Ribosome Inactivating Toxins

Ribosome Inactivating Toxins

Special Issue Editors

Julien Barbier
Daniel Gillet

MDPI • Basel • Beijing • Wuhan • Barcelona • Belgrade

MDPI

Special Issue Editors
Julien Barbier
CEA—Université Paris-Saclay
France

Daniel Gillet
CEA—Université Paris-Saclay
France

Editorial Office
MDPI
St. Alban-Anlage 66
Basel, Switzerland

This is a reprint of articles from the Special Issue published online in the open access journal *Toxins* (ISSN 2072-6651) from 2016 to 2018 (available at: http://www.mdpi.com/journal/toxins/special_issues/ribosome)

For citation purposes, cite each article independently as indicated on the article page online and as indicated below:

LastName, A.A.; LastName, B.B.; LastName, C.C. Article Title. *Journal Name* **Year**, *Article Number*, Page Range.

ISBN 978-3-03897-248-8 (Pbk)
ISBN 978-3-03897-249-5 (PDF)

Cover image courtesy of Julien Barbier and Daniel Gillet.

Contents

Preface to "Ribosome Inactivating Toxins" vii

Julien Barbier and Daniel Gillet
Ribosome Inactivating Proteins: From Plant Defense to Treatments against Human Misuse
or Diseases
Reprinted from: *Toxins* **2018**, *10*, 160, doi: 10.3390/toxins10040160 . . . . **1**

Maria Serena Fabbrini, Miku Katayama, Ikuhiko Nakase and Riccardo Vago
Plant Ribosome-Inactivating Proteins: Progesses, Challenges and Biotechnological Applications
(and a Few Digressions)
Reprinted from: *Toxins* **2017**, *9*, 314, doi: 10.3390/toxins9100314 . . . . **5**

Jeroen De Zaeytijd and Els J. M. Van Damme
Extensive Evolution of Cereal Ribosome-Inactivating Proteins Translates into Unique Structural
Features, Activation Mechanisms, and Physiological Roles
Reprinted from: *Toxins* **2017**, *9*, 123, doi: 10.3390/toxins9040123 . . . . **38**

Ludger Johannes
Shiga Toxin—A Model for Glycolipid-Dependent and Lectin-Driven Endocytosis
Reprinted from: *Toxins* **2017**, *9*, 340, doi: 10.3390/toxins9110340 . . . . **64**

Björn Becker, Tina Schnöder and Manfred J. Schmitt
Yeast Reporter Assay to Identify Cellular Components of Ricin Toxin A Chain Trafficking
Reprinted from: *Toxins* **2016**, *8*, 366, doi: 10.3390/toxins8120366 . . . . **75**

Xiao-Ping Li and Nilgun E. Tumer
Differences in Ribosome Binding and Sarcin/Ricin Loop Depurination by Shiga and
Ricin Holotoxins
Reprinted from: *Toxins* **2017**, *9*, 133, doi: 10.3390/toxins9040133 . . . . **89**

Wei-Wei Shi, Yun-Sang Tang, See-Yuen Sze, Zhen-Ning Zhu, Kam-Bo Wong and
Pang-Chui Shaw
Crystal Structure of Ribosome-Inactivating Protein Ricin A Chain in Complex with the
C-Terminal Peptide of the Ribosomal Stalk Protein P2
Reprinted from: *Toxins* **2016**, *8*, 296, doi: 10.3390/toxins8100296 . . . . **101**

Annie Villysson, Ashmita Tontanahal and Diana Karpman
Microvesicle Involvement in Shiga Toxin-Associated Infection
Reprinted from: *Toxins* **2017**, *9*, 376, doi: 10.3390/toxins9110376 . . . . **113**

Jun-Young Park, Yu-Jin Jeong, Sung-Kyun Park, Sung-Jin Yoon, Song Choi, Dae Gwin Jeong,
Su Wol Chung, Byung Joo Lee, Jeong Hun Kim, Vernon L. Tesh, Moo-Seung Lee and
Young-Jun Park
Shiga Toxins Induce Apoptosis and ER Stress in Human Retinal Pigment Epithelial Cells
Reprinted from: *Toxins* **2017**, *9*, 319, doi: 10.3390/toxins9100319 . . . . **135**

Christina C. Tam, Thomas D. Henderson II, Larry H. Stanker, Xiaohua He and
Luisa W. Cheng
Abrin Toxicity and Bioavailability after Temperature and pH Treatment
Reprinted from: *Toxins* **2017**, *9*, 320, doi: 10.3390/toxins9100320 . . . . **155**

Christina C. Tam, Luisa W. Cheng, Xiaohua He, Paul Merrill, David Hodge and
Larry H. Stanker
A Monoclonal–Monoclonal Antibody Based Capture ELISA for Abrin
Reprinted from: *Toxins* **2017**, *9*, 328, doi: 10.3390/toxins9100328 167

Yoav Gal, Ohad Mazor, Reut Falach, Anita Sapoznikov, Chanoch Kronman and Tamar Sabo
Treatments for Pulmonary Ricin Intoxication: Current Aspects and Future Prospects
Reprinted from: *Toxins* **2017**, *9*, 311, doi: 10.3390/toxins9100311 180

Yoav Gal, Anita Sapoznikov, Reut Falach, Sharon Ehrlich, Moshe Aftalion,
Chanoch Kronman and Tamar Sabo
Total Body Irradiation Mitigates Inflammation and Extends the Therapeutic Time Window for
Anti-Ricin Antibody Treatment against Pulmonary Ricinosis in Mice
Reprinted from: *Toxins* **2017**, *9*, 278, doi: 10.3390/toxins9090278 208

Sarah J. C. Whitfield, Gareth D. Griffiths, Dominic C. Jenner, Robert J. Gwyther,
Fiona M. Stahl, Lucy J. Cork, Jane L. Holley, A. Christopher Green and Graeme C. Clark
Production, Characterisation and Testing of an Ovine Antitoxin against Ricin; Efficacy, Potency
and Mechanisms of Action
Reprinted from: *Toxins* **2017**, *9*, 329, doi: 10.3390/toxins9100329 224

Amanda Y. Poon, David J. Vance, Yinghui Rong, Dylan Ehrbar and Nicholas J. Mantis
A Supercluster of Neutralizing Epitopes at the Interface of Ricin's Enzymatic (RTA) and Binding
(RTB) Subunits
Reprinted from: *Toxins* **2017**, *9*, 378, doi: 10.3390/toxins9120378 243

Gregory Hall, Shinichiro Kurosawa and Deborah J. Stearns-Kurosawa
Shiga Toxin Therapeutics: Beyond Neutralization
Reprinted from: *Toxins* **2017**, *9*, 291, doi: 10.3390/toxins9090291 259

Aleksander Rust, Lynda J. Partridge, Bazbek Davletov and Guillaume M. Hautbergue
The Use of Plant-Derived Ribosome Inactivating Proteins in Immunotoxin Development: Past,
Present and Future Generations
Reprinted from: *Toxins* **2017**, *9*, 344, doi: 10.3390/toxins9110344 277

Letizia Polito, Daniele Mercatelli, Massimo Bortolotti, Stefania Maiello, Alice Djemil,
Maria Giulia Battelli and Andrea Bolognesi
Two Saporin-Containing Immunotoxins Specific for CD20 and CD22 Show Different Behavior
in Killing Lymphoma Cells
Reprinted from: *Toxins* **2017**, *9*, 182, doi: 10.3390/toxins9060182 292

Ka-Yee Au, Wei-Wei Shi, Shuai Qian, Zhong Zuo and Pang-Chui Shaw
Improvement of the Pharmacological Properties of Maize RIP by Cysteine-Specific PEGylation
Reprinted from: *Toxins* **2016**, *8*, 298, doi: 10.3390/toxins8100298 310

Preface to "Ribosome Inactivating Toxins"

Ribosome inactivating proteins (RIPs) form a vast family of hundreds of toxins from plants, fungi, algae and bacteria. RIP activities have also been detected in animal tissues. They exert an N-glycosydase catalytic activity that is targeted to a single adenine of a ribosomal RNA, thereby blocking protein synthesis and leading intoxicated cells to apoptosis. In many cases they perform additional depurinating activities that act against other nucleic acids, such as viral RNA and DNA, or genomic DNA. Although their role remains only partially understood, their functions may be related to plant defense against predators and viruses, plant senescence or bacterial pathogenesis.

Most RIPs are no threat to human or animal health. However, several bacterial RIPs are major virulence factors involved in severe epidemic diseases such as dysentery or the hemolytic uremic syndrome that may occur in patients suffering from Shiga toxin-producing entero hemorrhagic Escherichia coli infection. A few RIPs synthesized in plant seeds have been involved in accidental or criminal poisonings, political intimidation or bio-suicides. Tremendous progress has been made in their detection, identification and characterization. However, the pathophysiologies of these intoxications seem much more complicated than being solely linked to cell death and are still far from being fully understood. There are no commercially available products to specifically prevent or block RIP action, although research progress has been made in the development of antibodies, small molecule inhibitors and vaccines.

Finally, RIPs have been engineered into immunotoxins by conjugating them to antibodies or other targeting moieties. Numerous clinical trials have shown great promise, also with regard to the difficulties in developing such therapies to destroy cancer cells.

This Special Issue of Toxins presents the most recent data on all aspects of RIPs: new RIPs, structure, function, mechanism of action, pathophysiology, anti-RIP drug development and RIP engineering into anticancer treatments.

Julien Barbier, Daniel Gillet
Special Issue Editors

Editorial

Ribosome Inactivating Proteins: From Plant Defense to Treatments against Human Misuse or Diseases

Julien Barbier and Daniel Gillet *

Service d'Ingénierie Moléculaire des Protéines (SIMOPRO), CEA, Université Paris-Saclay, LabEx LERMIT, 91191 Gif-sur-Yvette, France; julien.barbier@cea.fr
* Correspondence: Daniel.gillet@cea.fr; Tel.: +33-1-69-08-76-46

Received: 11 April 2018; Accepted: 13 April 2018; Published: 18 April 2018

Ribosome inactivating proteins (RIPs) form a vast family of hundreds of toxins from plants, fungi, algae, and bacteria. RIP activities have also been detected in animal tissues. They exert an N-glycosydase catalytic activity that is targeted to a single adenine of a ribosomal RNA, thereby blocking protein synthesis and leading intoxicated cells to apoptosis. In many cases, they have additional depurinating activities that act against other nucleic acids, such as viral RNA and DNA, or genomic DNA. Although their role remains only partially understood, their functions may be related to plant defense against predators and viruses, plant senescence, or bacterial pathogenesis.

In this Special Issue, a review by Fabbrini and colleagues [1] addresses our current knowledge about the function and mechanisms of action of plant type I and type II RIPs. In particular, they emphasize the diversity found in their mechanisms of action, although they share sequence and structural identities in their catalytic chain. In another review, De Zaeytijd and Van Damme focus on the heterogeneity of cereal RIPs from an evolutionary perspective, their differences from non-cereal RIPs, and their variety of roles in addition to defense against pathogens and insects [2].

Most RIPs are no threat to human or animal health. However, several bacterial RIPs are major virulence factors involved in severe epidemic diseases such as dysentery or the hemolytic uremic syndrome that may occur in patients suffering from Shiga toxin-producing entero-hemorrhagic *Escherichia coli* infection. A few RIPs synthesized in plant seeds such as ricin toxin, abrin, or sarcin have been or may be involved in accidental or criminal poisonings, political intimidation, or bio-suicides. In this Special Issue, four contributions address the most recent advances in understanding the three major steps of the intoxication process of cells by Shiga toxins, ricin, and/or sarcin: receptor-binding and triggering of endocytosis, the components of the intracellular trafficking machinery involved in intoxication and binding, and depurination of the ribosome. Johannes describes how the pentavalent binding of Shiga toxins to Gb3 gangliosides in lipid rafts induces membrane structural changes and stress leading to the internalization of the toxin-receptor complexes [3]. Becker and colleagues set up an elegant screening method in yeast, enabling them to not only confirm the importance of the GARP complex and other protein partners in ricin A chain intracellular trafficking, but also to identify seven new proteins involved along the pathway [4]. This method can now be applied to identify trafficking components used by other RIPs. Li and Tumer analyze the differences in ribosome binding and catalytic activities of the non-activated and activated Shiga and ricin holotoxins, showing opposite behaviors for these two toxins [5]. Finally, Shi et al. describe the structure of the complex of ricin A chain with the C-terminal peptide of the ribosome stalk protein P2 [6]. They discuss the differences in this interaction with that of other RIPs with the same ribosomal target.

The pathophysiology of intoxication by the most dangerous RIPs, such as Shiga toxins and ricin toxin, seem much more complicated than a sole link to circulation in the bloodstream and cell death, and is still far from being understood. Diana Karpman and her group review their recent work that brings important progress in the understanding of the mechanisms underlying the hemolytic uremic syndrome provoked by Shiga toxins [7]. They show that Shiga toxins are internalized by red blood cells and then released in microvesicles. It is these toxin-containing microvesicles that participate in the prothrombotic lesions, hemolysis, and transfer of the toxin from the circulation into the kidney, that are characteristic of this deadly syndrome. Furthermore, a research article from the groups of Lee and Park describes for the first time the apoptotic processes induced by Shiga toxins in human retinal pigment epithelial cells, suggesting the mechanisms leading to blindness in severe cases of hemolytic and uremic syndrome [8]. Gal et al. extend their characterization of the crucial role of inflammation in ricin toxin pathogenesis by showing that total body irradiation of mice decreases inflammation markers and extends time to death [9]. Second after ricin toxin, the plant RIP abrin is considered an increasing risk of malevolent and suicidal use. Thus, there is a need for detection and decontamination tools. Tam et al. set up a two-monoclonal antibody-based ELISA that can detect as low as 1 ng/mL of abrin and that shows no false positive detection of other plant RIPs [10]. The same team showed that while various pH treatments of the toxin did not affect its activity, heating above 74 °C completely inactivated its capacity to kill cells and mice. However, they showed that this treatment affects the lectin part of the toxin rather than its catalytic chain [11]. Interestingly, this article sets a correspondence between cytotoxicity testing and the mouse bioassay, which should help reduce the use of the mouse model for the evaluation of abrin and other RIPs.

Due to sporadic but recurrent cases of biosuicide and biothreats with ricin toxin, there is an urgent need for a treatment of human intoxication. The group of Kronman gives us a thorough review of existing data on potential countermeasures and treatment strategies, although none are approved for medical use [12]. While antibodies represent the most realistic approach in the case of early post-intoxication intervention, the review stresses the importance of not only eliminating the toxin but also downregulating the explosive inflammatory response triggered by the toxin, as additionally described in a research article by the same group [9].

Interestingly, Whitfield and colleagues describe in detail an F(ab')₂ polyclonal ovine antitoxin and its performance in a mouse model of ricin inhalation that is intended to be pharmaceutically qualified for human use [13]. Protection is mediated both by reducing the amount of circulating toxin and blocking its intracellular trafficking to the Golgi apparatus.

Many studies in the past showed that a fraction only of the antibodies generated in the course of an immune response against ricin toxin was neutralizing. Here, the group of Nicholas Mantis identifies the presence of a supercluster of neutralizing epitopes at the interface between the A and B chains of the toxin by analyzing a series of $V_H H$ camelid antibody fragments from a phage library generated against ricin toxin [14]. Interestingly, these antibodies do not interfere with the binding of the toxin to the galactose and N-acetyl-galactosamine residues of cell surface glycosylation. This is a step forward in understanding the basis for antibody-mediated protection against this toxin.

Hall et al. review the attempts to develop antibodies or other antitoxin strategies to treat the hemolytic and uremic syndrome caused by Shiga toxins, none of which have reached approval [15]. They suggest that the rarity of this disease is a major limit to achieving the necessary clinical trials. Then they advocate the development of drugs targeting the unfolded protein response and the ribotoxic stress response triggered by Shiga toxins as these pathways are involved in many other conditions, which may decrease the barriers to commercial development.

The final aspect of research on RIPs covered by this Special Issue concerns their use in the engineering of immunotoxins to target cancer or cells infected by Human immunodeficiency virus (HIV) by conjugation of antibodies or other targeting moieties. Two reviews by Fabbrini et al. and Rust et al. discuss the difficulties that have been encountered in the development of several generations of immunotoxins, none of which have been approved after clinical trials [1,16]. They also present the future trends of immunotoxin development. Two examples of the complexities of such development are given in the articles of Polito et al. and Au et al. The former analyzes the difference in the mechanism of killing of two closely related saporin-containing immunotoxins targeting different markers on B-cell lymphomas, CD20 and CD22 [17]. The latter addresses the effect of PEGylation on the pharmacology, biological activity, and antibody induction of a TAT-maize RIP construction designed to target HIV-infected cells [18].

Overall, this Special Issue of *Toxins* presents the most recent data on all aspects of RIPs, including function, diversity, evolution, as well as mechanism, pathophysiology, medical countermeasures, and engineering into anticancer drugs.

Acknowledgments: D.G. and J.B. are supported by CEA, the Joint Ministerial Program of R&D against CBRNE risks, ANR Anti-HUS grant ANR-14-CE16-0004, and The Swedish Research Council grant K2015-99X-22877-01-6.

Conflicts of Interest: The authors declare no conflict of interest.

References

1. Fabbrini, M.S.; Katayama, M.; Nakase, I.; Vago, R. Plant ribosome-inactivating proteins: Progesses, challenges and biotechnological applications (and a few digressions). *Toxins (Basel)* **2017**, *9*, 314. [CrossRef] [PubMed]
2. De Zaeytijd, J.; Van Damme, E.J. Extensive evolution of cereal ribosome-inactivating proteins translates into unique structural features, activation mechanisms, and physiological roles. *Toxins (Basel)* **2017**, *9*, 123. [CrossRef] [PubMed]
3. Johannes, L. Shiga toxin-a model for glycolipid-dependent and lectin-driven endocytosis. *Toxins (Basel)* **2017**, *9*, 340. [CrossRef] [PubMed]
4. Becker, B.; Schnoder, T.; Schmitt, M.J. Yeast reporter assay to identify cellular components of ricin toxin a chain trafficking. *Toxins (Basel)* **2016**, *8*, 366. [CrossRef] [PubMed]
5. Li, X.P.; Tumer, N.E. Differences in ribosome binding and sarcin/ricin loop depurination by shiga and ricin holotoxins. *Toxins (Basel)* **2017**, *9*, 133. [CrossRef] [PubMed]
6. Shi, W.W.; Tang, Y.S.; Sze, S.Y.; Zhu, Z.N.; Wong, K.B.; Shaw, P.C. Crystal structure of ribosome-inactivating protein ricin a chain in complex with the c-terminal peptide of the ribosomal stalk protein p2. *Toxins (Basel)* **2016**, *8*, 296. [CrossRef] [PubMed]
7. Villysson, A.; Tontanahal, A.; Karpman, D. Microvesicle involvement in shiga toxin-associated infection. *Toxins (Basel)* **2017**, *9*, 376. [CrossRef] [PubMed]
8. Park, J.Y.; Jeong, Y.J.; Park, S.K.; Yoon, S.J.; Choi, S.; Jeong, D.G.; Chung, S.W.; Lee, B.J.; Kim, J.H.; Tesh, V.L.; et al. Shiga toxins induce apoptosis and er stress in human retinal pigment epithelial cells. *Toxins (Basel)* **2017**, *9*, 319. [CrossRef] [PubMed]
9. Gal, Y.; Sapoznikov, A.; Falach, R.; Ehrlich, S.; Aftalion, M.; Kronman, C.; Sabo, T. Total body irradiation mitigates inflammation and extends the therapeutic time window for anti-ricin antibody treatment against pulmonary ricinosis in mice. *Toxins (Basel)* **2017**, *9*, 278. [CrossRef] [PubMed]
10. Tam, C.C.; Cheng, L.W.; He, X.; Merrill, P.; Hodge, D.; Stanker, L.H. A monoclonal-monoclonal antibody based capture elisa for abrin. *Toxins (Basel)* **2017**, *9*, 328. [CrossRef] [PubMed]
11. Tam, C.C.; Henderson, T.D.; Stanker, L.H.; He, X.; Cheng, L.W. Abrin toxicity and bioavailability after temperature and ph treatment. *Toxins (Basel)* **2017**, *9*, 320. [CrossRef] [PubMed]
12. Gal, Y.; Mazor, O.; Falach, R.; Sapoznikov, A.; Kronman, C.; Sabo, T. Treatments for pulmonary ricin intoxication: Current aspects and future prospects. *Toxins (Basel)* **2017**, *9*, 311. [CrossRef] [PubMed]
13. Whitfield, S.J.C.; Griffiths, G.D.; Jenner, D.C.; Gwyther, R.J.; Stahl, F.M.; Cork, L.J.; Holley, J.L.; Green, A.C.; Clark, G.C. Production, characterisation and testing of an ovine antitoxin against ricin; efficacy, potency and mechanisms of action. *Toxins (Basel)* **2017**, *9*, 329. [CrossRef] [PubMed]
14. Poon, A.Y.; Vance, D.J.; Rong, Y.; Ehrbar, D.; Mantis, N.J. A supercluster of neutralizing epitopes at the interface of ricin's enzymatic (RTA) and binding (RTB) subunits. *Toxins (Basel)* **2017**, *9*, 378. [CrossRef] [PubMed]
15. Hall, G.; Kurosawa, S.; Stearns-Kurosawa, D.J. Shiga toxin therapeutics: Beyond neutralization. *Toxins (Basel)* **2017**, *9*, 291. [CrossRef] [PubMed]
16. Rust, A.; Partridge, L.J.; Davletov, B.; Hautbergue, G.M. The use of plant-derived ribosome inactivating proteins in immunotoxin development: Past, present and future generations. *Toxins (Basel)* **2017**, *9*, 344. [CrossRef] [PubMed]

17. Polito, L.; Mercatelli, D.; Bortolotti, M.; Maiello, S.; Djemil, A.; Battelli, M.G.; Bolognesi, A. Two saporin-containing immunotoxins specific for cd20 and cd22 show different behavior in killing lymphoma cells. *Toxins (Basel)* **2017**, *9*, 182. [CrossRef] [PubMed]

18. Au, K.Y.; Shi, W.W.; Qian, S.; Zuo, Z.; Shaw, P.C. Improvement of the pharmacological properties of maize rip by cysteine-specific pegylation. *Toxins (Basel)* **2016**, *8*, 298. [CrossRef] [PubMed]

toxins

MDPI

Review

Plant Ribosome-Inactivating Proteins: Progesses, Challenges and Biotechnological Applications (and a Few Digressions)

Maria Serena Fabbrini [1], Miku Katayama [2,3], Ikuhiko Nakase [2] and Riccardo Vago [4,5,*]

[1]   MIUR, Italian Ministry of Instruction, University and Research, 20090 Monza, Italy; msfabbrini@gmail.com
[2]   NanoSquare Research Institution, Research Center for the 21st Century,
      Organization for Research Promotion, Osaka Prefecture University, 1-2, Gakuen-cho, Naka-ku,
      Osaka 599-8570, Japan; sxc04031@edu.osakafu-u.ac.jp (M.K.); i-nakase@21c.osakafu-u.ac.jp (I.N.)
[3]   Graduate School of Science, Osaka Prefecture University, 1-1, Gakuen-cho, Naka-ku, Osaka 599-8531, Japan
[4]   Urological Research Institute, Division of Experimental Oncology, IRCCS San Raffaele Hospital,
      20132 Milan, Italy
[5]   University Vita-Salute San Raffaele, 23132 Milan, Italy
*   Correspondence: vago.riccardo@hsr.it; Tel.: +39-02-2643-5664

Academic Editors: Julien Barbier and Daniel Gillet
Received: 31 August 2017; Accepted: 3 October 2017; Published: 12 October 2017

Abstract: Plant ribosome-inactivating protein (RIP) toxins are EC3.2.2.22 *N*-glycosidases, found among most plant species encoded as small gene families, distributed in several tissues being endowed with defensive functions against fungal or viral infections. The two main plant RIP classes include type I (monomeric) and type II (dimeric) as the prototype ricin holotoxin from *Ricinus communis* that is composed of a catalytic active A chain linked via a disulphide bridge to a B-lectin domain that mediates efficient endocytosis in eukaryotic cells. Plant RIPs can recognize a universally conserved stem-loop, known as the α-sarcin/ ricin loop or SRL structure in 23S/25S/28S rRNA. By depurinating a single adenine (A4324 in 28S rat rRNA), they can irreversibly arrest protein translation and trigger cell death in the intoxicated mammalian cell. Besides their useful application as potential weapons against infected/tumor cells, ricin was also used in bio-terroristic attacks and, as such, constitutes a major concern. In this review, we aim to summarize past studies and more recent progresses made studying plant RIPs and discuss successful approaches that might help overcoming some of the bottlenecks encountered during the development of their biomedical applications.

Keywords: plant ribosome inactivating proteins; ER-stress; saporin; targeted drug delivery; nanovectors

1. Prologue

The first time I heard the term ribosome-inactivating protein "RIP" was in 1987 when we were attending "GENE87" at the Milan University and one of the invited speakers was Prof. Fiorenzo Stirpe from the University "Alma Mater" of Bologna [1]. The speech was fascinating to all of us coming to attend the symposium from a Plant Biology institute. I had just started my own experimental thesis and it was even more intriguing that Prof. Stirpe was coming from a Medical School and not from a Botanical Institute. The bright idea of using plant-derived toxins to eliminate transformed cells was pioneered at that time. The two seminal papers by Endo and Tsurugi on the mechanism of action of ricin and type I RIPs acting on eukaryotic ribosomes were published this very same year [2,3]. Curiously, some researchers from an Italian pharmaceutical company came to our lab to get some advice on how to achieve the cloning of a RIP cDNA from *Saponaria officinalis* tissues. The dry seeds they were trying to use for preparing the cDNA library stored plenty of saponins that during the mashing procedures were producing huge amounts of bubbles (L. Benatti, personal communication).

This is the main reason why the first saporin cDNA was then cloned starting from fresh leaves [4], allowing me just by chance to meet the person with whom we still are sharing our lives. To end these digressions, we must certainly acknowledge the great amount of experimental work done by the group of Mike Lord and Lynne Roberts in Warwick while studying ricin, the prototype type II RIP, one of the most potent poisons known at that time, which was strikingly used to assassinate in a "rocambolesque" way a dissident in London during the heavy years of the cold war. Plant ribosome-inactivating proteins may be viewed as very special tools from the Plant Kingdom that allowed us to shed light on certain peculiar intracellular pathways, such as the retrograde transport along the secretory route or more recent findings about some RIP signal peptide(s) acting as stress-sensors. Still intracellular pathways of delivery need to be elucidated in detail to allow in the future more efficient uses in targeted anticancer therapy.

2. Biochemical and Structural Considerations

Several plant species belonging to 17 different families, among them Cucurbitaceae, Euphorbiaceae, and Poaceae, and families belonging to the superorder Caryophyllales, produce plant Ribosome-Inactivating Proteins (RIPs), although many others do not, including the plant type model *Arabidopsis thaliana* [5]. They are found in most plant species as gene families, reflecting their differential distribution in plant tissues (roots, leaves and seeds) and may share among major functions the protection against viral or fungal infections and possibly be relevant for the physiologic responses during plant senescence or following stress inducers [6,7]. RIPs belong to the N-glycosidase family of toxins (EC3.2.2.22) able to specifically and irreversibly inactivate the large ribosomal subunits depurinating a specific adenine base (A4324 in the rat 28S ribosomal rRNA) located in a universally conserved GAGA-tetraloop, also known as the α-sarcin/ricin loop, present in 23S/26S/28S rRNA. Plant RIPs can be divided into three main classes: type I like saporin from *Saponaria officinalis* are composed of a single polypeptide chain of approximately 30 KDa, type II as ricin from *Ricinus communis* [8] are heterodimers consisting of an A chain, functionally equivalent to the type I polypeptide linked via a disulphide bridge to a B subunit endowed with lectin-binding properties [9]. For a long time, all type 2 RIPs were considered to be highly potent toxins, but, so far, there are also known type II RIPs, which are not or only less toxic in vivo, and therefore they are denominated as non-toxic type II RIPs [10,11]. Finally, type III RIPs are polypeptides, which are synthesized as inactive precursors (ProRIPs) that will require proteolytic processing events to form an active RIP [12].

Residues that are highly conserved among RIPs (shown in Figure 1 with an asterisk), besides the main residues at the catalytic cleft (arrowed in Figure 1), are those belonging to the "N-glycosidase signature", which include Tyr80, Tyr123, and the key active site residues Glu177, Arg180, and Trp211 in RTA (Figure 2) and a few others surrounding this active site. The protein sequence identities between ricin A chain (RTA) and type I RIPs (Figure 1) are generally low and found to be respectively: saporin 22%, Gelonin 30%, pokeweed antiviral protein (PAP) 29%, thricosanthin 35%, dianthin 19%, bouganin 29%, and momordin / momorcharin, 33%.

Despite the differences in amino acid sequences, their overall three-dimensional fold is well conserved as estimated by the superimposition of the 3D structures of several type I RIPs with the one of RTA (Figure 3), which clearly demonstrates that RTA and type I RIPs all share a common "RIP fold" characterized by the presence of two major domains: an N-terminal domain, which is mainly beta-stranded, and a C-terminal domain that is predominantly alpha–helical.

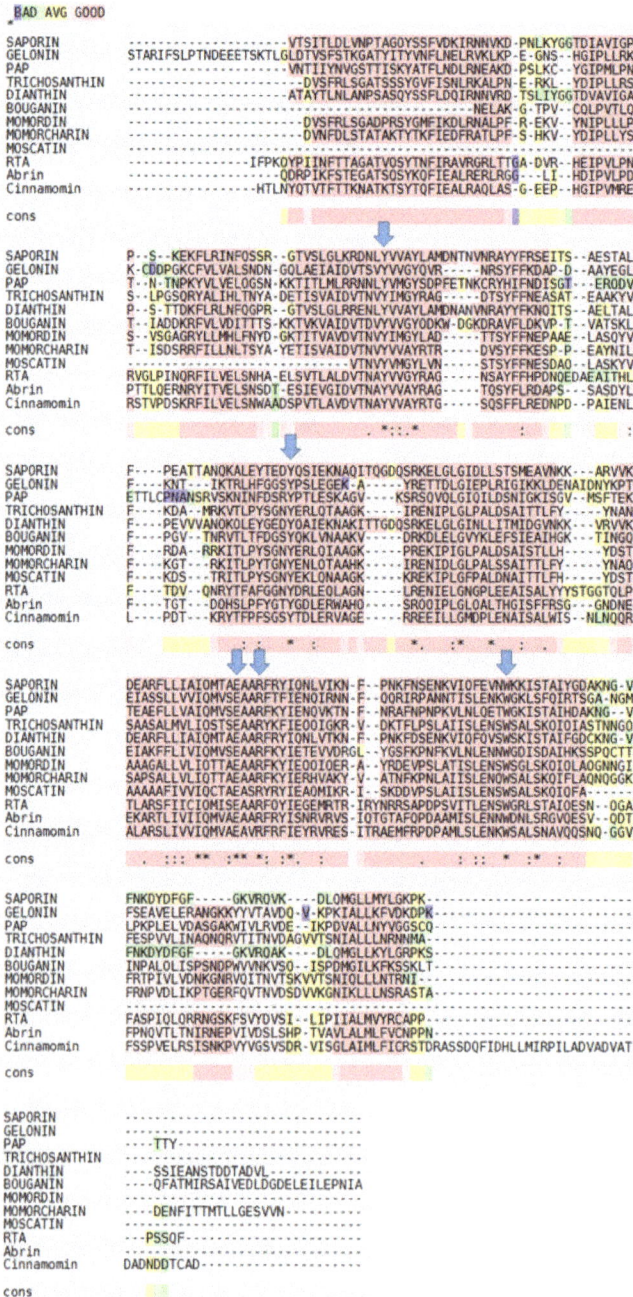

Figure 1. Amino acid sequence alignment of different type I RIPs compared to ricin (RTA), abrin and cinnamomin catalytic A chains by T-Coffee. Color shades indicate levels of amino acid homology between the aligned sequences. Conserved amino acids are identified with an asterisk and residues crucial for the catalytic activity are arrowed: Tyr72, Tyr120, Glu176, Arg179 and Trp208 in the sequence of saporin.

7

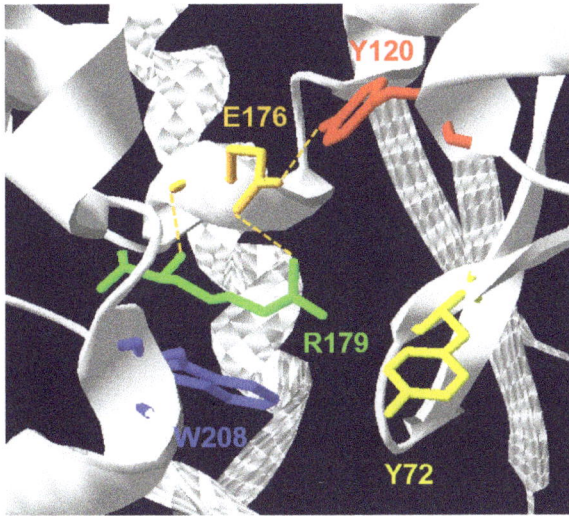

Figure 2. Three-dimensional reconstruction of catalytic cleft of saporin obtained by Swiss PDB Viewer (v4.0.4, SIB—Swiss Institute of Bioinformatics, Lausanne, Switzerland). Conserved residues crucial for the RIP signature are colored: Tyr72 (yellow), Tyr120 (red), Glu176 (orange), Arg179 (green) and Trp208 (blue). Hydrogen bonds among key residues are shown in orange.

Insertions and deletions, as compared to PAP, momordin from *Momordica charantia L.* and RTA were found to lay mainly in random coil regions. Glu177 and Arg180 in RTA (as Glu176 and Arg179 in saporin) are directly involved in the mechanism of catalysis. However, while a RTA Glu177 mutant was 20-fold less active than wild-type A chain in inhibiting translation in a reticulocyte lysate, the Arg179 saporin mutant was found 200-fold less active [13]. Double mutants at the catalytic site have been investigated for several type I RIPs, since for heterologous expression studies less active site mutants were needed to allow elucidating their biosynthesis. A loss of in vitro and in vivo saporin cytotoxicity can be achieved when Glu176 and Arg179 are both mutated to lysine and glutamine residues, respectively. This double saporin mutant (termed KQ) is, indeed, devoid of all the detrimental effects associated with RIP expression in several hosts [14–17]. Interestingly, mutation of Trp208 in saporin did not impair its in vitro enzymatic activity nor cytotoxicity [18], but this same residue has been demonstrated to be crucial for PAP structural integrity [19]. A negative electrostatic potential, arising from both the negatively charged phosphodiester backbone and conserved solvent-exposed acidic patches on the ribosomal proteins, covers much of the ribosomal surface [20]. The net positive charge of saporin and its high content in basic residues (around 10% lysine residues) are likely to be critical for the recognition of the ribosomal surface. In RTA, a set of arginine residues around the active site are involved in electrostatic interactions with the phosphodiester backbone of the α-sarcin/ricin loop [21,22]. Both RTA- and saporin-catalyzed rRNA modification shows a net dependence on salt and ion concentrations, indicating that these toxins can exploit multiple electrostatic interactions with their target ribosomes [23]. Extra enzymatic activities have been putatively ascribed to RIPs, but apart the Polynucleotide: adenosineglycosidase (PAG) activity documented as a DNA multiple depurination, DNAse-like and RNAse-like activities seem to be due to cross-contaminations of the protein preparation [15,24–27]. Similar observations were also recently made [28], when comparing saporin-6 to the leaf isoform L1/L3 (which behaves differently from all other isoforms studied to date), they showed that saporin-6 enzymatic activity released two adenines from ribosomes, the major fraction being the one deriving from the N-glycosidase activity while L1/L3 was able to "multidepurinate" ribosomes. Characterization of the kinetic parameters indicated that poly(A)

RNA depurination proceeds with a Km of 639 $\pm$ 32 μM and a kcat of 61 $\pm$ 1 min^{-1} at pH 7.8 and 25 °C. The catalytic efficiency of L1 on this substrate appears therefore to be considerably lower if compared to the action of a typical RIP, such as ricin A chain, on intact rat ribosomes which has been reported to occur with a Km of 2.6 μM and a kcat of 1777 min^{-1} [29]. The biological significance of the activity against rRNA at sites different from SRL, and on substrates other than the ribosomes (DNA, viral RNA, poly(A) mRNA) remains to be firmly established. When rRNA was extracted from mammalian cells exposed either to seed saporin or ricin, rRNA was found to be depurinated at a single site presumably corresponding to the one targeted by ricin [30]. Analysis of the in vivo activity of L1/L3 saporin will be required to assess whether multiple depurination plays any major role in the intoxication process and to clarify the mechanism of 80S ribosome inactivation by L1. The strong spatial similarities between type I RIPs, as shown in Figure 3, might suggest that different specificities/enzymatic activities could only reside in a restricted polypeptide area while assessment of whether these regions may contribute to altered activity would require either site-directed mutagenesis or protein domain swapping experiments.

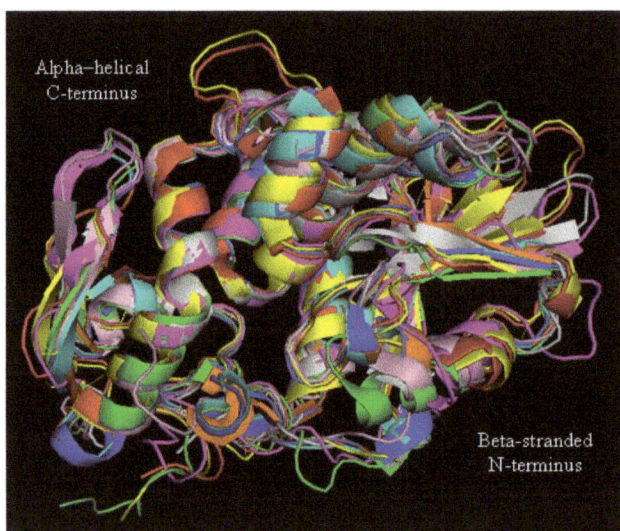

Figure 3. Three-dimensional structure of different type I RIPs and ricin A chain (RTA). Superimposition of secondary structure elements of Saporin (*red*, PDB code 1QI7), Gelonin (*pink*, 3KU0), PAP (*magenta*, 1GIK), Trichosanthin (*cyan*, 1QD2), Dianthin (*yellow*, 1RL0), Bouganin (*grey*, 3CTK), Momordin (*orange*, 1 MOM), Momorcharin (*blue*, 1AHA), RTA (*green*, 1J1 M), modified from [31].

Native ricin A chain carries two N-glycosylation (Asn-X-Ser/Thr) sites and is very poor in lysines (only two residues), a feature that has been linked to the cytosolic entry route of ricin A chains (see Section 5) [32]. Mature saporin has no oligosaccharide side chains similar to most type I RIPs that are hypothesized to be internalized by animal cells mainly passively by fluid-phase pinocytosis, with some relevant exceptions (see Section 5). Recently, a variant PAP from asiatic *Phytolacca acinosa*, PAP-S1aci, was resolved at 1.7 Angstroms (sharing 95% identity with PAP-S, one seed isoform of *Phytolacca americana*) and was found to have a proline substitution in the active site (Pro174) together with a rare type of N-glycosylation consisting of N-acetyl-D-glucosamine residues linked to the Asn10- Asn44- Asn255 sites, corresponding to putative rRNA-binding regions, mapped through computer modelling studies based on their structural data [33]. The authors hypothesize that these GlcNAc modifications may have evolved to exploit Mannose/GlcNAc-receptor-mediated endocytosis to enhance cytotoxic activity against seed predators. Their computer simulation studies suggest that

PAP-S1aci depurination activity would be adversely affected by larger, more typical oligosaccharide chains. The presence of short mannosyl residues in PAP-S1aci may thus confer an advantage to these seeds without compromising their RIP catalytic activity. The absence of carbohydrate side chains on PAP-I, a cell-wall protein may reflect its specialized antiviral role for "local suicide" of the virus-infected cells (see below).

3. Biotechnological Application in Agriculture

To explore the potential antimicrobial activity, different RIPs, including pokeweed antiviral protein (PAP), trichosanthin (TC) from *Trichosanthes kirilowii* Maxim and the antiviral protein from *Phytolacca insularis* Nakai have been expressed in transgenic plants successfully leading to plant resistance against various viral and/or fungal proteins [34–37]. Transformed tobacco plants with the nontoxic C-terminal deletion mutant PAPW237* of *Phytolacca americana* and the active site mutant PAPE176V showed both that the ribosomes from these transgenic tobacco plants were not found depurinated. Interestingly, extracts from transgenic plants expressing PAPW237* protected tobacco plants from Potato Virus X infection, while the plant extracts from catalytic site mutant PAPE176V did not. PAP proteins have been studied since the nineties by the group headed by Tumer focusing on their potential applications to agriculture [38]. They also found that transgenic *Arabidopsis* (Line 512) plant expressing PAPW237* displayed an enhanced resistance to a strain of Tobacco Etch virus (TEV) and detected several genes upregulated, including auxin-responsive genes (more than 4-fold vs. basal), as well as genes involved in immunity and plant defense when performing a transcript profiling analysis using the Affymetrix *Arabidopsis* microarray. Ribosomal RNA is not the only substrate for PAP enzymatic activity, and capped mRNAs and uncapped RNAs are subject to inactivation by PAP. Additionally, double-stranded (ds) supercoiled DNA could be cleaved by this toxin [38]. The catalytic site required for rRNA depurination is the same required for DNA cleavage, as the PAPE176V mutant also could not cleave dsDNA, which, when treated with PAPwt, contained apurinic/apyrimidinic (AP) sites due to the removal of adenines. This same PAG activity has also been reported for different other type I RIPs including Gelonin, momordin I, PAP-S and saporin [26], but not for the ricin A chain. However, its role in the intoxication process by RIPs has yet to be elucidated.

Due to the observation that several monomeric RIPs are active also towards "conspecific" ribosomes, and, because of this observation, inefficient endoplasmic reticulum (ER) translocation or mistargeting of RIPs potentially can lead to cell self-intoxication. Cells would need to prevent any unregulated accumulation of the catalytically active enzymes in the cytosolic compartment, implying that this class of enzymes could block the spread of certain pathogens by causing local suicide of the infected cells [39,40]. Following this hypothesis [41], plant cells undergoing plasma membrane breaching by a virus would allow entry of apoplast-located toxin(s). This localized cell death would concomitantly block viral replication and thus eventually block the systemic spread of the virus load throughout the plant. Although this has not always been the case, as when the Iris RIPs were expressed in transgenic tobacco plants [42], where only a local protection could be observed. A recent study showed that PAP formed homodimer complexes in the cytosol of pokeweed cells while its monomeric form was instead found in the apoplast [43]. PAP homodimers were much less active on rRNA as compared to the monomeric PAP. Thus, at least in the case of PAP, homodimerization is another mechanism to avoid this plant RIP depurinating its own rRNA. Since plants are often infected with multiple pathogens, it was suggested that engineering them with single RIP genes, such as the nontoxic PAP variant, which leaves intact host ribosomes and confer broad-spectrum disease resistance may be an advantageous application in agriculture to pursue.

4. ER-Stress Mediated Regulation of Type I RIPs' Signal Peptides

When heterologously expressed in tobacco protoplasts in the absence of the B chain, the catalytically active A chain of ricin is retained in the ER and then retrotranslocated to the cytosol where it inactivates tobacco ribosomes [6]. PAP has been suggested to follow a similar intoxication pathway when heterologously expressed in yeast cells [38]. To determine whether saporin expression was toxic to tobacco protoplasts, Marshall and colleagues used the sequence encoding the saporin precursor to study its fate in leaf protoplasts [16]. The experiments were performed using initially the non-toxic preproSAPKQ mutant, encoding 24 amino acid saporin signal peptide, 253-amino acid mature seed sequence isoform, and the C-terminal propeptide. In contrast to what observed with ricin A chain, inactive saporin polypeptides were found to be efficiently secreted into the incubation media [16]. Saporin wild type expression was, nevertheless, highly toxic to the protoplasts, indicating the newly synthesized polypeptides could have reached the cytosol (i) either because a few failed to be targeted to the ER or (ii) because they could retrotranslocate, as ricin A chain and PAP, or (iii) lastly, because a pool of endocytosed molecules would intoxicate them. However, when control protoplasts were co-incubated with saporin-expressing protoplasts, toxicity was found to be endogenously produced [16]. Mutated saporin variants at the catalytic site with an intermediate reduction in RIP activity were used, one termed SAPV (having Valine176 instead of Glutamic acid), as to still allow detection of saporin polypeptide biosynthesis. When specific mutations were introduced which abolished signal peptide cleavage in this background, considerably more mutant preproSAPV was recovered from tobacco protoplasts (as compared to preproSAPV) with their experiments confirming that when the former was co-expressed with a phaseolin reporter, toxicity was consistently much lower [16]. Saporin cytotoxicity apparently involved the action of toxin molecules that were at least first partially inserted into ER membranes. Retrotranslocation of ricin A chain from the ER to the cytosol has been shown to be dependent on the activity of an AAA-ATPase known as CDC48 in plant and yeast cells (p97 in mammalian cells) [44]. They thus further investigated whether saporin toxicity would also be dependent on CDC48 extraction. Unexpectedly, co-expression of dominant negative CDC48 QQ mutant, exacerbated saporin toxicity, as shown by an over 80% reduction in phaseolin reporter biosynthesis [16]. These observations were totally opposite to the effects of CDC48 QQ expression on RTA toxicity, where, on the contrary, a significant rescue in protein synthesis is observed [44]. Expression of CDC48 QQ in tobacco protoplasts causes an upregulation of several ER chaperones, including the Binding Immunoglobulin Protein (BiP), Calreticulin and Endoplasmin [44], due to induction of the unfolded protein response, caused by a disregulated accumulation of misfolded or orphan proteins [45]. ER stress could also be induced via treatment with the *N*-glycosylation inhibitor Tunicamycin showing saporin toxicity increased similarly to what seen upon co-expression with CDC48 QQ [16]. This observation was reminiscent of the "pre-emptive quality control pathway" a novel mechanism by which subcellular protein localization can be regulated in response to stress, as reported for the mammalian prion protein (PrP) [46,47]. The pathway was identified based on the observation of ER translocation of chaperones such as protein disulfide isomerase (PDI) or BiP, a critical facet of the ER stress response, is not attenuated even during maximal acute stress [46] and depends upon characteristics of the chaperone's signal peptide. Targeting of the saporin precursor to the ER carrying instead tobacco BiP's signal peptide prevented cytotoxicity, as expected [16]. This mechanism of toxicity modulation upon ER stress may have further implications for the role of type I RIPs in plant defense against pathogen attack.

Here, we wonder if other type I plant RIPs may undergo such an ER-stress response and show for a comparison the hydrophobicity plot of signal peptides of the chaperone BiP with those of PrP, saporin, dianthin, trichosanthin PAP, momorcharin and momordin (Figure 4).

Comparing the hydrophobicity plot of saporin and a few other type I RIP signal peptides indicate that they are similar in distribution to the signal peptide of Prion protein (PrP). At least for these six type I RIPs, the behavior in response to ER-stress might be expected to overlap with the one described for PrP and saporin. In contrast, Gelonin signal peptide displays a quite different behavior and is not expected to undergo this ER-stress regulated pre-emptive pathway (not shown). Interestingly, if we look at the phylogenetic tree of some type I (and few type II RIPs) (Figure 5), it shows that Gelonin, as well as bouganin constitute independent branches of this tree when compared to the other type

I RIPs investigated here that can be found grouped into two subgroups one constituted by saporin dianthin and PAP and the second one including trichosanthin, momordin and momorcharin.

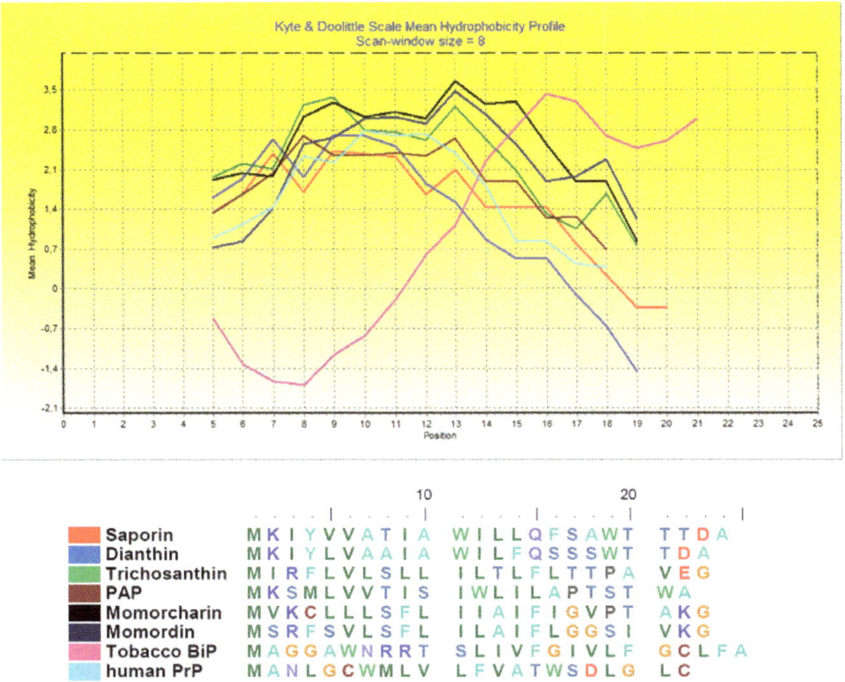

Figure 4. Kyte–Doolittle hydrophobicity Plot of the signal peptides of type I RIPs, tobacco BiP and PrP protein obtained by BioEdit software (version 7.0.9.0, Ibis Therapeutics, Carlsbad, CA, USA). Sequences used are shown under the plots.

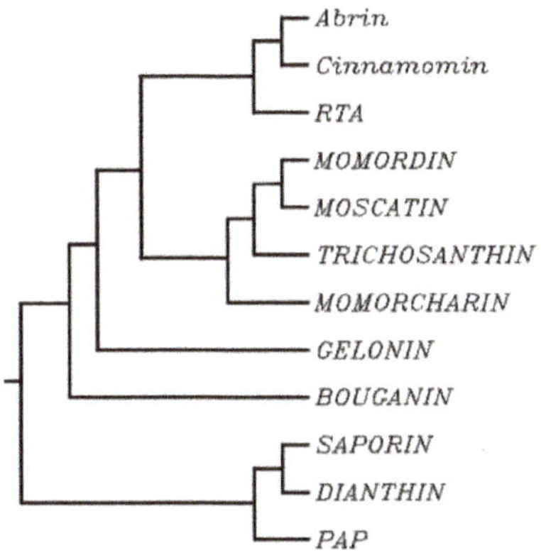

Figure 5. Phylogenetic tree of type I and type II RIPs obtained using the amino acidic sequences aligned with the Clustal Omega on-line software (EMBL-EBI, Hinxton, UK) with standard parameters. The branch length represents the number of changes that occurred in that branch.

5. Intoxication Routes in Mammalian Cells

Ricin holotoxin, as other type II RIPs, is able to exploit multiple receptors to enter mammalian cells thanks to the B-chain lectin properties binding cell surface galactoses/N-acetylgalactosamines and internalizing by (multiple) receptor-mediated endocytosis. Most of the holotoxin is then either recycled back to the cell surface or transported to endosomes where it is in part able to avoid proteolytic degradation in the endo-lysosomal compartment [48], while few molecules undergo retrograde transport reaching the Trans Golgi network (TGN) [49] and passing through the Golgi cisternae, where it was demonstrated that a modified RTA chain containing sulfation sites could be, indeed, found sulfated [50], to finally reach the ER, presumably thanks to interactions with recycling chaperones (calnexin/calreticulin) where the catalytic A chain (RTA) must be reductively separated from the cell-binding B chain [51] thanks to the action of PDI and thioredoxin-reductases [52,53]. Free RTA then mimics as a misfolded protein and is targeted to proteasomal degradation [48,54,55] by interacting with EDEM-1 (ER degradation-enhancing α-mannosidase I-like protein 1) and EDEM-2, which are key players in redirecting aberrant proteins for the ER-associated protein degradation (ERAD) (Figure 6).

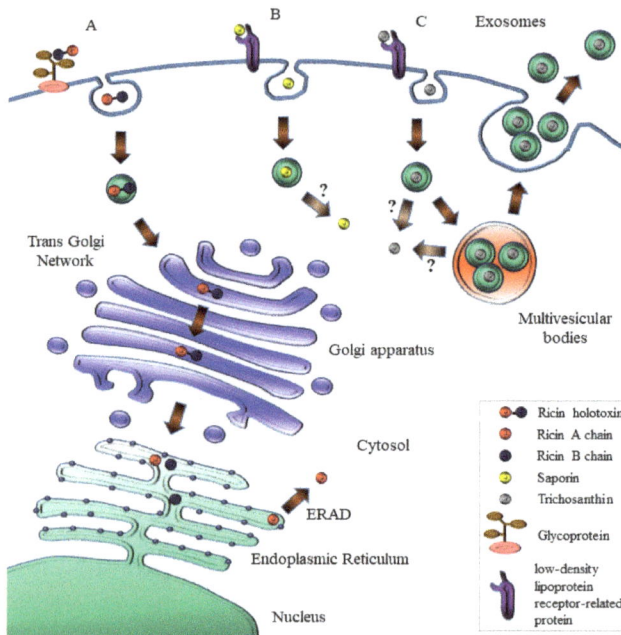

Figure 6. Schematic representation of the intoxication routes followed by ricin (**A**); saporin (**B**) and trichosanthin (**C**), modified from [56].

Both were shown to interact with ricin [57], but EDEM-2 may promote retrotranslocation out of the ER independently of translocon accessibility. Co-immunoprecipitation experiments demonstrated that ricin A chain interacted preferentially with EDEM-2. Ricin A chain used for these experiments does not contain carbohydrates (recombinantly expressed in *E. coli*) and therefore ricin interaction with the two EDEM proteins, in these experiments, represents nonglycan-mediated interplays. A single point mutation in ricin A-chain increases toxin degradation and inhibits EDEM-1-dependent ER retrotranslocation [58]. RTA contains a 12-residue hydrophobic C-terminal region that becomes exposed after the reduction of ricin in the ER. Indeed, mutation P250A increases also the endo-lysosomal degradation of the toxin, as well as reduces its transport out of the ER into the cytosol. The retrotranslocated RTA chain reaches the cytosol where it may refold rapidly, facilitated also by interaction with the target ribosomes [55]. RTA mimics ERAD substrates, escaping proteasomal degradation presumably also due to its paucity in lysine residues [32] and once in the cytosol, RTA interacts with Hsc70 chaperones, with its final destiny (refolding or degradation) depending on the presence of the co-chaperones that regulate Hsc70 activity [59]. Recently, with the aim of neutralizing ricin cytoxicity (which is an intensive field of investigation, see below) in ricin challenged mice they injected JJX12 a camelid bispecific antibody recognizing both an epitope on RTA (VH) and another one (VL) on the galactose-binding subunit that could passively protect mice against holotoxin lethal doses. JJX12 could affect dynamics of ricin uptake and trafficking also in human epithelial cells. Confocal microscopy and live cell imaging revealed that JJX12-toxin complexes formed at the cell surface could internalize via a pathway sensitive to amiloride, a known inhibitor of macropinocytosis. JJX12 interfered with retrograde transport of ricin to the TGN with the accumulation of the toxin in late endosomes found significantly enhanced [60].

The internalization pathway followed by type I RIPs in mammalian cells deserved much less attention, since it has been assumed that it should not rely on receptor-mediated endocytosis. Saporin cytotoxicity varies dramatically between different mammalian cell lines, with concentrations inhibiting protein synthesis by 50% (IC_{50}) ranging from the subnanomolar to the micromolar range, spanning at least three orders of magnitude [18,61]. In the case of saporin and trichosanthin, a family of closely related receptors has been identified to be involved in their internalization: the low-density lipoprotein receptor (LDLR) family that includes seven closely related family members: the very-low-density lipoprotein (VLDL) receptor, apoE receptor 2, multiple epidermal growth factor-like domains 7 (MEGF7), glycoprotein 330 (gp330/megalin/LRP2), lipoprotein receptor-related protein 1 (LRP1), and LRP1B. These proteins were shown to be promiscuous in ligand binding [62]. The α2-macroglobulin receptor/low-density lipoprotein receptor-related protein (LRP1) was shown to bind saporin in vitro [61,63] mediating saporin internalization in human monocytes and fibroblasts [64,65] (Figure 6). In human promyelocytic U937 cells, downregulation of LRP nicely paralleled resistance to saporin and to a urokinase–saporin conjugate [64]. Specific displacement of iodinated LRP1-receptor associated protein (RAP) with saporin in these cells was also independently demonstrated [66]. Mouse embryonic fibroblasts (MEF-2) derived from LRP1 knock-out mice were found 10-fold less sensitive to saporin as compared to MEF-1 control cells carrying both LRP1 and low-density lipoprotein receptor (LDLR) [30]. LB6 fibroblasts transfected with the human receptor for urokinase were used to study internalization of a human urokinase–saporin conjugate, which was initially designed to be internalized only following interaction of the urokinase domain with plasminogen activator inhibitors: surprisingly, saporin domain was instead able to trigger internalization of this conjugate. Confocal studies demonstrated a clear role for LRP1 in saporin–conjugate internalization [67]. Previously, a recombinant chimeric fusion in which only the amino-terminal fragment of human urokinase (unable to mediate internalization) was fused to saporin was found to be efficiently endocytosed by target cells [63]. While LRP1 is clearly involved in saporin uptake, at least in different cell lines, it did not appear to be essential for saporin cytotoxicity in Chinese hamster ovary (CHO) mutant cell line, CHO 13-5-1 [68], which had no detectable LRP mRNA. However, CHO cell lines defective for expression of either heparan sulphates or proteoglycans were also shown to be as sensitive to saporin isoforms as the parental CHO cells [65], suggesting that multiple receptors may be exploited by saporin in CHO. Trichosanthin (TC) is known to behave as an invasive toxin that targets syncythiotrophoblasts, macrophages, and T-cells and whose uptake was recently investigated [69]. TC binds cell surface receptors belonging to the LDL-related receptor family (Figure 6), and its well-known abortifaciens and renotoxic actions are caused presumably by LRP1-mediated uptake in trophoblasts and by LRP2/megalin-mediated uptake in proximal tubule epithelial cells [70]. Jurkatt-T cells, which do not express members of the LDL receptor family, are essentially resistant to free TC (as they are to saporin, our unpublished results), but turned extremely sensitive to the TC-loaded vesicles, which were secreted by at least two different target cell lines, JAR and K562, where part of the endocytosed TC was found incorporated into "pomegrenade" vesicles, deriving from multivesicular body (MVB) membranes, being subsequently secreted upon fusion of the MVB with the plasma membrane (Figure 6), thus targeting both syngeneic and allogeneic cells (Jurkatt-T cells). However, the possibility that other type I RIPs could exploit this exosome-mediated intercellular trafficking route remains to be investigated. Interestingly, Gelonin chimeric fusions were shown to be able to trigger mammalian cell death following instead of usual apoptotic an autophagy pathway [71,72].

The intracellular site(s) from which type I RIPs can escape into the cytosol remain unknown. However, several lines of evidence would exclude that the ricin Golgi-mediated retrograde transport to the ER could be used by other type I plant RIPs to exert their cytotoxicity. Brefeldin A, a fungal metabolite that causes disassembly of the Golgi complex, completely blocks ricin and RTA cytotoxicity [73,74], but fails to reduce saporin toxicity [30]. KDEL (single amino acid letter code) ER-retrieval sequence appended to saporin did not enhance saporin cytotoxicity [30], in contrast to its potentiating effects on RTA cytotoxicity [75]. ERAD mutants of CHO cells found resistant to ricin and *Pseudomonas* exotoxin A (PEA) remained as sensitive to saporin as parental CHO cells [76,77]. In addition, other biochemical essential features of RTA are not shared with some other type I RIPs. Lipid partitioning studies using Triton X-114 demonstrated that while RTA is predominantly found in the detergent phase, the ricin B chain, ricin holotoxin and several type I RIPs, including saporin, are instead found in the aqueous phase [78]. Most importantly, RTA has been shown to interact with negatively charged lipid vesicles and with ER membranes, undergoing a conformational change that makes it a better substrate for the ERAD system [79]. Saporin, in contrast, is a very stable protein, does not stably associate with negatively charged vesicles [32,65,79,80] and undergoes a peculiar stress-induced cytosolic delivery [16]. Altogether, the available data in literature would indicate that RTA compared to type I RIPs, such as saporin, may exploit different strategies to reach the cytosolic compartment in intoxicated cells.

6. Cell Death and Intracellular Signaling

Initially, it was assumed that mammalian cells exposed to the plant toxins would die simply because of the blockade of their protein translation machinery. However, it was shortly clear that besides inhibition of translation other players were to be involved in mediating cell-death. Griffiths and colleagues showed that intramuscular injection of ricin or abrin into rats resulted in the death of cells in rapidly dividing tissues, where the toxins were more concentrated. They observed that the morphology of dying cells in para-aortic lymph nodes and Peyers patches both in ricin and abrin injected-rats closely resembled that in cells undergoing apoptotic cell death [81,82].

Caspases are among the principal effectors of apoptosis, involved in pathways such as caspase-8-regulated extrinsic and caspase-9-regulated intrinsic pathways. The caspase-9 pathway links mitochondrial damage to caspase activation, and serves as an index of damage for the mitochondrial membrane function that loses its membrane potential [83]. Furthermore, co-downstream member caspase-3 is an executor of the DNA fragmentation and final steps leading to apoptosis, as exemplified by cleavage of poly(ADP-ribose) polymerase (PARP) during cell death [84].

Apoptotic death of U937 cells by ricin was evidenced by nuclear morphological changes, DNA fragmentation and an increased caspase-like activity. Komatsu and colleagues found also that an early event was that intracellular NAD(+) and ATP levels decreased in ricin-treated U937 cells, followed by the well-known ricin-mediated protein synthesis inhibition. The PARP inhibitor, 3-aminobenzamide (3-ABA), prevented depletion in NAD(+) and ATP levels, also protecting U937 cells from lysis. Overall, their results indicated that multiple apoptotic signaling pathways were triggered by ricin-treatment, one pathway leading to cell lysis via PARP activation while the other one lead to DNA fragmentation and caspase activation [85]. Toxicity with respect to apoptosis induction by various RIPs and the low molecular weight protein synthesis inhibitor, cycloheximide (CHX) was compared among various adherent and non-adherent cell lines, in particular focusing on the type II RIP abrin in one study [86].

Although there were no major differences observed between $IC_{50}s$ for protein synthesis inhibition in Jurkat-T cells or U937 cells using abrin, the extent of apoptosis was found to differ significantly. Most of the signaling cascades and pathways triggered by RIPs that are leading to apoptotic cell death were found funneled via mitochondria that are, indeed, major players in several stress-induced cell death pathways. Iordanov et al. show that ricin, α-sarcin and anisomycin were able to activate the SAP kinases or JNK1 in Rat-1 cells [87]. The damage to 28S rRNA by RIPs leads to a novel pathway of kinase activation known as "ribotoxic stress". Several type I RIPs, including TC, Gelonin and saporin have been demonstrated to activate these apoptotic pathways. In the latter case, the authors also suggested that initiation of apoptosis through mitochondrial cascade in U937 cells was an earlier event than the inhibition of translation by the plant RIP [88].

Another possible mode of action by plant toxins involves Reactive Oxygen Species (ROS) production, in response to stress and increase in intracellular calcium levels. Abrin was able to induce apoptosis by production of ROS, loss of mitochondrial membrane potential and release of cytochrome c into the cytosol. Shih et al. proposed that the apoptogenic activity of abrin A-chain also could be independent of its RNA N-glycosidase activity [86]. However, recently, a 225-fold lower abrin active-site mutant rABRA (R167L) was conjugated to the ricin B chain and its effects compared to a hybrid ricin B chain- abrin-conjugate in Jurkat-T cells. Rate of inhibition of protein translation, as well as the timing of activation of p38MAPK or caspase-3, were much slower by the mutated ricin B-rABRA (R167L), indicating that the lower abrin RIP potency correlated with the delay in activating cell death pathways, despite both conjugates used the same effectors in mediating cell death [89].

Trichosanthin induced high levels of ROS production in human choriocarcinoma cells [90]. Ricin holotoxin induces a rapid elevation in intracellular calcium levels in hepatoma cells [91]. Liver is one of the target organs for ricin holotoxin metabolism. An interesting study was performed in rats challenged with ricin: by using subcellular fractionation and immunoblotting procedures, the authors followed the fate of ricin in vivo in rat liver, focusing on endosome-associated events and on the induction of apoptosis [92]. Injected ricin rapidly accumulated in endosomes as an intact heterodimer (5 min–90 min) being later found (15 min–90 min) partially translocated to the cytosol as A- (and dissociated B-) chains. Strikingly, holotoxin could be found intact upon incubation with endosomal lysates and even in cell-free endosomes (pre-loaded with ricin in vivo). In the latter, a time-dependent translocation of ricin across the endosomal membrane could be observed depending on thioredoxin reductase-1, which was required to reduce the holotoxin disulphide bridge between A and B chain. Interestingly, ricin B chain alone, as a control, could induce cell death, as well. Holotoxin induced an intrinsic apoptotic pathway with increased cytochrome c release, activation of caspases-9 and -3 and DNA fragmentation. Reduced ricin or ricin B-chain caused cytochrome c release from mitochondria in vivo and in in vitro assays, suggesting ricin B-chain interaction with mitochondria contributes to the holotoxin-induced apoptosis [92].

The overall picture would point to a great level of complexity, suggesting once again that the pathways of apoptosis induced by RIPs are strictly dependent upon the cells lines investigated, as well as from the specific RIP utilized, and should be clarified when choosing to pursue an IT approach. In this context, when analyzing the effects of a targeted toxin, also the choice of the targeting moiety itself may also implicate different/alternate modalities in cell killing capabilities of the IT, as demonstrated by some studies by the group of Bolognesi: they compared cytotoxicities of two immunotoxins one obtained (by chemical conjugation of the plant toxin) conjugating seed saporin to Rituximab, an α-CD20 chimeric humanized antibody already approved by the United States (U.S.) Federal Drug Administration (FDA) to treat lymphomas, and the other one to OM124 an α-CD22 murine antibody. Both ITs were specific against Raji cells, being the α-CD22 murine immunotoxin (IT) two logs more efficient in cell killing Raji cells, presumably due to its faster internalization rate. Moreover, it induced a greater caspase activation compared to Rituximab IT. Cytotoxicity of both ITs could be in part prevented by pan-caspase inhibitor Z-VAD or necrostatin-1 a necroptosis inhibitor. Oxidative stress was involved in the cell-killing activity of Rituximab IT, as demonstrated by the protective role of the hydrogen peroxide scavenger catalase, but not in that of anti-CD22 IT [93]. Thus, depending on the targeting moiety, different pathways may be found activated. On the other hand, another study in L540 human Hodgkin's lymphoma cells showed at concentrations giving similar cellular inhibition of translation by ricin or saporin that ricin effects were more rapid when compared to those elicited by saporin. In addition, while the intrinsic apoptotic paths were equally activated, ricin activated the extrinsic caspase and effector caspase-3/7 pathways much more efficiently. Overall, the data indicated that different cell death mechanisms could be elicited by ricin and saporin with different timings and relative potency, hence, some of these differences might be also in part attributed to the concomitant pro-apoptotic effect of ricin B chain. The shutdown of intracellular signaling pathways may also contribute to the cytotoxicity of an IT as demonstrated by a recombinant IT made with recombinant Gelonin fused to a humanized single chain antibody (scFv, termed 4D5) deriving from Herceptin an anti Her2/neu [94].

In the case of solid tumors, it has been suggested that higher avidity and longer residence time of IgG-based immunoconjugates would outweigh improved tumor penetration of scFv-based constructs [95]. Since immunoconjugate development has been hampered by nonspecific toxicity and vascular leak syndrome (see below) the group of Rosemblum used Herceptin or its derived humanized single chain antibody (scFv, termed 4D5) to generate a Herceptin/rGel conjugate (bivalent) along with the two corresponding monovalent humanized recombinant ITs obtained using a classic flexible G4S-linker in the two orientations 4D5/rGel or rGel/4D5 [94]. The specific activity of engineered anti-Her2/neu single chain-immunotoxins fused to recombinant Gelonin (rGel) was compared to the activity of bivalent IgG-containing-immunoconjugate. The three constructs despite showing similar affinity to Her2/neu overexpressing ovarian cancer cells demonstrated significantly different antitumor activities. This study confirmed that monovalent fusion constructs can display virtually identical binding affinities as compared to their original IgG-counterpart [94]. However, rGel/4D5 orientation construct and Herceptin/rGel conjugate were found better performing than the 4D5/rGel construct both in in vitro and in vivo efficacy. Herceptin/rGel conjugate showed the most efficient and fastest internalization into target cells. The intracellular release of rGel after endocytosis of the various constructs was assessed after 4h exposure to 25 nM of each construct and the rGel moiety of all the immunotoxins could be observed by Western blots primarily in the cytosolic fraction. The rGel/4D5 displayed a comparatively higher and more prolonged intracellular concentration of toxin than 4D5/rGel. Extremely interesting were the data concerning the Her2/neu-related signaling events in SK-OV-3 cells. Both Herceptin/rGel conjugate and rGel/4D5 showed an impressive inhibition of phosphorylation of Her2/neu, EGFR, Akt and ERK, which are the most critical events for the Her2/neu signaling cascade. In contrast, 4D5/rGel showed a comparatively much reduced effect on these signaling pathways. Thus, the improved cytotoxicity coincided with an increased effect on cell signaling for recombinant rGel/4D5 [94].

7. Novel Potential Applications of RIPs

Historically, immunotoxins and chimeric fusions have been developed in parallel, with the former more extensively studied, following the introduction of monoclonal antibodies, which have gained great consideration for their selective specificity, being compared to the Paul Ehrlich's "magic bullet". On the other hand, a chimeric fusion Denileukin diftitox is the first, and until now the only, toxin-based formulation approved by the FDA employed in the treatment of cutaneous T-cell lymphomas [96] and tested for other clinical settings. This is a chimeric fusion formed by a truncated form of diphtheria toxin and the human cytokine interleukin-2, produced in bacteria. Toxin-based drugs have been mostly utilized in the treatment of hematological tumors because, being injected in the bloodstream, they can more easily reach target cancer cells. Numerous ITs formed by a toxic RIP domain like saporin, PAP, momordin and RTA, combined with a monoclonal antibody against markers of hematological malignancies such as CD2, CD19, CD22, CD25, CD30, CD38 and CD123 have been tested alone or in combination in animal models of lymphoma and leukemia [97–102] and some underwent early clinical phase studies. The anti-cancer activity of several ITs have been assayed also against solid tumors, but did not display the same successful results, likely due to the limited penetration within the tumor mass and relevant immunogenicity, the latter greatly reduced in hematological cancer patients. In the field of ligand-targeted chimerae, toxins have been in particular fused or conjugated to growth factors (EGF, [103], FGF [104] and VEGF [105]), but also with other polypeptides, such as uPa [106] and NCAM [107] as well, whose receptors have been found over-expressed on the cell surface of different cancer cells or in tumor vasculature, allowing to test the same therapeutics on several tumor models. For a comprehensive detailed description of ITs or fusion chimerae made with plant RIPs that have been also assayed using specific enhancers to promote toxin cytosolic delivery, refer to [108]. A historical description of ITs constructed using seed-extracted saporin (also termed saporin-S6) can be found in [109].

A major drawback of such biotechnological drugs is the triggering of immune responses after repeated administrations. Remarkably, plant toxin-based chimerae show less immunogenicity compared with those of bacterial origin. The development of ITs containing fully humanized antibodies or antibody fragments lacking Fc portions and the de-immunization of the toxin through the substitution of critical reactive amino acid residues contributed to significantly lessen the immunogenicity problems [31]. Furthermore, a serious side effect observed during IT clinical administration is the vascular leak syndrome (VLS), caused by the unspecific binding of the toxin domains to the vascular endothelial cells and characterized by interstitial edema, weight gain, and in most severe cases, pulmonary edema and hypotension. RTA-containing immunotoxins have been reported to elicit VLS [110]. An IT including a RTA mutated in a single amino acid flanking the consensus sequence identified as responsible of VLS exhibited a significantly reduced vascular damage in mouse models [111]. The clinical evaluation of the most promising targeted toxins is ongoing and, by exploiting novel discoveries and technologies in the biomedical fields, new toxin-related therapeutic options could be envisaged. Noteworthy examples are described below and summarized in Figure 7.

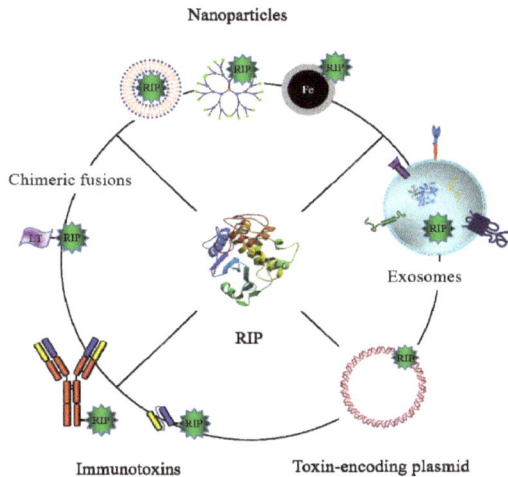

Figure 7. Present and future applications of RIPs. A schematic representation summarizes some of the potential application of a toxic plant RIP (green star and 3D-ribbon) starting with a classical Immunotoxin made by IgG chemically conjugated to native RIPs, which may be further transformed (left) into a fusion recombinant chimera or (right) into single chain variable fusions to RIP or even simply to DNA constructs encoding the plant RIP with all these different molecules being deliverable via targeted exosomes or engineered nanoparticles even for diagnostic purposes with $MnFe_2O_4$ nanoparticles (see text below).

7.1. Nanoparticles

Important drawbacks in cancer drug delivery approaches are often represented by (i) the scarce aqueous solubility of both natural and synthetic therapeutic agents; (ii) by their rapid clearance; (iii) by the dramatic side effects due to the lack of target selectivity and (iv) by the onset of resistance after repeated treatments.

In the last decades, many efforts have been directed to the development of delivery systems that may modify the drug biodistribution by conferring specific retention in the tumor site and favor tissue uptake. Nanotechnology has demonstrated great potential to overcome these limitations by increasing the circulation time of the drug, helping to bypass the hydrophobic nature of molecules by their incorporation, providing a number of additional administration routes and co-delivery of multiple drugs types and/or diagnostic agents for combined therapies. Furthermore, nanovectors can be arranged for a prompt controlled or sustained release of the drug, reducing the number of administrations and side effects or the risk of resistance [112]. Nanoparticle-based formulations can promote drug release with a selective activation within the target organ/tissue as a result of different stimuli, such as pH, enzymes, heat, light, etc. Liposomes, polymers and inorganic nanoparticles have been successfully employed for the delivery of protein toxins [113–115]. A library of cationic lipid-like materials (termed "lipidoids") was designed and employed to form nanocomplexes with proteins for intracellular delivery. Saporin and RNAse A were used as representative cytotoxic effector proteins, that when complexed with these lipidoids showed efficient cell internalization. A representative lipidoid EC16-1 was selected for protein delivery assay and, when administered to several cancer cells, saporin was demonstrated to be much more active in all the cell lines tested. Saporin-contained nanoparticles inhibited cell proliferation in vitro and suppressed tumor growth in a murine model of breast cancer [113]. A new transfection reagent, lipofectamine 3000, whose composition has not been displayed, has being tested for enhancing delivery of therapeutic proteins, including saporin [116]. In this recent report, they also show, being surprised that the J774.2 macrophage cell line already exhibits a great sensitivity to saporin, which is most likely due to the presence of LDL related receptor protein (LRP1) at the cell surface of macrophages, mediating uptake of the plant RIP. However, when they use cells that do not expose endocytic receptors for saporin entry, in the presence of lipofectamine 3000 saporin could potently inhibits growth of neuroblastoma N2a cells (EC50 = 1 nM in the presence of LF3000) but not in the presence of other transfection reagents such as lipofectamine LTX or Proteofectene. LF3000 increased the sensitivity of N2a cells to saporin by three orders of magnitude [116]. Stimuli-sensitive nanoconstructs have been investigated to enhance toxin penetration into the cytosol. Nanoparticles formed by polyglutamate (PGA), a sensitive polymer to the digestion by cathepsin B, a metalloproteinase overexpressed in the microenvironment of many tumors, have been loaded with indocyanine green (ICG), a FDA-approved near-infrared (NIR)-absorbing dye, and with saporin. Upon cleavage, ICG and saporin became bioavailable within the tumor context, facilitating their cellular uptake. Following irradiation, light causes endo/lysosome disruption with consequent drug release in the cytosol, where saporin exerts its toxic effect [114]. Another similar approach is the so-called photochemical internalization (PCI) to facilitate protein release to the cytosol after endocytosis of PCI-relevant photosensitizers. Saporin first was coupled to the polyamidoamine (PAMAM) dendrimer to improve its cellular uptake. Then, a further enhancement of the toxic activity was achieved on gingival and nasopharyngeal cancer cell lines after the combination with the PCI technology [115].

A peculiar approach has been tested by Su and collaborators, based on a two-stage delivery, where cancer cells pretreated with a Gelonin-based IT were killed following exposure to endosome-disrupting polymer nanoparticles. Nanoparticles were composed by the pH-sensitive poly(ethyleneoxide)-modified poly(beta-amino ester) (PBAE) encased within a biocompatible phospholipid shell and stabilized by the introduction of a phosphoethanolamine-conjugated polyethylene glycol (PEG) into the lipid coating. The immunotoxin comprised of a fusion of Gelonin to a fibronectin type III binding domain engineered to exhibit high affinity for the well-established cancer-associated EGF receptor. Co-administration allows co-internalization into common endosomes, followed by a swelling due to the "proton sponge effect" and subsequent IT cytosolic delivery. In vitro experiments indicate that such an approach may achieve a highly synergistic enhancement of antitumor activity [117]. The RIP-nanoparticle binomial can be used also as a diagnostic tool. For instance, the chimeric fusion VEGF121/rGel was conjugated to $MnFe_2O_4$ nanoparticles to be employed as a contrast agent, to monitor the efficacy of the antiangiogenic toxin-based therapy by magnetic resonance imaging (MRI). A satisfactory targeting capability to the vascular endothelial growth factor receptor 2 (VEGFR2) over-expressing cells was achieved providing the acquisition of clear neoangiogenic vascular distributions in orthotopic bladder tumor mice, a relevant information for an effective cancer therapy [118]. Despite successful attempts, the exploitation of nanoparticles (NP) for the delivery of toxins is quite underrepresented as compared to other cargos, like siRNAs or small molecules. Still ribosome inactivating proteins are mainly administrated as ITs or chimeric fusions.

Plant RIP-based NP have not only been employed as anti-cancer therapeutics, but also in the therapy of viral diseases, such as the acquired immunodeficiency syndrome (AIDS). An interesting approach to specifically target and kill cells activated early in the process of HIV production was developed. Ricin A chain was encapsulated in a polymer shell to form nanocapsules where a peptide cross-linker cleavable by HIV-1 protease was inserted. Once internalized in HIV infected cells, the cleaved nanocapsules released RTA, shutting down viral and cellular protein synthesis, leading to a prompt death of the producer cells [119].

7.2. Natural Vesicle-Mediated Delivery

Not only artificially prepared nanoparticles such as the liposomes as mentioned above, but also naturally produced extracellular vesicles like exosomes have focused attention as one of the next-generation intracellular carriers. Exosomes are secreted extracellular vesicles (diameters of 30–200 nm), containing biofunctional molecules such as microRNAs or enzymes [120–124]. Exosomes participate in cell-to-cell physiologic communication events, but also in disease progressions such as in tumors [120–124], by carrying the encapsulated signaling molecules to other cells/tissues in the body.

Our research group is currently developing saporin-encapsulated exosomes for specific receptor targeting and to promote effective cytosolic delivery by modification of functional peptides on exosomal membranes. Hereafter, we introduce the novel techniques for exosome-based saporin delivery with a detailed clarification of exosomal characteristics and peptide-based techniques.

Because of the pharmaceutical advantages of naturally occurring exosomes such as (1) effective knowledge of cell-to-cell communication extracellular routes; (2) absence of cytotoxicity; (3) controlled immunogenicity; (4) constitutive secretion; (5) encapsulation of additional biofunctional molecules; and (6) co-expression of functional proteins in membranes, they are expected to be next-generation therapeutic carriers [125,126]. Especially, a more sophisticated exosome-based delivery is considered making use of self-exosomes derived from single patients, to prevent any serious immunological-related problem. However, the high abundance of exosomes circulating in serum having negatively charged exosomal membranes [125,127–129], which repel negatively charged cellular membranes, would compete for their cellular uptake, and, therefore, methods for increasing cellular exosomal uptake efficacy must be developed to attain most effective intracellular delivery of exosomal contents.

Recently, our research group found that macropinocytosis (accompanied by actin reorganization, ruffling of the plasma membrane, and engulfment of large volumes of extracellular fluid) is an important entry route for the exosome uptake [130]. Stimulation of EGF receptor (EGFR) by exposing the receptor to natural ligands such as EGF induces signal transductions pathways leading to macropinocytosis via activation of the Rho family of small GTPase Rac [131,132]. When A431 cells (human epidermoid carcinoma, high expressor of EGFR) were treated with GFP-tagged exosomes (20 µg/mL) for 24 h at 37 °C, addition of the EGF (500 nM) significantly increased cellular uptake of the exosomes by approximately 27-fold [130]. Treatment with the macropinocytosis inhibitor,

5-(*N*-ethyl-*N*-isopropyl) amirolide (EIPA) instead reduced the internalization efficacy of exosomes. In a similar way, C-X-C chemokine receptor type 4 (CXCR4)-mediated macropinocytosis [133] was stimulated by stromal cell-derived factor 1α (SDF-1α), which enhanced cellular uptake of the exosomes (approximately 2.3-fold by co-treatment of SDF-1α (200 nM)) [130]. These results suggest that activation of the cancer-related receptors may induce an enhanced cellular uptake of exosomes by macropinocytotic entry route. Then, we examined delivery of saporin by encapsulation in exosomes. A431 cells were treated with saporin-encapsulated exosomes (4 µg/mL of the total exosomes, loaded by usage of electroporation) for 48 h at 37 °C, prior to WST-1 assay. Addition of EGF significantly enhanced cytotoxicity of saporin-encapsulated exosomes (in comparison to the absence of EGF), suggesting once again that induction of macropinocytosis by EGF treatment increases cellular uptake of saporin-encapsulated exosomes [130].

We further developed arginine-rich peptide-modified exosomes, which are able to actively induce macropinocytotic cellular uptake [134,135]. Oligoarginine has been shown to induce micropinocytosis via proteoglycans (e.g., syndecan-4) on plasma membranes [136,137]. Oligoarginine peptide (Rn; n = 4–16)-modified exosomes with a cross-linker (N-ε-maleimidocaproyl-oxysulfosuccinimide ester) were prepared by simple mixing of the peptides and target exosomes and the effects of peptide modification on cellular exosome uptake were evaluated [135]. Relative cellular uptake of Rn-conjugated GFP-tagged exosomes (20 µg/mL) was assayed in 10% serum-containing cell culture medium for 24 h at 37 °C into CHO-K1 cells. The modification of R8 peptides on exosomal membranes showed to greatly increase cellular uptake efficacy (29-fold higher than that of intact exosomes). Overall, our results also showed that the number of arginine residues in the peptide sequence modified on the exosomal membranes differently affected the induction of macropinocytosis and cellular uptake efficacy of the exosome. For example, modification of R16 peptides on exosomal membranes resulted in a slightly lower increased cellular uptake efficacy (18-fold higher than that of intact exosomes) [135]. However, when we artificially encapsulated saporin in exosomes (saporin-exosome), and compared to a saporin-exosome with modified R16 peptides, the latter showed higher biological activities than that of R8 peptide [135]. These results suggest that cytosolic release of saporin may differ when delivered by the exosomes.

We also developed a technique to obtain enhanced cytosolic release of exosomal contents in living cells by a simple method, namely using cationic lipids and pH-sensitive fusogenic peptide, GALA (amino acid sequence: WEAALAEALAEALAEHLAEALAEALEALAA) [125]. GALA peptide was designed to mimic viral fusion protein sequences that mediate escape of virus gene from acidic endosomes to cytosol [138]. We first studied the pH-sensitive fusogenic peptide GALA effects to enhance disruption of endosomal/exosomal membranes during endocytosis of exosomes. We already reported that cationic lipids/GALA complex is a useful system for efficient cytosolic delivery of protein loads [139,140]. In the current research, this system was applied to increase both cellular uptake and cytosolic release of exosomes. Cationic lipids were also used as an "adhesive" to paste the GALA peptide, which carries negative charges deriving from the seven glutamic acids [139,140], onto exosomal surface. As a model macromolecule, Texas red-labeled dextran was encapsulated into exosomes by electroporation, while addition of GALA to the complex of exosome and cationic lipids significantly enhanced cytosolic release of dextran inside cells, indicating that GALA may contribute to disruption of both endosomal and exosomal membranes [139,140]. Saporin-encapsulated exosomes in this combinatorial treatment, showed GALA peptide together with cationic lipids increased RIP activity of saporin, which was originally encapsulated in exosomes [125].

By combined usage of biologically functional peptides, we established simple techniques to enhance macropinocytotic uptake and favor cytosolic release of exosomal contents. Due to their pharmaceutical advantages, exosomes are expected to be the next-generation therapeutic carriers, with these experimental approaches being considered to strongly contribute to effective intracellular delivery of toxins in anti-cancer settings.

7.3. Suicide Gene Therapy

ITs with chimeric fusions represent the most diffused attempts to develop toxin-based targeted therapeutics. They display all major drawbacks common to proteinaceous biotechnological drugs, such as costly and time-consuming production and purification while potentially triggering immune responses. In addition, their cytotoxicity is linked to multiple parameters, such as accessibility and penetration inside the tumor mass, specific internalization into target cells, followed by an efficient release of the toxin in the cytoplasm, where it exerts the catalytic activity. Binding and internalization of the IT or chimeric toxin can be blocked through the loss or downregulation of targeted specific receptors by tumor cells, especially after repeated administrations [141]. To overcome some of these limitations, suicide gene therapy has been proposed as an alternative, making use of nucleic acids encoding the toxin domain under control of specific promoters. DNA holds several advantages: its production and management are cheaper and less time-consuming, as compared to manufacturing proteins; low doses are immediately effective, since few molecules produced in the cytosolic compartment are sufficient to achieve the killing activity. DNA-based therapeutics is both less immunogenic and less prone to induce drug resistance. The expression of the effector protein can be tightly controlled by inserting in the construct selected regulatory elements, like inducible or tumor-specific promoters to prevent detrimental expression of the toxin within healthy tissues. The most challenging step in the suicide gene therapy concerns an efficient and safe delivery of the DNA to cancer cells, and two main strategies have been developed for this purpose. Recombinant disarmed viruses or non-viral vectors have been investigated and both display some advantages and constrains. As long as an ideal system for all purposes has not been identified yet, the employment of the right vector has to be established according to the specific characteristics of the disease to be treated. To date, only a few toxins have been studied for cancer suicide gene therapy approaches with diphtheria toxin (DT) being considered a breakthrough, as it has been shown to be effective in numerous cellular and animal models [142]. Two phase I/II clinical trials have been performed where the DT expression was regulated by the H19 promoter, a well-known tumor-associated gene. The H19-DT construct was complexed with the synthetic polycation polyethylenimine (PEI) and administrated intratumorally in patients with recurrent, multiple non muscle invasive bladder tumors, resulting in a prolonged time to recurrence in responding patients (64% at 3 months, 40% at 2 years) [143]. Concerning plant RIPs, a direct intratumoral injection of a saporin-encoding plasmid under cytomegalovirus promoter complexed with the cationic lipid DOTAP in a melanoma mouse model determined a significant reduction of tumor growth with the antitumor effect being enhanced by repeated administrations. By replacing two key amino acids essential for the saporin catalytic activity (SAPKQ mutant), the anti-tumor effect was abolished, demonstrating that it was specifically due to the RIP N-glycosidase activity [14,144]. DNA constructs expressing ricin A chain or *Pseudomonas aeruginosa* exotoxin A genes under the control of the thyroid hormone promoter were delivered by retroviral vectors to glioblastoma cells and were able to eradicate brain tumors in rats [145]. So far, the therapeutic use of cationic lipid/polymers was mainly limited to intratumoral injections because their delivery capacity is prominently hampered by serum components and systemic toxicity occurs. Therefore, this approach is mostly confined to accessible tumors whose surface antigen markers have not been identified yet. Toxin-mediated gene therapy can be used to reduce tumor mass and allow surgery to better eradicate it, or to kill residual tumor cells after tumor surgery. To surmount these restrictions and extend the application of such an approach, masking agents like PEG were included in these formulations, increasing the circulation time in the bloodstream and reducing liposome toxicity. The selectivity towards target cells of toxin DNA constructs could be improved by concomitant association of antibodies or peptides. For instance, basic fibroblast growth factor 2, modified to carry a lysine stretch, was shown to effectively bind plasmid DNA encoding saporin and to deliver it to fibroblast growth factor receptor (FGFR) bearing cells, leading to cell death [146].

By encapsulating a DT-expressing plasmid in poly (lactic-co-glycolic-acid) (PLGA) nanoparticles modified with the carboxy-terminal–binding domain of the *Clostridium perfringens* enterotoxin

(CPE), the authors could confer them specificity towards the claudin-3/4 expressing cells [147]. The tumor-specific expression of DT was warranted by the transcriptional control of the p16 promoter. P16 protein is overexpressed in the majority of ovarian cancers correlating with tumor progression and poor prognosis. Nanoparticles were tested in chemotherapy-resistant ovarian cancer cells and within 12 h after intraperitoneal injection in mice harboring ovarian cancer xenografts, they caused a significant inhibition of tumor growth [147]. This is an approach coping therapeutic efficacy with a high safety profile, and could easily be extended to other tumors overexpressing claudin-3/4, such as pancreatic, breast, and prostate cancers. This represents a considerable step forward in the toxin-based suicide gene therapy for its high versatility.

7.4. Vaccines

Since Georgi Markov, the Bulgarian novelist and playwright, dissident of the communist regime was killed in London with a ricin pellet shot by a specially modified umbrella in the 1978 [148], type II RIPs have gained widespread attention as potential chemical weapons. Ricin is one of the deadliest poisons in nature; just a single ricin molecule entering the cytosol inactivates over 1.500 ribosomes per minute, causing immediate cell death whereas as little as 500 µg ricin may kill an adult. Initial symptoms in humans may occur within 6–8 h of exposure and clinical symptoms typically progress over 4 to 36 h. Death from ricin poisoning takes place within 36 to 72 h of exposure, depending on the route of exposure and the dose received [149]. People could be exposed to it through the air, food or water. It is unlikely to be absorbed through intact skin; however, the contact with ricin powder or ricin-based products may cause severe irritation and pain [149]. In the 1940s, ricin had been experimented with by U.S. military as a possible warfare agent and was reported in the 1980s as being a warfare agent used in Iraq. In November 2003, the U.S. Secret Service intercepted a letter addressed to the President and delivered to the White House, after it was found to contain ricin. Ten years later, an envelope that preliminarily tested positive for ricin has been again delivered to an U.S. President. Therefore, the need to quickly detect ricin in environmental samples, beverages or food matrices and especially to be able to neutralize or prevent its lethal effects in the case of a biological threat attack has become more and more urgent. Several technologies have been developed so far to allow for sensitive and rapid ricin detection assays. Radioimmunoassay, enhanced colorimetric and chemiluminescence ELISA, immuno-polymerase chain reaction, capillary electrophoresis/liquid chromatography coupled with mass spectrometry, array biosensors, microfluid chip-based immunoassay have been successfully tested on minute amount of ricin, showing sensitivity to power detection even at picomolar levels [149,150]. Medical treatment of patients after ricin exposures is largely symptomatic and currently there is no available antidote. Upon ricin inhalation, the likely symptoms are respiratory distress, fever, nausea, and tightness in the chest, followed by pulmonary edema, low blood pressure and respiratory failure, leading to death. The treatment is based on oxygen administration, bronchodilators, endotracheal intubation and supplemental positive end-expiratory pressure [149]. Significant efforts to develop a recombinant anti-ricin vaccine, as a countermeasure prevention of a biological attack, have been performed, ricin being classified as a biological threat by the U.S. Centers for Disease Control and Prevention. Therefore, several Phase I clinical trials have been launched [151] with two vaccines: RiVax™ and RVEc™. RiVax consists of a recombinant RTA carrying two mutations (V76M, Y80A) not affecting the tertiary structure of the protein, which abrogate both the enzymatic activity and the ability to induce vascular leak syndrome in humans. The protein has been expressed in bacteria and is immunogenic and not toxic in mice, rabbits and humans [152]. In a study on Rhesus Macaques, the parenteral immunization with RiVax elicited a serum antibody response that completely protected them from a lethal dose of aerosolized ricin [153]. The second vaccine candidate, RVEc™, has been developed by investigators at the United States Army Medical Research Institute of Infectious Diseases (USAMRIID) and consists in a truncated version of RTA lacking the C-terminal 68 residues as well as a small hydrophobic loop in the N-terminus, containing a potential protease-sensitive site on the protein surface [154]. Despite deletions, RVEc's 3D structure is superimposable onto RTA and it also elicits protective antibodies in rodents, New Zealand white rabbits, and African green monkeys [154–157]. A pilot clinical trial in humans has been carried with RiVax, by injecting three groups of five normal volunteers three times, at monthly intervals with 10, 33, or 100 μg of the vaccine. Ricin neutralizing antibodies were elicited in 1/5 individuals in the low-dose group, 4/5 in the intermediate-dose group, and in all in the high-dose group. Anti-RTA IgGs purified from pools of positive sera were effective in protecting mice from ricin intoxication. Side effects were limited in seven individuals to a modest increase in creatine phosphokinase (CPK) enzyme that was possibly correlated to the intramuscular (i.m.) injection, and a significant increase in only one individual. Nevertheless, the duration of the antibody responses was limited, spanning from 14 to 127 days after the third vaccination [152]. A second pilot phase I, dose escalation clinical trial using RiVax/alum, based on the results obtained on mice showing an enhanced response by approximately 10-fold and protective for at least a year [158,159]. The good manufacturing practice (GMP)-produced vaccine was adsorbed to aluminum salt adjuvant and three dose levels (1, 10 or 100 μg) were administrated to four or five volunteers per group. RiVax/alum induced higher titers of both total and neutralizing antibodies and in some volunteers, the titers were long lasting (five of eight were positive after one year). All volunteers experienced one or more toxicities associated with the i.m. injection [159]. The next steps will include tests on non-human primates and larger clinical trials to confirm the induction of a longer-lasting response. Concomitantly, a phase 1 escalating, multiple-dose study evaluating the safety and immunogenicity of RVEc^{TM} in healthy adults was performed showing that the vaccine was well tolerated and immunogenic at 20 and 50 μg dose levels. Significant ricin-specific ELISA antibody levels were seen in 100% of volunteers after the third dose [157]. The frequencies and types of adverse events experienced after RVEc^{TM} administration were similar to those reported for RiVax^{TM} vaccine, likely reflecting muscle damage associated to i.m. injections, as also reported for other approved vaccines [157]. Two phase I clinical trials are currently withdrawing volunteers (NCT02385825 and NCT02386150) to determine the safety and immunogenicity of a series of three primary vaccinations and a booster vaccination of RVEC^{TM}, administered at 10, 50, or 75 μg intramuscularly or at 10 or 20 μg intradermally to healthy adults. These studies are aimed at evaluating if the vaccine will display an acceptable safety profile and if it will elicit anti-ricin antibodies and ricin toxin-neutralizing antibodies in vaccine recipients.

One of the major hurdles to evaluating the vaccine efficacy in human beings is represented by the fact that the titer of ricin-specific serum antibodies, as well as the toxin-neutralizing antibody levels, not being accepted as definitive predictors of protective immunity [160].

7.5. Strategies to Promote Endosomal Escape

The number of targeted toxins under investigation in the tumor treatment has considerably risen over time, but in spite of their elevated toxicity, their efficacy in animal models remains far below expectations—a major reason being their limited reaching of cytosolic target ribosomes, due to either cell surface recycling after receptor-mediated internalization or to proteolytic degradation in lysosomes. Several strategies have been developed to improve the toxin performance by facilitating their release into the cytosol, where the toxic activity occurs. Different chemicals, cell-penetrating or fusogenic peptides and light-inducible techniques have been explored to destabilize the endosomal membrane integrity, finally leading to pore formation and content release [161,162]. By increasing the efficiency of endosomal escape, lower doses of drugs and a reduced number of administrations could be employed, decreasing immunogenicity and side effects, while retaining therapeutic effectiveness.

Cell penetrating peptides (CPPs) are 10–16 amino acids peptides originating from parts of viral, bacterial, insect and mammalian proteins. They are mainly structurally different from each other with no consensus sequence [163]. Among the most widely studied is the protein transduction domain TAT from the human HIV1 transcriptional activator protein Tat, *Drosophila* penetratin, herpes simplex virus structural protein VP22, a short amphipathic peptide Pep-1 and the PreS2-domain of hepatitis B virus surface antigen and a membrane translocation sequence from the human Kaposi. Three different CPPs were fused to dianthin and two of them increased the cytotoxicity of a toxin-based conjugate [164]. A fusion protein between the PreS2-domain and saporin exhibited an enhanced anti-tumor effect in a breast cancer mouse model [165]. Another CPP when appended to Gelonin, the resulting conjugate exhibited significantly improved tumoricidal effects and was shown to inhibit tumor growth in a xenograft mouse model of adenocarcinoma [166]. On the other hand, we found that arginine residues in CPP amino acid sequence significantly affect tumor accumulation in in vivo imaging [167], and further optimized methodology for CPP-based RIP delivery to tumor should be established.

In addition to CPPs, other peptides that support drug activation, release and accumulation can be combined with protein toxins for tumor-specific activation. Proteinase recognition sequence [168], endosomal translocation motif derived from diphtheria toxin [169], and cellular protein retention signals such as KDEL have been used to improve ricin A chain and PEA cytotoxicity [170].

Another way investigated to disrupt endosomal membranes favoring the drug escape is represented by the viral-derived fusogenic peptide GALA, a 30 amino acid peptide formed by a glutamic acid-alanine-leucine-alanine repeat. When pH drops to 5 in the acidified endosome, a conformational change occurs in GALA peptide leading to the formation of an amphipathic alpha-helix that binds to the membrane, forms a transmembrane pore and causes bilayer disassembling [171]. Moreover, non-peptidic substances able to buffer protons and swell when protonated, such as tertiary amine groups, cationic polymers, polyamidoamine dendrimers can be utilized to obtain endosome leakage through the so-called "proton sponge effect" [172]. To maintain the acidic pH, endosomes replace protons captured by cargo by generating an influx of protons and therefore of chloride ions and water, resulting in an osmotic rupture of the membrane [173]. Our research group developed a novel L17E endosomolytic peptides derived from the cationic and membrane-lytic spider venom peptide M-lycotoxin, showing that a combinatorial treatment of the peptide with saporin significantly enhanced saporin biological activity [174].

Saponins, from Latin soap, are plant-derived compounds that can be mainly divided into two groups depending on their chemical structure, steroids and triterpenoids, the latter bringing linear, as well as, branched sugar chains. Most saponins show hemolytic and membrane permeabilizing activity, which are employed by plants as defensive strategy from predators. Glycosylated triterpenoid saponins were demonstrated to substantially increase the effect of plant RIPs saporin and dianthin in murine mammary adenocarcinoma and colon carcinoma, respectively, by improving the transfer of the toxic compound into the cytosol upon co-administration [175–177]. Side effects like loss of body weight, altered blood parameters and immune response were only moderate and commonly reversible [178]. However, a main problem with the therapeutic use of saponins, as well as for other endosomal escape enhancers, is that they behave independently compared to the targeted toxins when they are co-administrated in terms of absorption, biodistribution, metabolism and excretion, aside from cellular internalization and release [179]. Hence, the major challenge is to synchronize the enhancer and the toxin effect to warrant that both are present simultaneously at the site of interaction. Several studies investigated the timing of co-administration of saponins and targeted toxins to obtain the utmost combined effect, avoiding off-target effects.

Another way to promote escape from endo-lysosomal compartments consists of light activation of an amphiphilic sensitizer taken up by the target cells, known as photochemical internalization (PCI) [180]. PCI was established as a drug delivery technology for the release of molecules that are sequestered in endosomes after endocytosis leading to the formation of reactive oxygen species (ROS), thus causing lipid peroxidation, protein oxidation and finally resulting in the rupture of endo-lysosomal vesicles and thus cytosolic release of the content [181]. This PCI technique has been successfully applied in vivo. For enhanced endosomal escape of the CD133-targeting immunotoxin AC133 mAb-saporin conjugate (PCI$_{AC133-saporin}$), Bostad et al. developed and showed that AC133-saporin co-localizes with the PCI-photosensitizer TPCS$_{2a}$, which upon light exposure induces cytosolic release of AC133-saporin. In vivo studies, targeting of CD133+ WiDr tumors with PCI$_{AC133-saporin}$ after one cycle of systemic injection and light exposure showed significant delay in tumor growth [182]. An experimental study is under way in Rosemblum's lab to test the enhancement effects of PCI on VEGF121-rGelonin fusion [183], since, contrary to the triterpenoid saponins, which are selective in their interaction for few, type I RIPs, this technique may apply to promoting endosomal efflux of any endocytosed molecule. PCI of EGF-Gelonin resulted in inhibitory effects on squamous cell carcinoma in a mouse xenograft model and reduction of tumor perfusion and necrosis induction in head and neck squamous cell carcinoma tumors [184].

Recently, a mutant saporin, where a cysteine has been introduced instead of an Ala 157 in an external loop, has been generated to allow conjugation with a small inhibitor molecule targeting integrins $\alpha v\beta 3/\alpha v\beta 5$. This conjugate was demonstrated to be highly cytotoxic and selective in breast cancer and melanoma cell lines. In addition, a tricomponent conjugate in which another cytotoxic agent, auristatin F, a potent microtubule inhibitor, was inserted between saporin and the small integrin inhibitor was tested showing it still maintained high activity and preserved specificity, but was not more effective in vitro as the first conjugate tested, deserving further investigation in xenograft tumor models [185]. This small-molecule RIP bioconjugate approach is highly versatile since it could be broadly applied using other small molecules towards different cancer-associated targets that are already available in large numbers [186].

7.6. Employment of Plant Type I RIPs in Other Pathological Models

Instead of anti-cancer treatment, RIPs have been also used to eliminate the cholinergic inputs to hippocampus [187], monocyte-derived inflammatory dendritic cells [188], and alloreactive CD4$^+$ and CD8$^+$ T-cells through CD137-mediated internalization [189]. The cholinergic projections to ventral subiculum were selectively eliminated using 192 IgG-saporin conjugates, and this model is useful for investigation of the cholinergic modulation of subicular theta-gamma activity on spatial learning and memory functions in rats. Eliminations of cholinergic inputs to ventral subiculum significantly reduced the subicular theta and enhanced the gamma activity during active wake and REM sleep states. The spatial learning was also severely impaired following cholinergic elimination of ventral subiculum [190]. Monocytes play crucial roles for the regulation of tissue homeostasis and the control of pathogens. However, accumulation of monocyte-derived inflammatory dendritic cells participates in the pathogenesis and persistence of inflammatory diseases including psoriasis and Crohn's disease. Alonso et al. developed a strategy for avoidance of the monocyte intermediates to deplete inflammatory dendritic cells through anti-CD209 antibody conjugated to saporin. Mice with an abundance of inflammatory dendritic cells as a consequence of lipopolysaccharide exposure were treated with anti-CD209 antibody conjugated to saporin, resulting in depletion of CD209 positive dendritic cells [188]. Graft-versus-host disease (GVHD) shows a major limitation to allogeneic hematopoietic stem cell transplantation. CD137 signaling is considered to be involved in multiple stages of GVHD development. Lee et al. developed a method to selectively eliminate alloreactive CD4$^+$ and CD8$^+$ T-cells through CD137-mediated internalization of anti-CD137 mAbs conjugated with saporin [189]. CD137-expressing cells were killed by the treatment with the conjugate, and transfer of donor T-cells after allodepletion showed no evident GVHD [189].

Substance P-saporin (SP-SAP) deserves a special mention. This is a neuropeptide-toxin conjugate aimed at selectively destroying neurons expressing the pain-related receptor for substance P (neurokinin-1 receptor, NK-1R). It is currently in Phase I clinical study recruiting terminal cancer patients with intractable pain (NCT02036281). Even if nociception is a complex process and substance P is not the only factor involved, intrathecal administration of SP-SAP has been shown to cause a robust change in a variety of pain states in animal models, including dogs [191,192]. In addition, the lumbar delivery of SP-SAP, but not SAP alone, in dogs resulted in a specific, dose-dependent reduction of superficial NK-1R bearing lumbar neurons. Notably, adverse effects in animal behavior, chronic motor functions disturbance, changes in blood pressure or heart rate, progression of the injury or any evidence of additional loss of neurons or function were not detected. Pharmacokinetic studies demonstrated that SP-SAP is cleared relatively rapidly from the cerebrospinal fluid with a clearance of over 50% occurring within 30–60 min [193].

8. Conclusions and Perspectives

A few hundred RIP-based ITs and chimeric fusions have been developed so far, but only a bacterial-derived toxin, Denileukin diftitox, has been approved by the FDA as a cancer therapy agent. Most of the clinical studies did not progress beyond Phase 1 study, due to the triggering of immune responses and considerable side effects, such as VLS. Hematologic malignancy patients have experienced on average somewhat less of these problems being immunocompromised; however, after a cycle of IT administration, half to almost 100% of patients treated having solid tumors develop anti-toxin antibodies. Efforts focused to reduce such negative effects are aimed to increase tumor penetration capability in solid tumors, while maintaining the target selectivity, by employing humanized single chain antibody fragments, peptides or small molecules and by masking antigens through PEGylation approaches. Biotechnology approaches as to limit VLS and RIP immune responses with de-immunization strategies, as with the Bouganvilleae de-bouganin toxin, should be pursued and prioritized. Nanotechnology is contributing in this direction by providing nanovectors able to increase the half-life of drugs, to bypass the hydrophobic nature of some molecules, to deliver multiple therapeutics and/or diagnostic agents and to reduce administration by controlled or sustained release of the drug. Along these lines, exosomes are emerging as suitable carriers of biological materials including toxins, providing a safe delivery route. The target selectivity can be achieved by modifying the vesicles, but also by engineering producing cells, increasing the number of possible combinations for the preparation of therapeutic agents. RIPs keep their potential unaltered with these new technologies, altogether opening further to clinical development opportunities.

Acknowledgments: This work was supported by the Italian Ministry of Health (GR-2011-02351220 to R.V.). We are grateful to the Recombinant Immunotoxin Collaborative Group members for excellent scientific support.

Author Contributions: M.S.F. and R.V. conceived the review. M.S.F., M.K., I.N. and R.V. wrote, revised and approved the manuscript.

Conflicts of Interest: The authors declare no conflict of interest.

References

1. Stirpe, F. Ribosome-inactivating proteins: From toxins to useful proteins. *Toxicon* **2013**, *67*, 12–16. [CrossRef] [PubMed]
2. Endo, Y.; Tsurugi, K. RNA N-glycosidase activity of ricin A-chain. Mechanism of action of the toxic lectin ricin on eukaryotic ribosomes. *J. Biol. Chem.* **1987**, *262*, 8128–8130. [PubMed]
3. Endo, Y.; Mitsui, K.; Motizuki, M.; Tsurugi, K. The mechanism of action of ricin and related toxic lectins on eukaryotic ribosomes. The site and the characteristics of the modification in 28 S ribosomal RNA caused by the toxins. *J. Biol. Chem.* **1987**, *262*, 5908–5912. [PubMed]
4. Benatti, L.; Saccardo, M.B.; Dani, M.; Nitti, G.; Sassano, M.; Lorenzetti, R.; Lappi, D.A.; Soria, M. Nucleotide sequence of cDNA coding for saporin-6, a type-1 ribosome-inactivating protein from Saponaria officinalis. *Eur. J. Biochem.* **1989**, *183*, 465–470. [CrossRef] [PubMed]
5. Arabidopsis Genome, I. Analysis of the genome sequence of the flowering plant Arabidopsis thaliana. *Nature* **2000**, *408*, 796–815. [CrossRef] [PubMed]
6. Di Cola, A.; Frigerio, L.; Lord, J.M.; Ceriotti, A.; Roberts, L.M. Ricin A chain without its partner B chain is degraded after retrotranslocation from the endoplasmic reticulum to the cytosol in plant cells. *Proc. Natl. Acad. Sci. USA* **2001**, *98*, 14726–14731. [CrossRef] [PubMed]
7. McLaughlin, J.E.; Boyer, J.S. Sugar-responsive gene expression, invertase activity, and senescence in aborting maize ovaries at low water potentials. *Ann. Bot.* **2004**, *94*, 675–689. [CrossRef] [PubMed]
8. Lord, J.M.; Roberts, L.M.; Robertus, J.D. Ricin: Structure, mode of action, and some current applications. *FASEB J.* **1994**, *8*, 201–208. [CrossRef] [PubMed]

9. Olsnes, S.; Pihl, A. Different biological properties of the two constituent peptide chains of ricin, a toxic protein inhibiting protein synthesis. *Biochemistry* **1973**, *12*, 3121–3126. [CrossRef] [PubMed]

10. Schrot, J.; Weng, A.; Melzig, M.F. Ribosome-inactivating and related proteins. *Toxins (Basel)* **2015**, *7*, 1556–1615. [CrossRef] [PubMed]

11. Jimenez, P.; Tejero, J.; Cordoba-Diaz, D.; Quinto, E.J.; Garrosa, M.; Gayoso, M.J.; Girbes, T. Ebulin from dwarf elder (Sambucus ebulus L.): A mini-review. *Toxins (Basel)* **2015**, *7*, 648–658. [CrossRef] [PubMed]

12. Walsh, T.A.; Morgan, A.E.; Hey, T.D. Characterization and molecular cloning of a proenzyme form of a ribosome-inactivating protein from maize. Novel mechanism of proenzyme activation by proteolytic removal of a 2.8-kilodalton internal peptide segment. *J. Biol. Chem.* **1991**, *266*, 23422–23427. [PubMed]

13. Pittaluga, E.; Poma, A.; Tucci, A.; Spano, L. Expression and characterisation in E. coli of mutant forms of saporin. *J. Biotechnol.* **2005**, *117*, 263–266. [CrossRef] [PubMed]

14. Zarovni, N.; Vago, R.; Solda, T.; Monaco, L.; Fabbrini, M.S. Saporin as a novel suicide gene in anticancer gene therapy. *Cancer Gene Ther.* **2007**, *14*, 165–173. [CrossRef] [PubMed]

15. Lombardi, A.; Bursomanno, S.; Lopardo, T.; Traini, R.; Colombatti, M.; Ippoliti, R.; Flavell, D.J.; Flavell, S.U.; Ceriotti, A.; Fabbrini, M.S. Pichia pastoris as a host for secretion of toxic saporin chimeras. *FASEB J.* **2010**, *24*, 253–265. [CrossRef] [PubMed]

16. Marshall, R.S.; D'Avila, F.; Di Cola, A.; Traini, R.; Spano, L.; Fabbrini, M.S.; Ceriotti, A. Signal peptide-regulated toxicity of a plant ribosome-inactivating protein during cell stress. *Plant J.* **2011**, *65*, 218–229. [CrossRef] [PubMed]

17. Errico Provenzano, A.; Posteri, R.; Giansanti, F.; Angelucci, F.; Flavell, S.U.; Flavell, D.J.; Fabbrini, M.S.; Porro, D.; Ippoliti, R.; Ceriotti, A.; et al. Optimization of construct design and fermentation strategy for the production of bioactive ATF-SAP, a saporin based anti-tumoral uPAR-targeted chimera. *Microb. Cell Fact.* **2016**, *15*, 194. [CrossRef] [PubMed]

18. Bagga, S.; Seth, D.; Batra, J.K. The cytotoxic activity of ribosome-inactivating protein saporin-6 is attributed to its rRNA N-glycosidase and internucleosomal DNA fragmentation activities. *J. Biol. Chem.* **2003**, *278*, 4813–4820. [CrossRef] [PubMed]

19. Rajamohan, F.; Pugmire, M.J.; Kurinov, I.V.; Uckun, F.M. Modeling and alanine scanning mutagenesis studies of recombinant pokeweed antiviral protein. *J. Biol. Chem.* **2000**, *275*, 3382–3390. [CrossRef] [PubMed]

20. Baker, N.A.; Sept, D.; Joseph, S.; Holst, M.J.; McCammon, J.A. Electrostatics of nanosystems: Application to microtubules and the ribosome. *Proc. Natl. Acad. Sci. USA* **2001**, *98*, 10037–10041. [CrossRef] [PubMed]

21. Monzingo, A.F.; Robertus, J.D. X-ray analysis of substrate analogs in the ricin A-chain active site. *J. Mol. Biol.* **1992**, *227*, 1136–1145. [CrossRef]

22. Marsden, C.J.; Fulop, V.; Day, P.J.; Lord, J.M. The effect of mutations surrounding and within the active site on the catalytic activity of ricin A chain. *Eur. J. Biochem.* **2004**, *271*, 153–162. [CrossRef] [PubMed]

23. Korennykh, A.V.; Correll, C.C.; Piccirilli, J.A. Evidence for the importance of electrostatics in the function of two distinct families of ribosome inactivating toxins. *RNA* **2007**, *13*, 1391–1396. [CrossRef] [PubMed]

24. Day, P.J.; Lord, J.M.; Roberts, L.M. The deoxyribonuclease activity attributed to ribosome-inactivating proteins is due to contamination. *Eur. J. Biochem.* **1998**, *258*, 540–545. [CrossRef] [PubMed]

25. Valbonesi, P.; Barbieri, L.; Bolognesi, A.; Bonora, E.; Polito, L.; Stirpe, F. Preparation of highly purified momordin II without ribonuclease activity. *Life Sci.* **1999**, *65*, 1485–1491. [CrossRef]

26. Barbieri, L.; Valbonesi, P.; Righi, F.; Zuccheri, G.; Monti, F.; Gorini, P.; Samori, B.; Stirpe, F. Polynucleotide: Adenosine glycosidase is the sole activity of ribosome-inactivating proteins on DNA. *J. Biochem.* **2000**, *128*, 883–889. [CrossRef] [PubMed]

27. Peumans, W.J.; Hao, Q.; Van Damme, E.J. Ribosome-inactivating proteins from plants: More than RNA N-glycosidases? *FASEB J.* **2001**, *15*, 1493–1506. [CrossRef] [PubMed]

28. Sturm, M.B.; Tyler, P.C.; Evans, G.B.; Schramm, V.L. Transition state analogues rescue ribosomes from saporin-L1 ribosome inactivating protein. *Biochemistry* **2009**, *48*, 9941–9948. [CrossRef] [PubMed]

29. Endo, Y.; Tsurugi, K. The RNA N-glycosidase activity of ricin A-chain. *Nucleic Acids Symp. Ser.* **1988**, *19*, 139–142.

30. Vago, R.; Marsden, C.J.; Lord, J.M.; Ippoliti, R.; Flavell, D.J.; Flavell, S.U.; Ceriotti, A.; Fabbrini, M.S. Saporin and ricin A chain follow different intracellular routes to enter the cytosol of intoxicated cells. *FEBS J.* **2005**, *272*, 4983–4995. [CrossRef] [PubMed]

31. Vago, R.; Ippoliti, R.; Fabbrini, M.S. Current status & Biomedical applications of Ribosome-inactivating proteins. In *Antitumor Potential and other Emerging Medicinal Properties of Natural Compounds*; Fang, E., Ng, T., Eds.; Springer: Berlin, Germany, 2013; pp. 145–179.

32. Deeks, E.D.; Cook, J.P.; Day, P.J.; Smith, D.C.; Roberts, L.M.; Lord, J.M. The low lysine content of ricin A chain reduces the risk of proteolytic degradation after translocation from the endoplasmic reticulum to the cytosol. *Biochemistry* **2002**, *41*, 3405–3413. [CrossRef] [PubMed]

33. Hogg, T.; Mendel, J.T.; Lavezo, J.L. Structural analysis of a type 1 ribosome inactivating protein reveals multiple LasparagineNacetylDglucosamine monosaccharide modifications: Implications for cytotoxicity. *Mol. Med. Rep.* **2015**, *12*, 5737–5745. [CrossRef] [PubMed]

34. Lodge, J.K.; Kaniewski, W.K.; Tumer, N.E. Broad-spectrum virus resistance in transgenic plants expressing pokeweed antiviral protein. *Proc. Natl. Acad. Sci. USA* **1993**, *90*, 7089–7093. [CrossRef] [PubMed]

35. Lam, Y.H.; Wong, Y.S.; Wang, B.; Wong, R.N.S.; Yeung, H.W.; Shaw, P.C. Use of trichosanthin to reduce infection by turnip mosaic virus. *Plant Sci.* **1996**, *114*, 111–117. [CrossRef]

36. Moons, A.; Gielen, J.; Vandekerckhove, J.; Van der Straeten, D.; Gheysen, G.; Van Montagu, M. An abscisic -acid- and salt-stress-responsive rice cDNA from a novel plant gene family. *Planta* **1997**, *202*, 443–454. [CrossRef] [PubMed]

37. Taylor, S.; Massiah, A.; Lomonossoff, G.; Roberts, L.M.; Lord, J.M.; Hartley, M. Correlation between the activities of five ribosome-inactivating proteins in depurination of tobacco ribosomes and inhibition of tobacco mosaic virus infection. *Plant J.* **1994**, *5*, 827–835. [CrossRef] [PubMed]

38. Di, R.; Tumer, N.E. Pokeweed antiviral protein: Its cytotoxicity mechanism and applications in plant disease resistance. *Toxins (Basel)* **2015**, *7*, 755–772. [CrossRef] [PubMed]

39. Prestle, J.; Schonfelder, M.; Adam, G.; Mundry, K.W. Type 1 ribosome-inactivating proteins depurinate plant 25S rRNA without species specificity. *Nucleic Acids Res.* **1992**, *20*, 3179–3182. [CrossRef] [PubMed]

40. Bonness, M.S.; Ready, M.P.; Irvin, J.D.; Mabry, T.J. Pokeweed antiviral protein inactivates pokeweed ribosomes; implications for the antiviral mechanism. *Plant J.* **1994**, *5*, 173–183. [CrossRef] [PubMed]

41. Kataoka, J.; Habuka, N.; Miyano, M.; Masuta, C.; Koiwai, A. Adenine depurination and inactivation of plant ribosomes by an antiviral protein of Mirabilis jalapa (MAP). *Plant Mol. Biol.* **1992**, *20*, 1111–1119. [CrossRef] [PubMed]

42. Vandenbussche, F.; Peumans, W.J.; Desmyter, S.; Proost, P.; Ciani, M.; Van Damme, E.J. The type-1 and type-2 ribosome-inactivating proteins from Iris confer transgenic tobacco plants local but not systemic protection against viruses. *Planta* **2004**, *220*, 211–221. [CrossRef] [PubMed]

43. Tourlakis, M.E.; Karran, R.A.; Desouza, L.; Siu, K.W.; Hudak, K.A. Homodimerization of pokeweed antiviral protein as a mechanism to limit depurination of pokeweed ribosomes. *Mol. Plant Pathol.* **2010**, *11*, 757–767. [CrossRef] [PubMed]

44. Marshall, R.S.; Jolliffe, N.A.; Ceriotti, A.; Snowden, C.J.; Lord, J.M.; Frigerio, L.; Roberts, L.M. The role of CDC48 in the retro-translocation of non-ubiquitinated toxin substrates in plant cells. *J. Biol. Chem.* **2008**, *283*, 15869–15877. [CrossRef] [PubMed]

45. Vitale, A.; Boston, R.S. Endoplasmic reticulum quality control and the unfolded protein response: Insights from plants. *Traffic* **2008**, *9*, 1581–1588. [CrossRef] [PubMed]

46. Kang, S.W.; Rane, N.S.; Kim, S.J.; Garrison, J.L.; Taunton, J.; Hegde, R.S. Substrate-specific translocational attenuation during ER stress defines a pre-emptive quality control pathway. *Cell* **2006**, *127*, 999–1013. [CrossRef] [PubMed]

47. Orsi, A.; Fioriti, L.; Chiesa, R.; Sitia, R. Conditions of endoplasmic reticulum stress favor the accumulation of cytosolic prion protein. *J. Biol. Chem.* **2006**, *281*, 30431–30438. [CrossRef] [PubMed]

48. van Deurs, B.; Tonnessen, T.I.; Petersen, O.W.; Sandvig, K.; Olsnes, S. Routing of internalized ricin and ricin conjugates to the Golgi complex. *J. Cell Boil.* **1986**, *102*, 37–47. [CrossRef]

49. Amessou, M.; Fradagrada, A.; Falguieres, T.; Lord, J.M.; Smith, D.C.; Roberts, L.M.; Lamaze, C.; Johannes, L. Syntaxin 16 and syntaxin 5 are required for efficient retrograde transport of several exogenous and endogenous cargo proteins. *J. Cell Sci.* **2007**, *120 Pt 8*, 1457–1468. [CrossRef]

50. Wesche, J.; Rapak, A.; Olsnes, S. Dependence of ricin toxicity on translocation of the toxin A-chain from the endoplasmic reticulum to the cytosol. *J. Biol. Chem.* **1999**, *274*, 34443–34449. [CrossRef] [PubMed]

51. Spooner, R.A.; Watson, P.D.; Marsden, C.J.; Smith, D.C.; Moore, K.A.; Cook, J.P.; Lord, J.M.; Roberts, L.M. Protein disulphide-isomerase reduces ricin to its A and B chains in the endoplasmic reticulum. *Biochem. J.* **2004**, *383 Pt 2*, 285–293. [CrossRef]

52. Bellisola, G.; Fracasso, G.; Ippoliti, R.; Menestrina, G.; Rosen, A.; Solda, S.; Udali, S.; Tomazzolli, R.; Tridente, G.; Colombatti, M. Reductive activation of ricin and ricin A-chain immunotoxins by protein disulfide isomerase and thioredoxin reductase. *Biochem. Pharmacol.* **2004**, *67*, 1721–1731. [CrossRef] [PubMed]

53. Bassik, M.C.; Kampmann, M.; Lebbink, R.J.; Wang, S.; Hein, M.Y.; Poser, I.; Weibezahn, J.; Horlbeck, M.A.; Chen, S.; Mann, M.; et al. A systematic mammalian genetic interaction map reveals pathways underlying ricin susceptibility. *Cell* **2013**, *152*, 909–922. [CrossRef] [PubMed]

54. Rapak, A.; Falnes, P.O.; Olsnes, S. Retrograde transport of mutant ricin to the endoplasmic reticulum with subsequent translocation to cytosol. *Proc. Natl. Acad. Sci. USA* **1997**, *94*, 3783–3788. [CrossRef] [PubMed]

55. Lord, J.M.; Spooner, R.A. Ricin trafficking in plant and mammalian cells. *Toxins (Basel)* **2011**, *3*, 787–801. [CrossRef] [PubMed]

56. de Virgilio, M.; Lombardi, A.; Caliandro, R.; Fabbrini, M.S. Ribosome-inactivating proteins: From plant defense to tumor attack. *Toxins (Basel)* **2010**, *2*, 2699–2737. [CrossRef] [PubMed]

57. Slominska-Wojewodzka, M.; Pawlik, A.; Sokolowska, I.; Antoniewicz, J.; Wegrzyn, G.; Sandvig, K. The role of EDEM2 compared with EDEM1 in ricin transport from the endoplasmic reticulum to the cytosol. *Biochem. J.* **2014**, *457*, 485–496. [CrossRef] [PubMed]

58. Sokolowska, I.; Walchli, S.; Wegrzyn, G.; Sandvig, K.; Slominska-Wojewodzka, M. A single point mutation in ricin A-chain increases toxin degradation and inhibits EDEM1-dependent ER retrotranslocation. *Biochem. J.* **2011**, *436*, 371–385. [CrossRef] [PubMed]

59. Spooner, R.A.; Hart, P.J.; Cook, J.P.; Pietroni, P.; Rogon, C.; Hohfeld, J.; Roberts, L.M.; Lord, J.M. Cytosolic chaperones influence the fate of a toxin dislocated from the endoplasmic reticulum. *Proc. Natl. Acad. Sci. USA* **2008**, *105*, 17408–17413. [CrossRef] [PubMed]

60. Herrera, C.; Klokk, T.I.; Cole, R.; Sandvig, K.; Mantis, N.J. A Bispecific Antibody Promotes Aggregation of Ricin Toxin on Cell Surfaces and Alters Dynamics of Toxin Internalization and Trafficking. *PLoS ONE* **2016**, *11*, e0156893. [CrossRef] [PubMed]

61. Cavallaro, U.; Nykjaer, A.; Nielsen, M.; Soria, M.R. Alpha 2-macroglobulin receptor mediates binding and cytotoxicity of plant ribosome-inactivating proteins. *Eur. J. Biochem.* **1995**, *232*, 165–171. [CrossRef] [PubMed]

62. Lillis, A.P.; Van Duyn, L.B.; Murphy-Ullrich, J.E.; Strickland, D.K. LDL receptor-related protein 1: Unique tissue-specific functions revealed by selective gene knockout studies. *Physiol. Rev.* **2008**, *88*, 887–918. [CrossRef] [PubMed]

63. Fabbrini, M.S.; Carpani, D.; Bello-Rivero, I.; Soria, M.R. The amino-terminal fragment of human urokinase directs a recombinant chimeric toxin to target cells: Internalization is toxin mediated. *FASEB J.* **1997**, *11*, 1169–1176. [PubMed]

64. Conese, M.; Nykjaer, A.; Petersen, C.M.; Cremona, O.; Pardi, R.; Andreasen, P.A.; Gliemann, J.; Christensen, E.I.; Blasi, F. alpha-2 Macroglobulin receptor/Ldl receptor-related protein(Lrp)-dependent internalization of the urokinase receptor. *J. Cell Boil.* **1995**, *131 Pt 1*, 1609–1622. [CrossRef]

65. Fabbrini, M.S.; Rappocciolo, E.; Carpani, D.; Solinas, M.; Valsasina, B.; Breme, U.; Cavallaro, U.; Nykjaer, A.; Rovida, E.; Legname, G.; et al. Characterization of a saporin isoform with lower ribosome-inhibiting activity. *Biochem. J.* **1997**, *322 Pt 3*, 719–727. [CrossRef]

66. Rajagopal, V.; Kreitman, R.J. Recombinant toxins that bind to the urokinase receptor are cytotoxic without requiring binding to the alpha(2)-macroglobulin receptor. *J. Biol. Chem.* **2000**, *275*, 7566–7573. [CrossRef] [PubMed]

67. Ippoliti, R.; Lendaro, E.; Benedetti, P.A.; Torrisi, M.R.; Belleudi, F.; Carpani, D.; Soria, M.R.; Fabbrini, M.S. Endocytosis of a chimera between human pro-urokinase and the plant toxin saporin: An unusual internalization mechanism. *FASEB J.* **2000**, *14*, 1335–1344. [CrossRef] [PubMed]

68. Bagga, S.; Hosur, M.V.; Batra, J.K. Cytotoxicity of ribosome-inactivating protein saporin is not mediated through alpha2-macroglobulin receptor. *FEBS Lett.* **2003**, *541*, 16–20. [CrossRef]

69. Zhang, F.; Sun, S.; Feng, D.; Zhao, W.L.; Sui, S.F. A novel strategy for the invasive toxin: Hijacking exosome-mediated intercellular trafficking. *Traffic* **2009**, *10*, 411–424. [CrossRef] [PubMed]

70. Chan, W.L.; Shaw, P.C.; Tam, S.C.; Jacobsen, C.; Gliemann, J.; Nielsen, M.S. Trichosanthin interacts with and enters cells via LDL receptor family members. *Biochem. Biophys. Res. Commun.* **2000**, *270*, 453–457. [CrossRef] [PubMed]

71. Rosenblum, M.G.; Cheung, L.H.; Liu, Y.; Marks, J.W., 3rd. Design, expression, purification, and characterization, in vitro and in vivo, of an antimelanoma single-chain Fv antibody fused to the toxin Gelonin. *Cancer Res.* **2003**, *63*, 3995–4002. [PubMed]

72. Vallera, D.A.; Oh, S.; Chen, H.; Shu, Y.; Frankel, A.E. Bioengineering a unique deimmunized bispecific targeted toxin that simultaneously recognizes human CD22 and CD19 receptors in a mouse model of B-cell metastases. *Mol. Cancer Ther.* **2010**, *9*, 1872–1883. [CrossRef] [PubMed]

73. Yoshida, T.; Chen, C.C.; Zhang, M.S.; Wu, H.C. Disruption of the Golgi apparatus by brefeldin A inhibits the cytotoxicity of ricin, modeccin, and Pseudomonas toxin. *Exp. Cell Res.* **1991**, *192*, 389–395. [CrossRef]

74. Simpson, J.C.; Roberts, L.M.; Lord, J.M. Free ricin A chain reaches an early compartment of the secretory pathway before it enters the cytosol. *Exp. Cell Res.* **1996**, *229*, 447–451. [CrossRef] [PubMed]

75. Wales, R.; Roberts, L.M.; Lord, J.M. Addition of an endoplasmic reticulum retrieval sequence to ricin A chain significantly increases its cytotoxicity to mammalian cells. *J. Biol. Chem.* **1993**, *268*, 23986–23990. [PubMed]

76. Teter, K.; Holmes, R.K. Inhibition of endoplasmic reticulum-associated degradation in CHO cells resistant to cholera toxin, Pseudomonas aeruginosa exotoxin A, and ricin. *Infect. Immunity* **2002**, *70*, 6172–6179. [CrossRef]

77. Geden, S.E.; Gardner, R.A.; Fabbrini, M.S.; Ohashi, M.; Phanstiel Iv, O.; Teter, K. Lipopolyamine treatment increases the efficacy of intoxication with saporin and an anticancer saporin conjugate. *FEBS J.* **2007**, *274*, 4825–4836. [CrossRef] [PubMed]

78. Day, P.J.; Pinheiro, T.J.; Roberts, L.M.; Lord, J.M. Binding of ricin A-chain to negatively charged phospholipid vesicles leads to protein structural changes and destabilizes the lipid bilayer. *Biochemistry* **2002**, *41*, 2836–2843. [CrossRef] [PubMed]

79. Mayerhofer, P.U.; Cook, J.P.; Wahlman, J.; Pinheiro, T.T.; Moore, K.A.; Lord, J.M.; Johnson, A.E.; Roberts, L.M. Ricin A chain insertion into endoplasmic reticulum membranes is triggered by a temperature increase to 37 °C. *J. Biol. Chem.* **2009**, *284*, 10232–10242. [CrossRef] [PubMed]

80. Santanche, S.; Bellelli, A.; Brunori, M. The unusual stability of saporin, a candidate for the synthesis of immunotoxins. *Biochem. Biophys. Res. Commun.* **1997**, *234*, 129–132. [CrossRef] [PubMed]

81. Griffiths, G.D.; Leek, M.D.; Gee, D.J. The toxic plant proteins ricin and abrin induce apoptotic changes in mammalian lymphoid tissues and intestine. *J. Pathol.* **1987**, *151*, 221–229. [CrossRef] [PubMed]

82. Hughes, J.N.; Lindsay, C.D.; Griffiths, G.D. Morphology of ricin and abrin exposed endothelial cells is consistent with apoptotic cell death. *Hum. Exp. Toxicol.* **1996**, *15*, 443–451. [CrossRef] [PubMed]

83. Krajewski, S.; Krajewska, M.; Ellerby, L.M.; Welsh, K.; Xie, Z.; Deveraux, Q.L.; Salvesen, G.S.; Bredesen, D.E.; Rosenthal, R.E.; Fiskum, G.; et al. Release of caspase-9 from mitochondria during neuronal apoptosis and cerebral ischemia. *Proc. Natl. Acad. Sci. USA* **1999**, *96*, 5752–5757. [CrossRef] [PubMed]

84. Boulares, A.H.; Yakovlev, A.G.; Ivanova, V.; Stoica, B.A.; Wang, G.; Iyer, S.; Smulson, M. Role of poly(ADP-ribose) polymerase (PARP) cleavage in apoptosis. Caspase 3-resistant PARP mutant increases rates of apoptosis in transfected cells. *J. Biol. Chem.* **1999**, *274*, 22932–22940. [CrossRef] [PubMed]

85. Komatsu, N.; Nakagawa, M.; Oda, T.; Muramatsu, T. Depletion of intracellular NAD(+) and ATP levels during ricin-induced apoptosis through the specific ribosomal inactivation results in the cytolysis of U937 cells. *J. Biochem.* **2000**, *128*, 463–470. [CrossRef] [PubMed]

86. Narayanan, S.; Surendranath, K.; Bora, N.; Surolia, A.; Karande, A.A. Ribosome inactivating proteins and apoptosis. *FEBS Lett.* **2005**, *579*, 1324–1331. [CrossRef] [PubMed]

87. Iordanov, M.S.; Pribnow, D.; Magun, J.L.; Dinh, T.H.; Pearson, J.A.; Chen, S.L.; Magun, B.E. Ribotoxic stress response: Activation of the stress-activated protein kinase JNK1 by inhibitors of the peptidyl transferase reaction and by sequence-specific RNA damage to the alpha-sarcin/ricin loop in the 28S rRNA. *Mol. Cell. Biol.* **1997**, *17*, 3373–3381. [CrossRef] [PubMed]

88. Sikriwal, D.; Ghosh, P.; Batra, J.K. Ribosome inactivating protein saporin induces apoptosis through mitochondrial cascade, independent of translation inhibition. *Int. J. Biochem. Cell Biol.* **2008**, *40*, 2880–2888. [CrossRef] [PubMed]

89. Shih, S.F.; Wu, Y.H.; Hung, C.H.; Yang, H.Y.; Lin, J.Y. Abrin triggers cell death by inactivating a thiol-specific antioxidant protein. *J. Biol. Chem.* **2001**, *276*, 21870–21877. [CrossRef] [PubMed]

90. Zhang, C.; Gong, Y.; Ma, H.; An, C.; Chen, D.; Chen, Z.L. Reactive oxygen species involved in trichosanthin-induced apoptosis of human choriocarcinoma cells. *Biochem. J.* **2001**, *355*, 653–661. [CrossRef] [PubMed]

91. Hu, R.; Zhai, Q.; Liu, W.; Liu, X. An insight into the mechanism of cytotoxicity of ricin to hepatoma cell: Roles of Bcl-2 family proteins, caspases, Ca(2+)-dependent proteases and protein kinase C. *J. Cell. Biochem.* **2001**, *81*, 583–593. [CrossRef] [PubMed]

92. Authier, F.; Djavaheri-Mergny, M.; Lorin, S.; Frenoy, J.P.; Desbuquois, B. Fate and action of ricin in rat liver in vivo: Translocation of endocytosed ricin into cytosol and induction of intrinsic apoptosis by ricin B-chain. *Cell Microbiol.* **2016**, *18*, 1800–1814. [CrossRef] [PubMed]

93. Polito, L.; Mercatelli, D.; Bortolotti, M.; Maiello, S.; Djemil, A.; Battelli, M.G.; Bolognesi, A. Two Saporin-Containing Immunotoxins Specific for CD20 and CD22 Show Different Behavior in Killing Lymphoma Cells. *Toxins (Basel)* **2017**, *9*, 182. [CrossRef] [PubMed]

94. Cao, Y.; Marks, J.W.; Liu, Z.; Cheung, L.H.; Hittelman, W.N.; Rosenblum, M.G. Design optimization and characterization of Her2/neu-targeted immunotoxins: Comparative in vitro and in vivo efficacy studies. *Oncogene* **2014**, *33*, 429–439. [CrossRef] [PubMed]

95. Mazor, Y.; Noy, R.; Wels, W.S.; Benhar, I. chFRP5-ZZ-PE38, a large IgG-toxin immunoconjugate outperforms the corresponding smaller FRP5(Fv)-ETA immunotoxin in eradicating ErbB2-expressing tumor xenografts. *Cancer Lett.* **2007**, *257*, 124–135. [CrossRef] [PubMed]

96. Olsen, E.; Duvic, M.; Frankel, A.; Kim, Y.; Martin, A.; Vonderheid, E.; Jegasothy, B.; Wood, G.; Gordon, M.; Heald, P.; et al. Pivotal phase III trial of two dose levels of denileukin diftitox for the treatment of cutaneous T-cell lymphoma. *J. Clin. Oncol.* **2001**, *19*, 376–388. [CrossRef] [PubMed]

97. Kreitman, R.J.; Wilson, W.H.; White, J.D.; Stetler-Stevenson, M.; Jaffe, E.S.; Giardina, S.; Waldmann, T.A.; Pastan, I. Phase I trial of recombinant immunotoxin anti-Tac(Fv)-PE38 (LMB-2) in patients with hematologic malignancies. *J. Clin. Oncol.* **2000**, *18*, 1622–1636. [CrossRef] [PubMed]

98. Blakey, D.C.; Watson, G.J.; Knowles, P.P.; Thorpe, P.E. Effect of chemical deglycosylation of ricin A chain on the in vivo fate and cytotoxic activity of an immunotoxin composed of ricin A chain and anti-Thy 1.1 antibody. *Cancer Res.* **1987**, *47*, 947–952. [PubMed]

99. Pasqualucci, L.; Wasik, M.; Teicher, B.A.; Flenghi, L.; Bolognesi, A.; Stirpe, F.; Polito, L.; Falini, B.; Kadin, M.E. Antitumor activity of anti-CD30 immunotoxin (Ber-H2/saporin) in vitro and in severe combined immunodeficiency disease mice xenografted with human CD30+ anaplastic large-cell lymphoma. *Blood* **1995**, *85*, 2139–2146. [PubMed]

100. Bolognesi, A.; Tazzari, P.L.; Olivieri, F.; Polito, L.; Lemoli, R.; Terenzi, A.; Pasqualucci, L.; Falini, B.; Stirpe, F. Evaluation of immunotoxins containing single-chain ribosome-inactivating proteins and an anti-CD22 monoclonal antibody (OM124): In vitro and in vivo studies. *Br. J. Haematol.* **1998**, *101*, 179–188. [CrossRef] [PubMed]

101. Flavell, D.J.; Noss, A.; Pulford, K.A.; Ling, N.; Flavell, S.U. Systemic therapy with 3BIT, a triple combination cocktail of anti-CD19, -CD22, and -CD38-saporin immunotoxins, is curative of human B-cell lymphoma in severe combined immunodeficient mice. *Cancer Res.* **1997**, *57*, 4824–4829. [PubMed]

102. ten Cate, B.; de Bruyn, M.; Wei, Y.; Bremer, E.; Helfrich, W. Targeted elimination of leukemia stem cells; a new therapeutic approach in hemato-oncology. *Curr. Drug Targets* **2010**, *11*, 95–110. [CrossRef] [PubMed]

103. Chandler, L.A.; Sosnowski, B.A.; McDonald, J.R.; Price, J.E.; Aukerman, S.L.; Baird, A.; Pierce, G.F.; Houston, L.L. Targeting tumor cells via EGF receptors: Selective toxicity of an HBEGF-toxin fusion protein. *Int. J. Cancer* **1998**, *78*, 106–111. [CrossRef]

104. Beitz, J.G.; Davol, P.; Clark, J.W.; Kato, J.; Medina, M.; Frackelton, A.R., Jr.; Lappi, D.A.; Baird, A.; Calabresi, P. Antitumor activity of basic fibroblast growth factor-saporin mitotoxin in vitro and in vivo. *Cancer Res.* **1992**, *52*, 227–230. [PubMed]

105. Veenendaal, L.M.; Jin, H.; Ran, S.; Cheung, L.; Navone, N.; Marks, J.W.; Waltenberger, J.; Thorpe, P.; Rosenblum, M.G. In vitro and in vivo studies of a VEGF121/rGelonin chimeric fusion toxin targeting the neovasculature of solid tumors. *Proc. Natl. Acad. Sci. USA* **2002**, *99*, 7866–7871. [CrossRef] [PubMed]

106. Cavallaro, U.; del Vecchio, A.; Lappi, D.A.; Soria, M.R. A conjugate between human urokinase and saporin, a type-1 ribosome-inactivating protein, is selectively cytotoxic to urokinase receptor-expressing cells. *J. Biol. Chem.* **1993**, *268*, 23186–23190. [PubMed]

107. Bussolati, B.; Grange, C.; Tei, L.; Deregibus, M.C.; Ercolani, M.; Aime, S.; Camussi, G. Targeting of human renal tumor-derived endothelial cells with peptides obtained by phage display. *J. Mol. Med. (Berl.)* **2007**, *85*, 897–906. [CrossRef] [PubMed]

108. Gilabert-Oriol, R.; Weng, A.; Mallinckrodt, B.; Melzig, M.F.; Fuchs, H.; Thakur, M. Immunotoxins constructed with ribosome-inactivating proteins and their enhancers: A lethal cocktail with tumor specific efficacy. *Curr. Pharm. Des.* **2014**, *20*, 6584–6643. [CrossRef] [PubMed]

109. Polito, L.; Bortolotti, M.; Mercatelli, D.; Battelli, M.G.; Bolognesi, A. Saporin-S6: A useful tool in cancer therapy. *Toxins (Basel)* **2013**, *5*, 1698–1722. [CrossRef] [PubMed]

110. Baluna, R.; Coleman, E.; Jones, C.; Ghetie, V.; Vitetta, E.S. The effect of a monoclonal antibody coupled to ricin A chain-derived peptides on endothelial cells in vitro: Insights into toxin-mediated vascular damage. *Exp. Cell Res.* **2000**, *258*, 417–424. [CrossRef] [PubMed]

111. Smallshaw, J.E.; Ghetie, V.; Rizo, J.; Fulmer, J.R.; Trahan, L.L.; Ghetie, M.A.; Vitetta, E.S. Genetic engineering of an immunotoxin to eliminate pulmonary vascular leak in mice. *Nat. Biotechnol.* **2003**, *21*, 387–391. [CrossRef] [PubMed]

112. Vago, R.; Collico, V.; Zuppone, S.; Prosperi, D.; Colombo, M. Nanoparticle-mediated delivery of suicide genes in cancer therapy. *Pharmacol. Res.* **2016**, *111*, 619–641. [CrossRef] [PubMed]

113. Wang, M.; Alberti, K.; Sun, S.; Arellano, C.L.; Xu, Q. Combinatorially designed lipid-like nanoparticles for intracellular delivery of cytotoxic protein for cancer therapy. *Angew. Chem.* **2014**, *53*, 2893–2898. [CrossRef] [PubMed]

114. Tarassoli, S.P.; de Pinillos Bayona, A.M.; Pye, H.; Mosse, C.A.; Callan, J.F.; MacRobert, A.; McHale, A.P.; Nomikou, N. Cathepsin B-degradable, NIR-responsive nanoparticulate platform for target-specific cancer therapy. *Nanotechnology* **2017**, *28*, 055101. [CrossRef] [PubMed]

115. Lai, P.S.; Pai, C.L.; Peng, C.L.; Shieh, M.J.; Berg, K.; Lou, P.J. Enhanced cytotoxicity of saporin by polyamidoamine dendrimer conjugation and photochemical internalization. *J. Biomed. Mater. Res. Part A* **2008**, *87*, 147–155. [CrossRef] [PubMed]

116. Rust, A.; Hassan, H.H.; Sedelnikova, S.; Niranjan, D.; Hautbergue, G.; Abbas, S.A.; Partridge, L.; Rice, D.; Binz, T.; Davletov, B. Two complementary approaches for intracellular delivery of exogenous enzymes. *Sci. Rep.* **2015**, *5*, 12444. [CrossRef] [PubMed]

117. Su, X.; Yang, N.; Wittrup, K.D.; Irvine, D.J. Synergistic antitumor activity from two-stage delivery of targeted toxins and endosome-disrupting nanoparticles. *Biomacromolecules* **2013**, *14*, 1093–1102. [CrossRef] [PubMed]

118. Cho, E.J.; Yang, J.; Mohamedali, K.A.; Lim, E.K.; Kim, E.J.; Farhangfar, C.J.; Suh, J.S.; Haam, S.; Rosenblum, M.G.; Huh, Y.M. Sensitive angiogenesis imaging of orthotopic bladder tumors in mice using a selective magnetic resonance imaging contrast agent containing VEGF121/rGel. *Investig. Radiol.* **2011**, *46*, 441–449. [CrossRef] [PubMed]

119. Wen, J.; Yan, M.; Liu, Y.; Li, J.; Xie, Y.; Lu, Y.; Kamata, M.; Chen, I.S. Specific Elimination of Latently HIV-1 Infected Cells Using HIV-1 Protease-Sensitive Toxin Nanocapsules. *PLoS ONE* **2016**, *11*, e0151572. [CrossRef] [PubMed]

120. van Dommelen, S.M.; Vader, P.; Lakhal, S.; Kooijmans, S.A.; van Solinge, W.W.; Wood, M.J.; Schiffelers, R.M. Microvesicles and exosomes: Opportunities for cell-derived membrane vesicles in drug delivery. *J. Control. Release* **2012**, *161*, 635–644. [CrossRef] [PubMed]

121. Tan, A.; Rajadas, J.; Seifalian, A.M. Exosomes as nano-theranostic delivery platforms for gene therapy. *Adv. Drug Deliv. Rev.* **2013**, *65*, 357–367. [CrossRef] [PubMed]

122. Vlassov, A.V.; Magdaleno, S.; Setterquist, R.; Conrad, R. Exosomes: Current knowledge of their composition, biological functions, and diagnostic and therapeutic potentials. *Biochim. Biophys. Acta* **2012**, *1820*, 940–948. [CrossRef] [PubMed]

123. Subra, C.; Grand, D.; Laulagnier, K.; Stella, A.; Lambeau, G.; Paillasse, M.; De Medina, P.; Monsarrat, B.; Perret, B.; Silvente-Poirot, S.; et al. Exosomes account for vesicle-mediated transcellular transport of activatable phospholipases and prostaglandins. *J. Lipid Res.* **2010**, *51*, 2105–2120. [CrossRef] [PubMed]

124. Conde-Vancells, J.; Rodriguez-Suarez, E.; Embade, N.; Gil, D.; Matthiesen, R.; Valle, M.; Elortza, F.; Lu, S.C.; Mato, J.M.; Falcon-Perez, J.M. Characterization and comprehensive proteome profiling of exosomes secreted by hepatocytes. *J. Proteome Res.* **2008**, *7*, 5157–5166. [CrossRef] [PubMed]

125. Nakase, I.; Futaki, S. Combined treatment with a pH-sensitive fusogenic peptide and cationic lipids achieves enhanced cytosolic delivery of exosomes. *Sci. Rep.* **2015**, *5*, 10112. [CrossRef] [PubMed]

126. Kooijmans, S.A.A.; Schiffelers, R.M.; Zarovni, N.; Vago, R. Modulation of tissue tropism and biological activity of exosomes and other extracellular vesicles: New nanotools for cancer treatment. *Pharmacol. Res.* **2016**, *111*, 487–500. [CrossRef] [PubMed]

127. Sokolova, V.; Ludwig, A.K.; Hornung, S.; Rotan, O.; Horn, P.A.; Epple, M.; Giebel, B. Characterisation of exosomes derived from human cells by nanoparticle tracking analysis and scanning electron microscopy. *Colloids Surfaces. B Biointerfaces* **2011**, *87*, 146–150. [CrossRef] [PubMed]

128. Hood, J.L.; San, R.S.; Wickline, S.A. Exosomes released by melanoma cells prepare sentinel lymph nodes for tumor metastasis. *Cancer Res.* **2011**, *71*, 3792–3801. [CrossRef] [PubMed]

129. Takahashi, Y.; Nishikawa, M.; Shinotsuka, H.; Matsui, Y.; Ohara, S.; Imai, T.; Takakura, Y. Visualization and in vivo tracking of the exosomes of murine melanoma B16-BL6 cells in mice after intravenous injection. *J. Biotechnol.* **2013**, *165*, 77–84. [CrossRef] [PubMed]

130. Nakase, I.; Kobayashi, N.B.; Takatani-Nakase, T.; Yoshida, T. Active macropinocytosis induction by stimulation of epidermal growth factor receptor and oncogenic Ras expression potentiates cellular uptake efficacy of exosomes. *Sci. Rep.* **2015**, *5*, 10300. [CrossRef] [PubMed]

131. Araki, N.; Hamasaki, M.; Egami, Y.; Hatae, T. Effect of 3-methyladenine on the fusion process of macropinosomes in EGF-stimulated A431 cells. *Cell Struct. Funct.* **2006**, *31*, 145–157. [CrossRef] [PubMed]

132. Dise, R.S.; Frey, M.R.; Whitehead, R.H.; Polk, D.B. Epidermal growth factor stimulates Rac activation through Src and phosphatidylinositol 3-kinase to promote colonic epithelial cell migration. American journal of physiology. *Gastrointest. Liver Physiol.* **2008**, *294*, G276–G285. [CrossRef] [PubMed]

133. Tanaka, G.; Nakase, I.; Fukuda, Y.; Masuda, R.; Oishi, S.; Shimura, K.; Kawaguchi, Y.; Takatani-Nakase, T.; Langel, U.; Graslund, A.; et al. CXCR4 stimulates macropinocytosis: Implications for cellular uptake of arginine-rich cell-penetrating peptides and HIV. *Chem. Biol.* **2012**, *19*, 1437–1446. [CrossRef] [PubMed]

134. Nakase, I.; Noguchi, K.; Fujii, I.; Futaki, S. Vectorization of biomacromolecules into cells using extracellular vesicles with enhanced internalization induced by macropinocytosis. *Sci. Rep.* **2016**, *6*, 34937. [CrossRef] [PubMed]

135. Nakase, I.; Noguchi, K.; Aoki, A.; Takatani-Nakase, T.; Fujii, I.; Futaki, S. Arginine-rich cell-penetrating peptide-modified extracellular vesicles for active macropinocytosis induction and efficient intracellular delivery. *Sci. Rep.* **2017**, *7*, 1991. [CrossRef] [PubMed]

136. Nakase, I.; Tanaka, G.; Futaki, S. Cell-penetrating peptides (CPPs) as a vector for the delivery of siRNAs into cells. *Mol. Biosyst.* **2013**, *9*, 855–861. [CrossRef] [PubMed]

137. Nakase, I.; Osaki, K.; Tanaka, G.; Utani, A.; Futaki, S. Molecular interplays involved in the cellular uptake of octaarginine on cell surfaces and the importance of syndecan-4 cytoplasmic V domain for the activation of protein kinase Calpha. *Biochem. Biophys. Res. Commun.* **2014**, *446*, 857–862. [CrossRef] [PubMed]

138. Subbarao, N.K.; Parente, R.A.; Szoka, F.C., Jr.; Nadasdi, L.; Pongracz, K. pH-dependent bilayer destabilization by an amphipathic peptide. *Biochemistry* **1987**, *26*, 2964–2972. [CrossRef] [PubMed]

139. Kobayashi, S.; Nakase, I.; Kawabata, N.; Yu, H.H.; Pujals, S.; Imanishi, M.; Giralt, E.; Futaki, S. Cytosolic targeting of macromolecules using a pH-dependent fusogenic peptide in combination with cationic liposomes. *Bioconj. Chem.* **2009**, *20*, 953–959. [CrossRef] [PubMed]

140. Nakase, I.; Kogure, K.; Harashima, H.; Futaki, S. Application of a fusogenic peptide GALA for intracellular delivery. *Methods Mol. Biol.* **2011**, *683*, 525–533. [PubMed]

141. Vago, R. Ribosome Inactivating Proteins: Exploiting Plant Weapons to Fight Human Cancer. *Genet. Syndr. Gene Therapy* **2015**, *6*, 272. [CrossRef]

142. Glinka, E.M. Eukaryotic expression vectors bearing genes encoding cytotoxic proteins for cancer gene therapy. *Plasmid* **2012**, *68*, 69–85. [CrossRef] [PubMed]

143. Gofrit, O.N.; Benjamin, S.; Halachmi, S.; Leibovitch, I.; Dotan, Z.; Lamm, D.L.; Ehrlich, N.; Yutkin, V.; Ben-Am, M.; Hochberg, A. DNA based therapy with diphtheria toxin-A BC-819: A phase 2b marker lesion trial in patients with intermediate risk nonmuscle invasive bladder cancer. *J. Urol.* **2014**, *191*, 1697–1702. [CrossRef] [PubMed]

144. Zarovni, N.; Vago, R.; Fabbrini, M.S. Saporin suicide gene therapy. *Methods Mol. Biol.* **2009**, *542*, 261–283. [PubMed]

145. Martin, V.; Cortes, M.L.; de Felipe, P.; Farsetti, A.; Calcaterra, N.B.; Izquierdo, M. Cancer gene therapy by thyroid hormone-mediated expression of toxin genes. *Cancer Res.* **2000**, *60*, 3218–3224. [PubMed]

146. Hoganson, D.K.; Chandler, L.A.; Fleurbaaij, G.A.; Ying, W.; Black, M.E.; Doukas, J.; Pierce, G.F.; Baird, A.; Sosnowski, B.A. Targeted delivery of DNA encoding cytotoxic proteins through high-affinity fibroblast growth factor receptors. *Hum. Gene Ther.* **1998**, *9*, 2565–2575. [CrossRef] [PubMed]

147. Cocco, E.; Deng, Y.; Shapiro, E.M.; Bortolomai, I.; Lopez, S.; Lin, K.; Bellone, S.; Cui, J.; Menderes, G.; Black, J.D.; et al. Dual-Targeting Nanoparticles for In Vivo Delivery of Suicide Genes to Chemotherapy-Resistant Ovarian Cancer Cells. *Mol. Cancer Ther.* **2017**, *16*, 323–333. [CrossRef] [PubMed]

148. Knight, B. Ricin-a potent homicidal poison. *Br. Med. J.* **1979**, *1*, 350–351. [PubMed]

149. Musshoff, F.; Madea, B. Ricin poisoning and forensic toxicology. *Drug Test. Anal.* **2009**, *1*, 184–191. [CrossRef] [PubMed]

150. Mu, X.; Tong, Z.; Huang, Q.; Liu, B.; Liu, Z.; Hao, L.; Dong, H.; Zhang, J.; Gao, C. An Electrochemiluminescence Immunosensor Based on Gold-Magnetic Nanoparticles and Phage Displayed Antibodies. *Sensors* **2016**, *16*, 308. [CrossRef] [PubMed]

151. Phase I Clinical Trials. Available online: www.clinicaltrials.gov (accessed on 8 October 2017).

152. Vitetta, E.S.; Smallshaw, J.E.; Coleman, E.; Jafri, H.; Foster, C.; Munford, R.; Schindler, J. A pilot clinical trial of a recombinant ricin vaccine in normal humans. *Proc. Natl. Acad. Sci. USA* **2006**, *103*, 2268–2273. [CrossRef] [PubMed]

153. Roy, C.J.; Brey, R.N.; Mantis, N.J.; Mapes, K.; Pop, I.V.; Pop, L.M.; Ruback, S.; Killeen, S.Z.; Doyle-Meyers, L.; Vinet-Oliphant, H.S.; et al. Thermostable ricin vaccine protects rhesus macaques against aerosolized ricin: Epitope-specific neutralizing antibodies correlate with protection. *Proc. Natl. Acad. Sci. USA* **2015**, *112*, 3782–3787. [CrossRef] [PubMed]

154. Olson, M.A.; Carra, J.H.; Roxas-Duncan, V.; Wannemacher, R.W.; Smith, L.A.; Millard, C.B. Finding a new vaccine in the ricin protein fold. *Protein Eng. Des. Sel.* **2004**, *17*, 391–397. [CrossRef] [PubMed]

155. Carra, J.H.; Wannemacher, R.W.; Tammariello, R.F.; Lindsey, C.Y.; Dinterman, R.E.; Schokman, R.D.; Smith, L.A. Improved formulation of a recombinant ricin A-chain vaccine increases its stability and effective antigenicity. *Vaccine* **2007**, *25*, 4149–4158. [CrossRef] [PubMed]

156. McLain, D.E.; Lewis, B.S.; Chapman, J.L.; Wannemacher, R.W.; Lindsey, C.Y.; Smith, L.A. Protective effect of two recombinant ricin subunit vaccines in the New Zealand white rabbit subjected to a lethal aerosolized ricin challenge: Survival, immunological response, and histopathological findings. *Toxicol. Sci.* **2012**, *126*, 72–83. [CrossRef] [PubMed]

157. Pittman, P.R.; Reisler, R.B.; Lindsey, C.Y.; Guerena, F.; Rivard, R.; Clizbe, D.P.; Chambers, M.; Norris, S.; Smith, L.A. Safety and immunogenicity of ricin vaccine, RVEc, in a Phase 1 clinical trial. *Vaccine* **2015**, *33*, 7299–7306. [CrossRef] [PubMed]

158. Marconescu, P.S.; Smallshaw, J.E.; Pop, L.M.; Ruback, S.L.; Vitetta, E.S. Intradermal administration of RiVax protects mice from mucosal and systemic ricin intoxication. *Vaccine* **2010**, *28*, 5315–5322. [CrossRef] [PubMed]

159. Vitetta, E.S.; Smallshaw, J.E.; Schindler, J. Pilot phase IB clinical trial of an alhydrogel-adsorbed recombinant ricin vaccine. *Clin. Vaccine Immunol.* **2012**, *19*, 1697–1699. [CrossRef] [PubMed]

160. Vance, D.J.; Mantis, N.J. Progress and challenges associated with the development of ricin toxin subunit vaccines. *Expert Rev. Vaccines* **2016**, *15*, 1213–1222. [CrossRef] [PubMed]

161. Fuchs, H.B.; Bachran, C.; Flavell, D.J. Diving through Membranes: Molecular Cunning to Enforce the Endosomal Escape of Antibody-Targeted Anti-Tumor Toxins. *Antibodies* **2013**, *2*, 209–235. [CrossRef]

162. Zhang, D.; Wang, J.; Xu, D. Cell-penetrating peptides as noninvasive transmembrane vectors for the development of novel multifunctional drug-delivery systems. *J. Control. Release* **2016**, *229*, 130–139. [CrossRef] [PubMed]

163. Sawant, R.; Torchilin, V. Intracellular transduction using cell-penetrating peptides. *Mol. Biosyst.* **2010**, *6*, 628–640. [CrossRef] [PubMed]

164. Lorenzetti, I.; Meneguzzi, A.; Fracasso, G.; Potrich, C.; Costantini, L.; Chiesa, E.; Legname, G.; Menestrina, G.; Tridente, G.; Colombatti, M. Genetic grafting of membrane-acting peptides to the cytotoxin dianthin augments its ability to de-stabilize lipid bilayers and enhances its cytotoxic potential as the component of transferrin-toxin conjugates. *Int. J. Cancer* **2000**, *86*, 582–589. [CrossRef]

165. Fuchs, H.; Bachran, C.; Li, T.; Heisler, I.; Durkop, H.; Sutherland, M. A cleavable molecular adapter reduces side effects and concomitantly enhances efficacy in tumor treatment by targeted toxins in mice. *J. Control. Release* **2007**, *117*, 342–350. [CrossRef] [PubMed]

166. Shin, M.C.; Zhang, J.; David, A.E.; Trommer, W.E.; Kwon, Y.M.; Min, K.A.; Kim, J.H.; Yang, V.C. Chemically and biologically synthesized CPP-modified Gelonin for enhanced anti-tumor activity. *J. Control. Release* **2013**, *172*, 169–178. [CrossRef] [PubMed]

167. Nakase, I.; Konishi, Y.; Ueda, M.; Saji, H.; Futaki, S. Accumulation of arginine-rich cell-penetrating peptides in tumors and the potential for anticancer drug delivery in vivo. *J. Control. Release* **2012**, *159*, 181–188. [CrossRef] [PubMed]

168. Shapira, A.; Gal-Tanamy, M.; Nahary, L.; Litvak-Greenfeld, D.; Zemel, R.; Tur-Kaspa, R.; Benhar, I. Engineered toxins "zymoxins" are activated by the HCV NS3 protease by removal of an inhibitory protein domain. *PLoS ONE* **2011**, *6*, e15916. [CrossRef] [PubMed]

169. O'Hare, M.; Brown, A.N.; Hussain, K.; Gebhardt, A.; Watson, G.; Roberts, L.M.; Vitetta, E.S.; Thorpe, P.E.; Lord, J.M. Cytotoxicity of a recombinant ricin-A-chain fusion protein containing a proteolytically-cleavable spacer sequence. *FEBS Lett.* **1990**, *273*, 200–204. [CrossRef]

170. Wales, R.; Chaddock, J.A.; Roberts, L.M.; Lord, J.M. Addition of an ER retention signal to the ricin A chain increases the cytotoxicity of the holotoxin. *Exp. Cell Res.* **1992**, *203*, 1–4. [CrossRef]

171. Li, W.; Nicol, F.; Szoka, F.C., Jr. GALA: A designed synthetic pH-responsive amphipathic peptide with applications in drug and gene delivery. *Adv. Drug Deliv. Rev.* **2004**, *56*, 967–985. [CrossRef] [PubMed]

172. Boussif, O.; Lezoualc'h, F.; Zanta, M.A.; Mergny, M.D.; Scherman, D.; Demeneix, B.; Behr, J.P. A versatile vector for gene and oligonucleotide transfer into cells in culture and in vivo: Polyethylenimine. *Proc. Natl. Acad. Sci. USA* **1995**, *92*, 7297–7301. [CrossRef] [PubMed]

173. Varkouhi, A.K.; Scholte, M.; Storm, G.; Haisma, H.J. Endosomal escape pathways for delivery of biologicals. *J. Control. Release* **2011**, *151*, 220–228. [CrossRef] [PubMed]

174. Akishiba, M.; Takeuchi, T.; Kawaguchi, Y.; Sakamoto, K.; Yu, H.H.; Nakase, I.; Takatani-Nakase, T.; Madani, F.; Graslund, A.; Futaki, S. Cytosolic antibody delivery by lipid-sensitive endosomolytic peptide. *Nat. Chem.* **2017**, *9*, 751–761. [CrossRef] [PubMed]

175. Thakur, M.; Mergel, K.; Weng, A.; von Mallinckrodt, B.; Gilabert-Oriol, R.; Durkop, H.; Melzig, M.F.; Fuchs, H. Targeted tumor therapy by epidermal growth factor appended toxin and purified saponin: An evaluation of toxicity and therapeutic potential in syngeneic tumor bearing mice. *Mol. Oncol.* **2013**, *7*, 475–483. [CrossRef] [PubMed]

176. von Mallinckrodt, B.T.; Thakur, M.; Weng, A.; Gilabert-Oriol, R.; Dürkop, H.; Brenner, W.; Lukas, M.; Beindorff, N.; Melzig, M.F.; Fuchs, H. Dianthin-EGF is an effective tumor targeted toxin in combination with saponins in a xenograft model for colon carcinoma. *Future Oncol.* **2014**, *10*, 2161–2175. [CrossRef] [PubMed]

177. Fuchs, H.; Niesler, N.; Trautner, A.; Sama, S.; Jerz, G.; Panjideh, H.; Weng, A. Glycosylated Triterpenoids as Endosomal Escape Enhancers in Targeted Tumor Therapies. *Biomedicines* **2017**, *5*, 14. [CrossRef] [PubMed]

178. Bachran, C.; Durkop, H.; Sutherland, M.; Bachran, D.; Muller, C.; Weng, A.; Melzig, M.F.; Fuchs, H. Inhibition of tumor growth by targeted toxins in mice is dramatically improved by saponinum album in a synergistic way. *J. Immunother.* **2009**, *32*, 713–725. [CrossRef] [PubMed]

179. Fuchs, H.; Weng, A.; Gilabert-Oriol, R. Augmenting the Efficacy of Immunotoxins and Other Targeted Protein Toxins by Endosomal Escape Enhancers. *Toxins (Basel)* **2016**, *8*, 200. [CrossRef] [PubMed]

180. Selbo, P.K.; Weyergang, A.; Hogset, A.; Norum, O.J.; Berstad, M.B.; Vikdal, M.; Berg, K. Photochemical internalization provides time- and space-controlled endolysosomal escape of therapeutic molecules. *J. Control Release* **2010**, *148*, 2–12. [CrossRef] [PubMed]

181. Berg, K.; Selbo, P.K.; Prasmickaite, L.; Tjelle, T.E.; Sandvig, K.; Moan, J.; Gaudernack, G.; Fodstad, O.; Kjolsrud, S.; Anholt, H.; Rodal, G.H.; Rodal, S.K.; Hogset, A. Photochemical internalization: A novel technology for delivery of macromolecules into cytosol. *Cancer Res.* **1999**, *59*, 1180–1183. [PubMed]

182. Bostad, M.; Olsen, C.E.; Peng, Q.; Berg, K.; Hogset, A.; Selbo, P.K. Light-controlled endosomal escape of the novel CD133-targeting immunotoxin AC133-saporin by photochemical internalization—A minimally invasive cancer stem cell-targeting strategy. *J. Control. Release* **2015**, *206*, 37–48. [CrossRef] [PubMed]

183. Weyergang, A.; Cheung, L.H.; Rosenblum, M.G.; Mohamedali, K.A.; Peng, Q.; Waltenberger, J.; Berg, K. Photochemical internalization augments tumor vascular cytotoxicity and specificity of VEGF(121)/rGel fusion toxin. *J. Control. Release* **2014**, *180*, 1–9. [CrossRef] [PubMed]

184. Berstad, M.B.; Cheung, L.H.; Berg, K.; Peng, Q.; Fremstedal, A.S.; Patzke, S.; Rosenblum, M.G.; Weyergang, A. Design of an EGFR-targeting toxin for photochemical delivery: In vitro and in vivo selectivity and efficacy. *Oncogene* **2015**, *34*, 5582–5592. [CrossRef] [PubMed]

185. Roy, S.; Axup, J.Y.; Forsyth, J.S.; Goswami, R.K.; Hutchins, B.M.; Bajuri, K.M.; Kazane, S.A.; Smider, V.V.; Felding, B.H.; Sinha, S.C. SMI-Ribosome inactivating protein conjugates selectively inhibit tumor cell growth. *Chem. Commun.* **2017**, *53*, 4234–4237. [CrossRef] [PubMed]

186. Krall, N.; Pretto, F.; Decurtins, W.; Bernardes, G.J.; Supuran, C.T.; Neri, D. A small-molecule drug conjugate for the treatment of carbonic anhydrase IX expressing tumors. *Angew. Chem.* **2014**, *53*, 4231–4235. [CrossRef] [PubMed]

187. Wiley, R.G.; Oeltmann, T.N.; Lappi, D.A. Immunolesioning: Selective destruction of neurons using immunotoxin to rat NGF receptor. *Brain Res.* **1991**, *562*, 149–153. [CrossRef]

188. Alonso, M.N.; Gregorio, J.G.; Davidson, M.G.; Gonzalez, J.C.; Engleman, E.G. Depletion of inflammatory dendritic cells with anti-CD209 conjugated to saporin toxin. *Immunol. Res.* **2014**, *58*, 374–377. [CrossRef] [PubMed]

189. Lee, S.C.; Seo, K.W.; Kim, H.J.; Kang, S.W.; Choi, H.J.; Kim, A.; Kwon, B.S.; Cho, H.R.; Kwon, B. Depletion of Alloreactive T-Cells by Anti-CD137-Saporin Immunotoxin. *Cell Transplant.* **2015**, *24*, 1167–1181. [CrossRef] [PubMed]

190. Rastogi, S.; Unni, S.; Sharma, S.; Laxmi, T.R.; Kutty, B.M. Cholinergic immunotoxin 192 IgG-SAPORIN alters subicular theta-gamma activity and impairs spatial learning in rats. *Neurobiol. Learn. Memory* **2014**, *114*, 117–126. [CrossRef] [PubMed]

191. Wiese, A.J.; Rathbun, M.; Butt, M.T.; Malkmus, S.A.; Richter, P.J.; Osborn, K.G.; Xu, Q.; Veesart, S.L.; Steinauer, J.J.; Higgins, D.; et al. Intrathecal substance P-saporin in the dog: Distribution, safety, and spinal neurokinin-1 receptor ablation. *Anesthesiology* **2013**, *119*, 1163–1177. [CrossRef] [PubMed]

192. Brown, D.C.; Agnello, K. Intrathecal substance P-saporin in the dog: Efficacy in bone cancer pain. *Anesthesiology* **2013**, *119*, 1178–1185. [CrossRef] [PubMed]

193. Allen, J.W.; Mantyh, P.W.; Horais, K.; Tozier, N.; Rogers, S.D.; Ghilardi, J.R.; Cizkova, D.; Grafe, M.R.; Richter, P.; Lappi, D.A.; Yaksh, T.L. Safety evaluation of intrathecal substance P-saporin, a targeted neurotoxin, in dogs. *Toxicol. Sci.* **2006**, *91*, 286–298. [CrossRef] [PubMed]

toxins

MDPI

Review

Extensive Evolution of Cereal Ribosome-Inactivating Proteins Translates into Unique Structural Features, Activation Mechanisms, and Physiological Roles

Jeroen De Zaeytijd and Els J. M. Van Damme *

Lab Biochemistry and Glycobiology, Department of Molecular Biotechnology, Ghent University,
Coupure links 653, B-9000 Ghent, Belgium; Jeroen.DeZaeytijd@ugent.be
* Correspondence: ElsJM.VanDamme@ugent.be; Tel.: +32-9264-6086; Fax: +32-9264-6219

Academic Editors: Julien Barbier and Daniel Gillet
Received: 27 February 2017; Accepted: 25 March 2017; Published: 29 March 2017

Abstract: Ribosome-inactivating proteins (RIPs) are a class of cytotoxic enzymes that can depurinate rRNAs thereby inhibiting protein translation. Although these proteins have also been detected in bacteria, fungi, and even some insects, they are especially prevalent in the plant kingdom. This review focuses on the RIPs from cereals. Studies on the taxonomical distribution and evolution of plant RIPs suggest that cereal RIPs have evolved at an enhanced rate giving rise to a large and heterogeneous RIP gene family. Furthermore, several cereal RIP genes are characterized by a unique domain architecture and the lack of a signal peptide. This advanced evolution of cereal RIPs translates into distinct structures, activation mechanisms, and physiological roles. Several cereal RIPs are characterized by activation mechanisms that include the proteolytic removal of internal peptides from the N-glycosidase domain, a feature not documented for non-cereal RIPs. Besides their role in defense against pathogenic fungi or herbivorous insects, cereal RIPs are also involved in endogenous functions such as adaptation to abiotic stress, storage, induction of senescence, and reprogramming of the translational machinery. The unique properties of cereal RIPs are discussed in this review paper.

Keywords: b-32; cereals; JIP60; Ribosome-inactivating proteins; RIP

1. Introduction

Ribosome-inactivating proteins (RIPs) are enzymes that can irreversibly inhibit protein translation by depurinating the ribosomal RNA. Despite the fact that RIP genes are also found in fungi, bacteria, and even in some insects, they are thought to have originated in plants [1,2]. The history of RIP research was recently summarized by Bolognesi et al. [3]. Although the term "ribosome-inactivating proteins" was only introduced in 1982 [4] and the biological activities of these proteins were elucidated in 1987 [5] research on RIPs dates back to the nineteenth century when the toxicity of seeds from the plants *Ricinus communis* and *Abrus precatorius* was attributed to the presence of the proteins ricin and abrin, respectively. These two toxins were the first RIPs discovered and the scientific community was highly interested in their toxic properties. The proteins were first labeled as haemagglutinins because of their haemagglutination activity [6,7]. Olsnes and Pihl [8,9] discovered that both ricin and abrin consist of two polypeptides linked together by a disulfide bridge. The agglutination activity of the proteins was attributed to the C-terminal B-chain which corresponds to a galactose-specific lectin domain. Endo and Tsurugi [5] showed that the N-terminal A-chain of the proteins possesses an enzymatic activity which enables them to remove a specific adenine residue from the conserved sarcin/ricin loop (SRL) of the 28S rRNA, and consequently these proteins were referred to as rRNA N-glycosidases. Later it was reported that these proteins depurinated ribosomal RNA at multiple sites [10] and also removed adenine residues from other substrates such as herring sperm DNA, poly(A) and Tobacco

Mosaic Virus (TMV) RNA [11]. For these reasons RIPs were considered as polynucleotide:adenosine glycosidases (PAG) [11–13]. Some authors also attributed other more exotic enzymatic activities to RIPs such as DNAse [14], RNAse [15], chitinase [16], phosphatase [17], lipase [18], and superoxide dismutase activities [19]. It cannot be excluded that these presumed activities were due to contaminants in the RIP preparations as reviewed in Peumans et al. [20]. However, more recent studies suggest that at least some RIPs from several *Phytolacca* species and *Beta vulgaris* possess endonuclease activity and/or superoxide dismutase activities. In these studies the necessary precautions were taken to exclude the possibility that the reported activities were due to contaminants [21–26]. Therefore the reported activities can be attributed to a particular structural configuration in these RIPs but cannot be considered a general feature of RIPs [22].

Following the elucidation of the enzymatic activity of ricin and abrin, RIPs were reported in a lot of plant species (reviewed by Schrot et al. [27]). Interestingly, some of these proteins did not possess a B-chain and consisted solely of the A-chain with RIP activity. This finding led to the classification of RIPs. Type-1 RIPs only consist of a RIP domain whereas the term type-2 RIPs refers to proteins composed of an enzymatic domain linked to a lectin domain.

Following the discovery of the cereal RIPs JIP60 (barley) and b-32 (maize) the term "type-3" was introduced by some authors [20,28,29] while others just considered these RIPs as atypical type-1 RIPs [30]. Alternatively, based on the domain architecture and evolutionary background type-1 and type-2 RIPs are called type-A RIPs and type-AB RIPs, respectively [2].

Type-A and type-AB RIPs differ in their cytotoxicity. In general, type-AB RIPs are toxic to animal cells since these RIPs can bind sugar moieties on the cell surface through their lectin domain, promoting their uptake in the cell. The absence of this glycoconjugate-binding activity explains the lower toxicity of RIPs devoid of a lectin domain. However, when type-A RIPs succeed to access the cytosol they also show cytotoxicity [31]. The existence of non-toxic type-AB RIPs emphasizes the fact that there are large differences in the toxicity of RIPs [32]. Overall, the toxicity is the result of the combined effects of: (i) Efficient binding to the cell surface; (ii) Uptake; (iii) Ability of the protein to reach the cytosol of the host cell; and (iv) Intracellular stability.

Because of their toxic effects some RIPs have a long history of being exploited for medicinal purposes. For example, trichosanthin (TCS) has long been used as a natural abortificient. The use of RIP producing plants in traditional medicine has recently been reviewed by Polito et al. [33]. Also in modern medicine there is a lot of interest in the toxic properties of RIPs. This interest increased after it was shown in 1970 that ricin and abrin were more toxic to tumor cells than to normal cells [34]. Ever since a lot of effort has been invested in exploring how to use RIPs to combat cancer. Initially some promising progress was made with immunotoxins where a RIP domain was coupled to an antibody, targeting the toxin specifically to tumorous cells (reviewed by de Virgilio et al. [35]). In some cases the RIP domain was not coupled to an antibody but to other targeting molecules, such as for example transferrin, since malignant cells overexpress the transferrin receptor [35]. Although some of these RIP based toxins showed promising results in clinical trials, a lot of problems arose concerning immunogenicity and aspecific toxicity of the toxins. The use of biomaterials such as gold nanoparticles, polymer/lipid nanoparticles, and nanocapsules allowed better control of the immunogenicity and delivery of these RIP based toxins as recently reviewed by Pizzo and Di Maro [36].

Several lines of evidence suggest that RIPs are involved in plant defense. First, the expression of several RIP genes is regulated by both abiotic and biotic stress factors. For example, the JIP60 gene from barley is only expressed in senescing leaves or leaves treated with methyl jasmonate (MeJA) [29,37]. Jiang et al. [38] showed that RIP genes from rice are stress responsive. Furthermore, overexpression of the OsRIP18 gene yielded rice plants that were more tolerant to salt and drought stress [39]. Second, feeding experiments with artificial diets as well as experiments with transgenic plants overexpressing RIPs showed the antiviral, antifungal, and insecticidal properties of RIPs as recently reviewed [40–42].

Though RIPs are present in a lot of different plant species, they are not ubiquitous, as evidenced by the absence of RIP genes from the *Arabidopsis* genome. Schrot et al. [27] recently provided an updated list of all known RIPs from plants. In the present review we will focus on RIPs from Poaceae, since this group of proteins differs in many aspects from RIPs of other plant species. Cereal RIPs have evolved at a higher rate resulting in some unique structures, domain architectures, activation mechanisms, subcellular localization, and physiological roles. To our knowledge only two other reviews have been published dealing with cereal RIPs in general [43,44]. Both reviews mainly focus on the b-32 RIP from maize. In view of the new data and insights generated in recent years an updated review on cereal RIPs can be justified.

2. Evolution and Unique Domain Architecture of Cereal RIPs

Initial studies on the evolution and taxonomical distribution of RIPs were merely based on the knowledge gathered from proteins that had been purified and characterized. Consequently only RIPs that are present in sufficient amounts to be purified are included in these phylogenetic analyses. Our research group performed some extensive in silico analyses based on both BLASTp and tBLASTn searches in genome as well as transcriptome databases. This allowed the mapping of the taxonomical distribution of RIPs and the elaboration of an evolutionary model for the RIP domain [1,2]. In 2013, Lapadula et al. [45] suggested a slightly different evolutionary model. The main difference between the two models is the explanation given for the occurrence of RIPs over different taxa like plants, bacteria, fungi, and some insects. While our research group is convinced that the presence of RIPs in non-plant taxa can be explained through horizontal gene transfer from the plant RIP lineage, Lapadula et al. [45] favor the idea that some paralogous genes exist in the common ancestor of bacteria and eukaryotes. The main weakness of this hypothesis is the absence of RIP genes in Archaea [45]. In the interest of this review we will only discuss the evolution of RIPs in the plant kingdom. The main findings from our in silico analyses indicate that: (i) RIPs are not ubiquitous since RIP sequences are absent from 24 out of 42 completed plant genomes (including *Arabidopsis*); (ii) The existing list of RIPs needs to be extended horizontally (more species) and vertically (more RIP genes per species); (iii) Extended RIP gene families are common in Poaceae/cereal species. The RIP genes from this family show some remarkable differences with other plants in terms of domain architecture of the RIP genes and complexity of the RIP gene family, suggesting a unique, quite recent, evolutionary process within the Poaceae/cereal species [2,46].

RIPs are classically divided into type-1 RIPs, that only consist of an A-chain with N-glycosidase activity, and type-2 RIPs in which the A-chain is linked to a C-terminal lectin B-chain. Special cases like the b-32 maize RIP (which can be considered a type-1 RIP that undergoes proteolytic activation) and the JIP60 from barley (where an *N*-terminal RIP-domain is fused to a C-terminal domain that shows high sequence similarity to the eukaryotic translation initiation factor 4E (eIF4E)) were classified as "type-3 RIPs". New insights into the evolution of RIP genes and the complexity of several Poaceae/cereal RIPs in terms of domain architecture no longer fit this nomenclature and give impetus to readdress the classification system. Proteins/genes consisting of a single N-glycosidase domain will be referred to as the "A type" and the chimeric forms as the "AN type" whereby N designates the different (unknown) C-terminal domains. $A^{\Delta N}$ RIPs are RIPs consisting only of an N-glycosidase domain resulting from the deletion of the N domain from the AN chimeric, whereas A^U RIPs have an unknown origin [2].

According to the evolutionary model (Figure 1), an ancestral RIP domain was developed in plants. Next, it is hypothesized that a fusion with a bacterial B-chain (lectin) by horizontal gene transfer gave rise to the classical lineage of typical AB RIPs like ricin. This assumption is based on the widespread distribution of B-chain encoding sequences in all major prokaryotic and eukaryotic taxa [1,2]. Furthermore, sequence comparisons revealed that the B-chain of plant type AB RIPs shares a higher similarity with the carbohydrate-binding part of an β-glycosidase-like glycosyl hydrolase and an α-L-arabinofuranosidase B family protein from the Actinomycete bacterium *Catenulispora acidiphila* than it does with B-chains of other eukaryotes [1]. Domain deletions gave rise to the classical $A^{\Delta B}$ RIPs. In addition to these typical RIP lineages a second lineage of chimeric plant RIPs, called the AX type, was established. This lineage is the result of a domain fusion with an X domain of unknown origin. Apart from the nature of the *C*-terminal domain, there is a very important difference between the AB and AX lineages. While classical AB RIPs are synthesized on the rough endoplasmic reticulum and are subsequently secreted or sequestered to the vacuole or cell wall, the RIPs descending from the AX lineage are synthesized without a signal peptide on free ribosomes, and hence remain in the nucleocytoplasmic space, where they can make contact with the host ribosomes. This difference in location between both AB and AX RIP lineages reflects a different function within plants and suggests an in planta activity of the AX RIPs in contrast to the generally accepted role of the AB RIPs in plant defense. $A^{\Delta B}$ and $A^{\Delta X}$ RIPs normally have the same subcellular localization as their parent chimeric forms although some exceptions occur. Based on the taxonomical distribution of AX and $A^{\Delta X}$ RIPs, the ancestral AX gene is suggested to have originated before the separation of monocots and dicots. However, the exact timing of this event is hard to predict [2,46].

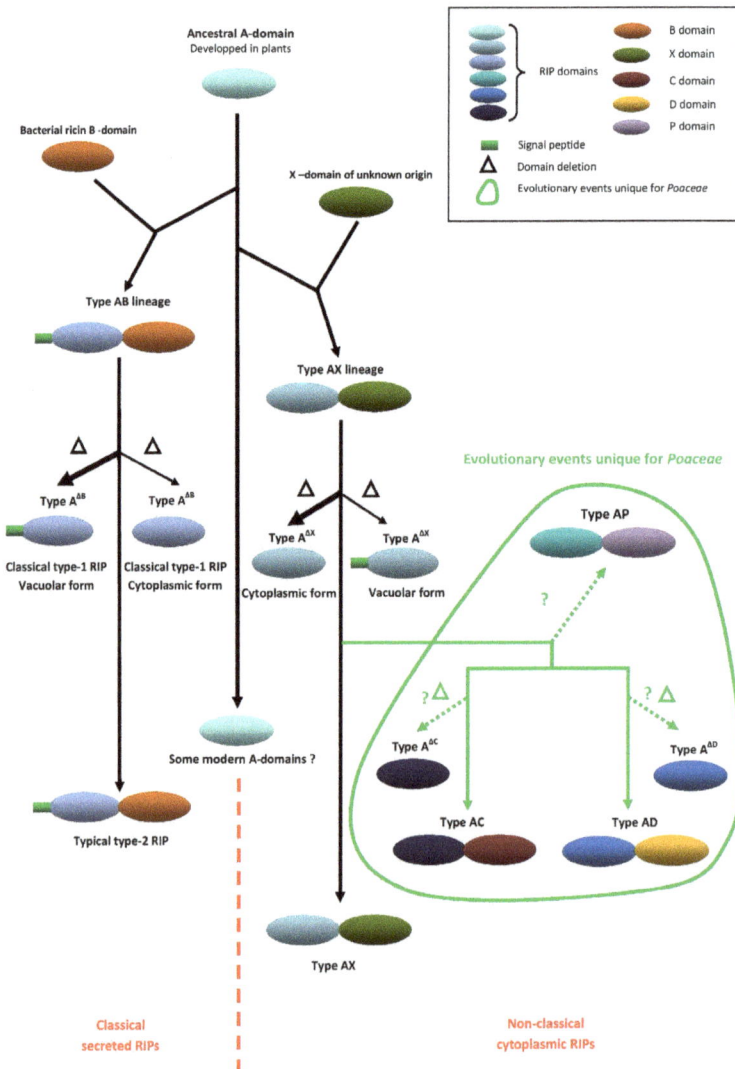

Figure 1. Evolutionary model for ribosome-inactivating protein (RIP) genes in plants.

Next to the AB and AX forms, other chimeric RIP forms such as the AC, AD and AP forms were found in the genomes of some Poaceae/cereal species. The so-called "type-3 RIP", JIP60 from barley, is probably the archetype of the AC group but these AC forms also occur in other cereal species. The AD type RIPs possess a C-terminal domain of yet another origin. In the AP forms an A domain is linked to a peptidase domain resulting in proteins with possible dual enzymatic activities. These forms are present in both *Oryza* and *Triticum*. Table 1 shows the occurrence of the different RIP forms in major cereal species [2,46].

Table 1. Different ribosome-inactivating protein (RIP) forms in cereal species.

Species	RIP Gene Architectures Reported
Oryza sativa (rice)	A^u, AC, AP
Avena barbata (oat)	A^u, AC
Hordeum vulgare (barley)	A^u, AC
Triticum aestivum (wheat)	A^u, AB, AP
Sorghum bicolor (sorghum)	A^u, AB
Zea mays (maize)	A^u, AB, AC, AD

The AC, AD, and AP forms presumably have the same subcellular localization as the AX forms. Low sequence identity exists between AC, AD, and AX forms suggesting that these forms evolved from an ancestral AX lineage. In the Poaceae family there are a lot of A-form RIP genes for which the origin cannot be traced. They are denominated as type-A^U RIPs. Possibly the AC and AD forms also gave rise to secondary $A^{\Delta C}$ and $A^{\Delta D}$ forms, though this assumption is rather hypothetical. An alternative hypothesis is that the A^U RIPs directly evolved from an ancestral A-form [2,46].

In summary, the RIP gene family evolved at an enhanced rate in Poaceae/cereal species. This assumption is supported by the large size and the heterogeneity of the RIP gene family within these species and the occurrence of chimeric forms that are not seen outside this clade. Because of the complexity it is very hard to reconstruct the exact evolution of the RIP gene family within the Poaceae/cereals. For example, in the *Oryza sativa* genome more than 30 genes with a RIP domain were identified [38]. Our preliminary screenings indicated that the RIP families in *Triticum aestivum* and *Hordeum vulgare* are of a similar complexity as in rice, whereas those of *Sorghum bicolor* and *Zea mays* are less complex. In future, more and better genomic and transcriptomic sequencing data can hopefully solve the complex evolutionary puzzle of the RIP gene family in Poaceae/cereal species. In parallel it is also very important to conduct biochemical analyses and to characterize the biological activity of these interesting RIPs. Together this will bring us closer to clarifying the physiological function of this unique cereal RIP family.

3. Biological Activity of Cereal RIP Domains

3.1. Structure of Non-Cereal and Cereal RIP Domains

3.1.1. Structure and Active Site of Classical Non-Cereal RIPs

RIPs are N-glycosidases that can depurinate a specific adenine residue of the conserved SRL in the rRNA of prokaryotic or eukaryotic ribosomes. In order to fulfill their function, these proteins have evolved to near enzymatic perfection [47]. All RIPs are characterized by a typical three dimensional (3D) fold. Ricin, the RIP from *Ricinus communis*, was the first RIP for which the 3D structure was resolved at atomic level [48]. Ricin is also by far the best studied RIP and can be considered as a representative model for the RIP family. To date the 3D structures of more than 20 RIPs, both A and AB forms, have been resolved and studied in detail [49,50].

The structure of the A domain is highly conserved, irrespective of the domain architecture of the protein. While some authors suggest that the N-glycosidase domain consists of three domains [51], it is generally accepted that the A domain contains two subdomains [50,52,53]. The large *N*-terminal subdomain is composed of six α-helices and a six-stranded mixed β-sheet, and the small C-terminal subdomain consists of an anti-parallel β-sheet and an α-helix with a bend in the middle [49].

The specificity pocket of RIPs, responsible for binding of the substrate adenine of the SRL, is generally found in the cleft between the N-terminal and C-terminal subdomains [50]. The residues in the specificity pocket that are important for the RNA N-glycosidase activity are absolutely invariant (Table 2). The Tyr residues are involved in the stacking of the substrate adenine. Residue Y80 (in RTA) can adopt different conformations depending on the liganded state. For example, in the native ricin structure Y80 blocks the active site cleft and a ligand must displace the side chain to enable binding into the pocket [50]. Glu and Arg are the key residues for catalysis. Arg protonates the N3 of the ribose ring hereby forming a ribose carbocation which is stabilized by Glu [54]. Next to its stabilizing role, it is also proposed that Glu serves as a general base with nucleophilic water molecules for the depurination reaction [47,55]. The nucleophilic attack by the activated water molecule breaks up the N9-C1 glycosidic bond, releasing the adenine [35]. Many RIPs from dicotyledonous plants possess a second glutamate (Glu) residue in the active site that serves as a back-up when the catalytic Glu is mutated [47,56,57].

Table 2. Comparison of the active site residues of RTA, maize RIP1 and barley bRIP1.

RIP	Active Site Residues				
RTA	Y80	Y123	E177	R180	W211
Maize RIP1 = b-32	Y94	Y130	E207	R210	W241
Barley bRIP1	Y87	Y118	E175	R178	W213

3.1.2. Structure and Active Site of Cereal RIPs

Thus far, the crystal structures of two cereal RIPs have been determined, in particular the maize RIP1 and bRIP1 from barley [58,59]. Comparison of the overall structures of these cereal RIPs to RIPs from other plants revealed some differences. The α-helix B and β-strand 6 in the N-terminal subdomain are absent in the maize RIP1 and the anti-parallel β-strands 7 and 8 in the C-terminal subdomain are replaced by a short α-helix in maize RIP1 [58]. Unique for the maize RIP1 is the presence of an inactivation region (see below). The bRIP1 from barley possesses three unique 3_{10}-helices G1/G2/G3 and an extra C-terminal α9-helix. In contrast, α-helix 2 and 2 β-strands present in the A domain of ricin (RTA) and other RIPs are absent in bRIP1 [59]. Especially the additional G2 helix and the C-terminal α9 helix are of importance and will be discussed in more detail below.

An important feature of cereal RIPs is that they do not seem to have a back-up glutamate like non-cereal RIPs. This observation was made for the structures of maize RIP1 and bRIP1 [58,59] but sequence comparison suggests that this phenomenon is also observed in other known RIPs of the family Poaceae, such as rice RIP (GenBank: BAB85659) and JIP60 (GenBank: AAB33361), and raises questions about the importance of the backup glutamate residue [58]. The active site residues for the toxic ricin A-chain (RTA), maize RIP1 and barley bRIP1 are identical (Table 2).

3.1.3. Proteolytic Activation Mechanisms of Cereal RIPs

An important feature in the synthesis of cereal RIPs is the occurrence of unique activation mechanisms that have not been reported in RIPs from species outside the grass family. Walsh et al. [28] first reported that the maize RIP1 (pI of 6) is expressed and stored as an inactive precursor (proRIP1) in the endosperm. During germination it is proteolytically processed by endogenous proteases that remove 16 residues from the N-terminus, 25 residues from the central domain, and 12 residues from the C-terminus of the RIP rendering a two-chain active form (αβ-form) with a pI of 9. The two chains of 16.5 and 8.5 kDa in the active form are tightly associated but are not covalently linked (Figure 2). The Pro-RIP was previously described as the b-32 endosperm protein [60–63]. Although the processing of proRIP1 is achieved in vivo through the action of endogenous proteases, the activation can also be performed in vitro by a variety of nonspecific proteases such as papain and subtilisin Carlsberg. Purified proRIP1 is about 10,000 times less active than the processed αβ-form [28]. Activity assays with deletion mutants representing the different naturally occurring processing events allowed the contribution of different peptides in suppressing the activity of the maize RIP1 to be checked. These experiments clearly showed that the 25-amino acid internal insertion was the primary inactivating element of the pro-RIP. Removal of the N- or C-terminal sequence only increases the activity by 6- or 5-fold, respectively, whereas MOD, an active mutant of maize RIP1 in which the central internal inactivation region of proRIP1 is replaced by a 2-residue short linker, was reported to be about 650 times more active than proRIP1 in an in vitro ribosome inactivating activity assay based on a rabbit reticulocyte lysate [28,58,64]. However, proteolytically activated proRIP1 is still more active than the active MOD mutant [65]. The excision of an internal peptide to render an active form resembles the processing of certain human hormones, such as insulin, and represents a unique activation mechanism that is not described for other non-cereal RIPs [66]. Bass et al. [67] first postulated that the internal inactivation region might exert its function by disrupting the protein fold and affecting key residues in the active site [67], but this is unlikely since the inactive proRIP1 shows a well-structured fold and the inactivation region is located more than 15 Å apart from the active site cleft. Rather the inactivation region is proposed to disrupt the interaction with ribosomal proteins [68].

The maize b-32 RIP is not a standalone case. Bass et al. [67] reported a second maize RIP, referred to as RIP2, that is not predominantly located to the endosperm. Although gene expression for maize RIP2 is regulated in a significantly different manner, maize RIP2 showed a similar activation mechanism as reported for maize RIP1. ProRIP2 also contains an internal 19-amino acid peptide (in a similar region as in proRIP1) that is very rich in acidic residues, suggesting that this stretch of amino acids would also serve as an inactivation region and should be removed proteolytically for activation of the protein. Indeed, addition of the unprocessed form did not inhibit translation in a cell-free system based on a rabbit reticulocyte lysate. However, pretreatment of the proRIP2 with papain resulted in inhibition of translation. Since the proRIP2 is not specifically expressed in seed tissue as is the case with proRIP1, the question remains whether activation in vivo is caused by exposure to endogenous proteases (as shown for proRIP1 during germination) or by exposure to proteases introduced by invading pests or pathogens [67]. Chuang et al. [69] reported that proRIP2 can be processed in the midgut of fall armyworm into the active form.

The proteolytic processing of inactive RIP precursors into their active forms is not only restricted to maize. Hey et al. [64] analyzed seed extracts for different members of the subfamily Panicoideae by immunoblotting using antisera against the α and β fragments of the maize RIP1 [28]. The species *Z. mays parviglumis* (three accessions), *Z. luxurians*, *Z. mays mexicana*, *T. dactyloides*, and *Sorghum bicolor* were tested and immunoreactivity was observed for all extracts. However, a more distantly related species *Coix lachryma-jobi* showed no cross reactivity. Interestingly, in addition to the polypeptides of 32–34 kDa corresponding to the full length precursor protein the Western blot analyses also yielded polypeptides around 11–16 kDa, possibly corresponding to the separate α- and β fragments, suggesting proteolytic processing.

The proteolytic activation of RIP precursors can be extended beyond the Panicoideae subfamily since Chaudhry et al. [37] suggested that the JIP60 precursor from barley is processed in vivo. As described earlier, JIP60 is a 60 kDa jasmonate-inducible chimeric protein that consists of an N-terminal RIP domain and a C-terminal domain which displays similarity to eukaryotic translation initiation factor 4E [37,70]. Like maize RIP1, JIP60 differs from other RIPs in containing internal acidic peptides of 20–25 amino acids that will be removed during maturation of the protein. Western blot analyses confirmed that JIP60 was processed in extracts of jasmonate-treated leaves as well as after incubation with papain. It was suggested that at least two processing steps were necessary for rendering a fully active RIP. The N-terminal RIP domain should be separated from the C-terminal domain by proteolytic cleavage and an internal peptide in the RIP domain has to be removed, similar to the case of maize RIP1 (Figure 2). Full length JIP60 did not significantly inhibit protein translation in a system based on a reticulocyte lysate, whereas the *N*-terminal domain did inhibit translation to a certain extent. When the linker peptide in the *N*-terminal domain was removed, translation was inhibited almost completely [37].

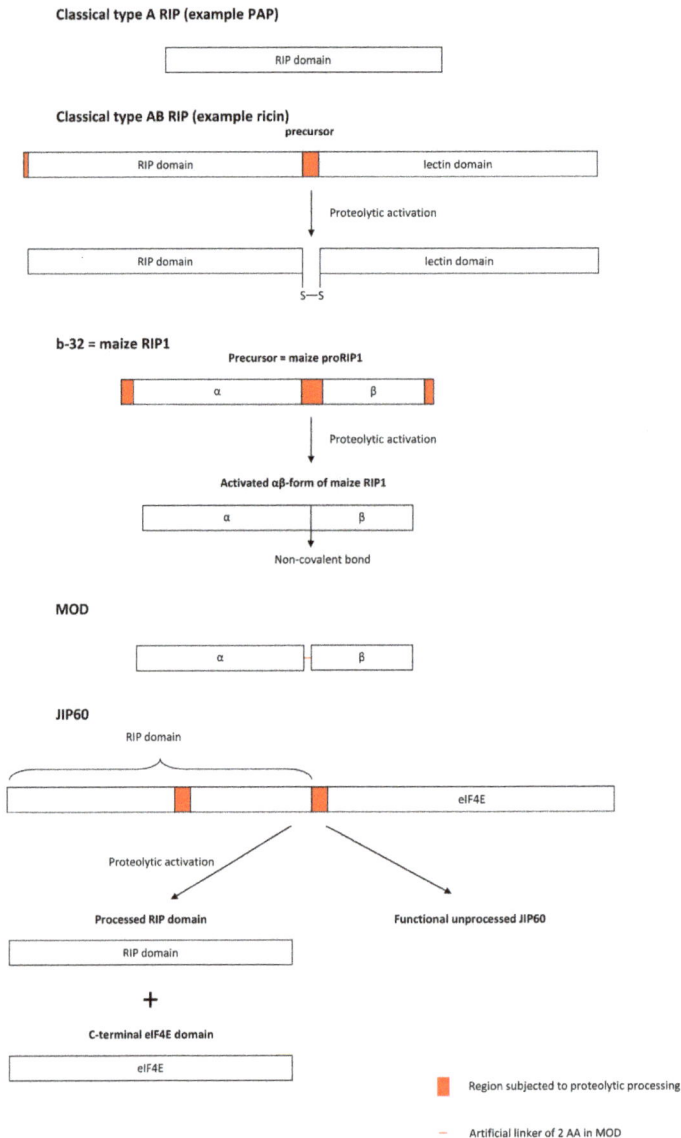

Figure 2. Activation mechanisms of the maize RIP1 (b-32) and the barley JIP60 in comparison with typical "Type-A" and "Type-AB" RIPs.

In contrast, Reinbothe et al. [29] reported that unprocessed JIP60 did inhibit translation of plant mRNAs in rabbit reticulocyte lysates. However, they attributed this arrest of translation to the inhibition of the initiation step rather than the elongation step of translation as is the case with typical RIPs. Instead of acting as a typical RIP and blocking the elongation step, the unprocessed JIP60 caused the decay of the rabbit polysomes into their subunits, preventing the initiation step of translation. Unprocessed JIP60 also had this effect on barley ribosomes isolated from leaves exposed to MeJA for longer than 24 h, or from osmotically stressed or desiccated leaves. Polysomes from leaves treated with a water control were not cleaved by unprocessed JIP60.

Rustgi et al. [70] expressed artificial proteins in *E. coli* corresponding to the full length JIP60, the unprocessed N-terminal RIP domain and the processed N-terminal RIP domain. In a rabbit reticulocyte lysate, full length JIP60 and the processed RIP domain completely abolished translation, whereas the unprocessed RIP domain did not. A comparative analysis of the polysomal profiles allowed the mode of action of translation inhibition to be analyzed. In the case of unprocessed JIP60, there was a drastic decrease in ribosome binding to the transcripts which was not the case with the processed RIP-domain. These experiments confirmed the hypothesis of Reinbothe et al. [29] that unprocessed JIP60 can act as a ribosome dissociation factor. However, when proteolytically cleaved, the released processed N-terminal domain can act as a classical RIP as postulated by Chaudhry et al. [37]. It remains unclear why the latter research group did not observe any effect of unprocessed JIP60 on translation.

3.1.4. "Switch Region" of Barley RIP

In 2012 the 3D structures of RIP1 from barley (bRIP1) in free form as well as in complex with adenine and AMP were determined [59]. In general the structure of bRIP1 shows high similarity with other RIP structures and most of the secondary structures are well superimposed. However, there are also some marked differences. The bRIP1 structure contains three unique 3_{10} helices G1, G2, and G3. Especially the G2 helix is of particular interest because this region displays two different conformations within the bRIP1 structure: it occurs as a 3_{10} helix or as a loop. As with other RIPs the geometry of Tyr84 (Tyr80 in RTA) in the active site can represent an "open" or a "closed" form depending on the liganded state. However, in bRIP1 the "open" or "closed" conformation of Tyr84 is correlated with a structural transition between a 3_{10} helix or a loop in this unique G2 region. When the G2 helix is formed, movement of a loop between the β-8 strand and the α-2 helix is blocked, thus preventing the transition from the open to the closed form needed for catalysis. On the contrary, when the region between the β-8 strand and the α-2 helix adopts a loop structure, the active site can close and enzyme catalysis can occur. Because of its putative role in the regulation of catalysis, this region was defined as a "switch region" [59].

Interestingly, this switch region is located at an equivalent position to the proteolytic cleavage site in the maize RIPs. These structural differences between apo and adenine-bound structures are not found in other RIP structures from dicotyledonous plants, which suggests a unique set of activation mechanisms in at least some cereal RIPs [59]. Since this region is also important for some RIPs (such as maize RIP1) for the interaction with ribosomes it might represent a major regulatory element of these proteins.

Barley bRIP1 is characterized by the presence of an additional C-terminal α-helix which is not found in maize RIP1 and RIPs from non-cereal species. However, several other cereal RIPs like those from barley, rice, and wheat have a strong sequence conservation in this C-terminal region [59]. At present it is unclear what the functional implications of this C-terminal α helix might be.

3.2. Interaction of RIPs with Ribosomes and Substrate Specificity

3.2.1. Interaction of Classical Non-Cereal RIPs with Ribosomes

The substrate of RIPs for the depurination is the SRL of the rRNA. Depurination of naked rRNA is about 80,000 fold slower compared to depurination of whole ribosomes [71]. Though the SRL is highly conserved among all ribosomes, a lot of plant RIPs are able to depurinate eukaryotic ribosomes but not prokaryotic ribosomes. Therefore it was suggested that ribosomal proteins contribute to the susceptibility of rRNA to depurination by RIPs as well as to the differential sensitivity of ribosomes to RIPs [72–74]. Indeed, the fact that some RIPs such as TCS and ricin can only inactivate eukaryotic ribosomes while the RIP from *Phytolacca americana* called "pokeweed antiviral protein" (PAP), can inactivate both eukaryotic and prokaryotic ribosomes can be explained by the different interaction with ribosomal proteins. It was shown that PAP binds ribosomal protein L3 and that this interaction is required for depurination activity [72]. L3 is highly conserved between eukaryotic and prokaryotic ribosomes which explains the interaction of PAP with both types of ribosomes [75]. In contrast to PAP, a lot of other RIPs like TCS and ricin were shown to interact with the acidic ribosomal stalk proteins (P-proteins) [49,73,76]. The ribosomal stalk is part of the GTPase activation center [77] which plays a role in the binding of elongation factors [78–80]. Thus these RIPs might have evolved to hijack the translation-factor-recruiting function of the ribosomal stalk in reaching their target site on the rRNA [81]. Unlike L3, the ribosomal stalk proteins are not conserved between prokaryotes and eukaryotes. The stalk of bacterial ribosomes is formed by ribosomal protein L10 in complex with two or three copies of L12 homodimers [82,83] while the eukaryotic stalk is composed of a pentameric P-complex containing P0 and two P1/P2 heterodimers [84,85]. Prokaryotic ribosomes are insensitive to TCS, but when the stalk proteins were replaced by their eukaryotic counterparts, the hybrid ribosomes became susceptible to TCS. These data suggest that the difference in ribosomal stalk proteins explains why RIPs like TCS and RTA are not active against prokaryotic ribosomes compared to PAP.

TCS and RTA interact with the P2 peptide, a conserved C-terminal 11 residue-sequence "SDDDMGFGLFD" of the stalk proteins and this property was suggested to be a general theme for many RIPs [49,73,86]. The binding mechanism of TCS to the P2-peptide was first elucidated and revealed that the basic residues K173, R174 and K177 in TCS form favorable electrostatic interactions with the acidic DDD motif of the P2 peptide while a hydrophobic pocket accommodates the C-terminal LF residues of the P2 peptide [73,87]. Shi et al. [88] suggested a different but similar mode of interaction based on both electrostatic and hydrophobic interactions for RTA. Residues R189 and R193 of RTA bind the SDDD motif electrostatically and the LF motif is inserted into a hydrophobic pocket [88]. However, Fan et al. [89] reported that the DDDM motif contributes little to the interaction of P2 with RTA, and that the binding is the result of hydrophobic interactions and the formation of some hydrogen bonds between the F114 and D115 of P2 and R235 of RTA [89]. For both TCS and RTA hydrophobic forces dominate the interaction with the P proteins.

3.2.2. Interaction of Cereal RIPs with Ribosomes

Maize RIP1 was also reported to interact with the P2 peptide of the acidic stalk proteins [58,68,90]. However, the interaction between maize RIP1 and the P2 peptide is significantly different compared to TCS and RTA. As mentioned previously, the β-strands 7 and 8 of the C-terminal domain that form the P2 binding site on TCS, are replaced by an α-helix in maize RIP. Therefore it was suggested that the P2 binding site in maize RIP1 might be different compared to other RIPs [58]. Using chemical-shift perturbation NMR (CSP-NMR) analysis the P2 binding site of maize RIP1 was mapped to a stretch of four lysine residues (K143–146) in the N-terminal part of the maize RIP1 [68]. Surprisingly this site is located next to the inactivation loop of maize proRIP1, reinforcing the hypothesis that the inactivation region is involved in sterically hindering ribosome binding rather than changing the conformation of the activity cleft. This assumption was confirmed by analyses whereby P2 was immobilized on a Sepharose column and subsequently challenged with both MOD and proRIP1. Only MOD could bind to P2 whereas proRIP1 could not. A MOD [K158A–K161A] mutant also did not bind to the column, confirming that binding is indeed established by the lysine stretch. Further pull down assays on the P2 column using buffers with different ionic strength proved that the interactions between maize RIP1 and P2 are electrostatic in nature, since binding of maize RIP1 was decreased with increasing salt concentrations. This was not the case for RTA or TCS. Furthermore removal of the FGLFD motif of the P2 peptide did not reduce interactions with maize RIP1 as it did for RTA and TCS [90]. Taken together these data suggest that maize RIP1 binds to P2, only after activation, and does so with a binding region that is very different from those of other RIPs. The binding is also solely dependent on electrostatic interactions and does not rely on hydrophobic interactions as classically seen with other RIPs [58,68,90].

Since maize RIP1 is the only cereal RIP thus far for which the interaction mechanism with ribosomal proteins was revealed, it is hard to predict if this unique mode of binding is valid for all cereal RIPs or this is a standalone case. Lapadula et al. [91] showed that the KKKK motif found in maize RIP1 is absent in RIP sequences from species like *Hordeum vulgare*, *Oryza sativa*, and *Triticum aestivum*. It was only present in some sequences from *Zea* species [91]. Since other cereal RIPs like JIP60 and barley RIP1 share some structural similarities with maize RIP1 in that they are also processed in a similar way to maize RIPs (as shown for JIP60 [37]) or contain a 'switch region' in the region corresponding to the inactivation loop in maize RIP1 (as is the case in barley RIP1) [59] it will be interesting to study the binding of these cereal RIPs with the ribosomal proteins.

3.2.3. Substrate Specificity of Classical Non-Cereal RIPs

Although interaction with different ribosomal proteins can account for the fact that some RIPs can inhibit both prokaryotic and eukaryotic ribosomes (like PAP) while others cannot, this does not immediately explain the fact that a lot of RIPs inhibit ribosomes of different eukaryotic sources with variable efficiency. The "SDDDMGFGLFD" motif of P2, that was proven to be essential in binding of a lot of RIPs to ribosomes, is conserved among all eukaryotes. So if P2 interaction is the only mechanism required for binding to ribosomes, RIPs should be capable of inhibiting ribosomes of different eukaryotic sources with the same efficiency, including plant ribosomes. However, this is not the case. For example, ricin is most active against mammalian ribosomes but yeast ribosomes are less sensitive [92]. RTA is around 23,000 times less active on plant ribosomes compared to mammalian ribosomes [93]. Type-2 RIPs in general are several-thousand-fold less active on plant ribosomes than on mammalian ribosomes [93,94] while type-1 RIPs can efficiently act on plant ribosomes [95]. To circumvent the depurination of host ribosomes, plants protect themselves from the effect of their RIPs. Most RIPs are synthesized with *N*-terminal signal sequences that target them to the endomembrane system, away from the ribosomes [96]. Additional carboxy terminal extensions are found in several RIP sequences that subsequently target them to the vacuole and prevent these proteins from becoming active before they reach the storage organelles [97–99]. Other RIPs like PAP are secreted and stored in the cell wall [100].

3.2.4. Substrate Specificity of Cytoplasmic RIPs from Cereals

Interestingly, RIPs from cereals seem to lack signal peptides, suggesting they should reside in the cytosol where they can make contact with the host ribosomes [28,37,101,102]. This raises some interesting questions concerning their mode of action. Are cereal ribosomes insensitive to these RIPs or do these RIPs exert their physiological role by inhibiting conspecific ribosomes? Different studies show that both options are possible. Both the pro-form as well as the activated form of the maize RIP1 are not very active on conspecific ribosomes [64]. This also suggests that the removal of the inactivation loop is not a measure to protect the host ribosomes. No results are available for the effect of the second maize RIP [67] on conspecific ribosomes.

Tritin, the RIP found in wheat germ, was reported to be inactive against wheat ribosomes [103,104], while other studies suggested it could inhibit conspecific ribosomes at least to some extent [102,105]. Massiah and Hartley [104] reported the existence of a second RIP in wheat that was not found in the seeds but in the leaves. Therefore they referred to the originally discovered tritin as tritin-S while the leaf form was given the name tritin-L. The authors showed that the S-form was inactive against conspecific ribosomes while the L-form could depurinate wheat ribosomes. Furthermore they reported that the S-form needed ATP as a cofactor in order to be enzymatically active while this does not hold true for the L-form. Sawasaki et al. [106] showed that tritin mediated the programmed senescence of the wheat coleoptiles by cleaving conspecific ribosomes. Although the authors did not mention explicitly which tritin form they focused on, our investigation of the gene sequence revealed that it encodes the tritin-S form. Hence this result contradicts the findings from Massiah and Hartley [104].

While barley seed RIPs are not active against barley ribosomes [101], the JIP60 found in barley leaves is a special interesting case. It was mentioned earlier that full length JIP60 can act as a ribosome dissociation factor [29,70]. However, only ribosomes isolated from leaves that had been exposed to MeJA for more than 24 h, or ribosomes from osmotically stressed or desiccated leaves were susceptible to the unprocessed JIP60. When processed, the *N*-terminal RIP domain of JIP60 can act as a genuine RIP [37,70]. Dunaeva et al. [107] reported N-glycosidase activity of (processed) JIP60 in tobacco and barley, suggesting that also the processed form is active on plant ribosomes.

In general, the interaction with ribosomal proteins seems to play a very important role in determining the substrate specificity of RIPs. However, there can be important differences depending on the RIP under study. PAP, RTA, TCS and maize RIP1 all show different modes of interaction with ribosomal proteins. More studies should be conducted to elucidate the binding of RIPs to ribosomal proteins as well as their substrate specificity. This is especially the case for the interaction between plant ribosomes and cereal RIPs given their unique structure, activation mechanisms, and presumed cytosolic localization in the plant cell.

4. Physiological Roles of Cereal RIPs

RIPs are supposed to help plants in safeguarding them from pathogenic fungi, viruses, or insects. These classical defense functions were also attributed to cereal RIPs. However, there is evidence that several cereal RIPs also mediate some very unique in planta functions. A summary of the physiological roles reported for cereal RIPs is given in Table 3.

Table 3. Summary of different physiological roles reported for cereal RIPs.

Species	RIP	Tissue	Role in Defense	In Planta Function
Zea mays	Maize RIP1 (= b-32)	Seeds	Antifungal, insecticidal	Storage function in seeds?
	Maize RIP2	Whole plant, except kernel	Expression upon herbivore attack, active against *Spodoptera frugiperda*	Involved in drought response?
Sorghum bicolor	Sorghum RIP	Seeds	Antifungal protein	Not reported
Oryza sativa	OsRIP18 = RA39	Tapetum	Not reported	Involved in drought and salt response. Involved in microspore maturation?
	Other rice RIPs	Variable	Expression of several genes induced by *Magnaporthe grisea* or *Xanthomonas oryzae*	Expression of several genes is enhanced after abiotic stress
Triticum aestivum	Tritin	Seed and leaf forms	Not reported	Involved in senescence
Hordeum vulgare	RIP30 and isoforms	Seeds	Antifungal protein	Not reported
	JIP60	Leaves	Re-organization of translational machinery in stress situations	Re-organization of translational machinery in stress situations

4.1. Role in Defense

4.1.1. Maize RIP1

Maize RIP1, also called RIP b-32, is by far the most intensively studied cereal RIP. It is exclusively found in the endosperm of maize kernels. The expression of the RIP1 gene is controlled by the Opaque-2 (O2) locus, which also controls the expression of other typical endosperm proteins such as the 22 kDa zeins [108]. The first indication that the b-32 protein is involved in protecting the maize seeds from biotic stress factors came from the observation that *o2* mutants were more susceptible to fungal attack [109] and insect feeding [110]. Since then, a lot of research has been conducted to clarify the role of maize RIP1 in plant defense.

The proteolytic activation mechanism of maize RIP1 has already been discussed in this review. The proRIP1 showed an inhibitory effect on the growth of *Rhizoctonia solani* in vitro [111]. However, other authors reported that proRIP1 had no effect on *Aspergillus flavus* and *Aspergillus nidulans*, while the proteolytically activated RIP1 reduced the growth of the pathogens in vitro. Furthermore when the RIP1 protein was mutated to abolish the RIP activity the protein did no longer inhibit the fungal pathogens, suggesting that RIP activity is essential for the antifungal activity [112]. It is not clear why activation of the proRIP1 seems to be necessary for antifungal activity in some experiments while this is not a prerequisite in other assays. One explanation could be that in those cases where the pro-form was active against fungi, the protein was actually activated by proteases secreted by the fungus.

Next to these in vitro bioassays, a lot of experiments with transgenic plants expressing the maize RIP1 were conducted in order to check the antifungal effect of the protein in planta. Transgenic rice plants expressing the maize RIP1 did not significantly reduce the disease severity caused by the fungal pathogens *Rhizoctonia solani* and *Magnaporthe griseae* [113]. However, when the protein was overexpressed in tobacco RIP1 it did confer increased protection against *Rhizoctonia solani* [111]. Similarly, transformed wheat plants expressing maize RIP1 were more resistant to infection by *Fusarium culmorum*, the causal agent of *Fusarium* head blight. When immature spikes were inoculated with spores, the number of infected spikelets per head at seven and 14 days after inoculation was significantly lower in transgenic plants compared to control plants [114]. Leaf tissue colonization bio-assays conducted with *Fusarium verticillioides* and leaves from transgenic maize ectopically overexpressing the b-32 protein clearly showed a reduction in mycelial growth compared to wild type leaves. Furthermore the level of resistance was correlated to the RIP1 content in the transgenic leaves [115]. The antifungal properties of maize RIP1 have been reviewed in detail by Motto and Lupotto [43] and Lanzanova et al. [116].

Besides antifungal properties, the maize RIP1 also acts against predatory insects. Dowd et al. [117] tested the effect of purified proRIP1 as well as papain activated RIP1 on a wide array of insects. Five different species of caterpillars were used for feeding assays and five different beetle species were used in choice assays. The caterpillars used in the feeding assays were: corn earworms (*Helicoverpa zea*), fall armyworms (*Spodoptera frugiperda*), European corn borers (*Ostrinia nubilalis*), cabbage loopers (*Trichoplusia ni*), and Indian meal moths (*Plodia interpunctella*). Only the papain activated RIP1 showed some significant effect on the caterpillars and the degree of toxicity was strongly dependent on the species tested. The toxic effects ranged from 70% mortality for the cabbage looper to no effect on the Indian meal moth. For the choice assays the following beetle species were used: freeman sap beetles (*Carpophilus freemani*), strawberry sap beetles (*Stelidota geminata*), maize weevils (*Sitophilus zeamais*), and dusky sap beetles (*Carpophilus lugubris*). In contrast to the effects on caterpillars, both proRIP1 and activated RIP1 significantly deterred feeding in choice assays with relative feeding rates being reduced up to 6-fold.

Efforts were made to confirm these effects in planta by overexpressing the maize RIP1 in plants and subsequently checking the effects on herbivorous insects. Tobacco plants expressing an activated form of the maize RIP1 showed enhanced resistance against the corn earworm (*Helicoverpa zea*). Insect feeding was reduced on transgenic plants compared to wild type plants and there was a higher mortality in the population fed on the transgenic plants. The survivors were characterized by a reduced bodyweight. The degree of damage to the plants caused by *Helicoverpa zea* was significantly though inversely correlated with RIP1 levels present in the plants [118]. Tobacco hybrids, obtained by crossing a maize RIP1 expressing line with a tobacco line overexpressing the tobacco anionic peroxidase also showed enhanced resistance against the corn earworm (*Helicoverpa zea*) and the cigarette beetle (*Lasioderma serricorne*). Transgenic plants were less prone to feeding by both insects compared to wild type plants. This effect was again inversely correlated with overexpression levels of both proteins in the tobacco plants [119]. Finally, maize plants ectopically overexpressing both maize RIP1 and a wheat germ agglutinin were more resistant to feeding by larvae of the fall armyworm (*Spodoptera frugiperda*) and of the corn earworm (*Helicoverpa zea*), and this level of resistance was correlated with the levels of maize RIP1 and the agglutinin [120]. Together, these data suggest that the maize RIP1 plays an important role in protecting developing maize seeds against attacks by fungal pathogens and herbivorous insects.

4.1.2. Maize RIP2

Compared to the maize RIP1, much less attention has been given to the second RIP identified from maize. As mentioned previously, similar to maize RIP1 the maize RIP2 is characterized by a proteolytic activation mechanism [67]. However, some important differences exist between the two proteins. The amino acid sequences of maize RIP1 and maize RIP2 share only 73% sequence similarity [121]. Furthermore the expression of RIP1 and RIP2 is differently regulated. While maize RIP1 is exclusively expressed in the kernel, maize RIP2 is expressed throughout the whole plant, except the kernel [67]. The RIP1 gene is found on chromosome 8 where it is under control of the opaque-2 regulatory locus and its expression is controlled developmentally. In contrast the RIP2 gene is located on chromosome 7 and RIP2 expression is responsive to environmental stimuli. Bass et al. [67] showed that RIP2 expression is higher upon drought stress, and a Chuang et al. [69] reported that RIP2 accumulates in maize leaves after caterpillar attack, suggesting a role for RIP2 in the defense against herbivory. This role of RIP2 was already hinted by the fact that the position of the RIP2 gene on chromosome 7 coincides with a strong QTL for caterpillar resistance [122]. Chuang et al. [69] reported that RIP2 expression strongly increased in leaves after feeding of the fall armyworm (*Spodoptera frugiperda*). Transcript levels for RIP2 as well as protein levels increased one hour after caterpillar attack and remained high during the 24 h monitoring period. Four days after initial feeding, RIP2 levels were still high in maize leaves. In contrast to caterpillar feeding, mechanical wounding did not alter RIP2 expression. Since several phytohormones like salicylic acid (SA), MeJA, and ethylene (ET) are known to be involved in the regulation of plant responses to biotic stress, the authors also investigated whether treatment with these hormones could trigger RIP2 expression. Only MeJA and ET induced RIP2 expression albeit only when combined with mechanical wounding. Treatment with ABA, a phytohormone involved in abiotic stress response, combined with mechanical wounding also induced RIP2 expression confirming the results of Bass et al. [67].

A second line of evidence for the role of RIP2 in herbivory defense came from the fact that the RIP2 protein survived digestion in the midgut of the fall armyworm. Defense proteins are often resistant to the proteases found in the midgut of caterpillars and can consequently be traced in the frass, while proteins not involved in defense mechanisms are easily degraded by the insect and are therefore not found in the frass. One of the predominant proteins found in the frass of fall armyworms fed on maize leaves was the maize RIP2. Furthermore, in the Western blot analyses polypeptides corresponding to the processed form of RIP2 were detected, suggesting that the protein is activated by insect proteases [69].

To confirm the presumed defensive role of the maize RIP2 against herbivorous insects, recombinant RIP2 was made and fed to larvae of fall armyworm. The concentration of recombinant protein in the artificial diet represented the amounts of RIP2 found in maize leaves under caterpillar attack. Both papain activated RIP2 as well as the unprocessed proRIP2 reduced the larval weight by 26% compared to larvae fed on BSA or buffer control. Immunoblotting of the frass of the fed larvae confirmed that proRIP2 was proteolytically activated in the midgut of the fall armyworm [69].

These data suggest that RIP2 is important in the inducible protection of the vegetative tissues of maize plants against fungal pathogens and insects, while RIP1 protects the developing seeds of the plant in a more constitutive manner.

4.1.3. Sorghum RIP

A protein similar to the maize RIP1 was also detected in sorghum seeds. Antibodies raised against the maize RIP1 cross reacted with this sorghum RIP and the occurrence of multiple reactive polypeptides suggested proteolytic processing of the sorghum RIP, as observed with the maize RIP1 [64]. The sorghum RIP is a member of the group of antifungal proteins (AFPs) found in sorghum caryopsis. Other AFPs include sormatin, chitinase, and glucanase. The presence of these AFPs is also characteristic for caryopsis of other cereals such as barley, maize, and wheat [63,123–125]. Seetharaman et al. [126] extracted and partially purified these sorghum AFPs and tested their biological activity against the grain molding pathogens *Fusarium moniliforme, Curvularia lunata,* and *Aspergillus flavus* in vitro. They reported that a mixture containing several AFPs was most inhibitory against the pathogens. RIP levels in sorghum caryopsis are highest at 15 days post anthesis and subsequently decreased, while the other AFP levels increase during development and only go down when grain maturity is reached [127]. Rodriguez-Herrera et al. [128] compared the levels of these AFPs in grain mold resistant (GMR) and susceptible (GMS) sorghum lines and noticed that in environments with grain mold incidence, the levels of RIP, sormatin, and chitinase were higher in GMR lines than in GMS lines. In a grain mold free environment the RIP levels were also higher in the GMR group. In the GMR lines, the levels of RIP, sormatin, and chitinase were higher in environments with grain mold incidence compared to environments without grain mold. These data suggest that GMR lines express more AFPs under grain mold pressure than GMS lines. In addition, RIP levels seem to be higher in GMR lines regardless of the presence of grain mold [128]. The sorghum RIP thus seems to play an important role in protecting the seeds against fungal attack.

4.1.4. Barley RIPs

Barley seeds contain three type-A RIPs, the most famous one being RIP30. The RIPs are very similar in amino acid sequence and can be considered isoforms [129]. Similar to the sorghum RIP, RIP30 is one of the antifungal proteins identified in barley seeds. Leah et al. [101] showed that RIP30 accumulates to high levels during late seed development, and inhibits the growth of the fungal pathogens *T. reesei* and *F. sporotrichioides* in vitro. However, the inhibitory effect was much more pronounced when a mixture containing the RIP and other AFPs such as glucanase and chitinase was used, suggesting that the effect of RIP30 is enhanced when the fungal cell walls are permeabilized by the action of these hydrolases [101]. Several experiments have been conducted to confirm the in planta antifungal effects of the barley RIP by overexpressing the protein in different plant species and subsequently examining the plant's resistance against fungal pathogens. In a study from Oldach et al. [130] transgenic wheat overexpressing the barley RIP did not significantly reduce formation of powdery mildew (*Erisyphe graminis*) or leaf rust (*Puccinia recondita*) colonies while wheat plants overexpressing chitinase or an antifungal protein from the fungus *Aspergillus giganteus* did reduce the disease symptoms. Transgenic wheat plants overexpressing an apoplastic targeted version of the barley seed RIP were not or only moderately protected against *Erysiphe graminis* but did show protection against the powdery mildew (*Blumeria graminis*). Co-expression of other AFPs such as barley seed chitinase and barnase did not significantly improve this level of protection [131,132]. More recent studies have proven that the overexpression of a combination of RIP30 and chitinase is successful. Transformed Indian mustard (*Brassica juncea*) expressing both barley RIP and chitinase showed 44% reduction in *Alternaria brassicae* hyphal growth in an in vitro assay and transgenic plants sprayed with an *A. brassicae* spore suspension showed reduced numbers, sizes, and expansion of lesions compared to wild type plants [133]. Similarly, potato plants overexpressing a combination of RIP and chitinase showed enhanced resistance to *Rhizoctonia solani* in a greenhouse assay [134]. Chopra and Saini [135] reported that transformed blackgram (*Vigna mungo*) plants expressing the same set of genes arrested the growth of *Corynespora cassiicola* in an in vitro antifungal assay and the plants also showed less disease symptoms compared to wild type plants after being sprayed with a spore suspension.

5. In planta Functions

A lot of cereal RIP genes are characterized by the absence of a signal peptide and therefore will most probably reside in the cytoplasm where they can make contact with the host ribosomes. This interaction can have implications regarding the possible endogenous functions of these RIPs.

5.1. Storage Function of Cereal Seed RIPs

Next to being used as antifungal agents, the maize RIP1 and barley seed RIPs are also thought to fulfill a nutritional role. It was previously mentioned that the expression of the maize RIP1 is controlled by the opaque-2 locus, which is known to regulate the expression of the zeins. It was also noticed that the maize RIP1 is high in lysine and methionine, amino acids that are underrepresented in the zeins [136]. Therefore the maize RIP1 could serve as a storage albumin that compensates for the lower methionine and lysine contents in the major seed storage proteins [43,137]. A similar nutritional function was suggested for the barley RIP30 isoforms that are expressed specifically in the starchy endosperm. The starchy endosperm cells differentiate during development and at maturity they are metabolically senesced. Leah et al. [101] hypothesized that the barley seed RIPs could play a role in this process.

5.2. Rice RIPs

A genome wide survey of the RIP family in *Oryza sativa* spp. *Japonica* conducted by Jiang et al. [38] identified 31 genes encoding proteins containing a RIP domain. All these putative RIPs were named type-1 RIPs. qPCR analyses showed that these RIP genes are differentially expressed in leaves, roots, and panicles and that the expression of several genes was stress responsive [38].

Analyses conducted by our group identified slightly different numbers of rice RIP genes in the genomes of *Oryza sativa* [2,46]. In silico analyses of expression data from RNAseq studies confirmed the differential expression of the RIP genes in different tissues and under different abiotic stresses. While Jiang et al. [38] only identified type-A RIPs, we believe that some of these RIPs have an AN type architecture. Both RIP sequences with and without signal peptide were found. The fact that some RIPs are differentially expressed after the application of abiotic stress suggests a role in the abiotic stress response.

The gene RA39 reported by Ding et al. [138] is a tapetum specific gene that is highly expressed in rice tapetal cells at the meiosis stage. RA39 encodes a secreted RIP and the recombinantly produced protein exhibited RNA N-glycosidase activity. During microspore development tapetal cells have a nutritional function and are degraded in order to free up nutrients for the maturing microspores. This tapetum degradation can be seen as a type of programmed cell death. Since the timing of this process coincides with the spatio-temporal regulation of RA39 expression, the authors hypothesized that the RIP might play a role in programmed cell death [138].

The RA39 studied by Ding et al. [138] is referred to as the OsRIP18 by Jiang et al. [38]. Expression of an OsRIP18 promoter-GUS construct confirmed the tapetum specific expression. Furthermore Jiang et al. [38] showed that the tapetal expression increased after polyethylene glycol and salinity stress, suggesting a role of the gene in drought and salinity tolerance. Rice plants ectopically overexpressing the RIP showed significantly increased tolerance to drought and high salinity stress, substantiating the possible function of OsRIP18 in abiotic stress tolerance [39]. Micro-analysis revealed that 128 genes were upregulated in rice plants overexpressing the RIP gene compared to wild type plants. Five of these genes are also upregulated in wild-type plants stressed by drought or salinity compared to non-stressed wild-type plants. The authors concluded that OsRIP18 might exert its function by re-organizing the translational machinery.

5.3. Maize RIP2

As mentioned earlier, the expression of maize RIP1 and RIP2 is differently regulated. Chuang et al. [69] showed that next to caterpillar feeding, a combination of mechanical wounding and ABA triggered maize RIP2 expression. Furthermore Bass et al. [67] reported that the level of RIP2 transcripts as well as the protein levels increased when plants were subjected to drought stress. These data suggest that besides protecting maize leaves against insect herbivory, maize RIP2 can also be involved in the abiotic stress response.

5.4. Tritin

Although there are some contradictions in the literature concerning the different isoforms of tritin found in wheat and their respective action on conspecific ribosomes [102–105], it is safe to say that at least one isoform of tritin serves an endogenous role in wheat by acting on wheat ribosomes. Sawasaki et al. [106] showed that tritin together with an RNA apurinic site-specific lyase (RALyase) mediates coleoptile senescence by cleaving the sarcin/ricin domain of conspecific ribosomes. RALyase is an enzyme that cleaves the phosphodiester bond at the depurination site of tritin. It was suggested that this enzyme completes the translational arrest caused by tritin, by reducing the residual translational elongation of the depurinated ribosomes [139]. Senescence is an important developmental process which allows the plant to redistribute nutrients from sinks to developing plant parts [140]. Programmed cell death is an active process during senescence [141]. Sawasaki et al. [106] noted that senescence of wheat coleoptiles was accompanied by morphological changes such as discoloration and withering. These changes were most obvious starting from 12 days after sowing of the wheat seeds. Interestingly, RT-PCR analyses showed that transcript levels for tritin emerged in wheat coleoptiles starting from day 10 and the expression dramatically increased from day 11. RALyase on the other hand apparently is constitutively expressed during senescence. Furthermore starting from day 10, the wheat ribosomes also appeared to be damaged by the combined effect of tritin and RALyase. The correlation between the expression pattern of tritin and RALyase, the onset of morphological changes and the ribosomal damage associated with tritin and RALyase activity, strongly suggest an important regulatory role for tritin in wheat coleoptile senescence. This hypothesis was confirmed by creating transgenic tobacco plants expressing tritin or RALyase under the control of the glucocorticoid-inducible promoter. When RIP expression was induced in seedlings, the plants started to show a senescent phenotype. However, this was not the case in seedlings where RALyase expression was induced, indicating that tritin is the driving force behind the regulation of senescence [106].

5.5. JIP60

The JIP60 protein has already been discussed in terms of its domain architecture and proteolytic processing. Similar to classical RIPs the processed *N*-terminal RIP domain inhibits translation by blocking the elongation step [37,70]. The N-glycosidase activity of the RIP domain was proven in vitro using rabbit reticulocyte ribosomes, as well as in planta [107]. It was shown that the unprocessed JIP60 can also abolish translation of both plant and rabbit ribosomes though this process occurs in a completely different way as with the *N*-terminal RIP part. The full length JIP60 acts as a ribosome dissociation factor and blocks the initiation of translation rather than the elongation step, by splitting the polysomes in their subunits. Remarkably, JIP60 only had this effect on ribosomes from leaves that had been exposed to MeJA or from osmotically stressed or desiccated leaves [29,70]. While these features were already discussed above, we did not yet clarify the role of the *C*-terminal eIF4E domain and how the processing contribute to the in planta function of JIP60.

Rustgi et al. [70] performed in vitro translation assays where transcripts of different genes were used as a template. These transcripts represented both housekeeping genes and genes involved in photosynthesis (RBSC, actin, LHCB2) as well as other jasmonate inducible genes (JIP23 and thionin). Derivatives of JIP60 were added to the reactions to study their effect on the translation of the different transcripts. It has already been mentioned that both the complete JIP60 protein as well as the processed RIP domain completely abolished translation, albeit in a different way. Interestingly the effect of the eIF4E domain derivative had no effect on the translation of the transcripts of housekeeping genes or genes involved in photosynthesis but an increased translation of the transcripts from jasmonate inducible genes was detected. The authors also observed that these JIP transcripts were increased in the polysome fraction. These findings suggest that the eIF4E domain resulting from the processing of the JIP60 selectively promotes the translation of other JIPs [70].

JIP60 is involved in the barley stress response. This is evidenced by the fact that its expression is dependent on MeJA, a phytohormone involved in plant defense. Rustgi et al. [70] also reported that the JIP60 gene is in close proximity to several QTLs for both biotic and abiotic stress resistance, confirming a probable role in stress response. JIP60 probably exerts its function by reprogramming the translational machinery in order to cope with unfavorable situations. Figure 3 represents a possible working model for JIP60.

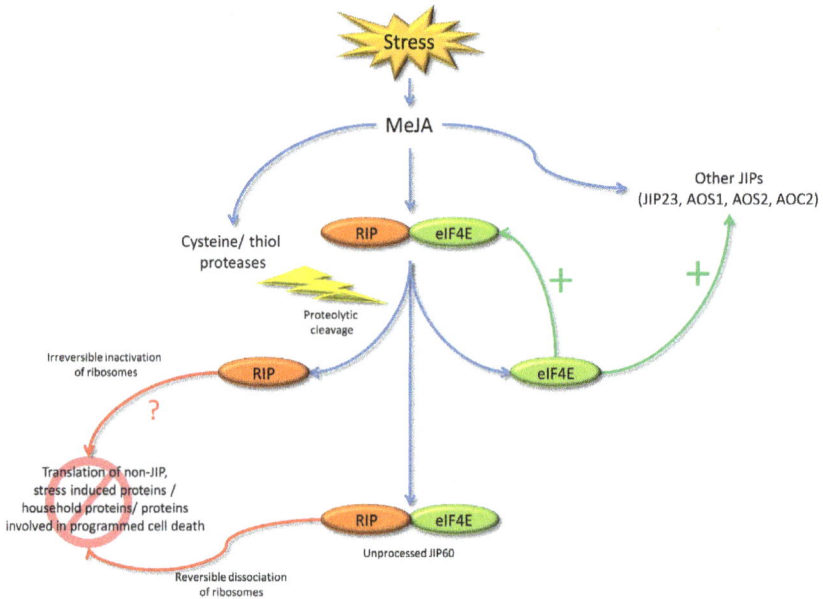

Figure 3. Working mechanism for JIP60 when the plant is subjected to stress (adapted from [70]). MeJA treatment affects the expression of several jasmonate-inducible proteins (JIPs) and proteases. Among the JIPs are JIP60, JIP23, and the enzymes involved in jasmonate biosynthesis:allene oxide synthases 1 and 2 (AOS1, AOS2) and allene oxide cyclase (AOC). JIP60 promotes the translation of JIPs through the action of the separate C-terminal domain, while unprocessed JIP60 as well as the separate N-terminal domain inhibit the translation of other proteins.

6. Concluding Remarks

In this review we discussed the characteristics of cereal RIPs which distinguish them from their counterparts in other plant species. These unique features of cereal RIPs probably result from the extensive evolution of cereal RIP genes. Several cereal RIPs are known for their proteolytic activation mechanisms. Famous examples are the maize RIP1 and the JIP60 from barley but evidence exists that homologs also occur in other species such as sorghum. Although the bRIP1 from barley does not contain an inactivation region that needs to be removed, a "switch region" was reported in bRIP1 at a location corresponding to the location of the inactivation loop of maize RIP1. This finding argues that structural regulation of RNA N-glycosidase activity might be a more general phenomenon for cereal RIPs than previously thought. As seen with the maize RIP1, this structural regulation might be achieved by affecting the interaction with ribosomal proteins rather than changing the active site conformation.

Another important difference between cereal RIPs and classical plant RIPs is the lack of signal peptides and consequently their presumed cytoplasmic localization. This has some important implications for the possible in planta function of cereal RIPs. There are some contradictions in the literature with regard to the effect of cereal RIPs on conspecific ribosomes. While certain RIPs were reported not to act on the host ribosomes, others like tritin clearly exert their function by inhibiting cereal ribosomes. Even more, JIP60 only dissociates barley ribosomes in certain stress situations suggesting that barley ribosomes are somehow 'marked' for cleavage in specific situations. This underlines the potential complexity of the interaction between cereal RIPs and (conspecific) ribosomes.

Studies on the physiological roles of cereal RIPs often focus on one specific action of an RIP. For example, the role of maize RIP2 in the defense against caterpillars was investigated. However, although the maize RIP2 gene was reported to be upregulated by drought stress, its possible role in drought tolerance has not yet been clarified. Possibly cereal RIPs, and especially the leaf forms, can have a dual function. On the one hand they can function as classical defense proteins protecting the plant against herbivorous insects or pathogenic fungi, while on the other hand they can be involved in regulating endogenous processes such as the abiotic stress response or developmental processes like senescence. Future research on cereal RIPs should focus on both aspects and can gain more insight into the importance of this unique class of proteins found in cereals. A better understanding of the physiological importance of RIPs can lead to future exploitation of their properties to create more resilient crops. Next to being useful for agricultural purposes, the unique properties of cereal RIPs could also be exploited in other research fields. An example is the use of maize RIP1 in the combat against human immunodeficiency virus (HIV). The unique activation mechanism of maize RIP1 has been exploited to enhance the specificity of the RIP against HIV infected cells by the introduction of HIV-1 protease recognition sequences in the internal inactivation region [142].

Acknowledgments: This research was funded by the Research Council of Ghent University (project BOF15/GOA/005) and the Fund for Scientific Research–Flanders (FWO Grant G007916N).

Conflicts of Interest: The authors declare no conflict of interest.

References

1. Peumans, W.J.; Van Damme, E.J.M. Evolution of plant ribosome-inactivating proteins. In *Toxic Plant Proteins*; Lord, J.M., Hartley, M.R., Eds.; Springer: Berlin/Heidelberg, Germany, 2010; Volume 18, pp. 1–26.
2. Peumans, W.; Shang, C.; Van Damme, E.J.M. Updated model of the molecular evolution of RIP genes. In *Ribosome-Inactivating Proteins: Ricin and Related Proteins*; Stirpe, F., Lappi, D., Eds.; John Wiley & Sons, Inc.: Somerset, NJ, USA, 2014; pp. 134–150.
3. Bolognesi, A.; Bortolotti, M.; Maiello, S.; Battelli, M.; Polito, L. Ribosome-inactivating proteins from plants: A historical overview. *Molecules* **2016**, *21*, 1627:1–1627:19.
4. Barbieri, L.; Stirpe, F. Ribosome-inactivating proteins from plants: Properties and possible uses. *Cancer Surv.* **1982**, *1*, 489–520. [CrossRef]
5. Endo, Y.; Tsurugi, K. RNA N-glycosidase activity of ricin A-chain. Mechanism of action of the toxic lectin ricin on eukaryotic ribosomes. *J. Biol. Chem.* **1987**, *262*, 8128–8130. [PubMed]
6. Ehrlich, P. Experimentelle Untersuchungen über Immunität. I. Ueber Ricin. *Med. Wochenschr.* **1891**, *17*, 976–979. [CrossRef]
7. Hellin, H. *Der giftige Eiweisskorper Abrin und seine Wirkung auf das Blut*; University of Dorpat: Tartu, Estonia, 1891.
8. Olsnes, S.; Pihl, A. Ricin—A potent inhibitor of protein synthesis. *FEBS Lett.* **1972**, *20*, 327–329. [CrossRef]
9. Olsnes, S.; Pihl, A. Isolation and properties of abrin: A toxic protein inhibiting protein synthesis: Evidence for different biological functions of its two constituent-peptide chains. *Eur. J. Biochem.* **1973**, *35*, 179–185. [CrossRef] [PubMed]
10. Barbieri, L.; Ferreras, J.M.; Barraco, A.; Ricci, P.; Stirpe, F. Some ribosome-inactivating proteins depurinate ribosomal RNA at multiple sites. *Biochem. J.* **1992**, *286*, 1–4. [CrossRef] [PubMed]
11. Barbieri, L.; Valbonesi, P.; Bonora, E.; Gorini, P.; Bolognesi, A.; Stirpe, F. Polynucleotide:Adenosine glycosidase activity of ribosome-inactivating proteins: Effect on DNA, RNA and poly(A). *Nucleic Acids Res.* **1997**, *25*, 518–522. [CrossRef] [PubMed]
12. Barbieri, L.; Gorini, P.; Valbonesi, P.; Castiglioni, P.; Stirpe, F. Unexpected activity of saporins. *Nature* **1994**. [CrossRef] [PubMed]
13. Barbieri, L.; Valbonesi, P.; Righi, F.; Zuccheri, G.; Monti, F.; Gorini, P.; Samori, B.; Stirpe, F. Polynucleotide:adenosine glycosidase is the sole activity of ribosome-inactivating proteins on DNA. *J. Biochem.* **2000**, *128*, 883–889. [CrossRef] [PubMed]

14. Ruggiero, A.; Chambery, A.; Di Maro, A.; Mastroianni, A.; Parente, A.; Berisio, R. Crystallization and preliminary X-ray diffraction analysis of PD-L1, a highly glycosylated ribosome inactivating protein with DNase activity. *Protein Pept. Lett.* **2007**, *14*, 407–709. [CrossRef] [PubMed]

15. Mock, J.W.Y.; Ng, T.B.; Wong, R.N.S.; Yao, Q.Z.; Yeung, H.W.; Fong, W.P. Demonstration of ribonuclease activity in the plant ribosome-inactivating proteins alpha- and beta-momorcharins. *Life Sci.* **1996**, *59*, 1853–1859. [CrossRef]

16. Shih, N.R.; McDonald, K.A.; Jackman, A.P.; Girbés, T.; Iglesias, R. Bifunctional plant defence enzymes with chitinase and ribosome inactivating activities from *Trichosanthes kirilowii* cell cultures. *Plant Sci.* **1997**, *130*, 145–150.

17. Chen, H.; Wang, Y.; Yan, M.G.; Yu, M.K.; Yao, Q.Z. The phosphatase activity of five ribosome-inactivating proteins. *Biochem. J.* **1996**, *12*, 125–129.

18. Lombard, S.; Helmy, M.E.; Piéroni, G. Lipolytic activity of ricin from *Ricinus sanguineus* and *Ricinus communis* on neutral lipids. *Biochem. J.* **2001**, *358*, 773–81. [CrossRef] [PubMed]

19. Li, X.D.; Chen, W.F.; Liu, W.Y.; Wang, G.H. Large-scale preparation of two new ribosome-inactivating proteins: Cinnamomin and camphorin from the seeds of *Cinnamomum camphora*. *Protein Expr. Purif.* **1997**, *10*, 27–31. [PubMed]

20. Peumans, W.J.; Hao, Q.; Van Damme, E.J.M. Ribosome-inactivating proteins from plants: More than RNA N-glycosidases? *FASEB J.* **2001**, *15*, 1493–1506. [CrossRef] [PubMed]

21. Iglesias, R.; Pérez, Y.; De Torre, C.; Ferreras, J.M.; Antolín, P.; Jiménez, P.; Rojo, M.Á.; Méndez, E.; Girbés, T. Molecular characterization and systemic induction of single-chain ribosome-inactivating proteins (RIPs) in sugar beet (*Beta vulgaris*) leaves. *J. Exp. Bot.* **2005**, *56*, 1675–1684. [CrossRef] [PubMed]

22. Iglesias, R.; Citores, L.; Di Maro, A.; Ferreras, J.M. Biological activities of the antiviral protein BE27 from sugar beet (*Beta vulgaris* L.). *Planta* **2015**, *241*, 421–433. [CrossRef] [PubMed]

23. Iglesias, R.; Citores, L.; Ragucci, S.; Russo, R.; Di Maro, A.; Ferreras, J.M. Biological and antipathogenic activities of ribosome-inactivating proteins from *Phytolacca dioica* L. *Biochim. Biophys. Acta Gen. Subj.* **2016**, *1860*, 1256–1264. [CrossRef] [PubMed]

24. Barbieri, L.; Polito, L.; Bolognesi, A.; Ciani, M.; Pelosi, E.; Farini, V.; Jha, A.K.; Sharma, N.; Vivanco, J.M.; Chambery, A.; Parente, A.; Stirpe, F. Ribosome-inactivating proteins in edible plants and purification and characterization of a new ribosome-inactivating protein from *Cucurbita moschata*. *Biochim. Biophys. Acta Gen. Subj.* **2006**, *1760*, 783–792. [CrossRef] [PubMed]

25. Ruggiero, A.; Di Maro, A.; Severino, V.; Chambery, A.; Berisio, R. Crystal structure of PD-L1, a ribosome inactivating protein from *Phytolacca dioica* L. leaves with the property to induce DNA cleavage. *Biopolym. Pept. Sci. Sect.* **2009**, *91*, 1135–1142. [CrossRef] [PubMed]

26. Aceto, S.; Di Maro, A.; Conforto, B.; Siniscalco, G.G.; Parente, A.; Delli Bovi, P.; Gaudio, L. Nicking activity on pBR322 DNA of ribosome inactivating proteins from *Phytolacca dioica* L. leaves. *Biol. Chem.* **2005**, *386*, 307–317. [CrossRef] [PubMed]

27. Schrot, J.; Weng, A.; Melzig, M.F. Ribosome-inactivating and related proteins. *Toxins (Basel)* **2015**, *7*, 1556–1615. [CrossRef] [PubMed]

28. Walsh, T.A.; Morgan, A.E.; Hey, T.D. Characterization and molecular cloning of a proenzyme form of a ribosome-inactivating protein from maize: Novel mechanism of proenzyme activation by proteolytic removal of a 2.8-kilodalton internal peptide segment. *J. Biol. Chem.* **1991**, *266*, 23422–23427. [PubMed]

29. Reinbothe, S.; Reinbothe, C.; Lehmann, J.; Becker, W.; Apel, K.; Parthier, B. JIP60, a methyl jasmonate-induced ribosome-inactivating protein involved in plant stress reactions. *Proc. Natl. Acad. Sci. USA* **1994**, *91*, 7012–7016. [CrossRef] [PubMed]

30. Stirpe, F.; Battelli, M.G. Ribosome-inactivating proteins: Progress and problems. *Cell. Mol. Life Sci.* **2006**, *63*, 1850–1866. [CrossRef] [PubMed]

31. Battelli, M.G. Cytotoxicity and toxicity to animals and humans of ribosome-inactivating proteins. *Mini Rev. Med. Chem.* **2004**, *4*, 513–521. [CrossRef] [PubMed]

32. Jiménez, P.; Gayoso, M.; Girbes, T. Non-toxic type 2 Ribosome-inactivating proteins. In *Ribosome-Inactivating Proteins: Ricin and Related Proteins*; Stirpe, F., Lappi, D., Eds.; John Wiley & Sons, Inc.: Somerset, NJ, USA, 2014; pp. 67–82.

33. Polito, L.; Bortolotti, M.; Maiello, S.; Battelli, M.G.; Bolognesi, A. Plants producing ribosome-inactivating proteins in traditional medicine. *Molecules* **2016**, *21*, 1560:1–1560:27. [CrossRef] [PubMed]

34. Lin, J.Y.; Tserng, K.Y.; Chen, C.C.; Lin, L.T.; Tung, T.C. Abrin and ricin: new anti-tumour substances. *Nature* **1970**, *227*, 292–293. [CrossRef] [PubMed]

35. De Virgilio, M.; Lombardi, A.; Caliandro, R.; Fabbrini, M.S. Ribosome-inactivating proteins: From plant defense to tumor attack. *Toxins (Basel)* **2010**, *2*, 2699–2737. [CrossRef] [PubMed]

36. Pizzo, E.; Di Maro, A. A new age for biomedical applications of ribosome inactivating proteins (RIPs): From bioconjugate to nanoconstructs. *J. Biomed. Sci.* **2016**. [CrossRef] [PubMed]

37. Chaudhry, B.; Müller-Uri, F.; Cameron-Mills, V.; Gough, S.; Simpson, D.; Skriver, K.; Mundy, J. The barley 60 kDa jasmonate-induced protein (JIP60) is a novel ribosome-inactivating protein. *Plant J.* **1994**, *6*, 815–824. [PubMed]

38. Jiang, S.Y.; Ramamoorthy, R.; Bhalla, R.; Luan, H.F.; Venkatesh, P.N.; Cai, M.; Ramachandran, S. Genome-wide survey of the RIP domain family in *Oryza sativa* and their expression profiles under various abiotic and biotic stresses. *Plant Mol. Biol.* **2008**, *67*, 603–614. [CrossRef] [PubMed]

39. Jiang, S.Y.; Bhalla, R.; Ramamoorthy, R.; Luan, H.F.; Venkatesh, P.N.; Cai, M.; Ramachandran, S. Over-expression of OSRIP18 increases drought and salt tolerance in transgenic rice plants. *Transgenic Res.* **2012**, *21*, 785–795. [CrossRef] [PubMed]

40. Vargas, L.R.B.; Carlini, C.R. Insecticidal and antifungal activities of ribosome-inactivating proteins. In *Ribosome-Inactivating Proteins: Ricin and Related Proteins*; Stirpe, F., Lappi, D., Eds.; John Wiley & Sons, Inc.: Somerset, NJ, USA, 2014; pp. 212–222.

41. Di, R.; Tumer, N.E. Pokeweed antiviral protein: Its cytotoxicity mechanism and applications in plant disease resistance. *Toxins (Basel)* **2015**, *7*, 755–772. [CrossRef] [PubMed]

42. Krivdova, G.; Neller, K.; Parikh, B.A.; Hudak, K.A. Antiviral and antifungal properties of RIPs. In *Ribosome-Inactivating Proteins: Ricin and Related Proteins*; Stirpe, F., Lappi, D., Eds.; John Wiley & Sons, Inc.: Somerset, NJ, USA, 2014; pp. 198–211.

43. Motto, M.; Lupotto, E. The genetics and properties of cereal ribosome-inactivating proteins. *Mini Rev. Med. Chem.* **2004**, *4*, 493–503. [PubMed]

44. Balconi, C.; Lanzanova, C.; Motto, M. Ribosome-inactivating proteins in cereals. In *Toxic Plant Proteins*; Lord, J., Hartley, M.R., Eds.; Springer: Berlin/Heidelberg, Germany, 2010; pp. 149–166.

45. Lapadula, W.J.; Sánchez Puerta, M.V.; Juri Ayub, M. Revising the taxonomic distribution, origin and evolution of ribosome inactivating protein genes. *PLoS ONE* **2013**, *8*, e72825:1–e72825:8. [CrossRef] [PubMed]

46. Shang, C.; Peumans, W.J.; Van Damme, E.J.M. Occurrence and taxonomical distribution of ribosome-inactivating proteins belonging to the Ricin/Shiga toxin superfamily. In *Ribosome-Inactivating Proteins: Ricin and Related Proteins*; Stirpe, F., Lappi, D., Eds.; John Wiley & Sons, Inc.: Somerset, NJ, USA, 2014; pp. 11–27.

47. Robertus, J.D.; Monzingo, A.F. The structure of ribosome inactivating proteins. *Mini Rev. Med. Chem.* **2004**, *4*, 477–486. [CrossRef] [PubMed]

48. Montfort, W.; Villafranca, J.E.; Monzingo, A.F.; Ernst, S.R.; Katzin, B.; Rutenber, E.; Xuong, N.H.; Hamlin, R.; Robertus, J.D. The three-dimensional structure of ricin at 2.8 A. *J. Biol. Chem.* **1987**, *262*, 5398–5403. [PubMed]

49. Shi, W.W.; Mak, A.N.S.; Wong, K.B.; Shaw, P.C. Structures and ribosomal interaction of ribosome-inactivating proteins. *Molecules* **2016**, *21*, 1588:1–1588:13. [CrossRef] [PubMed]

50. Robertus, J.D.; Monzingo, A.F. The structure and action of ribosome-inactivating proteins. In *Ribosome-Inactivating Proteins: Ricin and Related Proteins*; Stirpe, F., Lappi, D., Eds.; John Wiley & Sons, Inc.: Somerset, NJ, USA, 2014; pp. 111–133.

51. Tahirov, T.H.; Lu, T.H.; Liaw, Y.C.; Chen, Y.L.; Lin, J.Y. Crystal structure of abrin-a at 2.14 A. *J. Mol. Biol.* **1995**, *250*, 354–367. [CrossRef] [PubMed]

52. Katzin, B.J.; Collins, E.J.; Robertus, J.D. Structure of ricin A-chain at 2.5 A. *Proteins Struct. Funct. Bioinform.* **1991**, *10*, 251–259.

53. Mishra, V.; Bilgrami, S.; Sharma, R.S.; Kaur, P.; Yadav, S.; Krauspenhaar, R.; Betzel, C.; Voelter, W.; Babu, C.R.; Singh, T.P. Crystal structure of Himalayan mistletoe ribosome-inactivating protein reveals the presence of a natural inhibitor and a new functionally active sugar-binding site. *J. Biol. Chem.* **2005**, *280*, 20712–20721. [CrossRef] [PubMed]

54. Kim, Y.; Robertus, J.D. Analysis of several key active site residues of ricin a chain by mutagenesis and x-ray crystallography. *Protein Eng. Des. Sel.* **1992**, *5*, 775–779. [CrossRef]

55. Ho, M.-C.; Sturm, M.B.; Almo, S.C.; Schramm, V.L. Transition state analogues in structures of ricin and saporin ribosome-inactivating proteins. *Proc. Natl. Acad. Sci. USA* **2009**, *106*, 20276–20281. [CrossRef] [PubMed]

56. Kim, Y.; Mlsna, D.; Monzingo, A.F.; Ready, M.P.; Frankel, A.; Robertus, J.D. Structure of a ricin mutant showing rescue of activity by a noncatalytic residue. *Biochemistry* **1992**, *31*, 3294–3296. [CrossRef] [PubMed]

57. Wong, K.B.; Ke, Y.B.; Dong, Y.C.; Li, X.B.; Guo, Y.W.; Yeung, H.W.; Shaw, P.C. Structure/function relationship study of Gln156, Glu160 and Glu189 in the active site of trichosanthin. *Eur. J. Biochem.* **1994**, *221*, 787–791. [CrossRef] [PubMed]

58. Mak, A.N.S.; Wong, Y.T.; An, Y.J.; Cha, S.S.; Sze, K.H.; Au, S.W.N.; Wong, K.B.; Shaw, P.C. Structure-function study of maize ribosome-inactivating protein: Implications for the internal inactivation region and the sole glutamate in the active site. *Nucleic Acids Res.* **2007**, *35*, 6259–6267. [CrossRef] [PubMed]

59. Lee, B.G.; Kim, M.K.; Kim, B.W.; Suh, S.W.; Song, H.K. Structures of the ribosome-inactivating protein from barley seeds reveal a unique activation mechanism. *Acta Crystallogr. Sect. D Biol. Crystallogr.* **2012**, *68*, 1488–1500. [CrossRef] [PubMed]

60. Hartings, H.; Lazzaroni, N.; Marsan, P.A.; Aragay, A.; Thompson, R.; Salamini, F.; Di Fonzo, N.; Palau, J.; Motto, M. The b-32 protein from maize endosperm: Characterization of genomic sequences encoding two alternative central domains. *Plant Mol. Biol.* **1990**, *14*, 1031–1040. [CrossRef] [PubMed]

61. Di Fonzo, N.; Manzocchi, L.; Salamini, F.; Soave, C. Purification and properties of an endospermic protein of maize associated with the Opaque-2 and Opaque-6 genes. *Planta* **1986**, *167*, 587–594. [CrossRef] [PubMed]

62. Lohmer, S.; Maddaloni, M.; Motto, M.; Hartings, H.; Salamini, F.; Thompson, R.D. The maize regulatory locus Opaque-2 encodes a DNA-binding protein which activates the transcription of the b-32 gene. *EMBO* **1991**, *10*, 617–624.

63. Bass, H.W.; Webster, C.; OBrian, G.R.; Roberts, J.K.; Boston, R.S. A maize ribosome-inactivating protein is controlled by the transcriptional activator Opaque-2. *Plant Cell Online* **1992**, *4*, 225–234. [CrossRef]

64. Hey, T.D.; Hartley, M.; Walsh, T.A. Maize ribosome-inactivating protein (b-32). Homologs in related species, effects on maize ribosomes, and modulation of activity by pro-peptide deletions. *Plant Physiol.* **1995**, *107*, 1323–1332. [CrossRef] [PubMed]

65. Krawetz, J.E.; Boston, R.S. Substrate specificity of a maize ribosome-inactivating protein differs across diverse taxa. *Eur. J. Biochem.* **2000**, *267*, 1966–1974. [CrossRef] [PubMed]

66. Docherty, K.; Steiner, D.F. Post-translational proteolysis in polypeptide hormone biosynthesis. *Annu. Rev. Physiol.* **1982**, *44*, 625–638. [CrossRef] [PubMed]

67. Bass, H.W.; Krawetz, J.E.; Obrian, G.R.; Zinselmeier, C.; Habben, J.E.; Boston, R.S. Maize ribosome-inactivating proteins (RIPs) with distinct expression patterns have similar requirements for proenzyme activation. *J. Exp. Bot.* **2004**, *55*, 2219–2233. [CrossRef] [PubMed]

68. Yang, Y.; Mak, A.N.S.; Shaw, P.C.; Sze, K.H. Solution structure of an active mutant of maize ribosome-inactivating protein (MOD) and its interaction with the ribosomal stalk protein P2. *J. Mol. Biol.* **2010**, *395*, 897–907. [CrossRef] [PubMed]

69. Chuang, W.P.; Herde, M.; Ray, S.; Castano-Duque, L.; Howe, G.A.; Luthe, D.S. Caterpillar attack triggers accumulation of the toxic maize protein RIP2. *New Phytol.* **2014**, *201*, 928–939. [CrossRef] [PubMed]

70. Rustgi, S.; Pollmann, S.; Buhr, F.; Springer, A.; Reinbothe, C.; von Wettstein, D.; Reinbothe, S. JIP60-mediated, jasmonate- and senescence-induced molecular switch in translation toward stress and defense protein synthesis. *Proc. Natl. Acad. Sci. USA* **2014**, *111*, 14181–14186. [CrossRef] [PubMed]

71. Endo, Y.; Tsurugi, K. The RNA N-glycosidase activity of ricin A-chain. The characteristics of the enzymatic activity of ricin A-chain with ribosomes and with rRNA. *J. Biol. Chem.* **1988**, *263*, 8735–8739. [PubMed]

72. Hudak, K.A.; Dinman, J.D.; Tumer, N.E. Pokeweed antiviral protein accesses ribosomes by binding to L3. *J. Biol. Chem.* **1999**, *274*, 3859–3864. [CrossRef] [PubMed]

73. Chan, D.S.B.; Chu, L.O.; Lee, K.M.; Too, P.H.M.; Ma, K.W.; Sze, K.H.; Zhu, G.; Shaw, P.C.; Wong, K.B. Interaction between trichosanthin, a ribosome-inactivating protein, and the ribosomal stalk protein P2 by chemical shift perturbation and mutagenesis analyses. *Nucleic Acids Res.* **2007**, *35*, 1660–1672. [CrossRef] [PubMed]

74. McCluskey, A.J.; Poon, G.M.K.; Bolewska-Pedyczak, E.; Srikumar, T.; Jeram, S.M.; Raught, B.; Gariépy, J. The catalytic subunit of Shiga-like toxin 1 interacts with ribosomal stalk proteins and is inhibited by their conserved C-terminal domain. *J. Mol. Biol.* **2008**, *378*, 375–386. [PubMed]

75. Chiou, J.C.; Li, X.P.; Remacha, M.; Ballesta, J.P.G.; Tumer, N.E. The ribosomal stalk is required for ribosome binding, depurination of the rRNA and cytotoxicity of ricin a chain in *Saccharomyces cerevisiae*. *Mol. Microbiol.* **2008**, *70*, 1441–1452. [CrossRef] [PubMed]

76. Chan, S.H.; Hung, F.S.J.; Chan, D.S.B.; Shaw, P.C. Trichosanthin interacts with acidic ribosomal proteins P0 and P1 and mitotic checkpoint protein MAD2B. *Eur. J. Biochem.* **2001**, *268*, 2107–2112. [CrossRef] [PubMed]

77. Schmeing, T.M.; Ramakrishnan, V. What recent ribosome structures have revealed about the mechanism of translation. *Nature* **2009**, *461*, 1234–1242. [PubMed]

78. Bargis-Surgey, P.; Lavergne, J.P.; Gonzalo, P.; Vard, C.; Filhol-Cochet, O.; Reboud, J.P. Interaction of elongation factor eEF-2 with ribosomal P proteins. *Eur. J. Biochem.* **1999**, *262*, 606–611. [CrossRef] [PubMed]

79. Helgstrand, M.; Mandava, C.S.; Mulder, F.A.A.; Liljas, A.; Sanyal, S.; Akke, M. The ribosomal stalk binds to translation factors IF2, EF-Tu, EF-G and RF3 via a conserved region of the L12 C-terminal domain. *J. Mol. Biol.* **2007**, *365*, 468–479. [CrossRef] [PubMed]

80. Nomura, N.; Honda, T.; Baba, K.; Naganuma, T.; Tanzawa, T.; Arisaka, F.; Noda, M.; Uchiyama, S.; Tanaka, I.; Yao, M.; et al. Archaeal ribosomal stalk protein interacts with translation factors in a nucleotide-independent manner via its conserved C terminus. *Proc. Natl. Acad. Sci. USA* **2012**, *109*, 3748–3753. [CrossRef] [PubMed]

81. Choi, A.K.H.; Wong, E.C.K.; Lee, K.M.; Wong, K.B. Structures of eukaryotic ribosomal stalk proteins and its complex with trichosanthin, and their implications in recruiting ribosome-inactivating proteins to the ribosomes. *Toxins (Basel)* **2015**, *7*, 638–647. [CrossRef] [PubMed]

82. Diaconu, M.; Kothe, U.; Schlünzen, F.; Fischer, N.; Harms, J.M.; Tonevitsky, A.G.; Stark, H.; Rodnina, M.V.; Wahl, M.C. Structural basis for the function of the ribosomal L7/12 stalk in factor binding and GTPase activation. *Cell* **2005**, *121*, 991–1004. [CrossRef] [PubMed]

83. Subramanian, A.R. Copies of proteins L7 and L12 and heterogeneity of the large subunit of *Escherichia coli* ribosome. *J. Mol. Biol.* **1975**, *95*, 1–8. [CrossRef]

84. Lee, K.M.; Yu, C.W.H.; Chan, D.S.B.; Chiu, T.Y.H.; Zhu, G.; Sze, K.H.; Shaw, P.C.; Wong, K.B. Solution structure of the dimerization domain of ribosomal protein P2 provides insights for the structural organization of eukaryotic stalk. *Nucleic Acids Res.* **2010**, *38*, 5206–5216. [CrossRef] [PubMed]

85. Shimizu, T.; Nakagaki, M.; Nishi, Y.; Kobayashi, Y.; Hachimori, A.; Uchiumi, T. Interaction among silkworm ribosomal proteins P1, P2 and P0 required for functional protein binding to the GTPase-associated domain of 28S rRNA. *Nucleic Acids Res.* **2002**, *30*, 2620–2627. [CrossRef] [PubMed]

86. McCluskey, A.J.; Bolewska-Pedyczak, E.; Jarvik, N.; Chen, G.; Sidhu, S.S.; Gariépy, J. Charged and hydrophobic surfaces on the a chain of Shiga-Like toxin 1 recognize the C-terminal domain of ribosomal stalk proteins. *PLoS ONE* **2012**, *7*, e31191:1–e31191:11. [CrossRef] [PubMed]

87. Too, P.H.M.; Ma, M.K.W.; Mak, A.N.S.; Wong, Y.T.; Tung, C.K.C.; Zhu, G.; Au, S.W.N.; Wong, K.B.; Shaw, P.C. The C-terminal fragment of the ribosomal P protein complexed to trichosanthin reveals the interaction between the ribosome-inactivating protein and the ribosome. *Nucleic Acids Res.* **2009**, *37*, 602–610. [PubMed]

88. Shi, W.W.; Tang, Y.S.; Sze, S.Y.; Zhu, Z.N.; Wong, K.B.; Shaw, P.C. Crystal structure of ribosome-inactivating protein ricin a chain in complex with the C-terminal peptide of the ribosomal stalk protein P2. *Toxins (Basel)* **2016**, *8*, 296:1–296:12. [CrossRef] [PubMed]

89. Fan, X.; Zhu, Y.; Wang, C.; Niu, L.; Teng, M.; Li, X. Structural insights into the interaction of the ribosomal P stalk protein P2 with a type II ribosome-inactivating protein ricin. *Sci. Rep.* **2016**, *6*, 37803:1–37803:10.

90. Wong, Y.T.; Ng, Y.M.; Mak, A.N.S.; Sze, K.H.; Wong, K.B.; Shaw, P.C. Maize ribosome-inactivating protein uses Lys158-Lys161 to interact with ribosomal protein P2 and the strength of interaction is correlated to the biological activities. *PLoS ONE* **2012**, *7*, 1–10. [CrossRef] [PubMed]

91. Lapadula, W.J.; Sanchez-Puerta, M.V.; Juri Ayub, M. Convergent evolution led ribosome inactivating proteins to interact with ribosomal stalk. *Toxicon* **2012**, *59*, 427–432. [CrossRef] [PubMed]

92. Sturm, M.B.; Schramm, V.L. Detecting ricin: Sensitive luminescent assay for ricin A-chain ribosome depurination kinetics. *Anal. Chem.* **2009**, *81*, 2847–2853. [CrossRef] [PubMed]

93. Harley, S.M.; Beevers, H. Ricin inhibition of in vitro protein synthesis by plant ribosomes. *Proc. Natl. Acad. Sci. USA* **1982**, *79*, 5935–5938. [CrossRef] [PubMed]

94. Hartley, M.R.; Chacidock, J.A.; Bonness, M.S. The structure and function of ribosome-inactivating proteins. *Trends Plant Sci.* **1996**, *1*, 254–260. [CrossRef]

95. Prestle, J.; Schönfelder, M.; Adam, G.; Mundry, K.W. Type 1 ribosome-inactivating proteins depurinate plant 25S rRNA without species specificity. *Nucleic Acids Res.* **1992**, *20*, 3179–3182. [CrossRef] [PubMed]

96. Hartley, M.R.; Lord, J.M. Structure, function and applications of ricin and related cytotoxic proteins. In *Biosynthesis and Manipulation of Plant Products*; Grierson, D., Ed.; Springer: Houten, The Netherlands, 1993; pp. 210–239.

97. Neuhaus, J.M.; Rogers, J.C. Sorting of proteins to vacuoles in plant cells. In *Protein Trafficking in Plant Cells*; Soll, J., Ed.; Springer: Houten, The Netherlands, 1998; pp. 127–144.

98. Vitale, A.; Raikhel, N.V. What do proteins need to reach different vacuoles? *Trends Plant Sci.* **1999**, *4*, 149–155. [PubMed]

99. Wu, T.H.; Chow, L.P.; Lin, J.Y. Sechiumin, a ribosome-inactivating protein from the edible gourd, *Sechium edule* Swartz: Purification, characterization, molecular cloning and expression. *Eur. J. Biochem.* **1998**, *255*, 400–408. [CrossRef] [PubMed]

100. Ready, M.P.; Brown, D.T.; Robertus, J.D. Extracellular localization of pokeweed antiviral protein. *Proc. Natl. Acad. Sci. USA* **1986**, *83*, 5053–5056. [CrossRef] [PubMed]

101. Leah, R.; Tommerup, H.; Svendsen, I.; Mundy, J. Biochemical and molecular characterization of three barley seed proteins with antifungal properties. *J. Biol. Chem.* **1991**, *266*, 1564–1573. [PubMed]

102. Habuka, N.; Kataoka, J.; Miyano, M.; Tsuge, H.; Ago, H.; Noma, M. Nucleotide sequence of a genomic gene encoding tritin, a ribosome-inactivating protein from *Triticum aestivum*. *Plant Mol. Biol.* **1993**, *22*, 171–176. [CrossRef] [PubMed]

103. Taylor, B.E.; Irvin, J.D. Depurination of plant ribosomes by pokeweed antiviral protein. *FEBS Lett.* **1990**, *273*, 144–146. [CrossRef]

104. Massiah, A.J.; Hartley, M.R. Wheat ribosome-inactivating proteins: Seed and leaf forms with different specificities and cofactor requirements. *Planta* **1995**, *197*, 633–640. [CrossRef] [PubMed]

105. Kataoka, J.; Habuka, N.; Miyano, M.; Masuta, C.; Koiwai, A. Adenine depurination and inactivation of plant ribosomes by an antiviral protein of *Mirabilis jalapa* (MAP). *Plant Mol. Biol.* **1992**, *20*, 1111–1119. [CrossRef] [PubMed]

106. Sawasaki, T.; Nishihara, M.; Endo, Y. RIP and RALyase cleave the sarcin/ricin domain, a critical domain for ribosome function, during senescence of wheat coleoptiles. *Biochem. Biophys. Res. Commun.* **2008**, *370*, 561–565. [CrossRef] [PubMed]

107. Dunaeva, M.; Goebel, C.; Wasternack, C.; Parthier, B.; Goerschen, E. The jasmonate-induced 60 kDa protein of barley exhibits N-glycosidase activity in vivo. *FEBS Lett.* **1999**, *452*, 263–266. [CrossRef]

108. Soave, C.; Reggiani, R.; Di Fonzo, N.; Salamini, F. Clustering of genes for 20 kd zein subunits in the short arm of maize chromosome 7. *Genetics* **1981**, 363–377.

109. Loesch, P.J.; Foley, D.C.; Cox, D.F. Comparative resistance of opaque-2 and normal inbred lines of maize to ear-rotting pathogens. *Crop Sci.* **1976**, *16*, 841–842.

110. Gupta, S.C.; Asnani, V.L.; Khare, B.P. Effect of the opaque-2 gene in maize (*Zea mays* L.) on the extent of infestation by *Sitophilus oryzae* L. *J. Stored Prod. Res.* **1970**, *6*, 191–194. [CrossRef]

111. Maddaloni, M.; Forlani, F.; Balmas, V.; Donini, G.; Stasse, L.; Corazza, L.; Motto, M. Tolerance to the fungal pathogen *Rhizoctonia solani* AG4 of transgenic tobacco expressing the maize ribosome-inactivating protein b-32. *Transgenic Res.* **1997**, *6*, 393–402. [CrossRef]

112. Nielsen, K.; Payne, G.A.; Boston, R.S. Maize ribosome-inactivating protein inhibits normal development of *Aspergillus nidulans* and *Aspergillus flavus*. *Mol. Plant Microbe Interact.* **2001**, *14*, 164–172. [CrossRef] [PubMed]

113. Kim, J.K.; Duan, X.; Wu, R.; Seok, S.J.; Boston, R.S.; Jang, I.C.; Eun, M.Y.; Nahm, B.H. Molecular and genetic analysis of transgenic rice plants expressing the maize ribosome-inactivating protein b-32 gene and the herbicide resistance bar gene. *Mol. Breed.* **1999**, *5*, 85–94. [CrossRef]

114. Balconi, C.; Lanzanova, C.; Conti, E.; Triulzi, T.; Forlani, F.; Cattaneo, M.; Lupotto, E. *Fusarium* head blight evaluation in wheat transgenic plants expressing the maize b-32 antifungal gene. *Eur. J. Plant Pathol.* **2007**, *117*, 129–140. [CrossRef]

115. Lanzanova, C.; Giuffrida, M.G.; Motto, M.; Baro, C.; Donn, G.; Hartings, H.; Lupotto, E.; Careri, M.; Elviri, L.; Balconi, C. The *Zea mays* b-32 ribosome-inactivating protein efficiently inhibits growth of *Fusarium verticillioides* on leaf pieces in vitro. *Eur. J. Plant Pathol.* **2009**, *124*, 471–482. [CrossRef]

116. Lanzanova, C.; Torri, A.; Motto, M.; Balconi, C. Characterization of the maize b-32 ribosome inactivating protein and its interaction with fungal pathogen development. *Maydica* **2011**, *56*, 1709:1–1709:11.

117. Dowd, P.F.; Mehta, A.D.; Boston, R.S. Relative toxicity of the maize endosperm ribosome-inactivating protein to insects. *J. Agric. Food Chem.* **1998**, *46*, 3775–3779. [CrossRef]

118. Dowd, P.F.; Zuo, W.N.; Gillikin, J.W.; Johnson, E.T.; Boston, R.S. Enhanced resistance to *Helicoverpa zea* in tobacco expressing an activated form of maize ribosome-inactivating protein. *J. Agric. Food Chem.* **2003**, *51*, 3568–3574. [CrossRef] [PubMed]

119. Dowd, P.F.; Holmes, R.A.; Pinkerton, T.S.; Johnson, E.T.; Lagrimini, L.M.; Boston, R.S. Relative activity of a tobacco hybrid expressing high levels of a tobacco anionic peroxidase and maize ribosome-inactivating protein against *Helicoverpa zea* and *Lasioderma serricorne*. *J. Agric. Food Chem.* **2006**, *54*, 2629–2634. [CrossRef] [PubMed]

120. Dowd, P.F.; Johnson, E.T.; Price, N.P. Enhanced pest resistance of maize leaves expressing monocot crop plant-derived ribosome-inactivating protein and agglutinin. *J. Agric. Food Chem.* **2012**, *60*, 10768–10775. [CrossRef] [PubMed]

121. Bass, H.W.; Obrian, C.R.; Boston, R.S. Cloning and sequencing of a second ribosome-inactivating protein gene from maize. *Plant Physiol.* **1995**, *107*, 661–662. [CrossRef] [PubMed]

122. Brooks, T.D.; Willcox, M.C.; Williams, W.P.; Buckley, P.M. Quantitative trait loci conferring resistance to fall armyworm and southwestern corn borer leaf feeding damage. *Crop Sci.* **2005**, *45*, 2430–2434. [CrossRef]

123. Vigers, A.J.; Roberts, W.K.; Selitrennikoff, C.P. A new family of plant antifungal proteins. *Mol. Plant Microbe Interact.* **1991**, *4*, 315–323. [CrossRef] [PubMed]

124. Kumari, S.R.; Chandrashekar, A.; Shetty, H.S. Proteins in developing sorghum endosperm that may be involved in resistance to grain moulds. *J. Sci. Food Agric.* **1992**, *60*, 275–282. [CrossRef]

125. Darnetty; Leslie, J.F.; Muthukrishnan, S.; Swegle, M.; Vigers, A.J.; Selitrennikoff, C.P. Variability in antifungal proteins in the grains of maize, sorghum and wheat. *Physiol. Plant.* **1993**, *88*, 339–349. [CrossRef]

126. Seetharaman, K.; Whitehead, E.; Keller, N.P.; Waniska, R.D.; Rooney, L.W. In vitro activity of sorgum seed antifungal proteins against grain mold pathogens. *J. Agric. Food Chem.* **1997**, *45*, 3666–3671. [CrossRef]

127. Seetharaman, K.; Waniska, R.D.; Rooney, L.W. Physiological changes in sorghum antifungal proteins. *J. Agric. Food Chem.* **1996**, *44*, 2435–2441. [CrossRef]

128. Rodríguez-Herrera, R.; Waniska, R.D.; Rooney, W.L. Antifungal proteins and grain mold resistance in sorghum with nonpigmented testa. *J. Agric. Food Chem.* **1999**, *47*, 4802–4806. [CrossRef] [PubMed]

129. Asano, K.; Svensson, B.; Poulsen, F.M. Isolation and characterization of inhibitors of animal cell-free protein synthesis from barley seeds. *Carlsb. Res. Commun* **1984**, *49*, 619–626. [CrossRef]

130. Oldach, K.H.; Becker, D.; Lörz, H. Heterologous expression of genes mediating enhanced fungal resistance in transgenic wheat. *Mol. Plant Microbe Interact.* **2001**, *14*, 832–838. [CrossRef] [PubMed]

131. Bieri, S.; Potrykus, I.; Fütterer, S. Expression of active barley seed ribosome-inactivating protein in transgenic wheat. *Theor. Appl. Genet.* **2000**, *100*, 755–763. [CrossRef]

132. Bieri, S.; Potrykus, I.; Fütterer, J. Effects of combined expression of antifungal barley seed proteins in transgenic wheat on powdery mildew infection. *Mol. Breed.* **2003**, *11*, 37–48. [CrossRef]

133. Chhikara, S.; Chaudhury, D.; Dhankher, O.P.; Jaiwal, P.K. Combined expression of a barley class II chitinase and type I ribosome inactivating protein in transgenic *Brassica juncea* provides protection against *Alternaria brassicae*. *Plant Cell Tissue Organ Cult.* **2012**, *108*, 83–89. [CrossRef]

134. M'hamdi, M.; Chikh-Rouhou, H.; Boughalleb, N.; de Galarreta, J.R. resistance to *Rhizoctonia solani* by combined expression of chitinase and ribosome inactivating protein in transgenic potatoes (*Solanum tuberosum* L). *Span. J. Agric. Res.* **2012**, *10*, 778–785. [CrossRef]

135. Chopra, R.; Saini, R. Transformation of blackgram (*Vigna mungo* (L.) hepper) by barley chitinase and ribosome-inactivating protein genes towards improving resistance to *Corynespora* leaf spot fungal disease. *Appl. Biochem. Biotechnol.* **2014**, *174*, 2791–2800. [CrossRef] [PubMed]

136. Motto, M.; Thompson, R.; Salamini, F. Genetic regulation of carbohydrate and protein accumulation in seeds. In *Cellular and Molecular Biology of Plant Seed Development*; Larkins, B.A., Vasil, I.K., Eds.; Springer: Houten, The Netherlands, 1997; pp. 479–522.

137. Habben, J.E.; Kirleis, A.W.; Larkins, B.A. The origin of lysine-containing proteins in opaque-2 maize endosperm. *Plant Mol. Biol.* **1993**, *23*, 825–838. [CrossRef] [PubMed]

138. Ding, Z.J.; Wu, X.H.; Wang, T. The rice tapetum-specific gene RA39 encodes a type I ribosome-inactivating protein. *Sex. Plant Reprod.* **2002**, *15*, 205–212.

139. Ozawa, A.; Sawasaki, T.; Takai, K.; Uchiumi, T.; Hori, H.; Endo, Y. RALyase; A terminator of elongation function of depurinated ribosomes. *FEBS Lett.* **2003**, *555*, 455–458. [CrossRef]

140. Nam, H. The molecular genetic analysis of leaf senescence. *Curr. Opin. Biotechnol.* **1997**, *8*, 200–207. [CrossRef]
141. Pennell, R.; Lamb, C. Programmed cell death in plants. *Plant Cell* **1997**, *9*, 1157–1168. [CrossRef] [PubMed]
142. Law, S.K.Y.; Wang, R.R.; Mak, A.N.S.; Wong, K.B.; Zheng, Y.T.; Shaw, P.C. A switch-on mechanism to activate maize ribosome-inactivating protein for targeting HIV-infected cells. *Nucleic Acids Res.* **2010**, *38*, 6803–6812. [CrossRef] [PubMed]

toxins

MDPI

Review

Shiga Toxin—A Model for Glycolipid-Dependent and Lectin-Driven Endocytosis

Ludger Johannes

Cellular and Chemical Biology Department, Institut Curie, PSL Research University, U1143 INSERM, UMR3666 CNRS, 26 rue d'Ulm, 75248 Paris CEDEX 05, France; ludger.johannes@curie.fr

Academic Editors: Julien Barbier and Daniel Gillet
Received: 28 September 2017; Accepted: 20 October 2017; Published: 25 October 2017

Abstract: The cellular entry of the bacterial Shiga toxin and the related verotoxins has been scrutinized in quite some detail. This is due to their importance as a threat to human health. At the same time, the study of Shiga toxin has allowed the discovery of novel molecular mechanisms that also apply to the intracellular trafficking of endogenous proteins at the plasma membrane and in the endosomal system. In this review, the individual steps that lead to Shiga toxin uptake into cells will first be presented from a purely mechanistic perspective. Membrane-biological concepts will be highlighted that are often still poorly explored, such as fluctuation force-driven clustering, clathrin-independent membrane curvature generation, friction-driven scission, and retrograde sorting on early endosomes. It will then be explored whether and how these also apply to other pathogens, pathogenic factors, and cellular proteins. The molecular nature of Shiga toxin as a carbohydrate-binding protein and that of its cellular receptor as a glycosylated raft lipid will be an underlying theme in this discussion. It will thereby be illustrated how the study of Shiga toxin has led to the proposal of the GlycoLipid-Lectin (GL-Lect) hypothesis on the generation of endocytic pits in processes of clathrin-independent endocytosis.

Keywords: glycosphingolipid; globotriaosylceramide; Gb3; raft; galectin; integrin; CD44; cholera toxin; thermal Casimir-like force; spontaneous curvature

1. Introduction

Shiga toxin is a pathogenic protein that is produced by *Shigella dysenteriae*, while enterohemorrhagic strains of *Escherichia coli* secrete Shiga-like toxins, which are also termed verotoxins [1]. Notably, the verotoxins of *E. coli* strain O157:H7 are responsible for pathological manifestations that can lead to hemolytic-uremic syndrome (HUS), the leading cause for pediatric renal failure in the world [2]. These toxins are also a threat to adults as it became apparent in 2011, when an outbreak with *E. coli* strain O104:H4 in Germany and neighboring countries claimed dozens of adult victims, and thousands of adult patients who were hospitalized with severe symptoms. The most life-threatening extra-intestinal disease manifestations are renal failure and central nervous system complications. To date, no specific treatment options exist, and clinical management of HUS remains purely supportive [3].

In the cytosol of target cells, the A-subunits of Shiga/verotoxins catalyze the deadenylation of position 4324 of 28S ribosomal RNA, leading to protein biosynthesis inhibition and subsequently to cell death. The A-subunits alone cannot enter cells, however. For this, they need the non-covalent interaction with receptor-binding homopentameric B-subunits. All of these toxins share a common cellular receptor: the neutral glycosphingolipid (GSL) globotriaosylceramide (Gb3 or CD77) [1].

Toxin trafficking from the plasma membrane of target cells to cytosolic ribosomes has been particularly well studied for Shiga toxin, on which this review will be focused with an emphasis on recent literature. The principal steps are receptor binding and toxin clustering at the plasma membrane of target cells, the formation of membrane invaginations, and tubular endocytic pits, the scission of these invaginations to form endocytic carriers, and their intracellular trafficking to endosomes and the retrograde route (Figure 1). These steps are discussed in the following sections of this review.

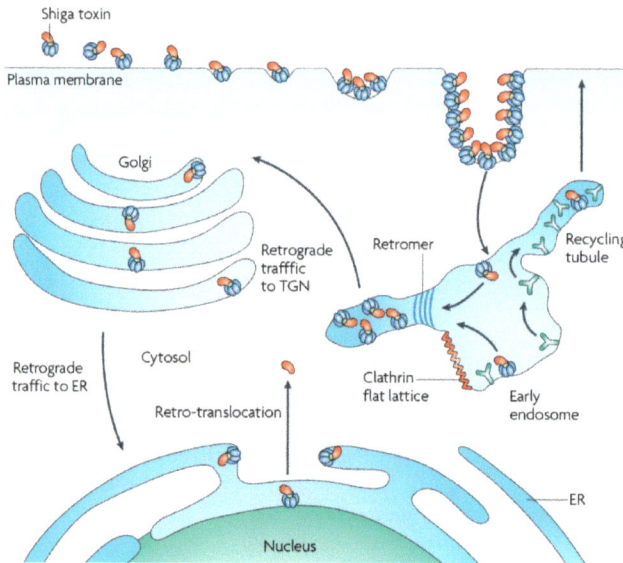

Figure 1. Intracellular trafficking of Shiga toxin. Reproduced from [1]. 2010, Nature Publishing Group.

2. Receptor Binding and Toxin Clustering

The receptor-binding B-subunit of Shiga toxin (STxB) is a homopentameric protein of 35 kDa. Each STxB pentamer displays a total of 15 Gb3 binding sites [4], and each of these binding sites has only millimolar affinity for the globotriose sugar moiety of Gb3 (reviewed in Reference [5]). Yet, Shiga toxin binds avidly and strongly to cells. This apparent high affinity is the result of multiple bond interactions, one STxB pentamer being in contact with several Gb3 molecules at a time. To release STxB from the membrane, all of these interactions have to be dissociated at the same time, which is statistically unlikely. Such avidity effect is a general hallmark of lectins. For example, the homopentameric receptor-binding B-subunit of cholera toxin from *Vibrio cholerae* has a relatively low binding pocket affinity in the micromolar range for its cellular receptor, the GSL GM1 (reviewed in References [5,6]). Yet, much like Shiga toxin, cholera toxin also binds to cells with apparent nanomolar affinity, due to avidity interactions.

The avidity binding effect would be favored if the GSLs that function as toxin receptors were themselves clustered, such that several receptor binding sites on toxin pentamers would be occupied straight away upon the initial contact with the plasma membrane of target cells. Using an advanced electron microscopy technique, nanoclustering has indeed been documented for the gangliosides GM1 and GM3 [7]. This resembles the situation described for another class of glycosylated lipids, the glycosylphosphatidylinositol (GPI)-anchored proteins, which were equally described to exist in membranes as nanoclusters of a few molecules [8].

For Shiga toxin, no direct interaction between toxins has been detected. Yet, on model membranes [9] and on cells [10], Shiga toxin clusters readily, suggesting that a membrane-mediated mechanism drives toxin molecules together. A recent study has investigated this aspect in further detail. Based on theoretical modeling, computer simulations, and experiments in model membrane systems and on cells, an original hypothesis has been proposed [10] according to which Shiga toxin molecules would suppress thermally excited membrane fluctuations not only at sites at which they bind, but also on the membrane patch between 2 adjacent toxin molecules, as long as these are not further apart than roughly the size of the toxin itself (Figure 2).

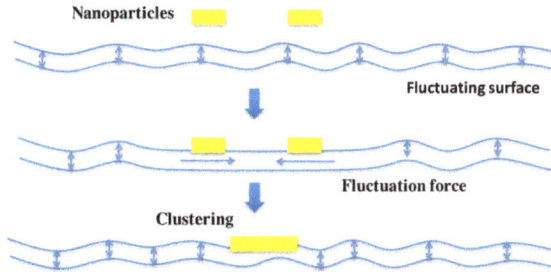

Figure 2. Hypothesis on fluctuation force-driven clustering. The represented nanoparticles could be Shiga toxin pentamers.

Unperturbed fluctuation of the membrane outside this toxin-delineated patch would push the toxin molecules together, even if these were not experiencing a direct attractive force. On theoretical grounds, such thermally induced membrane fluctuation forces (also termed thermal Casimir-like forces; References [11,12]; for a review, see Reference [13]) are expected to be as strong as electrostatic or van der Waals interactions, as long as the membrane inclusions that generate them are several nanometers in size [10]. However, as opposed to electrostatic or van der Waals interactions that operate at subnanometric distances, membrane fluctuation forces would have an effective radius of several nanometers, which corresponds to a gap in the interaction landscape of biological membranes between cytoskeleton-driven clustering that operates at tens to hundreds of nanometers [14], and the conventional subnanometric interaction forces. Of note, while membrane fluctuation forces are expected to apply to any tightly membrane-associated protein (or nanoparticle), other interaction forces would of course continue to be present in biological systems. Thereby, initial contacts that could be favored by the fluctuation forces would need to be further stabilized by other interactions for the generation of biologically meaningful outputs, and to avoid that all proteins (or nanoparticles) that are submitted to fluctuation forces coalesce into one big aggregate.

3. Formation of Membrane Invaginations and Tubular Endocytic Pits

Upon binding to Gb3, Shiga toxin does not directly reach through the plasma membrane to interact with the conventional cytosolic clathrin machinery. Yet, Shiga toxin has been seen in clathrin-coated pits [15], even if efficient inhibition of clathrin function causes at most a 35% reduction of Shiga toxin uptake [16,17]. How this localization into the clathrin pathway is operated remains to be further investigated. A key question here is whether toxin molecules "fall" into preexisting pits, or whether pit biogenesis is actually triggered by the toxin.

A sizable fraction of Shiga toxin still enters cells under conditions of clathrin pathway inhibition. In fact, protein toxins from plants and bacteria were the first cargo molecules for which clathrin-independent uptake into cells was suggested more than 35 years ago [18,19]. How membrane bending and endocytic pit formation are operated in these cases has remained a conundrum. Based on the observation that the formation of a Shiga toxin-Gb3 complex is necessary and sufficient to drive the formation of narrow tubular membrane invaginations [9], a molecular hypothesis has recently been suggested for how this might be achieved (Figure 3a). According to this model, the Shiga toxin-Gb3 complex would be endowed with curvature-active properties, i.e., the capacity to deform the membrane without the need of the cytosolic clathrin machinery. The clustering of several toxin molecules, likely favored by the fluctuation forces that were mentioned above, would then lead to the formation of deep and narrow invaginations as a first step towards the formation of tubular endocytic pits.

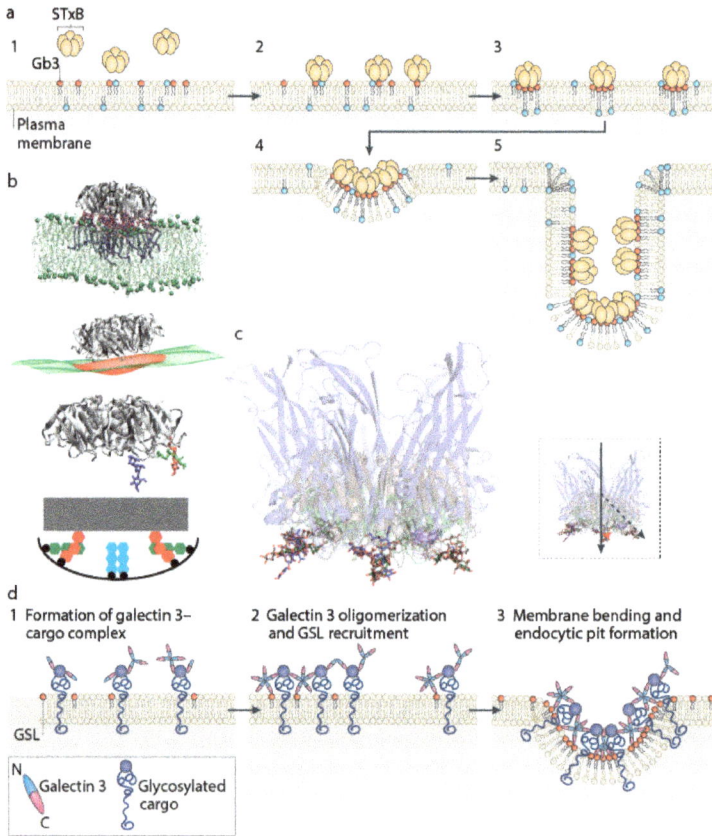

Figure 3. Lectin-driven construction of tubular endocytic pits. (**a**) Model for Shiga toxin B-subunit (STxB)-driven formation of membrane invaginations. (**b**) Molecular dynamics data on spontaneous curvature induced by STxB. The red and green binding sites represented in the lower part of the panel force the membrane to bend up at the edges of STxB pentamers. (**c**) Overlays of crystal structures of STxB (green, Reference [4]), cholera toxin B-subunit (red, Reference [20]), and VP1 capsid protein from SV40 (blue, Reference [21]). (**d**) GL-Lect hypothesis on the Gal3-driven, glycolipid-dependent formation of endocytic pits. Glycolipids are represented as red dots. Reproduced from [22]. 2015, Nature Publishing Group.

At first sight, this proposal appears counterintuitive. The asymmetric deposition of toxin molecules on the exoplasmic leaflet should lead to steric stress that would be expected to favor the buckling of the membrane to the outside of the cell or model membrane to which it binds, thereby generating positive curvature [23]. Obviously, the curvature-active properties of the Shiga toxin-Gb3 complex overcome this steric stress-induced buckling. How is this achieved? Results from molecular dynamics simulations have recently allowed to propose an explanation. It was shown that also in silico, the receptor-binding B-subunit of Shiga toxin (STxB) induces an increment of negative inward-oriented curvature when interacting with a patch of membrane that contains Gb3 receptor molecules [24] (Figure 3b). On average, 13 out of 15 Gb3 binding sites per STxB molecule were occupied. Because of the positioning of 10 of the Gb3 binding pockets at the rim of STxB molecules in a location slightly above the normal plane of the membrane, the latter must bend up to reach these sites, thereby generating an increment of negative, inward-oriented curvature.

This binding site geometry is preserved for the receptor-binding parts of cholera toxin and simian virus 40 (SV40) (Figure 3c), for which it was shown previously that they also have curvature-active properties, endowing them with the capacity to drive tubular membrane invaginations through interaction with their GSL receptor molecules [25], as observed for Shiga toxin [9]. Strikingly, these GSL-binding pathogenic lectins do not have any sequence similarity, which suggests that this binding site geometry might be the result of convergent evolution towards a common function: membrane mechanical work in relation to inward-oriented curvature generation for the construction of endocytic pits. Of note, cholera toxin and SV40 have indeed both been described to be efficiently internalized into cells in which the clathrin pathway is inhibited [18,26]. Further pathogens and pathogenic factors exist that also interact with GSLs in one way or another to get into cells (reviewed for gangliosides in Reference [27]), suggesting that this mechanism is used more widely.

An example of a cellular lectin that drives the GSL-dependent biogenesis of tubular endocytic pits for the uptake of endogenous cargo proteins via clathrin-independent carriers has indeed been described: galectin-3 (Gal3). Gal3 is a member of a family of 15 galectins, 12 of which are expressed in human [28]. This galactose-binding lectin has many physiological (cell migration, immune modulation, inflammation, signaling, etc.) and pathological (cancer, fibrosis, diabetes, etc.) functions, raising the question of how a single protein can manage all of this. The cell biology of Gal3 still remains very little explored. Interestingly, it has recently been shown that membrane-associated Gal3 drives the GSL-dependent formation of narrow tubular invaginations on model membranes [29], similar to Shiga toxin [9], cholera toxin, and SV40 [25]. On cells, Gal3 drives the GSL-dependent biogenesis of clathrin-independent carriers (CLICs), via which Gal3-interacting proteins such as the cell adhesion and migration factors CD44 and β1 integrin are internalized [29].

Based on the results published in Reference [29], the GlycoLipid-Lectin (GL-Lect) hypothesis was put forward [22]. According to this hypothesis, monomeric Gal3 in solution binds to glycosylated cargo proteins such as CD44 and β1 integrin on target cells (Figure 3d). This leads to the oligomerization of Gal3 and its capacity to interact with GSLs in a way such as to induce membrane bending and thereby, the biogenesis of endocytic pits, from which CLICs are then generated. According to this model, Gal3 functions like an endocytic adaptor that links glycosylated cargo proteins and membrane curvature-generating GSLs into the same compositional nanoenvironments from which tubular endocytic pits emerge. In contrast, the pathogenic lectins from Shiga toxin, cholera toxin, and SV40 are themselves cargoes, which are produced as stable pentamers (Figure 3c) such that they can immediately aim for the membrane mechanical part of the program, bypassing the cargo recognition step.

Several aspects of the GL-Lect model are worth further discussion. One concerns the question of why monomeric Gal3 fails to bind GSLs directly. Most likely, the affinity of the sugar binding pocket of Gal3 for carbohydrates on GSLs is weak, such that efficient binding becomes possible only upon the formation of multiple-bond interactions (avidity) of each Gal3 oligomer with several GSLs. Another key question concerns the membrane-bound structure of Gal3. The C-terminal carbohydrate recognition domain of the protein has been crystallized in solution, while the N-terminal domain turned out to be unstructured [30]. It is still unknown when and how the N-terminal domain folds, and what type of oligomeric configuration Gal3 adopts on membranes. It has been proposed that the protein may form pentamers [31], but also, that even larger oligomers may arise [32]. One may finally point out that GSLs are basic fabric of raft-type membrane domains [33]. The lectin-induced clusters of GSLs (or of other glycolipids such as GPI-anchored proteins) that are likely formed at sites of CLIC formation might therefore be viewed as stabilized raft-type membrane domains [34], whose intrinsic connectivity may be critical for tubular endocytic pit construction, CLIC formation, and subsequent intracellular sorting.

4. Membrane Scission, Targeting to Endosomes, and Retrograde Transport

Shiga toxin-induced membrane invaginations need to pinch off from the plasma membrane by fusion of the opposing walls of invaginated tubular endocytic pits. The conventional pinchase dynamin affects Shiga toxin uptake, but is not absolutely required [9,16]. The difficulty with coming to a conclusion for the involvement of dynamin is that feedback loops exist to actin [35], which also contributes to the scission of Shiga toxin-induced membrane invaginations [36]. It has indeed been demonstrated that these invaginations are prone to undergo domain formation, thereby generating domain boundary forces that drive the spontaneous line tension-driven squeezing of the tubule membrane leading to scission [37]. Actin was identified as one of the domain formation triggers [36], likely by directly or indirectly binding to lipids, thereby changing the entropy of the system [38].

Recent findings indicate that yet another scission modality might be operating on Shiga toxin-induced membrane invaginations. The model that is presented in Figure 3a predicts that the signal that is sent by the toxin to the cytosol might be mechanical: the formation of a highly curved membrane domain, which would then be recognized by cellular machinery for further processing (Figure 4).

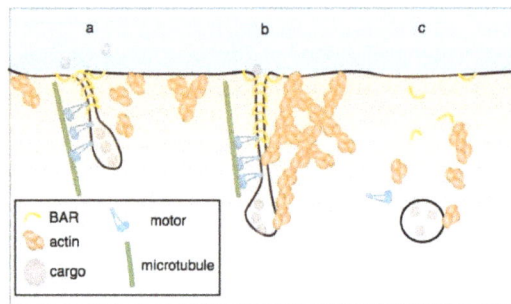

Figure 4. Friction-driven scission. See text for details. Reproduced from [39]. 2017, Cell Press.

Proteins of the Bin, amphiphysin, and Rvs (BAR) domain family are specialized in such curvature recognition [40], and by screening through a BAR domain protein library, endophilin-A2 (EndoA2) was identified as being functionally localized onto Shiga toxin-induced membrane invaginations in relation to the scission reaction [41]. Furthermore, it was demonstrated for Shiga [41] and cholera toxin [42] that these invaginations are pulled upon by microtubule-based dynein motor activity. Strikingly, if this dynein-mediated pulling force is exerted on tubular invaginations that are scaffolded by EndoA2, membrane scission ensues [41]. This is due to the fact that the EndoA2 scaffold exerts a friction force onto the membrane that limits lipid diffusion [39]. Upon pulling with speeds superior to 50 nm/s (such as that exerted by the dynein motor), the lack in lipid supply leads to tube thinning, and then to scission. Such pulling force or friction-driven scission may operate more generally in processes of clathrin and caveolin-independent endocytosis that are often little reliant on dynamin [43].

It might appear surprising that several scission modalities operate on Shiga toxin-induced membrane invaginations: dynamin [9], actin [36], and EndoA2/dynein [41]. It could be demonstrated that when interfering with each of these individually or in combination, the effects on scission appear to be roughly additive [41], suggesting that the overall probability of scission is the result of individual contributions.

Once formed, Shiga toxin-containing carriers are targeted to the endosomal system. This is favored by association with microtubules [44], which might be initiated right at the plasma membrane, as discussed above. Furthermore, the vSNARE proteins VAMP2, VAMP3, and VAMP8 are recruited to Shiga toxin-containing endocytic carriers for their targeting to the endosomal membrane system [45].

From endosomes, Shiga toxin is transported via the retrograde route to the TGN, the Golgi apparatus, and then on to the endoplasmic reticulum, from where the catalytic A-subunit is translocated to the cytosol [46]. The study of Shiga toxin has allowed to propose the existence of a trafficking interface between early endosomes and the TGN [47], which has since become a hotspot of membrane biology research. Shiga toxin trafficking at this interface has been described to depend on Rab6 [48], SNARE complexes involving syntaxin-16 [48,49] and syntaxin-5 [50], epsinR [17], Arl1 [51], OCRL [52], retromer [53–55], the tethering complex GARP [56] and the GARP interactor TSSC1 [57], the ARF1 GAP protein AGAP2 [58], GPP130 [59], the ERM proteins ezrin and moesin [60], annexins A1 and A2 [61], and UNC50 [62].

The early endosome-TGN trafficking interface has also become the target of small molecule inhibitors to protect against the Shiga toxin [63,64]. The Retro series compounds have been shown to protect cells and mice against Shiga toxin [65] and the plant toxin ricin that also uses the retrograde transport route to intoxicate cells [66]. In Retro compound-treated cells, Shiga toxin is retained in early endosomes, therefore reaching the endoplasmic reticulum and the cytosol less efficiently, leading to a reduced toxic effect. Efforts to identify the cellular target(s) of these compounds are still ongoing.

5. Conclusions

The study of Shiga toxin endocytosis has allowed the discovery of a wide range of novel cellular mechanisms that had previously gone undetected. Yet, much remains to be done. Notably the functional dissection of the Shiga toxin receptor, the GSL Gb3, is still in its infancy, due to the difficulty to manipulate the structure of this glycosylated lipid in a controlled manner within cells. This is true for GSLs in general, notably for the ones with long acyl chains that are found in naturally occurring species and that are very difficult to reconstitute into cells. Several of the mechanisms that were discovered with the help of Shiga toxin are themselves still poorly explored. To name two examples: Fluctuation forces for the clustering of tightly membrane-associated proteins (or nanoparticles) will need to be measured directly using appropriate force sensors, and the structural intricacies underlying the capacity of GL-Lect drivers such as galectin-3 (Gal3) to switch from glycoprotein to glycolipid recognition remain to be elucidated. Membrane biology research on or around Shiga toxin is thereby expected to remain a fruitful ground for discovery.

Acknowledgments: Weria Pezeshkian is acknowledged for help with Figure 2. Work in the Johannes team in the context of the theme of the current review is supported by grants from the Agence Nationale pour la Recherche (ANR-14-CE16-0004-03, ANR-14-CE14-0002-02, ANR-16-CE23-0005-02, ANR-16-CE23-0005-02), Human Frontier Science Program grant RGP0029-2014, European Research Council advanced grant (project 340485), European Union program H2020-MSCA-ITN-2014 BIOPOL, and the Swedish Research Council K2015-99X-22877-01-6. The Johannes team is member of Labex CelTisPhyBio (11-LBX-0038) and Idex Paris Sciences et Lettres (ANR-10-IDEX-0001-02 PSL).

Conflicts of Interest: The author declares no conflict of interest.

References

1. Johannes, L.; Römer, W. Shiga toxins—From cell biology to biomedical applications. *Nat. Rev. Microbiol.* **2010**, *8*, 105–116. [CrossRef] [PubMed]

2. Karch, H.; Denamur, E.; Dobrindt, U.; Finlay, B.B.; Hengge, R.; Johannes, L.; Ron, E.Z.; Tønjum, T.; Sansonetti, P.J.; Vicente, M. The enemy within us: Lessons from the 2011 european escherichia coli o104:H4 outbreak. *EMBO Mol. Med.* **2012**, *4*, 841–848. [CrossRef] [PubMed]

3. Tarr, P.I.; Gordon, C.A.; Chandler, W.L. Shiga-toxin-producing escherichia coli and haemolytic uraemic syndrome. *Lancet* **2005**, *365*, 1073–1086. [CrossRef]

4. Ling, H.; Boodhoo, A.; Hazes, B.; Cummings, M.D.; Armstrong, G.D.; Brunton, J.L.; Read, R.J. Structure of shiga-like toxin i b-pentamer complexed with an analogue of its receptor gb3. *Biochemistry* **1998**, *37*, 1777–1788. [CrossRef] [PubMed]

5. Pina, D.G.; Johannes, L. Cholera and shiga toxin b-subunits: Thermodynamic and structural considerations for function and biomedical applications. *Toxicon* **2005**, *45*, 389–393. [CrossRef] [PubMed]

6. Ewers, H.; Helenius, A. Lipid-mediated endocytosis. *Cold Spring Harb. Perspect. Biol.* **2011**, *3*, a004721. [CrossRef] [PubMed]

7. Fujita, A.; Cheng, J.; Hirakawa, M.; Furukawa, K.; Kusunoki, S.; Fujimoto, T. Gangliosides gm1 and gm3 in the living cell membrane form clusters susceptible to cholesterol depletion and chilling. *Mol. Biol. Cell* **2007**, *18*, 2112–2122. [CrossRef] [PubMed]

8. Goswami, D.; Gowrishankar, K.; Bilgrami, S.; Ghosh, S.; Raghupathy, R.; Chadda, R.; Vishwakarma, R.; Rao, M.; Mayor, S. Nanoclusters of gpi-anchored proteins are formed by cortical actin-driven activity. *Cell* **2008**, *135*, 1085–1097. [CrossRef] [PubMed]

9. Römer, W.; Berland, L.; Chambon, V.; Gaus, K.; Windschiegl, B.; Tenza, D.; Aly, M.R.; Fraisier, V.; Florent, J.-C.; Perrais, D.; et al. Shiga toxin induces tubular membrane invaginations for its uptake into cells. *Nature* **2007**, *450*, 670–675. [CrossRef] [PubMed]

10. Pezeshkian, W.; Gao, H.; Arumugam, S.; Becken, U.; Bassereau, P.; Florent, J.C.; Ipsen, J.H.; Johannes, L.; Shillcock, J. Mechanism of shiga toxin clustering on membranes. *ACS Nano* **2017**, *11*, 314–324. [CrossRef] [PubMed]

11. Bartolo, D.; Ajdari, A.; Fournier, J.B.; Golestanian, R. Fluctuations of fluctuation-induced casimir-like forces. *Phys. Rev. Lett.* **2002**, *89*, 230601. [CrossRef] [PubMed]

12. Yolcu, C.; Deserno, M. Membrane-mediated interactions between rigid inclusions: An effective field theory. *Phys. Rev. E Stat. Nonlinear Soft Matter Phys.* **2012**, *86*, 031906. [CrossRef] [PubMed]

13. Nguyen, V.D.; Dang, M.T.; Nguyen, T.A.; Schall, P. Critical casimir forces for colloidal assembly. *J. Phys. Condens. Matter* **2016**, *28*, 043001. [CrossRef] [PubMed]

14. Rao, M.; Mayor, S. Active organization of membrane constituents in living cells. *Curr. Opin. Cell Biol.* **2014**, *29*, 126–132. [CrossRef] [PubMed]

15. Sandvig, K.; Olsnes, S.; Brown, J.E.; Petersen, O.W.; van Deurs, B. Endocytosis from coated pits of shiga toxin: A glycolipid-binding protein from *shigella dysenteriae* 1. *J. Cell Biol.* **1989**, *108*, 1331–1343. [CrossRef] [PubMed]

16. Lauvrak, S.U.; Torgersen, M.L.; Sandvig, K. Efficient endosome-to-golgi transport of shiga toxin is dependent on dynamin and clathrin. *J. Cell Sci.* **2004**, *117*, 2321–2331. [CrossRef] [PubMed]

17. Saint-Pol, A.; Yélamos, B.; Amessou, M.; Mills, I.; Dugast, M.; Tenza, D.; Schu, P.; Antony, C.; McMahon, H.T.; Lamaze, C.; et al. Clathrin adaptor epsinr is required for retrograde sorting on early endosomal membranes. *Dev. Cell* **2004**, *6*, 525–538. [CrossRef]

18. Montesano, R.; Roth, J.; Robert, A.; Orci, L. Non-coated membrane invaginations are involved in binding and internalization of cholera and tetanus toxins. *Nature* **1982**, *296*, 651–653. [CrossRef] [PubMed]

19. Moya, M.; Dautry-Varsat, A.; Goud, B.; Louvard, D.; Boquet, P. Inhibition of coated pit formation in hep2 cells blocks the cytotoxicity of diphtheria toxin but not that of ricin toxin. *J. Cell Biol.* **1985**, *101*, 548–559. [CrossRef] [PubMed]

20. Zhang, R.G.; Westbrook, M.L.; Westbrook, E.M.; Scott, D.L.; Otwinowski, Z.; Maulik, P.R.; Reed, R.A.; Shipley, G.G. The 2.4 a crystal structure of cholera toxin b subunit pentamer: Choleragenoid. *J. Mol. Biol.* **1995**, *251*, 550–562. [CrossRef] [PubMed]

21. Neu, U.; Woellner, K.; Gauglitz, G.; Stehle, T. Structural basis of gm1 ganglioside recognition by simian virus 40. *Proc. Natl. Acad. Sci. USA* **2008**, *105*, 5219–5224. [CrossRef] [PubMed]

22. Johannes, L.; Wunder, C.; Shafaq-Zadah, M. Glycolipids and lectins in endocytic uptake processes. *J. Mol. Biol.* **2016**, *428*, 4792–4818. [CrossRef] [PubMed]

23. Stachowiak, J.C.; Schmid, E.M.; Ryan, C.J.; Ann, H.S.; Sasaki, D.Y.; Sherman, M.B.; Geissler, P.L.; Fletcher, D.A.; Hayden, C.C. Membrane bending by protein-protein crowding. *Nat. Cell Biol.* **2012**, *14*, 944–949. [CrossRef] [PubMed]

24. Pezeshkian, W.; Hansen, A.G.; Johannes, L.; Khandelia, H.; Shillcock, J.; Sunil Kumar, P.B.; Ipsen, J.H. Membrane invagination induced by shiga toxin b-subunit: From molecular structure to tube formation. *Soft Matter* **2016**, *12*, 5164–5171. [CrossRef] [PubMed]

25. Ewers, H.; Römer, W.; Smith, A.E.; Bacia, K.; Dmitrieff, S.; Chai, W.; Mancini, R.; Kartenbeck, J.; Chambon, V.; Berland, L.; et al. Gm1 structure determines sv40-induced membrane invagination and infection. *Nat. Cell Biol.* **2010**, *12*, 11–18. [CrossRef] [PubMed]

26. Damm, E.M.; Pelkmans, L.; Kartenbeck, J.; Mezzacasa, A.; Kurzchalia, T.V.; Helenius, A. Clathrin- and caveolin-1-independent endocytosis: Entry of simian virus 40 into cells devoid of caveolae. *J. Cell Biol.* **2005**, *168*, 477–488. [CrossRef] [PubMed]

27. Ravindran, M.S.; Tanner, L.B.; Wenk, M.R. Sialic acid linkage in glycosphingolipids is a molecular correlate for trafficking and delivery of extracellular cargo. *Traffic* **2013**, *14*, 1182–1191. [CrossRef] [PubMed]

28. Dumic, J.; Dabelic, S.; Flogel, M. Galectin-3: An open-ended story. *Biochim. Biophys. Acta* **2006**, *1760*, 616–635. [CrossRef] [PubMed]

29. Lakshminarayan, R.; Wunder, C.; Becken, U.; Howes, M.T.; Benzing, C.; Arumugam, S.; Sales, S.; Ariotti, N.; Chambon, V.; Lamaze, C.; et al. Galectin-3 drives glycosphingolipid-dependent biogenesis of clathrin-independent carriers. *Nat. Cell Biol.* **2014**, *16*, 595–606. [CrossRef] [PubMed]

30. Seetharaman, J.; Kanigsberg, A.; Slaaby, R.; Leffler, H.; Barondes, S.H.; Rini, J.M. X-ray crystal structure of the human galectin-3 carbohydrate recognition domain at 2.1-a resolution. *J. Biol. Chem.* **1998**, *273*, 13047–13052. [CrossRef] [PubMed]

31. Ahmad, N.; Gabius, H.J.; Andre, S.; Kaltner, H.; Sabesan, S.; Roy, R.; Liu, B.; Macaluso, F.; Brewer, C.F. Galectin-3 precipitates as a pentamer with synthetic multivalent carbohydrates and forms heterogeneous cross-linked complexes. *J. Biol. Chem.* **2004**, *279*, 10841–10847. [CrossRef] [PubMed]

32. Lepur, A.; Salomonsson, E.; Nilsson, U.J.; Leffler, H. Ligand induced galectin-3 self-association. *J. Biol. Chem.* **2012**, *287*, 21751–21756. [CrossRef] [PubMed]

33. Simons, K.; Ikonen, E. Functional rafts in cell membranes. *Nature* **1997**, *387*, 569–572. [CrossRef] [PubMed]

34. Simons, K.; Gerl, M.J. Revitalizing membrane rafts: New tools and insights. *Nat. Rev. Mol. Cell Biol.* **2010**, *11*, 688–699. [CrossRef] [PubMed]

35. Taylor, M.J.; Lampe, M.; Merrifield, C.J. A feedback loop between dynamin and actin recruitment during clathrin-mediated endocytosis. *PLoS Biol.* **2012**, *10*, e1001302. [CrossRef] [PubMed]

36. Römer, W.; Pontani, L.L.; Sorre, B.; Rentero, C.; Berland, L.; Chambon, V.; Lamaze, C.; Bassereau, P.; Sykes, C.; Gaus, K.; et al. Actin dynamics drive membrane reorganization and scission in clathrin-independent endocytosis. *Cell* **2010**, *140*, 540–553. [CrossRef] [PubMed]

37. Johannes, L.; Wunder, C.; Bassereau, P. Bending "on the rocks"—A cocktail of biophysical modules to build endocytic pathways. *Cold Spring Harb. Perspect. Biol.* **2014**, *6*, a016741. [CrossRef] [PubMed]

38. Liu, A.P.; Fletcher, D.A. Actin polymerization serves as a membrane domain switch in model lipid bilayers. *Biophys. J.* **2006**, *91*, 4064–4070. [CrossRef] [PubMed]

39. Simunovic, M.; Manneville, J.B.; Renard, H.F.; Evergren, E.; Raghunathan, K.; Bhatia, D.; Kenworthy, A.K.; Voth, G.A.; Prost, J.; McMahon, H.; et al. Friction mediates scission of membrane nanotubes scaffolded by bar proteins. *Cell* **2017**, *170*, 172–184. [CrossRef] [PubMed]

40. Rao, Y.; Haucke, V. Membrane shaping by the bin/amphiphysin/rvs (bar) domain protein superfamily. *Cell. Mol. Life Sci.* **2011**, *68*, 3983–3993. [CrossRef] [PubMed]

41. Renard, H.-F.; Simunovic, M.; Lemière, J.; Boucrot, E.; Garcia-Castillo, M.D.; Arumugam, S.; Chambon, V.; Lamaze, C.; Wunder, C.; Kenworthy, A.K.; et al. Endophilin-a2 functions in membrane scission in clathrin-independent endocytosis. *Nature* **2015**, *517*, 493–496. [CrossRef] [PubMed]

42. Day, C.A.; Baetz, N.W.; Copeland, C.A.; Kraft, L.J.; Han, B.; Tiwari, A.; Drake, K.R.; De Luca, H.; Chinnapen, D.J.; Davidson, M.W.; et al. Microtubule motors power plasma membrane tubulation in clathrin-independent endocytosis. *Traffic* **2015**, *16*, 572–590. [CrossRef] [PubMed]

43. Blouin, C.M.; Lamaze, C. Interferon gamma receptor: The beginning of the journey. *Front. Immunol.* **2013**, *4*, 267. [CrossRef] [PubMed]

44. Hehnly, H.; Sheff, D.; Stamnes, M. Shiga toxin facilitates its retrograde transport by modifying microtubule dynamics. *Mol. Biol. Cell* **2006**, *17*, 4379–4389. [CrossRef] [PubMed]

45. Renard, H.-F.; Garcia-Castillo, M.D.; Chambon, V.; Lamaze, C.; Johannes, L. Clathrin-independent endocytosis of vamp2/3/8 snare proteins and their function in shiga toxin trafficking into cells. *J. Cell Sci.* **2015**, *128*, 2891–2902. [CrossRef] [PubMed]

46. Spooner, R.A.; Lord, J.M. How ricin and shiga toxin reach the cytosol of target cells: Retrotranslocation from the endoplasmic reticulum. *Curr. Top. Microbiol. Immunol.* **2012**, *357*, 19–40. [PubMed]

47. Mallard, F.; Tenza, D.; Antony, C.; Salamero, J.; Goud, B.; Johannes, L. Direct pathway from early/recycling endosomes to the golgi apparatus revealed through the study of shiga toxin b-fragment transport. *J. Cell Biol.* **1998**, *143*, 973–990. [CrossRef] [PubMed]

48. Mallard, F.; Tang, B.L.; Galli, T.; Tenza, D.; Saint-Pol, A.; Yue, X.; Antony, C.; Hong, W.J.; Goud, B.; Johannes, L. Early/recycling endosomes-to-tgn transport involves two snare complexes and a rab6 isoform. *J. Cell Biol.* **2002**, *156*, 653–664. [CrossRef] [PubMed]

49. Ganley, I.G.; Espinosa, E.; Pfeffer, S.R. A syntaxin 10-snare complex distinguishes two distinct transport routes from endosomes to the trans-golgi in human cells. *J. Cell Biol.* **2008**, *180*, 159–172. [CrossRef] [PubMed]

50. Tai, G.; Lu, L.; Wang, T.L.; Tang, B.L.; Goud, B.; Johannes, L.; Hong, W. Participation of syntaxin 5/ykt6/gs28/gs15 snare complex in transport from the early/recycling endosome to the tgn. *Mol. Biol. Cell* **2004**, *15*, 4011–4022. [CrossRef] [PubMed]

51. Tai, G.; Lu, L.; Johannes, L.; Hong, W. Functional analysis of arl1 and golgin-97 in endosome-to-tgn transport using recombinant shiga toxin b fragment. *Methods Enzymol.* **2005**, *404*, 442–453. [PubMed]

52. Choudhury, R.; Diao, A.; Zhang, F.; Eisenberg, E.; Saint-Pol, A.; Williams, C.; Konstantakopoulos, A.; Lucocq, J.; Johannes, L.; Rabouille, C.; et al. Lowe syndrom protein ocrl1 interacts with clathrin and regulates protein trafficking between endosomes and the trans-golgi network. *Mol. Biol. Cell* **2005**, *16*, 3467–3479. [CrossRef] [PubMed]

53. Popoff, V.; Mardones, G.A.; Tenza, D.; Rojas, R.; Lamaze, C.; Bonifacino, J.S.; Raposo, G.; Johannes, L. The retromer complex and clathrin define a post-early endosomal retrograde exit site. *J. Cell Sci.* **2007**, *120*, 2022–2031. [CrossRef] [PubMed]

54. Bujny, M.V.; Popoff, V.; Johannes, L.; Cullen, P.J. The retromer component, sorting nexin-1, is required for efficient early endosome-to-trans golgi network retrograde transport of shiga toxin. *J. Cell Sci.* **2007**, *120*, 2010–2021. [CrossRef] [PubMed]

55. Utskarpen, A.; Slagsvold, H.H.; Dyve, A.B.; Skanland, S.S.; Sandvig, K. Snx1 and snx2 mediate retrograde transport of shiga toxin. *Biochem. Biophys. Res. Commun.* **2007**, *358*, 566–570. [CrossRef] [PubMed]

56. Perez-Victoria, F.J.; Mardones, G.A.; Bonifacino, J.S. Requirement of the human garp complex for mannose 6-phosphate-receptor-dependent sorting of cathepsin d to lysosomes. *Mol. Biol. Cell* **2008**, *19*, 2350–2362. [CrossRef] [PubMed]

57. Gershlick, D.C.; Schindler, C.; Chen, Y.; Bonifacino, J.S. Tssc1 is novel component of the endosomal retrieval machinery. *Mol. Biol. Cell* **2016**, *27*, 2867–2878. [CrossRef] [PubMed]

58. Shiba, Y.; Römer, W.; Mardones, G.A.; Burgos, P.V.; Lamaze, C.; Johannes, L. Agap2 regulates retrograde transport between early endosomes and the tgn. *J. Cell Sci.* **2010**, *123*, 2381–2390. [CrossRef] [PubMed]

59. Mukhopadhyay, S.; Linstedt, A.D. Manganese blocks intracellular trafficking of shiga toxin and protects against shiga toxicosis. *Science* **2012**, *335*, 332–335. [CrossRef] [PubMed]

60. Kvalvaag, A.S.; Pust, S.; Sundet, K.I.; Engedal, N.; Simm, R.; Sandvig, K. The erm proteins ezrin and moesin regulate retrograde shiga toxin transport. *Traffic* **2013**, *14*, 839–852. [CrossRef] [PubMed]

61. Tcatchoff, L.; Andersson, S.; Utskarpen, A.; Klokk, T.I.; Skanland, S.S.; Pust, S.; Gerke, V.; Sandvig, K. Annexin a1 and a2: Roles in retrograde trafficking of shiga toxin. *PLoS ONE* **2012**, *7*, e40429. [CrossRef] [PubMed]

62. Selyunin, A.S.; Iles, L.R.; Bartholomeusz, G.; Mukhopadhyay, S. Genome-wide sirna screen identifies unc50 as a regulator of shiga toxin 2 trafficking. *J. Cell Biol.* **2017**. [CrossRef] [PubMed]

63. Mukhopadhyay, S.; Linstedt, A.D. Retrograde trafficking of ab(5) toxins: Mechanisms to therapeutics. *J. Mol. Med.* **2013**, *91*, 1131–1141. [CrossRef] [PubMed]

64. Gupta, N.; Noel, R.; Goudet, A.; Hinsinger, K.; Michau, A.; Pons, V.; Abdelkafi, H.; Secher, T.; Shima, A.; Shtanko, O.; et al. Inhibitors of retrograde trafficking active against ricin and shiga toxins also protect cells from several viruses, leishmania and chlamydiales. *Chem. Biol. Interact.* **2017**, *267*, 96–103. [CrossRef] [PubMed]

65. Secher, T.; Shima, A.; Hinsinger, K.; Cintrat, J.C.; Johannes, L.; Barbier, J.; Gillet, D.; Oswald, E. Retrograde trafficking inhibitors of shiga toxins reduces morbidity and mortality of mice infected with enterohemorrhagic escherichia coli (stec). *Antimicrob. Agents Chemother.* **2015**, *59*, 5010–5013. [CrossRef] [PubMed]

66. Stechmann, B.; Bai, S.K.; Gobbo, E.; Lopez, R.; Merer, G.; Pinchard, S.; Panigai, L.; Tenza, D.; Raposo, G.; Beaumelle, B.; et al. Inhibition of retrograde transport protects mice from lethal ricin challenges. *Cell* **2010**, *141*, 231–242. [CrossRef] [PubMed]

toxins

Communication

Yeast Reporter Assay to Identify Cellular Components of Ricin Toxin A Chain Trafficking

Björn Becker, Tina Schnöder and Manfred J. Schmitt *

Molecular and Cell Biology, Department of Biosciences and Center of Human and Molecular Biology (ZHMB), Saarland University, Saarbrücken D-66123, Germany; b.becker@microbiol.uni-sb.de (B.B.); Tina.Schnoeder@med.uni-jena.de (T.S.)
* Correspondence: mjs@microbiol.uni-sb.de; Tel.: +49-681-302-4730; Fax: +49-681-302-4710

Academic Editor: Tomas Girbes
Received: 11 October 2016; Accepted: 30 November 2016; Published: 6 December 2016

Abstract: RTA, the catalytic A-subunit of the ribosome inactivating A/B toxin ricin, inhibits eukaryotic protein biosynthesis by depurination of 28S rRNA. Although cell surface binding of ricin holotoxin is mainly mediated through its B-subunit (RTB), sole application of RTA is also toxic, albeit to a significantly lower extent, suggesting alternative pathways for toxin uptake and transport. Since ricin toxin trafficking in mammalian cells is still not fully understood, we developed a GFP-based reporter assay in yeast that allows rapid identification of cellular components required for RTA uptake and subsequent transport through a target cell. We hereby show that Ypt6p, Sft2p and GARP-complex components play an important role in RTA transport, while neither the retromer complex nor COPIB vesicles are part of the transport machinery. Analyses of yeast knock-out mutants with chromosomal deletion in genes whose products regulate ADP-ribosylation factor GTPases (Arf-GTPases) and/or retrograde Golgi-to-ER (endoplasmic reticulum) transport identified Sso1p, Snc1p, Rer1p, Sec22p, Erv46p, Gea1p and Glo3p as novel components in RTA transport, suggesting the developed reporter assay as a powerful tool to dissect the multistep processes of host cell intoxication in yeast.

Keywords: *S. cerevisiae*; ricin toxin A chain (RTA); ribosome inactivating protein (RIP); retrograde protein transport; trans-Golgi network (TGN); endoplasmic reticulum (ER)

1. Introduction

Understanding intracellular ricin transport is important for the development of effective strategies against acute ricin intoxication. As a member of the A/B toxin family, ricin is a highly potent protein toxin from the seeds of the castor oil plant *Ricinus communis* that belongs to class-II ribosome inactivating proteins (RIPs) [1]. It consists of two disulfide-bonded polypeptides, amongst which the B-chain (RTB) serves as the cell binding subunit mediating toxin uptake by mammalian target cells [2]. The cytotoxic A-chain of ricin (RTA) acts as *N*-glycosidase that cleaves a specific adenine residue within a conserved sarcin/ricin loop of eukaryotic 28S rRNA, which subsequently causes a block in eEF-2 mediated translation initiation followed by rapid cell death [3].

Over the years, intensive analyses of ricin trafficking in mammalian cells identified a complex network of pathways that is parasitized by the toxin [4–9]. Although intoxication initiates by RTB binding to terminal galactose and/or *N*-acetylglucosamine residues in cell surface proteins or lipids, followed by ricin uptake through clathrin-dependent as well as -independent endocytosis and vesicular transport to early endosomes, RTA without RTB is likewise capable to kill yeast and mammalian cells (IC$_{50}$ of 50–100 μg/mL), though cell killing is much more efficient for ricin holotoxin (IC$_{50}$ of 2 ng/mL) [10–17]. Whilst most of the dimeric toxin is subsequently exocytosed and/or degraded in lysosomes, only 5% of the internalized toxin reaches the trans-Golgi network (TGN) [18] through retrograde transport by components such as the GARP-complex [5], Stx5 [6] and the GTPases Rab6A and Rab6A' [7], while neither retromer components [5] nor the GTPases Rab9 and Rab11 [11] are required for endosome–TGN transport. Retrograde toxin transport within the Golgi is mediated by the TRAPP-complex [5] followed by backward transport to the endoplasmic reticulum (ER). Since ricin itself does not contain an ER retention/targeting signal that could potentially mediate its interaction with KDEL receptors of the target cell, it has been proposed that RTB binds to resident luminal ER proteins, which indirectly allows toxin transport to the ER [19,20]. Within the ER, ricin is recognized by EDEM1 and EDEM2 [21,22] and subsequently retrotranslocates into the cytosol, most likely through the Sec61 translocon by using components of the ER-associated protein degradation (ERAD) machinery, including Derlins 1–3 [23]. Consequently, inhibition of retrograde transport, either by chemical inhibitors or via specific antibodies, efficiently protects mice and cell lines against ricin intoxication, indicating that detailed knowledge and understanding of intracellular toxin trafficking is a prerequisite for the development of a protective ricin antidote which, until now, is still not available [6,24,25].

Interestingly, intact yeast cells are phenotypically ricin resistant due to a lack of galactosylated RTB binding sites at their cell surface, while the same cells become ricin sensitive after RTA expression in the ER [23,26,27] or when exogenously applied to cell wall lacking spheroplasts [17]. This important observation turns yeast into an attractive model to study RTA uptake and intracellular transport in a lower eukaryote. Furthermore, as retrograde protein transport in its basic mechanisms is similar between yeast and mammalian cells (for reviews see e.g., [28–31]), we focused on the development of a yeast-based assay to identify cellular components of intracellular RTA transport and to address three major questions: (i) Which proteins are involved in RTA transport from the plasma membrane through the endosomal compartment to the ER? (ii) Are there any similarities and/or differences in RTA trafficking between yeast and mammalian cells? (iii) How useful is yeast as a model to analyze ricin transport pathways?

2. Results and Discussion

2.1. Fluorescence-Based Reporter Assay for RTA Toxicity in Yeast

To screen yeast for cellular components involved in retrograde toxin transport after external RTA application, a GFP-based reporter assay was designed in which RTA toxicity is measured indirectly through fluorescence emission after in vivo translation of a secreted GFP reporter containing an N-terminal signal peptide and ER import signal derived from the yeast viral K28 preprotoxin [32]. Expression of this reporter (K28SP-GFP) is driven from the inducible *GAL1* promoter and requires active protein biogenesis for fluorescence emission. Since RTA efficiently blocks mRNA translation by depurinating 28S rRNA [19,33], this assay is expected to allow the identification of both known and novel host cell proteins that are involved in intracellular RTA transport. In case that RTA uptake and/or compartmental transport is negatively affected in a particular yeast knock-out mutant, this should be in line with an increased in vivo fluorescence of the reporter. The general setup of this bioassay and its use in an *S. cerevisiae* wild-type strain and various isogenic knock-out mutants after transformation with the reporter plasmid is outlined in Figure 1a. To bypass the lack of specific RTA binding sites at the yeast cell surface, spheroplasts of yeast transformants are cultivated in microtiter plates under conditions of induced GFP expression in the presence or absence of externally applied RTA and fluorescence kinetics is measured in wild-type and mutant spheroplasts over a time window of 20 h (Figure 1b).

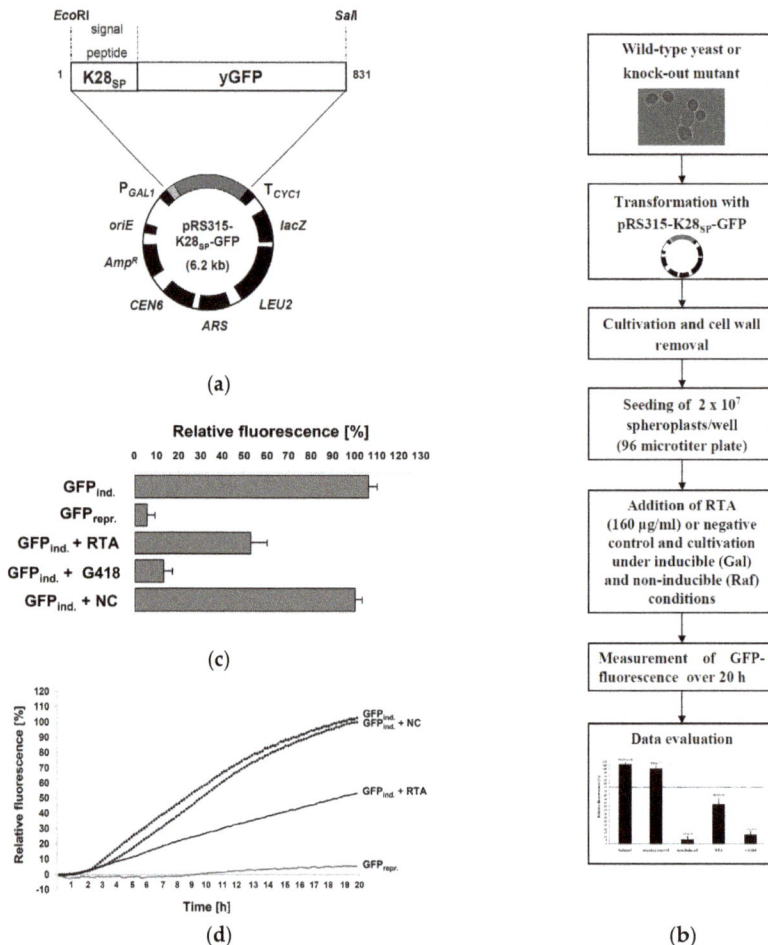

Figure 1. (a) Schematic overview of the reporter plasmid used to screen yeast for proteins involved in ricin toxin A chain (RTA) transport. Yeast enhanced GFP extended by an N-terminal signal peptide (SP) derived from K28 preprotoxin [34] is placed under transcriptional control of the *GAL1* promoter and *CYC1* terminator, allowing inducible expression in the presence of galactose. Indicated numbers correspond to amino acid position in GFP; (b) Experimental assay setup. After transformation of a particular yeast strain with pRS315-K28$_{SP}$-GFP and cell wall removal by zymolyase treatment, 2×10^7 spheroplasts were seeded in 96 microtiter plates and GFP expression was induced by the addition of 3% galactose. Simultaneously, RTA or the negative control sample was added and fluorescence development was measured over a time window of 20 h; (c) Relative fluorescence emission of 2×10^7 wild-type yeast spheroplasts expressing GFP from the reporter plasmid pRS315-K28$_{SS}$-GFP under induced (3% galactose, GFP$_{ind.}$) and non-induced (2% raffinose, GFP$_{repr.}$) culture conditions. Yeast spheroplasts were also incubated in the presence of RTA (5 μM), G418 (300 μg/mL) or the negative control (NC) over 20 h. Standard deviation is indicated; all measurements were repeated 6 to 12 times as independent experiments without technical replicates; (d) Time course of GFP fluorescence development of wild-type yeast spheroplasts expressing GFP from the reporter plasmid pRS315-K28$_{SS}$-GFP in the presence (RTA) or absence (NC) of 5 μM RTA.

In a first experiment, the effect of extracellular applied RTA on wild-type spheroplasts was analyzed; a Ni^{2+}/NTA-purified culture supernatant from *E. coli* carrying the empty vector was included as negative control, whereas cells treated with the aminoglycoside antibiotic and protein biosynthesis blocker geneticin (G418) served as positive control. As expected, G418-treated cells revealed only a slight but clearly detectable increase in GFP fluorescence after induction (13.4% ± 3.9%) while no fluorescence (5.5% ± 3.6%) was seen in non-induced cells (Figure 1c). Induced cells (105.9% ± 4.2%) showed a similar level in fluorescence as negative control cells (set to 100%), while fluorescence significantly declined in RTA-treated spheroplasts (52.8% ± 7.2%). Based on the cytotoxic effect of RTA against yeast, an IC_{50} of approximately 5 μM (160 μg/mL) can be calculated, which is remarkably close to the reported toxicity of RTA against mammalian cells with IC_{50} values ranging from 1.5 to 3 μM (50–100 μg/mL), while toxicity of ricin holotoxin, consisting of RTA and RTB, is several orders of magnitude higher with an IC_{50} of 30 pM (2 ng/mL) [16]. Thus, RTA seems two to three times less toxic against yeast in comparison to mammalian cells and approximately 10^5 times less toxic than ricin holotoxin (for which IC_{50} data are not available in yeast).

To get an idea about the timing of GFP expression and its translational inhibition by RTA in our experimental setup, relative GFP fluorescence was determined over a time period of 20 h. As illustrated in Figure 1d, GFP fluorescence became detectable and continuously increased 90 min after *GAL1*-induced expression, while the inhibitory effect of RTA on GFP translation and fluorescence became detectable after 210 min; this delay in RTA-mediated fluorescence decrease most likely reflects the kinetics of RTA uptake and intracellular transport to the cytosol. Although the simultaneous induction of GFP expression and RTA application caused an increase in the GFP background signal, the chosen 20 h end-point GFP signal nevertheless turned out suitable to detect significant differences in fluorescence inhibition between toxin-treated and control cells. In this way, a stringent correlation between in vivo reporter fluorescence and the level of protein biosynthesis after external RTA application could be demonstrated, confirming the suitability of this assay to detect the impact of RTA on in vivo protein biosynthesis in the genetic background of different knock-out mutants.

2.2. Assay Validation

To validate the yeast bioassay, the influence of external RTA application was determined in yeast mutants with chromosomal deletions in genes whose products are known to either affect (Δ*der*1, Δ*hrd*1) or not affect (Δ*yos*9, Δ*nup*120) RTA in vivo toxicity. In the case of ERAD components, Hrd1p and Der1p have already been demonstrated to be involved in ER-to-cytosol retrotranslocation of RTA in yeast, while siRNA mediated knock-down of the mammalian Der1p homologue, Derlin1, causes reduced sensitivity against ricin [5,35]. Thus, both mutants should show an increased fluorescence in the yeast reporter assay. As negative controls, deletion mutants Δ*yos*9 and Δ*nup*120 were included since neither a lack of the ER quality control lectin Yos9p nor a lack in the nuclear pore protein Nup20p affect RTA transport or in vivo toxicity [35]. As shown in Figure 2a, GFP fluorescence in RTA treated Δ*der*1 and Δ*hrd*1 cells was significantly increased to 82.1% ± 9.3% and 85.7% ± 6.9%, while fluorescence in Δ*yos*9 (55.7% ± 5%) and Δ*nup*120 (52% ± 3.2%) spheroplasts did not differ from wild-type cells. In a final validation, a threshold of significance (dotted line in Figure 2a,b) based on fluorescence emission in the positive controls was set to 75% (in comparison to the 52.8% signal for the average fluorescence of RTA-treated wild-type cells) to exclude false positive hits and to identify only those knock-out mutants whose defects have a strong impact on RTA trafficking.

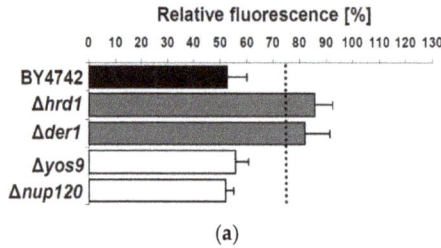

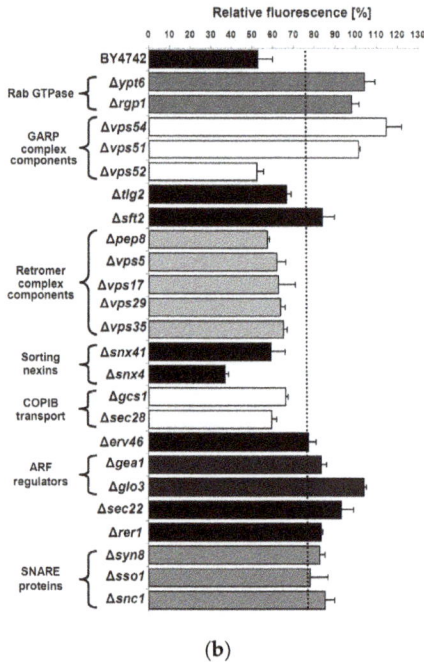

Figure 2. (a) Assay validation. Relative fluorescence of Δder1, Δhrd1, Δyos9 and Δnup120 knock-out mutant spheroplasts is indicated. Measurements were performed in the presence of RTA (5 µM) under induced culture conditions (3% galactose) over 20 h. Shown fluorescence values at 20 h were normalized to the fluorescence of spheroplasts treated with the negative control sample; (b) Impact of the indicated yeast deletion mutants on RTA trafficking. Fluorescence values of 2×10^7 yeast spheroplasts of each mutant were displayed after 20 h induction (3% galactose) in the presence of RTA (5 µM). Values shown in (a,b) were normalized to the fluorescence of negative control spheroplasts, mean values and standard deviations are indicated. All measurements were at least repeated 3 to 12 times as independent experiments without technical replicates. The dotted line in (a) and (b) indicates the chosen threshold of 75% for a positive hit.

2.3. Cellular Pathways Involved in Endosome-to-Golgi Transport of RTA

Having confirmed the general suitability of the reporter assay, we first focused on cellular components such as Ypt6p, Rgp1p and Ric1p, which are known to play a central role in retrograde protein transport from endosomes to the TGN [36,37]. In particular, Ypt6p is involved in the fusion of endosomal vesicles with the TGN, during which its GTPase activity is strictly regulated by two proteins, Rgp1p and Ric1p, acting in a complex as nucleotide exchange factor of Ypt6p [38]. As illustrated in Figure 2b, fluorescence of Δypt6 (104.1% ± 5.0%) and Δrgp1 cells (98.1% ± 3.5%) significantly exceeded the defined threshold and resulted in values that closely reflected reporter fluorescence in the negative controls (100.0% ± 3.2%), indicating that Ypt6p and its regulator Rgp1p are involved in RTA transport in yeast. This result nicely matches reports in mammalian cells in which ricin transport from endosomes to the Golgi has also been shown to be regulated by Rab6A and Rab6A′, the human homologues of yeast Ypt6p [5,7].

Since in yeast, Tlg2p and GARP complex components such as Vps51p, Vps52p, Vps53p and Vps54p are known to mediate vesicle docking at the TGN [28], we asked if a loss of any of these components affects RTA transport. Furthermore, Stx16—the mammalian homologue of Tlg2p—has been described to interact with the GARP-complex [39], and Stx18 in conjunction with GARP is required for intracellular ricin trafficking in mammalian cells [5]. As illustrated in Figure 2b, spheroplasts of both a Δ*vps*54 (114.5% ± 7.6%) and a Δ*vps*51 (101.3% ± 0.8%) knock-out showed a significant increase in GFP fluorescence compared to wild-type, while the corresponding values in Δ*vps*52 (52.5% ± 3.1%) and Δ*tlg*2 (66.9% ± 2.2%) cells were clearly below the threshold, suggesting that not all GARP complex components are required for RTA transport in yeast. It is widely accepted that GARP-complex stability depends on each subunit and mutations in any component lead to strong missorting and mislocation defects [40]. Unexpectedly, similar results were obtained in mammalian cells in which a siRNA mediated knock-down of the GARP-complex components Vps54 and Vps53, in contrast to Vps52, resulted in a significant decrease in ricin sensitivity [5]. Although the unaffected ricin sensitivity seen after Vps52 knock-down had been interpreted as a false negative result, our analogous observation in yeast indicates that the so far postulated model for GARP-complex assembly and function in the absence of Vps52 might indeed be correct, at least with respect to intracellular RTA transport. Furthermore, and in contrast to the documented role of Syn16 in ricin transport in mammalian cells [5,41], the present result on its homologue in yeast indicates Tlg2p—in contrast to Ypt6p—dispensable for endosome-to-Golgi transport of RTA.

As a further candidate, the impact of the membrane protein Sft2p on RTA transport was studied. In yeast, it has been postulated that Sft2p facilitates fusion of endosome-derived vesicles with the late Golgi. Furthermore, Sft2p genetically interacts with Sed5p, the yeast homologue of mammalian syntaxin 5 (Stx5), and Sft2p overexpression can suppress some *sed5* alleles [42]. Interestingly, mice treated with retro-2 to selectively prevent fusion of Stx5 containing vesicles with the TGN were shown to survive a lethal dose of ricin [6]. We now show that a yeast Δ*sft*2 knock-out shows a strong increase in GFP fluorescence (84.1% ± 5.8%) (Figure 2b), indicating that Sft2p is likewise required for efficient retrograde RTA transport in yeast. As the retromer complex is another essential component with a proposed function in retrograde endosome-to-Golgi transport [43], we tested the corresponding yeast proteins Vps5p, Vps17p, Vps35p, Pep8p/Vps26p and Vps29p [44] for their involvement in RTA trafficking. Although siRNA knock-down of several retromer complex components did not affect ricin sensitivity of mammalian cells [5], a yeast Δ*snx*4 knock-out resulted in a dramatic decrease in GFP fluorescence (37% ± 1.8%), while neither Δ*pep*8, Δ*vps*5, Δ*vps*17, Δ*vps*29 and Δ*vps*35 nor Δ*snx*41 cells significantly differed from wild-type after RTA-treatment (Figure 2b). Both Snx4p and Snx41p, forming a complex in vivo, belong to the family of sorting nexins which, in conjunction with the retromer complex, mediate various retrieval pathways from endosomes to the TGN [45]. Since Snx4p can form a complex with Snx42p/Atg20p (which was not tested here) that also plays a role in endosomal sorting [45], we cannot exclude that the observed absence of a protective effect against RTA in a Δ*snx*4 background is due to redundancy. Nevertheless, our data clearly indicate that the tested sorting nexins and retromer complex components are not important for intracellular RTA transport in yeast (Figure 2b), thus nicely matching similar observations in mammalian cells [5]. To analyze proteins involved in COPIB-mediated intra-Golgi transport [46], cells of a Δ*gcs*1 and a Δ*sec*28 knock-out mutant were tested. While Gcs1p is a regulatory protein in COPIB-transport, Sec28p is a structural COPIB component responsible for complex stability [47,48]. As shown in Figure 2b, reporter fluorescence was not significantly affected in either mutant (Δ*gcs*1, 66.3% ± 1.2%; Δ*sec*28, 59.6% ± 2.7%), suggesting that COPIB-vesicles are not involved in RTA-transport in yeast.

2.4. Impact of Golgi-to-ER Transport and/or Endocytosis Components on RTA Toxicity

Since recent studies in mammalian cells indicated that ERGIC2 (ER/Golgi intermediate compartment) is an important regulator in Golgi-to-ER trafficking of ricin and siRNA-mediated knock-down of ERGIC2 renders cells resistant against high doses of ricin [5], we tested various yeast mutants defective in Golgi-to-ER transport for potential effects on RTA in vivo activity. The yeast homologues of ERGIC2 (Erv41p) and ERGIC3 (Erv46p) form an active complex cycling between the ER and Golgi, which is important for membrane fusion in ER/Golgi transport [49,50]. When analyzed in the yeast assay (Figure 2b), cells of a Δerv46 knock-out mutant caused a significant increase (77.5% ± 3.8%) in fluorescence compared to wild-type. Since Erv46p is only functional when present in a complex with Erv41p, the result strongly indicates that the Erv41p/Erv46p complex participates in RTA trafficking and that Golgi-to-ER transport of RTA in yeast shows a striking similarity to ricin transport in mammalian cells. However, as it was recently reported that a direct expression of RTA in the yeast ER lumen requires a cycling between the ER and the Golgi as a pre-requisite for RTA dislocation from the ER to the cytosol [9,35], it cannot be excluded that yeast mutants defective in Golgi-to-ER transport might likewise disrupt the recycling of ER-localized RTA rather than its initial transport to the ER. Since recent studies in yeast also demonstrated that the Erv41p/Erv46p complex serves as retrograde receptor for the retrieval of non-HDEL-bearing ER residents [51], it is conceivable that RTA utilizes this retrieval pathway for retrograde transport from the Golgi to the ER.

To include central regulators of Arf GTPases in the present analysis, Glo3p and Gea1p were likewise examined. While Glo3p is an ADP-ribosylation factor and GTPase activating protein (ArfGAP), which regulates Golgi-ER transport, Gea1p represents a guanine nucleotide exchange factor for ADP-ribosylation factors (ARFs) [52,53]. Nucleotide exchange on ARFs is mediated by Gea1p and this mechanism is essential for in vivo Golgi-to-ER transport [54]. As shown in Figure 2b, spheroplasts of Δgea1 (83.5% ± 2.7%) and Δglo3 (104.1% ± 1.2%) mutants showed fluorescence values exceeding the threshold of 75% after RTA-treatment, indicating that both proteins are important for RTA trafficking from the Golgi to the ER. Considering that Glo3p and Gea1p are likewise known to be equally important for the regulation of Arf GTPases in yeast, we hypothesize that Arf GTPases might also be involved in retrograde RTA transport.

We next analyzed Sec22p and Rer1p for a potential role in RTA trafficking; Rer1p acts as a retrieval receptor in returning membrane proteins to the ER [55]. In contrast, Sec22p is an R-SNARE present in a complex with Bet1p, Bos1p and Sed5p [56] that constantly cycles between the Golgi and the ER and is responsible for both anterograde and retrograde transport [57]. Since the mammalian homologue of Sec22p, Sec22B, is also important for ricin toxicity [5], we tested the corresponding yeast knock-out mutants and thereby identified a strong increase in reporter fluorescence in Δsec22 (93.2% ± 5.8%) and Δrer1 (83.7% ± 0.6%) cells, indicating that Golgi-to-ER transport of RTA depends on the presence of both Sec22p and Rer1p. In support of the proposed function of Ypt6p in the retrograde transport and recycling of Sec22p from the Golgi to the ER [36], we now demonstrate that yeast cells lacking Ypt6p show an RTA-resistant phenotype (Figure 2b), indicating that Ypt6p as Rab GTPase might be involved in regulating endosome-to-TGN as well as TGN-to-ER transport. Furthermore, the data obtained here for Sec22p underlines the similarity in Golgi-to-ER transport of RTA in yeast and ricin trafficking in mammalian cells. For host cell intoxication by the RTA/RTB holotoxin, it has been proposed that RTB binds to KDEL-bearing proteins in the Golgi and thereby hijacks ER residents for retrograde transport to the ER [19,20,58]. Based on the data presented here, we assume the existence of several alternative pathways, including Rer1p, Sec22p and Arf proteins, to ensure efficient RTA transport from the Golgi to the ER. Although Rer1p and the Arf regulators Glo3p and Gea1p have, to our knowledge, not yet been described as important factors in ricin trafficking in mammalian cells, they might be promising novel candidates for ricin transport in higher eukaryotic cells.

To extend the analysis to gene products that potentially affect endocytotic RTA uptake from the plasma membrane, proteins such as Syn8p, Sso1p and Snc1p were selected as candidates for further analysis [59–62]. As illustrated in Figure 2b, a significant increase in fluorescence in the respective knock-out mutants strongly indicated that Syn8p (82.8% ± 2.6%), Sso1p (78.2% ± 8.6%) and Snc1p (85.4% ± 4.9%) have an impact on RTA trafficking. It is generally accepted that the v-SNARE Snc1p together with Tgl2p is required for both, secretory vesicle trafficking to the plasma membrane and retrograde vesicle transport from early endosomes to the TGN [59,63]. However, in the case of RTA transport, the results obtained here for Δ*snx*4 and Δ*snx*41 mutants do not argue for an involvement of Snc1p in endosome-to-Golgi transport as no increase in GFP fluorescence was observed after RTA-treatment in either mutant and both sorting nexins have been described to be involved in retrieving Snc1p from endosomes to the Golgi [45]. In addition, the significant impact observed in the Δ*sso*1 mutant fosters the assumption that Snc1p might indeed be involved in the endocytosis of RTA. It is known that v-SNAREs such as Snc1p confer the docking and fusion of two classes of secretory vesicles by forming a functional SNARE complex with the plasma membrane t-SNAREs Sso1, Sso2, and Sec9 [61,62,64,65]. Thus, yeast cells lacking Sso1p should be blocked in the fusion of Snc1p containing secretory vesicles which, in turn, would negatively affect efficient RTA endocytosis. At least for the depletion of Syn8p, an increased resistance against external applied RTA was observed. Syn8p forms a complex with Snc1p and it is proposed that both Syn8p and Snc1p play a role in plasma membrane-to-endosome transport [60]. It is thus conceivable that a knock-out of *SYN8* disturbs efficient retrograde RTA transport to early endosomes. However, Sso1p, which shows similarities to mammalian Syn1A and Syn1B, as well as the VAMP3 homologue Snc1p are important for RTA trafficking in yeast [66,67]. Furthermore, the yeast homologue of mammalian Syn8, Syn8p, is also part of this transport step. Whether these proteins are also involved in ricin endocytosis in mammalian cells is unknown.

Based on the proteins identified in the present study, we propose a refined model of intracellular RTA transport (Figure 3), which shows striking similarities to ricin holotoxin trafficking in mammalian cells, suggesting yeast as an attractive and powerful model to dissect single steps and pathways of host cell intoxication. We could demonstrate that retrograde transport pathways, including Sft2p and various GARP complex components, regulate intracellular toxin trafficking, while the retromer complex and COPIB vesicles are not part of the transport machinery. Proteins such as Sso1p, Snc1p, Rer1p, Sec22p, Erv46p, Gea1p and Glo3p were identified as novel components of retrograde RTA transport and, therefore, might represent promising novel candidates for future analyses in mammalian cells. Based on the reporter assay developed here, we intend to perform a genome-wide screen to identify additional novel gene products that are required for RTA uptake and subsequent transport through a target cell.

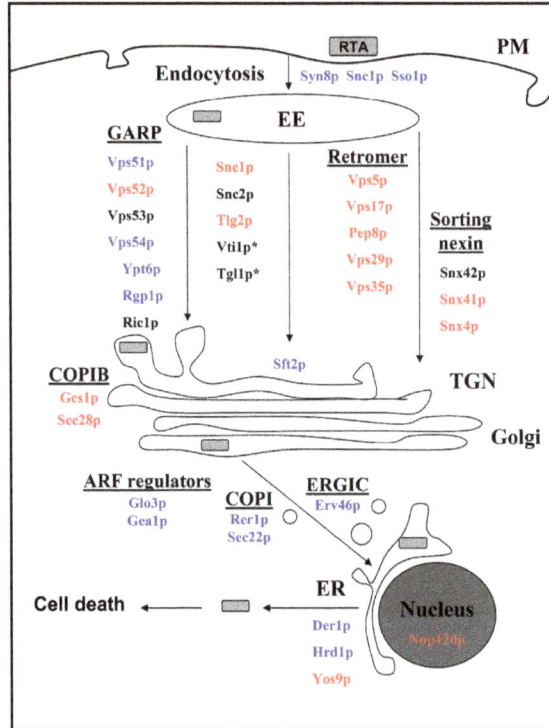

Figure 3. Overview of cellular components involved in RTA uptake and intracellular trafficking in yeast. Proteins involved in RTA transport are shown in blue, non-involved proteins in red; proteins marked with an asterisk (*) are essential for cell viability in yeast; proteins shown in black were not tested. EE, early endosome; PM, plasma membrane; TGN, trans-Golgi network; ER, endoplasmic reticulum; ERGIC, ER/Golgi intermediate compartment; COPI, coat protein I vesicle.

3. Materials and Methods

3.1. Escherichia coli *Strains, Plasmids, Culture Media and Genetic Techniques*

Standard molecular manipulations were performed as described [68]. *E. coli* TOP10 (*F'mcrA* Δ (*mrr-hsdRMS-mcrBC*) Φ80*lac-Z*ΔM15 Δ*lacX*74 *recA*1 *araD*139 Δ (*ara-leu*) 7697 *galU galK rpsL* (*StrR*) *endA*1 *nupG*) was used for cloning. Construction of the RTA expression plasmid, pET-RTA$_{\text{His}}$, was previously described [17]. Vectors pPGK-M28I [32] and pUG35 [69] were used as template to amplify K28$_{\text{SP}}$-GFP by SOE-PCR [70] and to obtain the final reporter plasmid pRS315-K28$_{\text{SP}}$-GFP. In the first amplification, the sequences of the K28 pptox signal peptide and yeast-enhanced GFP were amplified using HiFi polymerase (Roche) and the following primer pairs: 5'-K28$_{\text{SS}}$ (5'-CTCGAGGGATTCATGGACTTCAGTGCTGCTACTTGCGTA-CTGATG) and 3'-K28$_{\text{SS}}$-GFP-SOE (3'-CCTCGCCCTTGCTCACACCCCGTGCATATTTGAGATT) for K28$_{\text{SS}}$ as well as 5'-K28$_{\text{SS}}$-GFP-SOE (5'-AATCTCAAATATGCACGGGGTGTGAGCAAGGGCGAGG) and 3'-GFP (3'-AGATCTAAGCTTGTCGACTTACT TGTACAGCTCGTCCAT) for GFP. In a second step, both fragments were amplified by PCR in the presence of the 5'-K28$_{\text{SS}}$ and 3'-GFP primers. The primer introduced a 5' *Eco*RI and 3' *Sal*I cleavage site, respectively (underlined). After amplification, the gene fusion was subcloned into pYES2.1/V5-His-TOPO (Invitrogen) and routinely sequenced. The final fusion was cloned as *Eco*RI/*Sal*I fragment under transcriptional control of the *GAL1* promoter in a centromeric yeast expression vector derived from pRS315 [71] to obtain the expression vector pRS315-K28$_{SP}$-GFP.

3.2. Affinity Purification of RTA

Purification of RTA was performed as previously described [17]. After purification of (His)$_6$-tagged RTA by Ni^{2+}/NTA chromatography, eluted protein fractions were desalted and equilibrated in 0.8 M sorbitol. An equally purified cell lysate of *E. coli* expressing the empty vector pET24a$^{(+)}$ without RTA served as negative control. After concentration through 10 kDa cut-off spin columns (Sartorius, Viva Spin 20, Göttingen, Germany), purified proteins were filter sterilized and stored at 4 °C. Total protein content was determined using a BCA protein assay kit (Pierce, Waltham, MA, USA).

3.3. Yeast Strains, Transformation and Culture Media

The *S. cerevisiae* wild-type strain BY4742 (MATα *his3Δ1*, *leu2Δ0*, *lys2Δ0*, *ura3Δ0*) and its isogenic knock-out mutants were obtained from Open Biosystems (Lafayette, CO, USA). Yeast cultures were grown at 30 °C, either in YEPD media or in synthetic media containing 2% glucose (SD medium) or 2% raffinose (SR medium). Synthetic medium lacking leucine (SDL or SRL) was used for transformed yeast cells. Yeast transformation with the expression plasmid pRS315-K28$_{SP}$-GFP was achieved by the lithium acetate method [34]. Positive clones were selected on SDL agar. For spheroplast preparation, yeast cells carrying pRS315-K28$_{SP}$-GFP were grown in SRL medium at 30 °C to late exponential phase, harvested at 8000 rpm and washed twice with sterile water. Subsequently, 5×10^8 cells were resuspended in 50 mL spheroplasting buffer (0.8 M sorbitol, 10 mM Tris-HCl (pH 7.5), 10 mM CaCl$_2$, 2 mM DTT and 200 µg/mL zymolyase 20 T), incubated at 30 °C for 90 min at 100 rpm, harvested at 4 °C and 2000 rpm, washed twice with 0.8 M sorbitol stabilized SRL medium, and finally resuspended in 5 mL stabilized SRL medium. Subsequently, spheroplasts were used in the GFP-fluorescence assay. To control preparation efficacy, 2×10^7 spheroplasts were centrifuged for 10 min at 2000 rpm after spheroplast preparation, resuspended in H$_2$O distilled, shaken for 30 s and plated out on SDL agar for three days at 30 °C. For data evaluation, only samples with efficiency higher than 98% (i.e., less than 2% of non-spheroplasted cells) were used.

3.4. GFP-Fluorescence Assay

Yeast cells carrying the expression vector pRS315-K28$_{SP}$-GFP were spheroplasted as described above. Resuspended spheroplasts (2×10^7 cells in 200 µL) were seeded in 96 microtiter plates (Nunc, Roskilde, Denmark). For induction of GFP-expression, 30 µL of 30% galactose solution and 70 µL stabilized SRL medium containing purified RTA or the negative control were added, yielding a final RTA concentration of 5 µM, corresponding to 160 µg/mL RTA. GFP-fluorescence (485 nm/527 nm) was measured every 10 min over 20 h. Each experiment was performed at least three times ($n = 3$) at 30 °C, 120 rpm and a shaking diameter of 1 mm. Measurements were carried out in a fluorescence reader equipped with an integrated shaker (Fluoroskan Ascent, Labsystems, Vantaa, Finland). After 20 h incubation, relative GFP fluorescence in % was calculated for the 20 h time point according to the following equation:

$$\text{fluorescence} = \frac{\left(\text{fluorescence}_{\text{sample(20h)}} - \text{fluorescence}_{\text{sample(0h)}} \right)}{\left(\text{fluorescence}_{\text{negativecontrol(20h)}} - \text{fluorescence}_{\text{negativecontrol(0h)}} \right)}$$

Acknowledgments: This study was kindly supported by a grant from the Deutsche Forschungsgemeinschaft (SFB 1027, A6).

Author Contributions: B.B. and M.J.S. conceived and designed the experiments; B.B. and T.S. performed the experiments; B.B. analyzed the data; B.B. and M.J.S. wrote the paper.

Conflicts of Interest: The authors declare no conflict of interest.

References

1. Wei, G.Q.; Liu, R.S.; Wang, Q.; Liu, W.Y. Toxicity of two type II ribosome-inactivating proteins (cinnamomin and ricin) to domestic silkworm larvae. *Arch. Insect Biochem. Physiol.* **2004**, *57*, 160–165. [CrossRef] [PubMed]
2. Hartley, M.R.; Lord, J.M. Cytotoxic ribosome-inactivating lectins from plants. *Biochim. Biophys. Acta* **2004**, *1701*, 1–14. [CrossRef] [PubMed]
3. Holmberg, L.; Nygard, O. Depurination of A4256 in 28 S rRNA by the ribosome-inactivating proteins from barley and ricin results in different ribosome conformations. *J. Mol. Biol.* **1996**, *259*, 81–94. [CrossRef] [PubMed]
4. Kvalvaag, A.S.; Pust, S.; Sandvig, K. Vps11, a subunit of the tethering complexes HOPS and CORVET, is involved in regulation of glycolipid degradation and retrograde toxin transport. *Commun. Integr. Biol.* **2014**, *7*. [CrossRef] [PubMed]
5. Moreau, D.; Kumar, P.; Wang, S.C.; Chaumet, A.; Chew, S.Y.; Chevalley, H.; Bard, F. Genome-wide RNAi screens identify genes required for Ricin and PE intoxications. *Dev. Cell* **2011**, *21*, 231–244. [CrossRef] [PubMed]
6. Stechmann, B.; Bai, S.K.; Gobbo, E.; Lopez, R.; Merer, G.; Pinchard, S.; Panigai, L.; Tenza, D.; Raposo, G.; Beaumelle, B.; et al. Inhibition of retrograde transport protects mice from lethal ricin challenge. *Cell* **2010**, *141*, 231–242. [CrossRef] [PubMed]
7. Utskarpen, A.; Slagsvold, H.H.; Iversen, T.G.; Walchli, S.; Sandvig, K. Transport of ricin from endosomes to the Golgi apparatus is regulated by rab6a and rab6a'. *Traffic* **2006**, *7*, 663–672. [CrossRef] [PubMed]
8. Lord, M.J.; Jolliffe, N.A.; Marsden, C.J.; Pateman, C.S.; Smith, D.C.; Spooner, R.A.; Watson, P.D.; Roberts, L.M. Ricin. Mechanisms of cytotoxicity. *Toxicol. Rev.* **2003**, *22*, 53–64. [CrossRef] [PubMed]
9. Spooner, R.A.; Lord, J.M. Ricin trafficking in cells. *Toxins* **2015**, *7*, 49–65. [CrossRef] [PubMed]
10. Moya, M.; Dautry-Varsat, A.; Goud, B.; Louvard, D.; Boquet, P. Inhibition of coated pit formation in Hep2 cells blocks the cytotoxicity of diphtheria toxin but not that of ricin toxin. *J. Cell Biol.* **1985**, *101*, 548–559. [CrossRef] [PubMed]
11. Iversen, T.G.; Skretting, G.; Llorente, A.; Nicoziani, P.; van Deurs, B.; Sandvig, K. Endosome to Golgi transport of ricin is independent of clathrin and of the Rab9- and Rab11-GTPases. *Mol. Biol. Cell* **2001**, *12*, 2099–2107. [CrossRef] [PubMed]
12. Sandvig, K.; Olsnes, S. Entry of the toxic proteins abrin, modeccin, ricin, and diphtheria toxin into cells. II. Effect of pH, metabolic inhibitors, and ionophores and evidence for toxin penetration from endocytotic vesicles. *J. Biol. Chem.* **1982**, *257*, 7504–7513. [PubMed]
13. Moisenovich, M.; Tonevitsky, A.; Maljuchenko, N.; Kozlovskaya, N.; Agapov, I.; Volknandt, W.; Bereiter-Hahn, J. Endosomal ricin transport: Involvement of Rab4- and Rab5-positive compartments. *Histochem. Cell Biol.* **2004**, *121*, 429–439. [CrossRef] [PubMed]
14. Sandvig, K.; Bergan, J.; Kavaliauskiene, S.; Skotland, T. Lipid requirements for entry of protein toxins into cells. *Prog. Lipid Res.* **2014**, *54*, 1–13. [CrossRef] [PubMed]
15. Simpson, J.C.; Roberts, L.M.; Lord, J.M. Free ricin a chain reaches an early compartment of the secretory pathway before it enters the cytosol. *Exp. Cell Res.* **1996**, *229*, 447–451. [CrossRef] [PubMed]
16. Wales, R.; Roberts, L.M.; Lord, J.M. Addition of an endoplasmic reticulum retrieval sequence to ricin A chain significantly increases its cytotoxicity to mammalian cells. *J. Biol. Chem.* **1993**, *268*, 23986–23990. [PubMed]
17. Becker, B.; Schmitt, M.J. Adapting yeast as model to study ricin toxin a uptake and trafficking. *Toxins* **2011**, *3*, 834–847. [CrossRef] [PubMed]
18. Van Deurs, B.; Sandvig, K.; Petersen, O.W.; Olsnes, S.; Simons, K.; Griffiths, G. Estimation of the amount of internalized ricin that reaches the trans-Golgi network. *J. Cell Biol.* **1988**, *106*, 253–267. [CrossRef] [PubMed]
19. Lord, J.M.; Roberts, L.M.; Robertus, J.D. Ricin: Structure, mode of action, and some current applications. *FASEB J.* **1994**, *8*, 201–208. [CrossRef] [PubMed]
20. Day, P.J.; Owens, S.R.; Wesche, J.; Olsnes, S.; Roberts, L.M.; Lord, J.M. An interaction between ricin and calreticulin that may have implications for toxin trafficking. *J. Biol. Chem.* **2001**, *276*, 7202–7208. [CrossRef] [PubMed]

21. Slominska-Wojewodzka, M.; Gregers, T.F.; Walchli, S.; Sandvig, K. EDEM is involved in retrotranslocation of ricin from the endoplasmic reticulum to the cytosol. *Mol. Biol. Cell* **2006**, *17*, 1664–1675. [CrossRef] [PubMed]

22. Slominska-Wojewodzka, M.; Pawlik, A.; Sokolowska, I.; Antoniewicz, J.; Wegrzyn, G.; Sandvig, K. The role of EDEM2 compared with EDEM1 in ricin transport from the endoplasmic reticulum to the cytosol. *Biochem. J.* **2014**, *457*, 485–496. [CrossRef] [PubMed]

23. Simpson, J.C.; Roberts, L.M.; Romisch, K.; Davey, J.; Wolf, D.H.; Lord, J.M. Ricin a chain utilises the endoplasmic reticulum-associated protein degradation pathway to enter the cytosol of yeast. *FEBS Lett.* **1999**, *459*, 80–84. [CrossRef]

24. Herrera, C.; Klokk, T.I.; Cole, R.; Sandvig, K.; Mantis, N.J. A bispecific antibody promotes aggregation of ricin toxin on cell surfaces and alters dynamics of toxin internalization and trafficking. *PLoS ONE* **2016**, *11*. [CrossRef] [PubMed]

25. Yermakova, A.; Klokk, T.I.; O'Hara, J.M.; Cole, R.; Sandvig, K.; Mantis, N.J. Neutralizing monoclonal antibodies against disparate epitopes on ricin toxin's enzymatic subunit interfere with intracellular toxin transport. *Sci. Rep.* **2016**, *6*. [CrossRef] [PubMed]

26. Allen, S.C.; Moore, K.A.; Marsden, C.J.; Fulop, V.; Moffat, K.G.; Lord, J.M.; Ladds, G.; Roberts, L.M. The isolation and characterization of temperature-dependent ricin A chain molecules in *Saccharomyces cerevisiae*. *FEBS J.* **2007**, *274*, 5586–5599. [CrossRef] [PubMed]

27. Yan, Q.; Li, X.P.; Tumer, N.E. Wild type RTA and less toxic variants have distinct requirements for Png1 for their depurination activity and toxicity in *Saccharomyces cerevisiae*. *PLoS ONE* **2015**, *9*, e113719. [CrossRef] [PubMed]

28. Bonifacino, J.S.; Rojas, R. Retrograde transport from endosomes to the trans-Golgi network. *Nat. Rev. Mol. Cell Biol.* **2006**, *7*, 568–579. [CrossRef] [PubMed]

29. Lee, M.C.; Miller, E.A.; Goldberg, J.; Orci, L.; Schekman, R. Bi-directional protein transport between the ER and Golgi. *Annu. Rev. Cell Dev. Biol.* **2004**, *20*, 87–123. [CrossRef] [PubMed]

30. Duden, R. ER-to-Golgi transport: Cop I and Cop II function (review). *Mol. Membr. Biol.* **2003**, *20*, 197–207. [CrossRef] [PubMed]

31. Seaman, M.N. Endosome protein sorting: Motifs and machinery. *Cell. Mol. Life Sci.* **2008**, *65*, 2842–2858. [CrossRef] [PubMed]

32. Schmitt, M.J. Cloning and expression of a cDNA copy of the viral K28 killer toxin gene in yeast. *Mol. Gen. Genet.* **1995**, *246*, 236–246. [CrossRef] [PubMed]

33. Endo, Y.; Tsurugi, K. RNA N-glycosidase activity of ricin A-chain. Mechanism of action of the toxic lectin ricin on eukaryotic ribosomes. *J. Biol. Chem.* **1987**, *262*, 8128–8130. [PubMed]

34. Eiden-Plach, A.; Zagorc, T.; Heintel, T.; Carius, Y.; Breinig, F.; Schmitt, M.J. Viral preprotoxin signal sequence allows efficient secretion of green fluorescent protein by *Candida glabrata*, *Pichia pastoris*, *Saccharomyces cerevisiae*, and *Schizosaccharomyces pombe*. *Appl. Environ. Microbiol.* **2004**, *70*, 961–966. [CrossRef] [PubMed]

35. Li, S.; Spooner, R.A.; Allen, S.C.; Guise, C.P.; Ladds, G.; Schnoder, T.; Schmitt, M.J.; Lord, J.M.; Roberts, L.M. Folding-competent and folding-defective forms of ricin A chain have different fates after retrotranslocation from the endoplasmic reticulum. *Mol. Biol. Cell* **2010**, *21*, 2543–2554. [CrossRef] [PubMed]

36. Luo, Z.; Gallwitz, D. Biochemical and genetic evidence for the involvement of yeast Ypt6-GTPase in protein retrieval to different Golgi compartments. *J. Biol. Chem.* **2003**, *278*, 791–799. [CrossRef] [PubMed]

37. Li, B.; Warner, J.R. Mutation of the rab6 homologue of *Saccharomyces cerevisiae*, Ypt6, inhibits both early Golgi function and ribosome biosynthesis. *J. Biol. Chem.* **1996**, *271*, 16813–16819. [PubMed]

38. Siniossoglou, S.; Peak-Chew, S.Y.; Pelham, H.R. Ric1p and Rgp1p form a complex that catalyses nucleotide exchange on Ypt6p. *EMBO J.* **2000**, *19*, 4885–4894. [CrossRef] [PubMed]

39. Perez-Victoria, F.J.; Bonifacino, J.S. Dual roles of the mammalian GARP complex in tethering and SNARE complex assembly at the trans-Golgi network. *Mol. Cell. Biol.* **2009**, *29*, 5251–5263. [CrossRef] [PubMed]

40. Conibear, E.; Stevens, T.H. Vps52p, Vps53p, and Vps54p form a novel multisubunit complex required for protein sorting at the yeast late Golgi. *Mol. Biol. Cell* **2000**, *11*, 305–323. [CrossRef] [PubMed]

41. Amessou, M.; Fradagrada, A.; Falguieres, T.; Lord, J.M.; Smith, D.C.; Roberts, L.M.; Lamaze, C.; Johannes, L. Syntaxin 16 and syntaxin 5 are required for efficient retrograde transport of several exogenous and endogenous cargo proteins. *J. Cell Sci.* **2007**, *120*, 1457–1468. [CrossRef] [PubMed]

42. Conchon, S.; Cao, X.; Barlowe, C.; Pelham, H.R. Got1p and Sft2p: Membrane proteins involved in traffic to the Golgi complex. *EMBO J.* **1999**, *18*, 3934–3946. [CrossRef] [PubMed]

43. Johannes, L.; Popoff, V. Tracing the retrograde route in protein trafficking. *Cell* **2008**, *135*, 1175–1187. [CrossRef] [PubMed]

44. Seaman, M.N.; McCaffery, J.M.; Emr, S.D. A membrane coat complex essential for endosome-to-Golgi retrograde transport in yeast. *J. Cell Biol.* **1998**, *142*, 665–681. [CrossRef] [PubMed]

45. Hettema, E.H.; Lewis, M.J.; Black, M.W.; Pelham, H.R. Retromer and the sorting nexins Snx4/41/42 mediate distinct retrieval pathways from yeast endosomes. *EMBO J.* **2003**, *22*, 548–557. [CrossRef] [PubMed]

46. Day, K.J.; Staehelin, L.A.; Glick, B.S. A three-stage model of Golgi structure and function. *Histochem. Cell Biol.* **2013**, *140*, 239–249. [CrossRef] [PubMed]

47. Robinson, M.; Poon, P.P.; Schindler, C.; Murray, L.E.; Kama, R.; Gabriely, G.; Singer, R.A.; Spang, A.; Johnston, G.C.; Gerst, J.E. The Gcs1 Arf-GAP mediates Snc1,2 v-SNARE retrieval to the Golgi in yeast. *Mol. Biol. Cell* **2006**, *17*, 1845–1858. [CrossRef] [PubMed]

48. Eugster, A.; Frigerio, G.; Dale, M.; Duden, R. COP I domains required for coatomer integrity, and novel interactions with ARF and ARF-GAP. *EMBO J.* **2000**, *19*, 3905–3917. [CrossRef] [PubMed]

49. Otte, S.; Belden, W.J.; Heidtman, M.; Liu, J.; Jensen, O.N.; Barlowe, C. Erv41p and Erv46p: New components of COPII vesicles involved in transport between the ER and Golgi complex. *J. Cell Biol.* **2001**, *152*, 503–518. [CrossRef] [PubMed]

50. Orci, L.; Ravazzola, M.; Mack, G.J.; Barlowe, C.; Otte, S. Mammalian Erv46 localizes to the endoplasmic reticulum-Golgi intermediate compartment and to cis-Golgi cisternae. *Proc. Natl. Acad. Sci. USA* **2003**, *100*, 4586–4591. [CrossRef] [PubMed]

51. Shibuya, A.; Margulis, N.; Christiano, R.; Walther, T.C.; Barlowe, C. The Erv41-Erv46 complex serves as a retrograde receptor to retrieve escaped ER proteins. *J. Cell Biol.* **2015**, *208*, 197–209. [CrossRef] [PubMed]

52. Lewis, S.M.; Poon, P.P.; Singer, R.A.; Johnston, G.C.; Spang, A. The ArfGAP Glo3 is required for the generation of COPI vesicles. *Mol. Biol. Cell* **2004**, *15*, 4064–4072. [CrossRef] [PubMed]

53. Spang, A.; Herrmann, J.M.; Hamamoto, S.; Schekman, R. The ADP ribosylation factor-nucleotide exchange factors Gea1p and Gea2p have overlapping, but not redundant functions in retrograde transport from the Golgi to the endoplasmic reticulum. *Mol. Biol. Cell* **2001**, *12*, 1035–1045. [CrossRef] [PubMed]

54. Peyroche, A.; Paris, S.; Jackson, C.L. Nucleotide exchange on ARF mediated by yeast Gea1 protein. *Nature* **1996**, *384*, 479–481. [CrossRef] [PubMed]

55. Sato, K.; Sato, M.; Nakano, A. Rer1p, a retrieval receptor for endoplasmic reticulum membrane proteins, is dynamically localized to the Golgi apparatus by coatomer. *J. Cell Biol.* **2001**, *152*, 935–944. [CrossRef] [PubMed]

56. Liu, Y.; Flanagan, J.J.; Barlowe, C. Sec22p export from the endoplasmic reticulum is independent of snare pairing. *J. Biol. Chem.* **2004**, *279*, 27225–27232. [CrossRef] [PubMed]

57. Liu, Y.; Barlowe, C. Analysis of Sec22p in endoplasmic reticulum/Golgi transport reveals cellular redundancy in SNARE protein function. *Mol. Biol. Cell* **2002**, *13*, 3314–3324. [CrossRef] [PubMed]

58. Sandvig, K.; Torgersen, M.L.; Engedal, N.; Skotland, T.; Iversen, T.G. Protein toxins from plants and bacteria: Probes for intracellular transport and tools in medicine. *FEBS Lett.* **2010**, *584*, 2626–2634. [CrossRef] [PubMed]

59. Lewis, M.J.; Nichols, B.J.; Prescianotto-Baschong, C.; Riezman, H.; Pelham, H.R. Specific retrieval of the exocytic snare Snc1p from early yeast endosomes. *Mol. Biol. Cell* **2000**, *11*, 23–38. [CrossRef] [PubMed]

60. Lewis, M.J.; Pelham, H.R. A new yeast endosomal SNARE related to mammalian syntaxin 8. *Traffic* **2002**, *3*, 922–929. [CrossRef] [PubMed]

61. Aalto, M.K.; Ronne, H.; Keranen, S. Yeast syntaxins Sso1p and Sso2p belong to a family of related membrane proteins that function in vesicular transport. *EMBO J.* **1993**, *12*, 4095–4104. [PubMed]

62. Gurunathan, S.; Chapman-Shimshoni, D.; Trajkovic, S.; Gerst, J.E. Yeast exocytic v-SNAREs confer endocytosis. *Mol. Biol. Cell* **2000**, *11*, 3629–3643. [CrossRef] [PubMed]

63. Protopopov, V.; Govindan, B.; Novick, P.; Gerst, J.E. Homologs of the synaptobrevin/VAMP family of synaptic vesicle proteins function on the late secretory pathway in *S. cerevisiae*. *Cell* **1993**, *74*, 855–861. [CrossRef]

64. David, D.; Sundarababu, S.; Gerst, J.E. Involvement of long chain fatty acid elongation in the trafficking of secretory vesicles in yeast. *J. Cell Biol.* **1998**, *143*, 1167–1182. [CrossRef] [PubMed]

65. Rossi, G.; Salminen, A.; Rice, L.M.; Brunger, A.T.; Brennwald, P. Analysis of a yeast snare complex reveals remarkable similarity to the neuronal SNARE complex and a novel function for the C terminus of the SNAP-25 homolog, Sec9. *J. Biol. Chem.* **1997**, *272*, 16610–16617. [CrossRef] [PubMed]

66. Bennett, M.K.; Garcia-Arraras, J.E.; Elferink, L.A.; Peterson, K.; Fleming, A.M.; Hazuka, C.D.; Scheller, R.H. The syntaxin family of vesicular transport receptors. *Cell* **1993**, *74*, 863–873. [CrossRef]

67. Gerst, J.E.; Rodgers, L.; Riggs, M.; Wigler, M. SNC1, a yeast homolog of the synaptic vesicle-associated membrane protein/synaptobrevin gene family: Genetic interactions with the RAS and CAP genes. *Proc. Natl. Acad. Sci. USA* **1992**, *89*, 4338–4342. [CrossRef] [PubMed]

68. Sambrook, J.; Maniatis, T.; Fritsch, E.F. *Molecular Cloning: A Laboratory Manual*, 2nd ed.; Cold Spring Harbor Laboratory Press: Cold Spring Harbor, NY, USA, 1989.

69. Cormack, B.P.; Bertram, G.; Egerton, M.; Gow, N.A.; Falkow, S.; Brown, A.J. Yeast-enhanced green fluorescent protein (yEGFP)a reporter of gene expression in *Candida albicans*. *Microbiology* **1997**, *143 Pt 2*, 303–311. [CrossRef] [PubMed]

70. Horton, R.M.; Hunt, H.D.; Ho, S.N.; Pullen, J.K.; Pease, L.R. Engineering hybrid genes without the use of restriction enzymes: Gene splicing by overlap extension. *Gene* **1989**, *77*, 61–68. [CrossRef]

71. Sikorski, R.S.; Hieter, P. A system of shuttle vectors and yeast host strains designed for efficient manipulation of DNA in *Saccharomyces cerevisiae*. *Genetics* **1989**, *122*, 19–27. [PubMed]

Article

Differences in Ribosome Binding and Sarcin/Ricin Loop Depurination by Shiga and Ricin Holotoxins

Xiao-Ping Li * and Nilgun E. Tumer

Department of Plant Biology, Rutgers, The State University of New Jersey, New Brunswick, NJ 08901, USA;
tumer@aesop.rutgers.edu
* Correspondence: xpli@aesop.rutgers.edu; Tel.: +1-848-932-6355

Academic Editors: Julien Barbier and Daniel Gillet
Received: 3 January 2017; Accepted: 5 April 2017; Published: 11 April 2017

Abstract: Both ricin and Shiga holotoxins display no ribosomal activity in their native forms and need to be activated to inhibit translation in a cell-free translation inhibition assay. This is because the ribosome binding site of the ricin A chain (RTA) is blocked by the B subunit in ricin holotoxin. However, it is not clear why Shiga toxin 1 (Stx1) or Shiga toxin 2 (Stx2) holotoxin is not active in a cell-free system. Here, we compare the ribosome binding and depurination activity of Stx1 and Stx2 holotoxins with the A1 subunits of Stx1 and Stx2 using either the ribosome or a 10-mer RNA mimic of the sarcin/ricin loop as substrates. Our results demonstrate that the active sites of Stx1 and Stx2 holotoxins are blocked by the A2 chain and the B subunit, while the ribosome binding sites are exposed to the solvent. Unlike ricin, which is enzymatically active, but cannot interact with the ribosome, Stx1 and Stx2 holotoxins are enzymatically inactive but can interact with the ribosome.

Keywords: Shiga toxin 1; Shiga toxin 2; Stx1; Stx2; ricin; ribosome binding; depurination activity

---

1. Introduction

Shiga-toxin producing _E. coli_ (STEC) infections can cause life threatening complications such as hemolytic uremic syndrome (HUS) or hemorrhagic colitis (HC), and are the leading cause of death from foodborne bacterial infections in children [1]. Shiga toxins (Stxs) are the primary virulence factors of STEC and belong to a group of proteins called type II ribosome inactivating proteins (RIPs). There are two main Stx types, designated Shiga toxin 1 (Stx1) and Shiga toxin 2 (Stx2), and within each are many subtypes. Stx1 and Stx2 contain a catalytically active A subunit and five copies of cell binding B subunits [2,3]. Stx1 and Stx2 share 55% and 57% amino sequence identity in the A and B subunits, respectively, and have similar molecular structures. The A subunits remove a universally conserved adenine from the sarcin/ricin loop (SRL) of the large ribosomal RNA and inhibit translation. The B subunits bind to a globotriaosylceramide (Gb3 or CD77) receptor on the cell surface and facilitate endocytosis. Ricin produced by the castor bean is another type II RIP. The A subunit of ricin (RTA) has similar enzymatic function and molecular structure to the A subunits of Stxs, while the B subunit of ricin (RTB) is functionally similar but structurally different and binds to different cellular receptors [2]. Residues important for the enzymatic activity of the A subunits are conserved among all RIPs [4]. Although Stx and ricin holotoxins are very toxic to eukaryotic cells, they are not active towards the ribosome and are activated after separation of the B subunits from the A subunits [5,6].

Stx1 and Stx2 are AB$_5$ toxins consisting of one A subunit and a pentamer of B subunits. In _E. coli_, the genes encoding Stx1 or Stx2 A (A1 + A2) and B subunits are linked, but they are encoded by separate ORFs [7]. In ricin, the RTA and RTB are encoded by a single gene and synthesized as preproricin with a signal peptide preceding the A chain and a linker peptide between RTA and RTB [8]. The signal sequence and the linker peptide are cleaved during the maturation of ricin, but RTA and RTB remain linked together by a disulfide bond [9,10]. Stx1, Stx2 or ricin traffic from the endosome to the trans-Golgi network in a retrograde manner and reach the endoplasmic reticulum (ER) [5,6]. During this process, A1 and A2 chains of Stxs are cleaved by furin between R251 and M252 in Stx1 and between R250 and A251 in Stx2 (Figure 1) [11]. The furin cleavage sites are located in protease-sensitive loops. After furin cleavage, the A1 subunit and A2 chain remain connected by a disulfide bond and the B subunit remains connected to the A1 subunit through the A2 chain. In the ER, the disulfide bond is reduced by protein disulfide isomerase, releasing the A1 subunit of Stxs from the B subunit, and RTA from RTB. The A1 subunits of Stxs and RTA unfold and retrotranslocate out of the ER and refold into an active form in the cytosol [5,6].

A chain of Stx

```
KEFTLDFSTAKTYVD SLNVIRSAIGTPLQTISS GGTSLLMIDSGTGDNLFAVDVRGI
DPEEGRFNNLRLIVE RNNLYVTGFVNRTNNVFY RFADFS HVTFPGTTAVTL SGDSSY
TTLQRVAGI SRTGMQ INRHSL TTSYLD LMSHSGTSLTQS VARAMLRFVTVTAEALRF
RQIQRGFRTTLDDLS GRSYVMTAEDVD LTLNWGRLSSVL PDYHGQDSVRVGRISFGS
                          S-S
INAILGSVALILNCHHHASRVARMASDEFPSMCPADGRVRGITHNKILWDS STLGAI
LMRRTISS        242        ↑            261
                        Furin
```

A chain of Stx2

```
REFTIDFSTQQSYVS SLNSIRTEISTPLEHISQGTTSVSVINHTPPGSYFAVDIRGLD
VYQARFDHLRLIIEQNNLYVAGFVNTATNTFYRFSDFTHISVPGVTTVSMTTDSSYTT
LQRVAALERSGMQIS RHSLVS SYLALMEFSGNTMTRDASRAVLRFVTVTAEALRFRQI
QREFRQALSETAPVYTMTPGDVDLTLNWGRISNVLPEYRGEDGVRVGRISFNNISAIL
                      S-S
GTVAVILNCHHQGARSVRAVNEDSQPECQITGD RPVIKINNTLWESNTAAAFLNRKSQ
FLYTTGK  241        ↑              260
                 Furin
```

Figure 1. The sequences of mature A chains of *Shigella* Stx and Stx2. Furin recognition sites are shown in blue. The sequence before the furin cleavage site corresponds to the A1 subunits and the sequence after the furin cleavage site corresponds to the A2 subunits.

Stx2-producing *E. coli* are more often associated with HUS than Stx1-producing *E. coli* but the reason for this is unclear [12,13]. We showed that the A1 subunit of Stx2 (Stx2A1) bound ribosomes more tightly, had higher activity and was more toxic than the A1 subunit of Stx1 (Stx1A1) [14]. RTA, Stx1A1 and Stx2A1 interact with the C-terminus of the ribosomal P-protein stalk to access the SRL [15–19]. The eukaryotic ribosomal P-protein stalk is a pentameric complex located on the 60S subunit in close proximity to the SRL. It consists of two P1/P2 dimers, which bind to uL10 (previously P0) to form uL10-(P1/P2)$_2$ structure. The C-terminal sequence of the five P-proteins is exactly the same and is conserved among all eukaryotes. The ribosomal P-protein stalk together with the sarcin/ricin loop (SRL), are part of the GTPase associated center, which is responsible for the recruitment of translational GTPases and stimulation of factor-dependent GTP hydrolysis [20–23].

The X-ray crystal structures of Shiga toxin from *Shigella* (Stx), which differs only in one residue from *E. coli* Stx1, and *E. coli* Stx2 are resolved [24–28], but not the structures of the A1 subunits alone. Figure 2 compares the structure of RTA with the structures of StxA1 and Stx2A1 derived from their holotoxin structure. On the left, the residues in Stx and in RTA, which interact with the C-termini of ribosomal stalk P-proteins, are shown in magenta and in light blue. Rotating 180° along the y-axis, on the right, the active sites are shown in red. Cys242/241 in Stx/Stx2 A1, which forms a disulfide bond with Cys261/260 in Stx/Stx2 A2, and Cys259 in RTA, which forms a disulfide bond with Cys2 in RTB, are shown in yellow.

We showed that the RTA-ribosomal stalk interaction site is located on the opposite side of the active site [19]. In ricin holotoxin, the ribosome binding site of RTA was blocked by RTB [19]. Consequently, ricin holotoxin did not bind or depurinate the ribosome. However, the active site of ricin was not blocked by RTB and ricin holotoxin could depurinate the naked RNA [19].

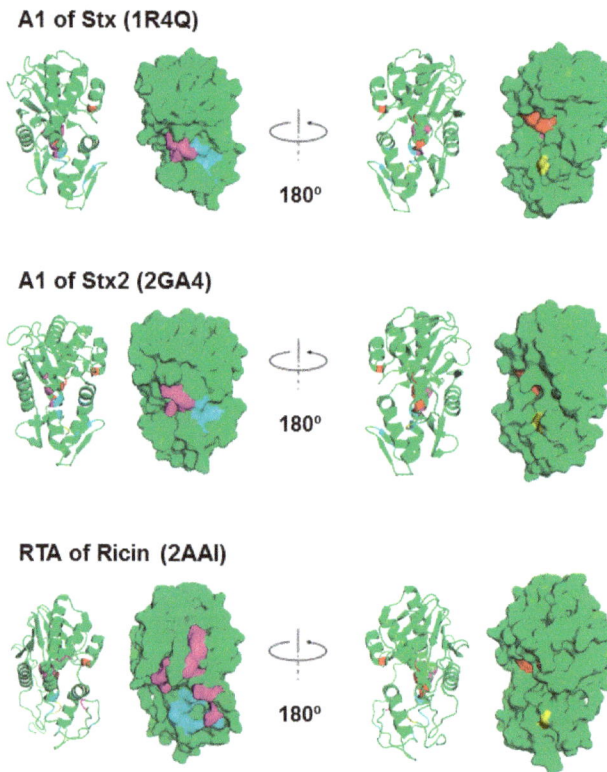

Figure 2. Structures of the StxA1, Stx2A1 and RTA shown as ribbon and as surface. The residues at the active site, Tyr76/Tyr77, Tyr114, Glu167 and Arg170 in Stx; Tyr80, Tyr123, Glu177 and Arg180 in RTA, are shown in red. The arginines that are critical for P protein interaction, Arg172, Arg176 and Arg179 in Stxs and Arg189, Arg191, Arg193, Arg196, Arg197, Arg234 and Arg235 in RTA, are shown in magenta. The residues that form a hydrophobic pocket and interact with the last six residues of the P-proteins, Gln173, Leu198, Ile224/Ile223, Phe226/Phe225, Leu233/Leu232, Ser235/Thr234 in Stx/Stx2 and Tyr183, Leu207, Phe240, Ile247, Pro250 and Ile251 in RTA, are shown in cyan [29]. Cys242/241 in Stx/Stx2 A1, which form a disulfide bond with Cys261/260 in Stx/Stx2 A2, and Cys259 in RTA, which forms a disulfide bond with Cys2 in RTB, are shown in yellow. Stx PDB ID: 1R4Q, Stx2 PDB ID: 2RA4 and ricin PDB ID: 2AA1. The structures were generated using PyMOL.

The X-ray crystal structures of a peptide mimic of the conserved C-terminus of ribosomal P-proteins in complex with trichosanthin, a type I RIP, and RTA have been resolved [29–31]. The last six amino acids (P6) of the P-protein peptide bind to a hydrophobic pocket on RTA, which is blocked by RTB in the holotoxin. Several ribosome binding residues on the A1 subunits of Stxs have been identified [32]. As shown in Figure 2, the ribosome binding sites of Stxs are also located on the opposite face of the active site. Stx1 holotoxin displays almost no ribosomal activity in its native form. Why Stx1 and Stx2 holotoxins are not active on ribosomes remains unknown. Here we compare ribosome binding and depurination activity of Stx1 and Stx2 holotoxins with the A1 subunits of Stx1 and Stx2 using intact yeast ribosomes or a 10-mer RNA mimic of the SRL. We demonstrate that the A2 chain and the B subunit block the active site of Stx1 and Stx2, but not their ribosome binding site. Unlike ricin holotoxin, the ribosome binding sites of Stx1 and Stx2 holotoxins are accessible to the ribosome. These differences have implications for the development of antidotes against STEC.

2. Results

2.1. Stx1 and Stx2 Holotoxins Can Bind the Ribosome

Computer modeling of Stx1A1 in complex with a peptide that mimics the last 11 amino acids of the ribosomal P-stalk proteins (P11) suggested that the ribosomal stalk binding site and the active site are located on opposite sides of the protein [30]. Our recent study showed that point mutations at Arg172 or Arg176 located on the opposite face of the active site (Figures 1 and 3) disrupted the ribosome and ribosomal stalk interaction of Stx1A1 and Stx2A1, but did not affect their depurination activity on RNA, indicating that their active sites were intact and were located on the opposite face of the ribosome binding site [32]. Figure 3 shows the structure of Stx holotoxins in comparison with ricin. In Stxs, five B-subunits form a doughnut shape and A2 chains insert into the doughnut hole. The position of the B subunits is different between Stxs and ricin. The pentameric B subunits are docked on the active site of StxA1 and Stx2A1 and block the active site, while the ribosome binding site is exposed in Stx and Stx2 holotoxins. However, in ricin holotoxin the RTB covers the ribosome binding site of RTA while the active site of RTA is exposed.

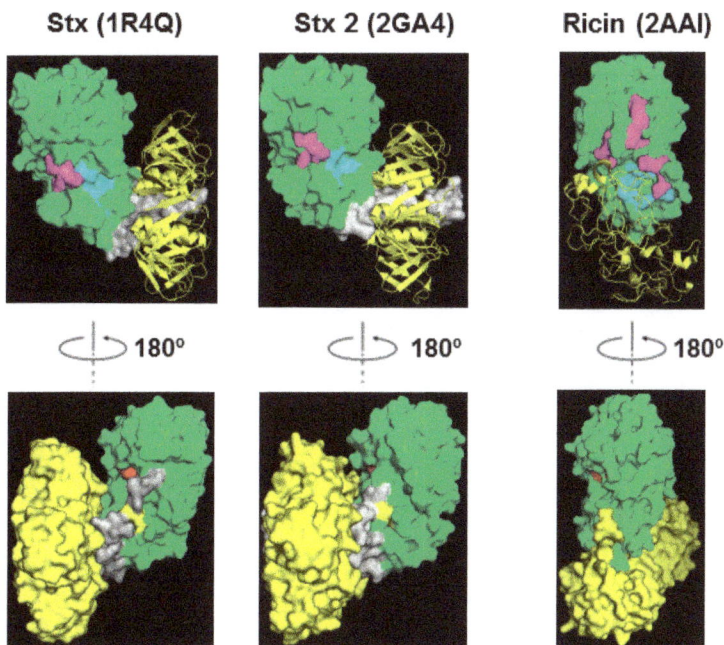

Figure 3. Structures of Stx, Stx2 and ricin holotoxin shown in ribbon and surface. The A1 subunits are shown in green, the A2 chains are shown in gray and the B subunits are shown in yellow. The labeling of the residues is the same as in Figure 1.

To determine if Stx1 and Stx2 holotoxins can bind ribosomes, we immobilized Stx1, Stx2 and ricin on a CM5 chip of Biacore T200 and passed yeast ribosomes over the surface. As shown in Figure 4, both Stx1 and Stx2 could interact with yeast ribosomes. The binding levels of the interaction were ribosome concentration-dependent. In order to observe the interaction signal of Stx1, we immobilized 37% more Stx1 (2541 RU for Stx1 and 1853 RU for Stx2) on the chip. When the same concentration of ribosomes was passed over the surface, the binding level of Stx2 was still about 90% higher than Stx1 at 2.5 nM and 66% higher than Stx1 at 40 nM ribosome concentration. These data indicate that Stx1 and Stx2 holotoxins can bind ribosomes, and Stx2 holotoxin has higher affinity for the ribosome than Stx1 holotoxin. Similar results were observed when Stx1A1 and Stx2A1 interacted with the ribosome [14]. There was no detectable interaction between the ribosome and ricin holotoxin, as previously observed [19].

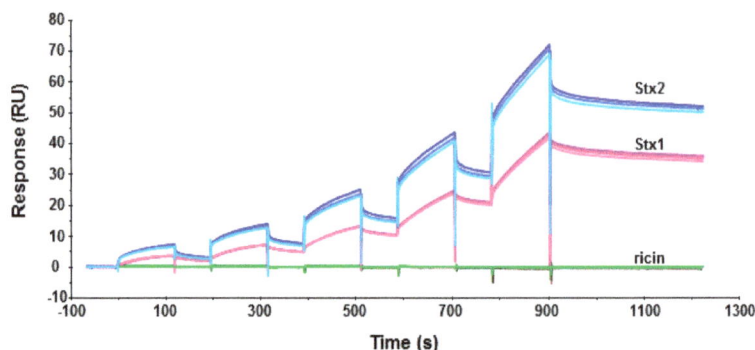

Figure 4. Interaction of yeast ribosomes with Stx1 (magenta), Stx2 (blue) and ricin (green) holotoxins. Toxins were immobilized on a CM5 chip. Ricin was immobilized on Fc2 at 1873 RU, Stx1 was immobilized on Fc3 at 2541 RU, and Stx2 was immobilized on Fc4 at 1853 RU. Fc1 was activated and blocked as a control. Ribosomes were passed over the surfaces at 2.5, 5, 10, 20 and 40 nM. The running buffer contained 10 mM Hepes, pH 7.4, 150 mM NaCl, 5 mM $MgCl_2$, 50 μM EDTA and 0.003% Surfactant P20. The surface was regenerated with one-minute-injection of 3 M NaCl for three times. The interaction was measured at 25 °C. The signals were normalized to the ligand level of 1000 RU. The experiment was replicated three times.

2.2. Stx Holotoxins Are Inactive Towards the Ribosome, but Are Active After Separation of the A1 from the A2 Chain and the B Subunit

We measured the depurination activities of Stx holotoxins before and after treatment with trypsin and TCEP (Tris (2-carboxyethyl) phosphine hydrochloride) to separate the A1 from the A2 chain and the B subunit. Stx1 and Stx2 were treated with TPCK (*N*-tosyl-L-phenylalanine chloromethyl ketone)-treated trypsin, which can cleave at the furin sites, with or without TCEP, to reduce the disulfide bond. Ricin was treated with TCEP only. After treatment, the toxins were separated on non-reducing SDS-PAGE. As shown in Figure 5, the A (A1 + A2) subunits migrated at 31 and 32 kDa for Stx1 and Stx2, respectively, and the B subunits migrated at about 8 kDa without TCEP treatment. After TCEP treatment, the A1 and A2 subunits were separated and the A1 subunits migrated at 28 kDa. The A2 subunits of 5 kDa were not observed on SDS-PAGE, possibly because their amounts were low. Ricin holotoxin migrated to about 60 kDa. The A and B subunits were separated after TCEP treatment with molecular weights around 30 and 32 kDa, respectively.

We previously measured the depurination kinetics of recombinant Stx1A1 and Stx2A1 from *E. coli* on ribosomes and on an SRL mimic RNA [14]. Both StxA1 and RTA can depurinate ribosomes at physiological pH but they depurinate RNA only at an acidic pH [14]. Using previously established conditions to measure depurination kinetics [14], we compared the depurination activities of Stx1 and Stx2 holotoxins with their activated forms where the catalytic A1 subunit was separated from the A2 chain and the B subunit. Stx1 and Stx2 holotoxins were not active towards the ribosome. However, after trypsin and TCEP treatment to release the A1 subunits, Stx1A1 and Stx2A1 were able to depurinate the yeast ribosome at 374-fold and 92-fold higher levels than Stx1 and Stx2 holotoxins, respectively, at 0.5 μM ribosome concentration (Figure 6a). A previous study examined the k_{cat} and K_m of Stx1 holotoxin on *Artemia salina* ribosomes before and after treatment with trypsin, urea and DTT [33]. Our results are consistent with this study in the requirement for activation, although this study used a different assay, different source of ribosomes, different method of activation, and much higher toxin concentrations [33].

Toxins	Stx1			Stx2			Ricin (glycosylated)	
Chains	A1	A2	B	A1	A2	B	A	B
Number of amino acids	251	42	69	250	47	70	267	262
MW (KD)	27.62	4.62	7.69	27.93	5.29	7.82	28*	32*

(a)

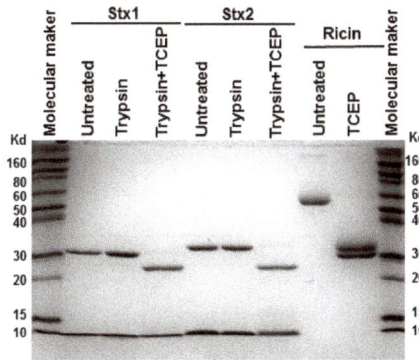

(b)

Figure 5. Molecular weight of the subunits of Stx1, Stx2 and ricin (**a**) and SDS-PAGE analysis (**b**); Stx1 and Stx2 (7 μg) were treated with trypsin with or without TCEP, and ricin (7 μg) was treated with TCEP at 25 °C for 2 h and separated on a 12% SDS-PAGE. The gel was stained with Coomassie blue. The protein marker is Invitrogen Novex®sharp pre-stained protein marker. * Molecular weight is affected by glycosylation in both A and B subunits of ricin.

We further examined the activity of Stx1 and Stx2 holotoxins with their activated forms toward a 10-mer RNA stem loop mimic of the SRL. Activated Stx holotoxins could depurinate the SRL mimic RNA, while the holotoxins were inactive (Figure 6b). These results are consistent with the structural data, indicating that the A2 chain and the B subunit block the active sites of Stx1 and Stx2 [24–28]. We have previously shown that ricin holotoxin could not depurinate the ribosome but can depurinate the naked RNA using a qRT-PCR assay to measure depurination [19]. Here, as a control, we measured the activities of ricin holotoxin toward the yeast ribosome and the SRL using a highly sensitive luminescent assay [34]. Consistent with the previously published data, ricin holotoxin could not depurinate yeast ribosomes unless it was reduced to separate the catalytic A subunit from the B subunit (Figure 6c). In contrast, both unreduced and reduced ricin holotoxin could depurinate the SRL mimic RNA with similar activity (Figure 6d). Stx1A1, Stx2A1 (pH 4.5) and ricin (pH 4.0) have different optimum catalytic pH on nucleic acid substrates [14,34,35]. We examined depurination of the SRL mimic RNA at the same pH (pH 4.5) in order to compare the activity of activated Stx1 and Stx2 holotoxins with ricin. At pH 4.5, A1 subunits of Stxs are about 100-fold more active than ricin holotoxin or RTA (Figure 6b,d). To determine if Stx1 and Stx2 holotoxins have any activity on the SRL mimic RNA at a very high concentration, we compared their activity with ricin holotoxin at 1 μM toxin concentration using the same conditions as in Figure 6b,d. As shown in Figure 6e, we could not detect any depurination activity of Stx1 and Stx2 holotoxins, even at this higher dose. In contrast, ricin holotoxin could depurinate the SRL at a similar level as that in Figure 6d. Ricin holotoxin could depurinate the SRL mimic RNA at 252-, 326-, and 364-fold higher levels than Stx1 and at 51-, 76-, and 114-fold higher levels than Stx2 at 1, 2 and 4 min, respectively (Figure 6e).

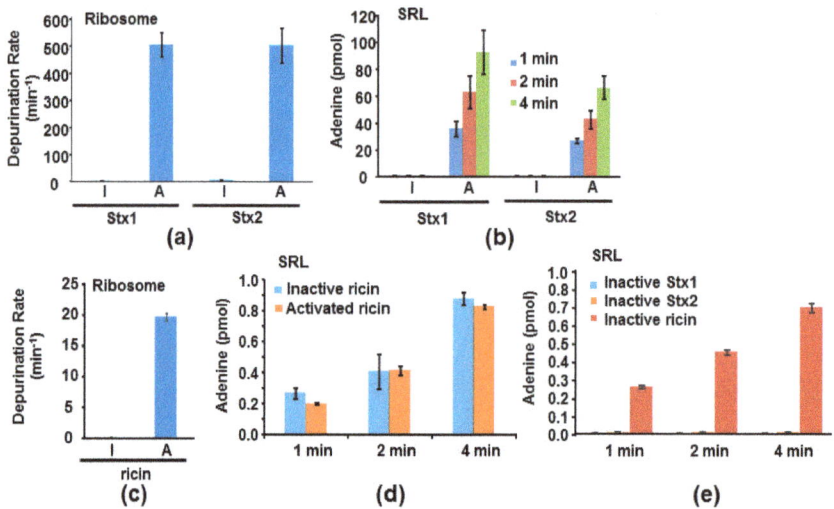

Figure 6. Depurination of yeast ribosomes (**a,c**) and SRL mimic RNA (**b,d,e**) by Stxs (**a,b,e**) and ricin (**c,d,e**). (**a,c**) Stx holotoxins (I for inactive) were activated by trypsin and TCEP (Tris (2-carboxyethyl) phosphine hydrochloride) treatment to release the A1 subunit from the A2 chain and the B subunit (A for active) and ricin holotoxin (I for inactive) was activated by TCEP treatment to release the A subunit from the B subunit (A for active). Ribosomes were used at 0.5 μM, Stxs were used at 0.2 nM and ricin was used at 1 nM. The reaction was incubated at 25 °C. The adenine released was measured by the enzyme coupled fluorescence assay for 15 min. The linear part of the reactions was used to calculate the activities. The depurination rate is shown as pmol of adenine released per pmol of toxin in one minute. (**b,d**) The SRL mimic RNA was used at 2 μM, Stxs were used at 5 nM and ricin was used at 40 nM. The reactions were incubated at 20 °C at pH 4.5. The depurination was stopped at 1, 2 and 4 min. The adenine released was measured by the enzyme coupled fluorescence assay. The amount of adenine released is shown as pmol of adenine released per pmol of toxins at the sampling time point. (**e**) The SRL mimic RNA concentration was 2 μM and the holotoxin concentrations were 1 μM. The reactions were incubated at 20 °C at pH 4.5. The depurination was stopped at 1, 2 and 4 min. The adenine released was measured by the enzyme coupled fluorescence assay. The amount of adenine released is shown as pmol of adenine released per pmol of toxins.

In summary, the A2 chain and the B subunit block the active site in the Stx holotoxins, but leave the ribosome binding sites open. As a result, Stx holotoxins could not depurinate either the ribosome or the 10-mer RNA stem loop mimic of the SRL, even though they could interact with the ribosome. In contrast, in ricin holotoxin RTB covers the ribosome binding side of RTA and leaves the active site of RTA exposed. Ricin holotoxin could depurinate the SRL mimic RNA at low pH, but could not interact with the ribosome or depurinate the ribosome. These results explain why Stx1, Stx2 and ricin holotoxins are inactive towards ribosomes and show that the molecular basis for the inactivity of Stx1/Stx2 holotoxins differs from that of the ricin holotoxin.

3. Discussion

We previously showed that ricin holotoxin is not active on ribosomes because the ribosome binding site of RTA is blocked by the B subunit [19]. However, the molecular basis for the inactivity of Stx1 and Stx2 holotoxins towards ribosomes remained unknown. Here, we compared the ribosome binding and depurination activity of Stx1 and Stx2 holotoxins with the A1 subunits of Stx1 and Stx2 and demonstrated that the active sites of Stx1 and Stx2 are blocked by the A2 chain and the B subunit, while the ribosome binding sites are exposed. Although Stx1 and Stx2 holotoxins are enzymatically inactive, they are able to interact with the ribosome. Previous results indicated that the A1 subunit of Stx2 interacts with the ribosome with higher affinity than the A1 subunit of Stx1 [14]. Conserved arginines at the P-protein stalk binding site did not contribute to the differences in the affinity of the A1 subunits for the ribosome [32]. We show here that Stx2 holotoxin also interacts with the ribosome with higher affinity than Stx1 holotoxin (Figure 4). Since the difference in ribosome binding previously observed for the A1 subunits [14] is also observed with the Stx1 and Stx2 holotoxins, the regions of the A1 subunits, which are covered by the A2 chain and the B subunit, are not likely responsible for the differences in the ribosome binding. These results suggest that residues at the active site of Stx2A1 and Stx1A1 do not contribute to the higher affinity of Stx2 for the ribosome. The surface charge and the distribution of the surface charge are different between Stx1A1 and Stx2A1 [32]. Our results suggest that surface charge differences in regions other that the active site or the stalk binding site contributes to the higher affinity of Stx2 for the ribosome.

Although Stx2A1 is more active in depurination of the ribosome and the RNA than Stx1A1 [32], we did not observe this difference when the holotoxins were activated by trypsin and TCEP (Figure 6). As previously observed [14], the holotoxin is not a good choice for comparison of the depurination activities of Stx1 and Stx2. Stx1 and Stx2 holotoxins have to be activated by trypsin and TCEP treatment to separate the A2 chains and the B subunits from the A1 subunits. It is impossible to digest Stx1 and Stx2 equally with trypsin. Since trypsin cleaves the peptide chains at the carboxyl side of lysine and arginine, trypsin can cleave the toxins at other sites besides the furin cleavage site between A1 and A2, and inactivate them. Furthermore, during toxin purification, some of the toxin molecules are already cleaved between A1 and A2 subunits and the holotoxins can be activated to up to 80% of their maximal activity after treatment with only TCEP. Therefore, the activity measured after trypsin plus TCEP (or DTT) treatment does not necessarily reflect the total activity of the A1 subunits in the Stx holotoxins, but the activity of the A1 subunits remaining after the trypsin digestion.

The X-ray crystal structure comparison of Stx2 holotoxin with *Shigella* Stx indicated some differences. While the active site of the A subunit was blocked by the A2 chain in the *Shigella* Stx, the active site of Stx2 holotoxin was accessible to a small substrate, adenosine [26]. However, it was predicted that while Stx2 holotoxin was active against adenosine, it was unlikely to bind to the SRL because the A2 chain and the B subunit would interfere with binding to the 28S rRNA [26]. Our results (Figure 6c,e) provide direct experimental evidence that the Stx1 and Stx2 holotoxins cannot depurinate the SRL or the ribosome because their active sites are blocked by the A2 chain and the B subunit. In contrast, ricin holotoxin could depurinate the SRL mimic RNA. However, depurination of the RNA occurs only at the low pH, but not at the physiologic pH. Ricin holotoxin could not interact with the ribosome or depurinate the ribosome due to the blockage of the ribosome binding site by RTB. These results demonstrate the molecular basis for the inactivity of ricin, Stx1 and Stx2 holotoxins towards ribosomes and suggest that deactivation of the toxins could be achieved by blocking the ribosome binding site as an alternative to the active site, since targeting the active site has not yielded an inhibitor with sufficiently high activity to date [36].

Although Stx1A1, Stx2A1 and RTA depurinate small RNA substrates at a low pH [14,34,35], at the physiologic pH, Stx1A1, Stx2A1 and RTA need to interact with the ribosomal P-proteins to catalyze depurination of the SRL. Stx1A1, Stx2A1 and RTA can depurinate the ribosome with high affinity. However, since the ribosome and the A subunits interact with each other electrostatically [32,37], the local concentration of the toxins around the ribosome can be high. In order to block ribosome interaction inhibitors that can bind either to the P-protein stalk interaction site or to the active site would require very high affinity. The molecular details of how Stx1A1 and Stx2A1 interact with ribosomes and the structures of the C-termini of P-proteins in complex with the toxins could lead to the identification of inhibitors that can disrupt ribosome interactions and thereby offer new opportunities for antidote development.

4. Materials and Methods

4.1. Materials

Stx1 and Stx2 holotoxins were from Phoenix Laboratory at the Tufts Medical Center, Boston, MA, USA. They are purified by receptor analog affinity chromatography [38]. Ricin holotoxin was from Vector Laboratory, Burlingame, CA, USA. It was purified from castor bean (catalog number L-1090). Yeast monomeric ribosomes were purified as described previously [19]. The 10-mer SRL mimic oligo (5'-rCrGrCrGrArGrArGrCrG-3') was purchased from Integrated DNA Technologies, Coralville, Iowa, IA, USA.

4.2. Toxin Structures

Stx (1R4Q), Stx2 (2GA4) and ricin (2AAI) holotoxins were downloaded from PDB. The A1 subunits of Stx and Stx2 or the A subunit of ricin were obtained from the holotoxins. The structures were visualized using PyMOL (The PyMOL Molecular Graphics System, Version 1.8 Schrödinger, LLC, Portland, OR, USA).

4.3. SDS-PAGE Analysis

Stx1 or Stx2 holotoxin (7 μg) was treated with 50 ng TPCK-treated trypsin or trypsin plus 50 mM of TCEP at 25 °C for 2 h in a total volume of 10 μL in trypsin reaction buffer (50 mM Tris-HCl pH 8.0, 20 mM $CaCl_2$). TPCK-treated trypsin was from Biolabs (Ipswich, MA, USA). Trypsin was reconstituted following the manufacturer's instructions and used within a week. Ricin holotoxin (7 μg) was treated with 50 mM of TCEP at 25 °C for 2 h in 10 μL in its original buffer (10 mM of phosphate pH 7.8, 150 mM NaCl, and 0.08% sodium azide). 5 μL of 5X SDS-PAGE loading buffer was added to the treated and untreated Stx1, Stx2 or ricin. The samples were heated at 95 °C for 5 min and separated on 12% SDS-PAGE. The gel was stained with Coomassie blue.

4.4. Holotoxin-Ribosome Interaction

Holotoxins were immobilized on a CM5 chip of a Biacore T200. Ricin was immobilized on Fc2 at 1873 RU, Stx1 on Fc3 at 2541 RU and Stx2 on Fc4 at 1853 RU. Fc1 was activated and blocked as a control. Yeast ribosomes at 2.5, 5, 10, 20 and 40 nM were passed over the surface using single kinetic injection method. The running buffer contained 10 mM Hepes, pH 7.4, 150 mM NaCl, 5 mM $MgCl_2$, 50 μM EDTA and 0.003% surfactant P20. The surface was regenerated by three times 1 min injection of 3 M NaCl. The interaction was measured at 25 °C.

4.5. Toxin Activation

The activation conditions were the same as used in the SDS-PAGE analysis described in Section 4.3. Stx1 and Stx2 were activated by treatment with trypsin and TCEP. At the end of the reaction 1 mM of PMSF was added to stop the reaction. The activated Stx1 or Stx2 was stored at 4 °C and used within two days. The activated ricin was also stored at 4 °C and used within two days.

4.6. Ribosome Depurination

Ribosome depurination was measured using continuous enzyme coupled luminescent assay [34]. Ribosomes were used at 0.5 μM, Stx1 and Stx2 holotoxins were used at 0.2 nM and ricin was used at 1 nM. The reaction was started by adding the toxin, and luminescence intensity was recorded continuously for 15 min using a BioTek Synergy 4 Microplate reader (Winooski, VT, USA). Adenine standards were assayed at same time. The rates were determined from the linear region of the luminescence intensity. The adenine generated from ribosomes alone was subtracted.

4.7. SRL Depurination

Stem-loop depurination was performed using a synthetic 10-mer RNA mimic of the SRL. The discontinuous luminescence assay described by Sturm and Schramm was used [34]. In the reaction mixture, the SRL was used at 2 µM, Stxs at 5 nM or 1 µM and ricin at 40 nM or 1 µM. The reaction was incubated at 20 °C in 50 µL of 10 mM potassium citrate-KOH and 1 mM EDTA at pH 4.5 for both Stxs and ricin. The reaction was started by adding the toxins. Samples were taken at 1, 2 and 4 min and mixed immediately with 2X coupling buffer to stop the reaction. Adenine standards were assayed at the same time. Luminescence intensity was measured using a BioTek Synergy 4 Microplate reader.

Acknowledgments: The authors thank Jennifer Kahn for reading the manuscript. The work was supported by National Institutes of Health grants, AI092011 and AI072425 to Nilgun E. Tumer.

Author Contributions: Both authors designed the experiments and wrote the paper.

Conflicts of Interest: The authors declare no conflict of interest.

References

1. Tesh, V.L.; O'Brien, A.D. The pathogenic mechanisms of Shiga toxin and the Shiga-like toxins. *Mol. Microbiol.* **1991**, *5*, 1817–1822. [CrossRef] [PubMed]
2. Olsnes, S. The history of ricin, abrin and related toxins. *Toxicon* **2004**, *44*, 361–370. [CrossRef] [PubMed]
3. Bergan, J.; Dyve Lingelem, A.B.; Simm, R.; Skotland, T.; Sandvig, K. Shiga toxins. *Toxicon* **2012**, *60*, 1085–1107. [CrossRef] [PubMed]
4. Lapadula, W.J.; Sanchez Puerta, M.V.; Juri Ayub, M. Revising the taxonomic distribution, origin and evolution of ribosome inactivating protein genes. *PLoS ONE* **2013**, *8*, e72825. [CrossRef] [PubMed]
5. Sandvig, K.; van Deurs, B. Endocytosis, intracellular transport, and cytotoxic action of Shiga toxin and ricin. *Physiol. Rev.* **1996**, *76*, 949–966. [PubMed]
6. Sandvig, K.; van Deurs, B. Delivery into cells: Lessons learned from plant and bacterial toxins. *Gene Ther.* **2005**, *12*, 865–872. [CrossRef] [PubMed]
7. Sato, T.; Shimizu, T.; Watarai, M.; Kobayashi, M.; Kano, S.; Hamabata, T.; Takeda, Y.; Yamasaki, S. Genome analysis of a novel Shiga toxin 1 (Stx1)-converting phage which is closely related to Stx2-converting phages but not to other Stx1-converting phages. *J. Bacteriol.* **2003**, *185*, 3966–3971. [CrossRef] [PubMed]
8. Lamb, F.I.; Roberts, L.M.; Lord, J.M. Nucleotide sequence of cloned cDNA coding for preproricin. *Eur. J. Biochem.* **1985**, *148*, 265–270. [CrossRef] [PubMed]
9. Frigerio, L.; Vitale, A.; Lord, J.M.; Ceriotti, A.; Roberts, L.M. Free ricin A chain, proricin, and native toxin have different cellular fates when expressed in tobacco protoplasts. *J. Biol. Chem.* **1998**, *273*, 14194–14199. [CrossRef] [PubMed]
10. Bolognesi, A.; Bortolotti, M.; Maiello, S.; Battelli, M.G.; Polito, L. Ribosome-inactivating proteins from plants: A historical overview. *Molecules* **2016**, *21*, 1627. [CrossRef] [PubMed]
11. Fagerquist, C.K.; Sultan, O. Top-down proteomic identification of furin-cleaved alpha-subunit of Shiga toxin 2 from *Escherichia coli* O157:H7 using MALDI-TOF-TOF-MS/MS. *J. Biomed. Biotechnol.* **2010**, *2010*, 123460. [CrossRef] [PubMed]
12. Nataro, J.P.; Kaper, J.B. Diarrheagenic *Escherichia coli*. *Clin. Microbiol. Rev.* **1998**, *11*, 142–201. [PubMed]
13. Manning, S.D.; Motiwala, A.S.; Springman, A.C.; Qi, W.; Lacher, D.W.; Ouellette, L.M.; Mladonicky, J.M.; Somsel, P.; Rudrik, J.T.; Dietrich, S.E.; et al. Variation in virulence among clades of *Escherichia coli* O157:H7 associated with disease outbreaks. *Proc. Natl. Acad. Sci. USA* **2008**, *105*, 4868–4873. [CrossRef] [PubMed]
14. Basu, D.; Li, X.P.; Kahn, J.N.; May, K.L.; Kahn, P.C.; Tumer, N.E. The A1 subunit of Shiga toxin 2 has higher affinity for ribosomes and higher catalytic activity than the A1 subunit of Shiga toxin 1. *Infect. Immun.* **2016**, *84*, 149–161. [CrossRef] [PubMed]

15. Chiou, J.C.; Li, X.P.; Remacha, M.; Ballesta, J.P.; Tumer, N.E. The ribosomal stalk is required for ribosome binding, depurination of the rRNA and cytotoxicity of ricin A chain in *Saccharomyces cerevisiae*. *Mol. Microbiol.* **2008**, *70*, 1441–1452. [CrossRef] [PubMed]

16. Chiou, J.C.; Li, X.P.; Remacha, M.; Ballesta, J.P.; Tumer, N.E. Shiga toxin 1 is more dependent on the P proteins of the ribosomal stalk for depurination activity than Shiga toxin 2. *Int. J. Biochem. Cell Biol.* **2011**, *43*, 1792–1801. [CrossRef] [PubMed]

17. McCluskey, A.J.; Poon, G.M.; Bolewska-Pedyczak, E.; Srikumar, T.; Jeram, S.M.; Raught, B.; Gariepy, J. The catalytic subunit of Shiga-like toxin 1 interacts with ribosomal stalk proteins and is inhibited by their conserved C-terminal domain. *J. Mol. Biol.* **2008**, *378*, 375–386. [CrossRef] [PubMed]

18. McCluskey, A.J.; Bolewska-Pedyczak, E.; Jarvik, N.; Chen, G.; Sidhu, S.S.; Gariepy, J. Charged and hydrophobic surfaces on the A chain of Shiga-like toxin 1 recognize the C-terminal domain of ribosomal stalk proteins. *PLoS ONE* **2012**, *7*. [CrossRef] [PubMed]

19. Li, X.P.; Kahn, P.C.; Kahn, J.N.; Grela, P.; Tumer, N.E. Arginine residues on the opposite side of the active site stimulate the catalysis of ribosome depurination by ricin A chain by interacting with the P-protein stalk. *J. Biol. Chem.* **2013**, *288*, 30270–30284. [CrossRef] [PubMed]

20. Wahl, M.C.; Moller, W. Structure and function of the acidic ribosomal stalk proteins. *Curr. Protein Pept. Sci.* **2002**, *3*, 93–106. [CrossRef] [PubMed]

21. Gonzalo, P.; Reboud, J.P. The puzzling lateral flexible stalk of the ribosome. *Biol. Cell* **2003**, *95*, 179–193. [CrossRef]

22. Ballesta, J.P.; Remacha, M. The large ribosomal subunit stalk as a regulatory element of the eukaryotic translational machinery. *Prog. Nucleic Acid Res. Mol. Biol.* **1996**, *55*, 157–193. [PubMed]

23. Tchorzewski, M. The acidic ribosomal P proteins. *Int. J. Biochem. Cell Biol.* **2002**, *34*, 911–915. [CrossRef]

24. Kozlov, Y.V.; Chernaia, M.M.; Fraser, M.E.; James, M.N. Purification and crystallization of Shiga toxin from *Shigella dysenteriae. J. Mol. Biol.* **1993**, *232*, 704–706. [CrossRef] [PubMed]

25. Fraser, M.E.; Fujinaga, M.; Cherney, M.M.; Melton-Celsa, A.R.; Twiddy, E.M.; O'Brien, A.D.; James, M.N. Structure of Shiga toxin type 2 (Stx2) from *Escherichia coli* O157:H7. *J. Biol. Chem.* **2004**, *279*, 27511–27517. [CrossRef] [PubMed]

26. Fraser, M.E.; Cherney, M.M.; Marcato, P.; Mulvey, G.L.; Armstrong, G.D.; James, M.N. Binding of adenine to Stx2, the protein toxin from *Escherichia coli* O157:H7. *Acta Crystallogr. Sect. F Struct. Biol. Cryst. Commun.* **2006**, *62*, 627–630. [CrossRef] [PubMed]

27. Robertus, J.D. Toxin structure. *Cancer Treat. Res.* **1988**, *37*, 11–24. [PubMed]

28. Katzin, B.J.; Collins, E.J.; Robertus, J.D. Structure of ricin A-chain at 2.5 Å. *Proteins* **1991**, *10*, 251–259. [CrossRef] [PubMed]

29. Shi, W.W.; Tang, Y.S.; Sze, S.Y.; Zhu, Z.N.; Wong, K.B.; Shaw, P.C. Crystal structure of ribosome-inactivating protein ricin A chain in complex with the C-terminal peptide of the ribosomal stalk protein P2. *Toxins* **2016**, *8*. [CrossRef] [PubMed]

30. Too, P.H.; Ma, M.K.; Mak, A.N.; Wong, Y.T.; Tung, C.K.; Zhu, G.; Au, S.W.; Wong, K.B.; Shaw, P.C. The C-terminal fragment of the ribosomal P protein complexed to trichosanthin reveals the interaction between the ribosome-inactivating protein and the ribosome. *Nucleic Acids Res.* **2009**, *37*, 602–610. [CrossRef] [PubMed]

31. Fan, X.; Zhu, Y.; Wang, C.; Niu, L.; Teng, M.; Li, X. Structural insights into the interaction of the ribosomal P stalk protein P2 with a type II ribosome-inactivating protein ricin. *Sci. Rep.* **2016**, *6*. [CrossRef] [PubMed]

32. Basu, D.; Kahn, J.N.; Li, X.P.; Tumer, N.E. Conserved arginines at the P-protein stalk binding site and the active site are critical for ribosome interactions of Shiga toxins but do not contribute to differences in the affinity of the A1 subunits for the ribosome. *Infect. Immun.* **2016**, *84*, 3290–3301. [CrossRef] [PubMed]

33. Brigotti, M.; Carnicelli, D.; Alvergna, P.; Mazzaracchio, R.; Sperti, S.; Montanaro, L. The RNA-*N*-glycosidase activity of Shiga-like toxin I: Kinetic parameters of the native and activated toxin. *Toxicon* **1997**, *35*, 1431–1437. [CrossRef]

34. Sturm, M.B.; Schramm, V.L. Detecting ricin: Sensitive luminescent assay for ricin A-chain ribosome depurination kinetics. *Anal. Chem.* **2009**, *81*, 2847–2853. [CrossRef] [PubMed]

35. Chen, X.Y.; Link, T.M.; Schramm, V.L. Ricin A-chain: Kinetics, mechanism, and RNA stem-loop inhibitors. *Biochemistry* **1998**, *37*, 11605–11613. [CrossRef] [PubMed]

36. Wahome, P.G.; Robertus, J.D.; Mantis, N.J. Small-molecule inhibitors of ricin and Shiga toxins. *Curr. Top. Microbiol. Immunol.* **2012**, *357*, 179–207. [PubMed]

37. Li, X.P.; Chiou, J.C.; Remacha, M.; Ballesta, J.P.; Tumer, N.E. A two-step binding model proposed for the electrostatic interactions of ricin A chain with ribosomes. *Biochemistry* **2009**, *48*, 3853–3863. [CrossRef] [PubMed]

38. Donohue-Rolfe, A.; Acheson, D.W.; Kane, A.V.; Keusch, G.T. Purification of Shiga toxin and Shiga-like toxins I and II by receptor analog affinity chromatography with immobilized P1 glycoprotein and production of cross-reactive monoclonal antibodies. *Infect. Immun.* **1989**, *57*, 3888–3893. [PubMed]

toxins

MDPI

Article

Crystal Structure of Ribosome-Inactivating Protein Ricin A Chain in Complex with the C-Terminal Peptide of the Ribosomal Stalk Protein P2

Wei-Wei Shi [†], Yun-Sang Tang [†], See-Yuen Sze, Zhen-Ning Zhu, Kam-Bo Wong and Pang-Chui Shaw *

Centre for Protein Science and Crystallography, School of Life Sciences, The Chinese University of Hong Kong, Shatin, N.T., Hong Kong, China; Shiww@cuhk.edu.hk (W.-W.S.); samtys0910@gmail.com (Y.-S.T.); seeyuen123@gmail.com (S.-Y.S.); janet.chuk@gmail.com (Z.-N.Z.); kbwong@cuhk.edu.hk (K.-B.W.)
* Correspondence: pcshaw@cuhk.edu.hk; Tel.: +852-3943-1363
† These authors contributed equally to this work.

Academic Editors: Julien Barbier and Daniel Gillet
Received: 24 August 2016; Accepted: 30 September 2016; Published: 13 October 2016

Abstract: Ricin is a type 2 ribosome-inactivating protein (RIP), containing a catalytic A chain and a lectin-like B chain. It inhibits protein synthesis by depurinating the N-glycosidic bond at α-sarcin/ricin loop (SRL) of the 28S rRNA, which thereby prevents the binding of elongation factors to the GTPase activation center of the ribosome. Here, we present the 1.6 Å crystal structure of Ricin A chain (RTA) complexed to the C-terminal peptide of the ribosomal stalk protein P2, which plays a crucial role in specific recognition of elongation factors and recruitment of eukaryote-specific RIPs to the ribosomes. Our structure reveals that the C-terminal GFGLFD motif of P2 peptide is inserted into a hydrophobic pocket of RTA, while the interaction assays demonstrate the structurally untraced SDDDM motif of P2 peptide contributes to the interaction with RTA. This interaction mode of RTA and P protein is in contrast to that with trichosanthin (TCS), Shiga-toxin (Stx) and the active form of maize RIP (MOD), implying the flexibility of the P2 peptide-RIP interaction, for the latter to gain access to ribosome.

Keywords: Ricin; ribosome-inactivating protein; ribosomal P stalk protein; ribosome

---

1. Introduction

Ricin is a well-known type 2 ribosome-inactivating protein (RIP) discovered in the late 19th century in the seeds of *Ricinus communis*. It agglutinates red blood cells and is one of the most potent and lethal substances known [1]. Ricin hydrolyzes the *N*-glycosidic bond at adenine−4324 in the 28S rRNA of eukaryotic ribosomes [2]. This adenine is located at a GAGA hairpin within the sarcin/ricin loop [3], which is highly conserved in all large ribosomal subunits and is essential for the proper assembly of the functional core of the large subunit [4]. In eukaryotes, removal of the specific adenine hinders the elongation factor 1-dependent binding of aminoacyl-tRNA and GTP-dependent binding of elongation factor (EF) 2 to ribosome. In prokaryotes, damaged ribosomes do not bind EF-Tu or EF-G. As a result, protein synthesis is arrested at the elongation step [5].

Ricin is more active on eukaryotic ribosome compared to naked 28S rRNA. The depurination activity for intact ribosome is over 80,000-fold higher than naked 28S rRNA, suggesting that the presence of ribosomal proteins facilitating the rRNA depurination by ricin [6]. These observations lead to a postulation that the interaction with ribosomal proteins is essential for RIPs to exert the conserved *N*-glycosidase activity. To date, it has been demonstrated that ricin A chain (RTA) interacts with the mammalian L9 and L10e (=P0) ribosomal proteins by crosslinking-trypsin digestion analysis [7] and

RTA fails to interact with ribosome mutants lacking P1 or P2 in surface plasmon resonance interaction analysis [8]. RTA-P protein interaction is essential for ribosome inactivating action as yeast mutants with deletion of P1 and/or P2 have less depurination [8]. In eukaryotic ribosome, P proteins are located at the stalk and is composed of a pentameric complex of acidic ribosomal proteins, with one P0 and two P1 and P2, all of which possess a conserved amino acid sequence rich in acidic residues in their C-termini [9].It was also suggested that RTA binds to human ribosomal stalk protein P2 both in vitro and in vivo [10]. The RIP-binding site in P2 proteins has been mapped to eleven highly conserved C-terminal residues SDDDMGFGLFD (C11-P) and peptides corresponding to this sequence are found to be involved in stalk activity [11] and interacts with several RIPs, including trichosanthin (TCS) [12,13], the active form of maize RIP (MOD) [14] and Shiga toxin (Stx) 1 [15,16].

Our group has previously revealed the crystal structure of TCS-C11-P complex at 2.2 Å and identified the three basic residues K173, R174 and K177 in TCS forming charge–charge interactions with the acidic DDD motif while a hydrophobic pocket lined by F166, L188 and L215 accommodates the LF motif [17]. Furthermore, based on our NMR structure of the stalk protein complex of P1/P2 heterodimer and biochemical analyses, we demonstrated that the flexible C-terminal tail of eukaryotic ribosome stalk can form an arm-like structure and extend with a radius up to ~125 Å [18–20], hinting that the complex can help recruiting RIPs towards the core of the ribosome upon RIP-P protein interaction. Apart from interacting with RIPs, several studies have reported that the conserved C-terminal motif of ribosomal P proteins is crucial for binding translation factors or interacting with elongation factors in bacterial [21,22], archaeal [11] and eukaryotic ribosomes [23]. All the experimental evidence suggest TCS or other RIPs might hijack the ribosomal stalk proteins by binding to their conserved C-termini to gain access to the SRL of rRNA [24].

RTA and TCS maintain a homologous structure topology and both have been reported to interact with the conserved C-terminus of P proteins. However, they share low similarity in amino acid sequences. Therefore, to decipher the molecular mechanism of RTA-P protein interaction will provide more insights on the interaction mode between RIPs and P proteins. Recently, four positively charged arginines R189, R193, R234 and R235 in RTA were suggested to be important in RIP-stalk binding [25], implying that RTA might adopt same interaction mode to binding to the conserved C-terminal tail of P proteins. In order to clarify the nature of binding between RTA and ribosomal stalk, we crystallized and solved the structure of RTA and the conserved C-terminal nonamer of the P-proteins. Here, we show that RTA adopts a novel binding mode to interact with the P protein via hydrophobic interactions rather than electrostatic interactions. Based on our structural analyses and comparison of the putative P2 binding pocket of TCS, RTA, Stx and MOD, we conclude that P protein may adopt different orientations when interacting with RIPs.

2. Results

2.1. RTA Interacts with the Conserved C-Terminal Region of Acidic Ribosomal Stalk Protein P

RTA can specifically recognize and bind to the C-terminus of human ribosomal stalk protein both in vitro and in vivo to facilitate its ribosome-inactivating activity [10,26]. The C-terminus of P-proteins share a distinctive feature consisting of a cluster of acidic and hydrophobic amino acids, especially the last 11 residues (SDDDMGFGLFD) (as shown in Figure 1a) are conserved. The C-terminus of the stalk protein is functionally significant since it can bind to translation factors both before and after GTP hydrolysis during protein synthesis [11]. The interaction between RTA and P2 was investigated by in vitro pull-down assay. In order to map the boundary sufficient for RTA binding, we fused the last 9, 11, 17 residues of P2 to the GST protein to create GST-C9, GST-C11, GST-C17 fusion proteins, and detected the interactions between RTA and these P2 truncations. Our data showed that RTA could be retained by all three GST-tagged proteins-coupled Sepharose (Figure 1b). The binding affinities were then measured by Isothermal titration calorimetry (ITC). The determined binding affinities of RTA toward C9 and C11 peptide are almost the same, with constant (K_D) at 3.4 µM and 2.3 µM, respectively (Figure 1c,d).

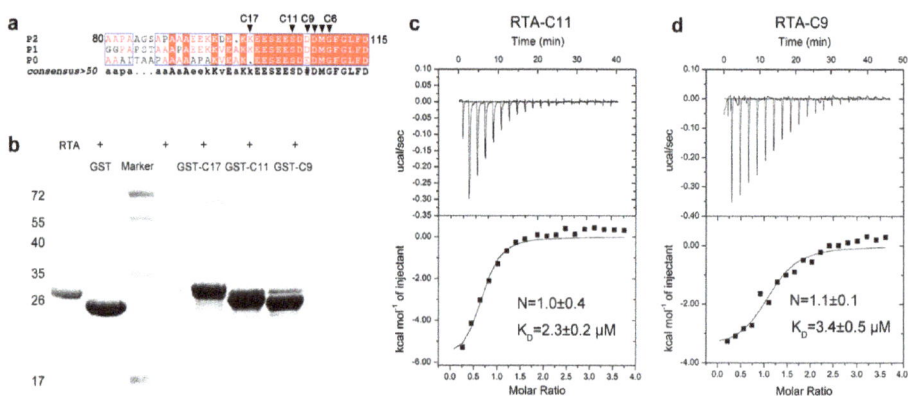

Figure 1. In vitro interaction assays of RTA and P2. (**a**) The C-terminal consensus sequence in the ribosomal P protein family. The sequence alignment was prepared by Multalin online program (http://multalin.toulouse.inra.fr/multalin/). The last residues of the C-terminus sequence of P2 are marked with black triangle; (**b**) GST pull-down assay of C-terminal truncations of P2 and RTA. Lane 1 represents the RTA mobility on 15% SDS-PAGE. RTA with GST (Lane 2) and RTA alone (Lane 4) were used as blank control; (**c**) ITC binding spectra for GST-C11 titrated to RTA; (**d**) ITC binding spectra for GST-C9 titrated to RTA. The K_D value is fitted to a one-site binding model using Origin software (Microcal, Northampton, MA, USA).

2.2. Structure of RTA-C9-P2 Complex

To further decipher the molecular interaction between RTA and P2, we co-crystallized RTA in complex with a synthetic peptide C11-P2 (SDDDMGFGLFD) and C9-P2 (DDMGFGLFD), respectively. The structure of the RTA-C9-P2 complex and RTA-C11-P2 complex were solved to a resolution of 1.6 Å (Table 1) and 2.3 Å (Table S1), respectively. These two datasets are almost identical, they share the same space group and the RTA molecules and P2 peptides in these two complexes can be superimposed well (Figure S1). Each asymmetric unit of the crystal structure contains two molecules of RTA with an overall root-mean-square deviation (RMSD) of 0.195 Å over 255 Cα atoms. The crystal packing results in a buried interface of 1260 Å^2 (630 Å^2/molecule), which was supported by 38 non-bonded contacts, two salt bridges and three hydrogen bonds. Refinement was first performed on RTA, and then C9-P2 peptide was manually built, each molecule of RTA forms a 1:1 complex with C9-P2 peptide (Figure 2a). RTA contains residues 6–260, displaying a highly similar fold as the previously reported structure [27], the overall main chain RMSD of RTA moiety in RTA-C9-P2 complex and wild-type RTA (PDB code: 1RTC) is 0.545 Å over 255 comparable Cα atoms. For C9-P2 and C11-P2 peptide, only the last 6 residues (GFGLFD, C6-P2) could be fitted to the observed electron density, while N-terminal residues DDM and SDDDM were not defined in the final model (Figure 2c). The positions of the two C6-P2 peptide chains are almost identical. The C6-P2 peptide binds to the C-terminal domain of RTA, which is composed of 11 key residues Q182, Y183, S203, L207, Q233, R234, R235, F240, I247, P250 and I251 (as shown in Figure 2b,d), providing a major hydrophobic interface via main-chain and side-chain interactions, while only R235 contributing one key hydrogen bond contact for C6-P2 peptide binding. To further confirm the C6-P2 is sufficient for binding to RTA, we repeated GST-pull down assays using GST-C6, GST-C7 and GST-C8 fusion proteins. By comparison with other P2 truncations, GST-C6, GST-C7 and GST-C8 have similar binding ability (Figure 3a). Using ITC, we determined the RTA binding affinity toward C6, which has a K_D of 18.8 µM (Figure 3b), decreasing eight-fold comparing to C11. Using structure modeling (Figure S2), we propose the two positively charged R189 and R193 probably participate in binding to the SDDD motif of P2. Our observations are in contrast to TCS binding wherein a C7 peptide was found unable to bind to TCS [12].

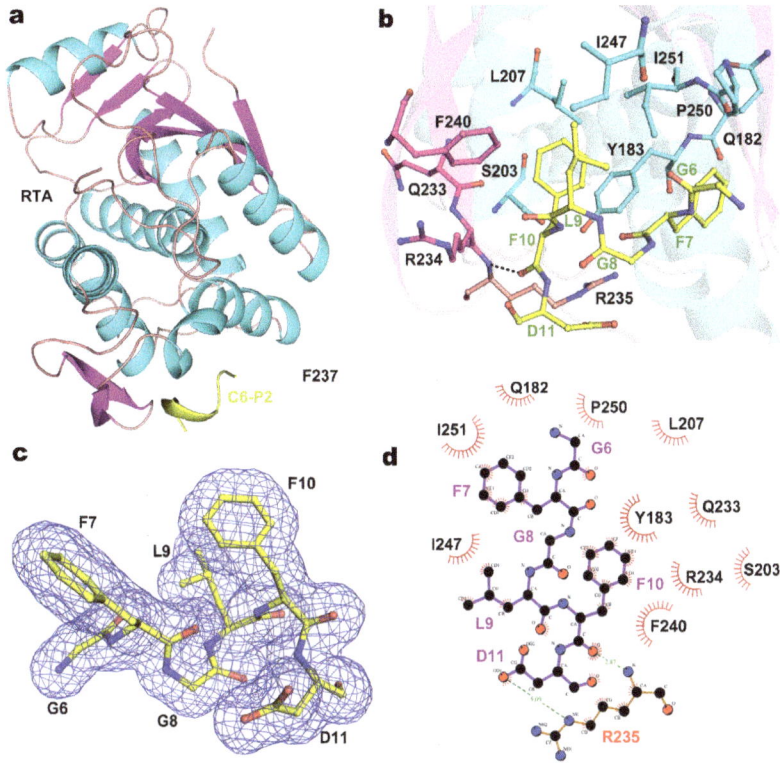

Figure 2. Structure and binding pocket of RTA-C6-P2 complex. (**a**) Overall structure of RTA-C6-P2 complex. The α-helixes, β-sheets and loops of RTA are colored in cyan, magenta and light pink, respectively. Hexameric peptide of P2 ribbon is shown in yellow; (**b**) The binding pocket of C6-P2 peptide. The binding residues are shown as sticks in the same color corresponding to the secondary elements they belong to. The residues of C6-P2 are shown as yellow sticks. The hydrogen bond is indicated by dashed lines; (**c**) 2Fo-Fc electron density map of the P2 peptide at the binding pocket of RTA. The density map is presented as a blue mesh at 1.5 σ level; (**d**) Ligplot analysis of the interactions between RTA and peptide C6-P2.

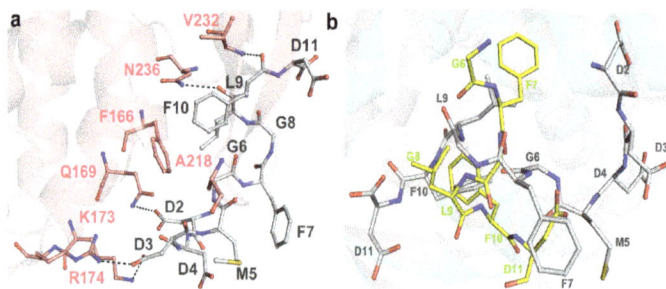

Figure 3. Structural comparison of P2 binding pockets of RTA and TCS. (**a**) P2 binding pocket of TCS-C11-P2 (PDB code: 2JDL). Key residues are colored in pink, and hydrogen bonds are highlighted with black dash; (**b**) Structural superposition of RTA-C6-P2 and TCS-C11-P2. RTA, C6-P2, TCS and C11-P2 are colored in cyan, yellow, pink and gray, respectively.

2.3. Structural Analysis and Comparison of P2 Binding Pockets of RTA and TCS

Our previous structural study showed that both the N- and C-terminal regions of C11-P (SDDDMGFGLFD) are involved in the interaction with TCS [12]. The negatively charged DDD motif was stabilized by three hydrogen bonds donated by two positively charged residues K173 and R174, and one non-polar residue Q169 of TCS. The C-terminal hydrophobic region MGFGLFD was stabilized by hydrophobic forces contributed by F166, A218, V232 and N236 (Figure 4a). Four positively charged arginines at the corresponding region in RTA, namely R189, R193, R234 and R235, were identified to be important in RIP-stalk binding [25]. Structural superposition of RTA-C6-P2 and TCS-C11-P2 complex showed RTA and TCS maintain a homologous structure topology (RMSD 1.21 Å over 233 Cα atom) even though they share low similarity in amino acid sequences [28]. However, C6-P2 binds to RTA in a distinguished manner, adopts a different orientation and is seriously distorted when compared with C11-P2 in the TCS-C11-P2 complex (Figure 4b). The P2 binding sites of RTA and TCS are distinctively different, in particularly K173 in TCS, which is the key residue for electrostatic interaction, becomes T190 in RTA. In addition, Q169, V232 and N236 in TCS, which use side-chains for forming hydrogen bonds to accommodate C11-P, become G186, I247 and I251 in RTA, respectively, making hydrogen bonding not plausible. The P2 binding pocket of RTA is hydrophobic and is composed of six hydrophobic residues (Y183, L207, F240, I247, P250 and I251) and five polar residues (Q182, S203, Q233, R234 and R235) to form non-bond contacts to accommodate C6-P2 peptide. Our structural and interaction studies suggested that the conserved hydrophilic DDDM motif at the C-terminus of P2 contributes to interacting with RTA, while the hydrophobic part GFGLFD is accommodated by the hydrophobic binding pocket of RTA. The previous proposed docking model of the RTA-C11 complex [17], has missed some important features, making it not accurate. In the RTA-C6- P2 structure, it revealed that R235 and R234 participate in interacting with F10 and D11 other than interacting with the M5 and G6 in the docking model (Figure S3). Also, most of the hydrophobic residues of RTA, such as Y183, L207, F240, I247, P250 and I251, which contribute to interactions with the hydrophobic part (GFGLF) of P2, were not identified in the docking model (Figure S3b). This also explains why P2 peptide has adopted a different conformation in interacting with RTA.

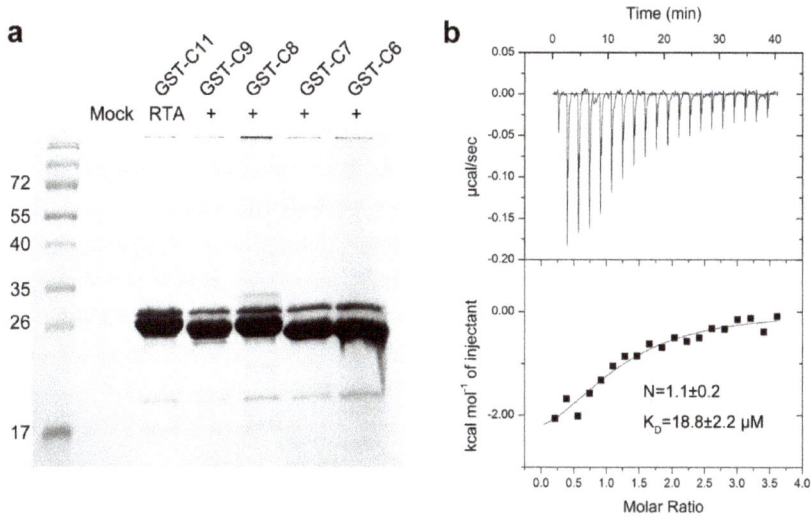

Figure 4. RTA interacts with the conserved C-terminal region of P2. (**a**) The interaction between C6-C8 of P2 and RTA was checked by in vitro GST-pull down assay; (**b**) The ITC spectra of GST-C6 to RTA.

2.4. P2 Protein May Adopt Different Ways of Binding to Different RIPs

From the above structural comparison of the P2 interaction modes and P2 binding site residues of RTA and TCS, we postulate P2 protein may adopt different conformations to interact with RIPs to facilitate RIPs gaining access to ribosome. This is supported by the different charge distribution patterns on RIPs. We superimposed the structures of MOD (PDB code: 2PQJ) and Stx (PDB code: 1R4Q) onto RTA and TCS individually to compare their putative P2 binding pockets. These RIPs are structurally homologous to each other, Stx and MOD possess a similar fold as RTA with RMSD 1.95 Å over 210 Cα atoms and 2.08 Å over 195 Cα atoms, respectively. Meanwhile, compared with TCS, Stx exhibits a lower RMSD 1.21 Å over 233 comparable Cα atoms than MOD with a RMSD 1.89 Å over 203 comparable Cα atoms. The overall landscape and polarity of the electrostatic surface potential around P2 binding pockets of RTA (Figure 5a) and TCS (Figure 5c) are quite different, which probably determined the orientation of P2 peptide upon binding. We attempted to predict P2 binding modes on Stx by superimposing C6-P2 and C11-P2 to Stx. Compared with TCS and RTA, the P2 binding pocket of Stx exhibited distinct shape and electrostatic charge distributions using C6-P2 and C11-P2 as reference (Figure 5b,d). Based on the complex structure of TCS-C11-P2 and RTA-C6-P2, we mapped the related residues of the hydrophobic patch within 4 Å from C11-P2/C6-P2 in TCS (F166, A184, L188, L215, N216, V223 and L225) and RTA (Y183, L207, F240, I247, P250 and I251). The amino acid composition and configuration were found to be different. At the corresponding position, we identified six residues S225, F226, G227, N230, A231 and G234 in Stx, which are also distinct from the corresponding residues in TCS and RTA. For the positively charged patch comparison, R176 and R179, K173 and R174, R189 and R193 were found in Stx, TCS and RTA, respectively. Although the distance between each pair is not the same, they may form charge-charge interactions with the DDD motif of P2. The different shape, electrostatic charge distribution of the P2 binding pocket and variant composition of hydrophobic patch suggested these RIPs possibly using a novel binding mode to interact with P2.

In addition, we previously determined that MOD interacts with the conserved C-terminal peptide of P2 without hydrophobic interactions [29], in a manner different to TCS or RTA. Four positive charged lysines K143-K146 were identified to be involved in matching the negative charged DDD motif on P2 [14]. They are located far from the positively charged region which corresponds to binding the acidic DDD motif on TCS (corresponding residues K173, R174 and K177). Comparison of primary sequences shows that these identified amino acid residues responsible for P protein interaction in Stx and MOD are not conserved in TCS and RTA (Figure 5e). The variability of composition of P2 binding site residues, pocket shape and charge distribution of RIPs, implying that C-terminal region of P protein adopts a flexible conformation to interact with different RIPs.

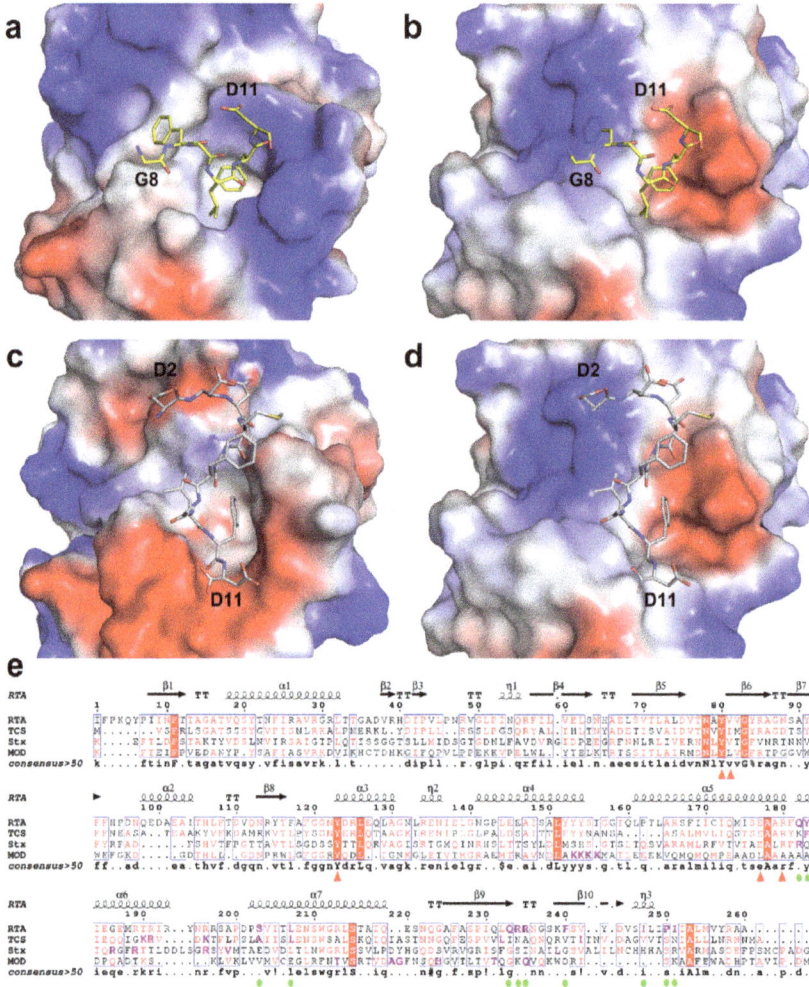

Figure 5. Comparison of surface polarity of P2 binding pocket between RTA and other typical RIPs. (**a**) Electrostatic surface potential map of C6-P2 binding pocket of RTA; (**b**) Electrostatic surface potential map of Stx (PDB code: 1R4Q) using C6-P2 as a reference; (**c**) Electrostatic surface potential map of TCS-C11-P2 (PDB code: 2JDL); (**d**) Electrostatic surface potential map of Stx using C11-P2 as a reference. The surface color represents electric potential, red color is negatively charged surface, blue color is positively charged surface. The neutral region is colored in white or lighter color shade. Electrostatic surface potential is generated by PyMol software; (**e**) Sequence alignment of RTA and other classical RIPs. The key residues of P2 binding pocket of RTA were marked with green dots, whereas the rRNA-glycosidase catalytic residues of RTA were labeled with red triangles. The top secondary structure elements are shown according to the crystal structure of RTA. The identified P2 binding site residues in the different RIPs are highlighted in purple.

Table 1. Crystal parameters, data collection, and structure refinement.

	RTA-C6-P2
Data collection	
Space group	*P 1 21 1*
Unit cell a, b, c (Å) α, β, γ (°)	67.45, 59.88, 67.50 90.00, 99.89, 90.00
Resolution range (Å)	25.00–1.55 (1.61–1.55) [a]
Unique reflections	76937
Completeness (%)	97.8 (96.4)
<I/σ(I)>	24.0 (3.5)
R$_{merge}$ [b] (%)	33.9 (4.6)
Average redundancy	3.6 (3.6)
Structure refinement	
Resolution range (Å)	24.65–1.55
R-factor [c]/R-free [d] (%)	25.6/21.7
Number of protein atoms	4366
Number of water atoms	234
RMSD [e] bond lengths (Å)	0.0261
RMSD bond angles (°)	1.500
Ramachandran plot [f] (residues, %)	-
Most favored (%)	95.8
Additional allowed (%)	4.2
Outliers (%)	0
PDB entry	5GU4

[a] The values in parentheses refer to statistics in the highest bin; [b] R$_{merge}$ = $\sum_{hkl}\sum_i |I_i(hkl) - <I(hkl)>| / \sum_{hkl}\sum_i I_i(hkl)$, where $I_i(hkl)$ is the intensity of an observation and <I(hkl)> is the mean value for its unique reflection; Summations are over all reflections; [c] R-factor = $\sum_h ||F_o(h)| - |F_c(h)|| / \sum_h |F_o(h)|$, where $|F_o|$ and $|F_c|$ are the observed and calculated structure-factor amplitudes, respectively; [d] R-free was calculated with 5% of the data excluded from the refinement; [e] Root mean square deviation from ideal values; [f] Categories were defined by Molprobity.

3. Discussion

In this study we have solved the structure of RTA complexed with C6-P2 peptide. In this structure, a hydrophobic pocket at the C-terminal domain of RTA donates a major interface to associate with the hydrophobic GFGLFD motif on the human ribosomal stalk P2 peptide. Nevertheless, our in vitro interaction assays also showed that the negatively charged DDD motif on P2 peptide does contribute to the binding of RTA (K$_D$ of RTA and C11-P2 is 2.3 μM, C6-P2 is 18.8 μM). Using structure modeling, we identified two arginines, R189, R193 probably interact with the N-terminal DDD motif of P2. This is consistent with previous biochemical data which identified four positively charged arginines R189, R193, R234 and R235 on RTA to be important in RIP-stalk binding [25].

Our structure of RTA-C9-P2 only captured the electron density of C6-P2; this may be caused by the flexibility of the N-terminal of P2, which was also observed in our previous structural study of TCS-C11-P2 complex. The asymmetric unit contains two molecules of TCS and two C11-P2 peptide (chain C and D). For chain C, C10-P2 (DDDMGFGLFD) could be fitted to the observed electron density and for chain D, only C9-P2 (DDMGFGLFD) was fitted [17]. A possible explanation is that RIPs form transient complexes with ribosomal stalk P proteins to facilitate their fast access to the targeted adenine on the SRL of the ribosome. It was previously evidenced that the interaction of RIP with ribosomal protein increased the efficiency of the rat ribosome inactivation activity of RTA, with a catalytic efficiency is 80,000 times faster than that on naked 28S rRNA [6]. The current structure may only represent a snapshot of a series of transient interactions in the course of SRL targeting. Even though we demonstrated hydrophobic interactions play an important role in mediating RTA and ribosomal stalk P protein interaction, electrostatic interactions might be involved in the subsequent proximity steps to the SRL of ribosome, as a previous study mentioned electrostatic interactions might be exploited by RTA to facilitate ribosome targeting in a similar mechanism of restrictocin [30].

According to the binding mode of P2 peptide to TCS and RTA, the shape and key residues of P2 binding pocket in the RIP may affect the P2 conformation. The structural differences implied that the C-terminus of P2 may as well adopt a flexible conformation for binding. Our previous studies also have shown that MOD interacts with the conserved C-terminal P2 peptide via charge-charge interactions but not hydrophobic interactions and the binding pocket in MOD may be far away from those of RTA and TCS [29]. Comparison of the primary sequences shows that the biochemically identified amino acid residues responsible for P protein interaction in Stx and MOD are not conserved in other RIPs (Figure 5e).

In sum, we demonstrate the amino acid composition, shape and polarity of P2 binding pocket of RIPs are versatile. This strengthened our hypothesis that P2 protein may adopt different conformation and orientation to bind RIP.

4. Materials and Methods

4.1. Cloning, Expression, and Purification of RTA and P2 Variants

RTA gene was cloned into PET28a with a hexahistidine tag at the N-terminus. The *Escherichia coli* BL21 (DE3) pLysS strain was used for the expression of recombinant protein. The transformed cells were grown at 37 °C in LB culture medium (Invitrogen, Life Technologies, Camarillo, CA, USA) containing appropriate antibiotics until the $OD_{600\ nm}$ reached about 0.8. Protein expression was then induced with 0.2 mM isopropyl 1-thio-β-D-galactopyranoside (Sigma-Aldrich, St. Louis, MO, USA) by another 20 h at 16 °C. Cells were harvested by centrifugation ($6000\times g$, 4 °C, 10 min) and resuspended in 40 mL of lysis buffer (50 mM Tris-Cl, pH 7.5, 150 mM NaCl, 5% (v/v) glycerol). After cell disruption and centrifugation, the supernatant containing the soluble target protein was purified with a HisTrap HP 5 mL column (GE Healthcare Biosciences, Pittsburgh, PA, USA) equilibrated with the binding buffer (20 mM Tris-Cl, pH 7.5, 150 mM NaCl). The target protein was eluted and further purified by Superdex 75 column (GE Healthcare Biosciences, Pittsburgh, PA, USA) equilibrated with 20 mM Tris-Cl, pH 7.5, 100 mM NaCl. Protein purity was evaluated by SDSPAGE, and concentrated to appropriate concentration by ultrafiltration (Millipore Amicon, Merk, Darmstadt, Germany). After liquid nitrogen freezing, protein samples were stored at −80 °C.

The last 6, 7, 8, 9, 11 and 17 residues of P2 were cloned into PGEX-4T-1 vector (GE Healthcare Life Sciences, Pittsburgh, PA, USA) with a GST-tag at the N-terminus. GST and the recombinant GST-tagged proteins were purified by glutathione sepharose chromatography (GE Healthcare Biosciences, Pittsburgh, PA, USA) and gel filtration individually using PBS pH 7.4 buffer (USB, Cleveland, OH, USA). Purity was assessed by SDS-PAGE and protein stored in the same manner as RTA.

4.2. Crystallization, Data Collection and Processing

Synthesized peptides C9-P2 and C11-P2 (GL Biochem, Shanghai, China) were added into RTA to final concentration 5 mM and incubated for 6 h at 4 °C before crystallization.

Commercially available Crystal Screen 1–2 and Index screen (Hampton Research) were used for crystallization trials in 96-well plates (XtalFinder, XtalQuest Inc., Beijing, China) at 16 °C. The crystals were obtained using the hanging drop vapor-diffusion method, by equilibrating 1 μL of 15 mg/mL RTA-C9-P2 mixture with an equal volume of the reservoir solution (2.8 M sodium acetate, tetrahydrate pH 7.0) (USB, Cleveland, OH, USA). Further optimization was carried out using Additive Screen kit (Hampton Research). Crystals which produced good diffraction quality were grown in 2.8 M sodium acetate tetrahydrate, pH 7.0, 30%–35% glucose. All the crystals were transferred to cryoprotectant (reservoir solution supplemented with 30% glycerol) and flash-cooled with liquid nitrogen. The data were collected at 100 K in a liquid nitrogen stream using beamline 13B1 with a Q315r CCD (ADSC, MAR Research, Norderstedt, Germany) at the Biological Crystallization Facility at National Synchrotron Radiation Research Center (NSRRC), Hsinchu, Taiwan. Data were scaled and merged with ScalePack installed with HKL2000 [31].

4.3. Structure Determination and Refinement

The structure of the RTA-C6-P2 complex was determined by molecular replacement with Phaser in CCP4 suite [32] using Protein Data Bank code 3PX8 as the search model. The initial model of the RTA was obtained and refined by REFMAC5 [33]. The C9-P2 peptide was manually built and refined in Coot [34]. The overall assessment of model quality was evaluated using the programs MOLPROBITY [35] and PROCHECK [36]. Sequence alignment was prepared using the online program Multalin (F.Corpet, INRA Toulouse, France) (http://multalin.toulouse.inra.fr/multalin/). The crystallographic parameters of the structure are listed in Table 1. All structure figures were prepared with PyMOL (DeLano Scientific, Palo Alto, CA, USA) [37].

4.4. In Vitro GST-Pull-Down Assays

To assess the interaction of RTA with GST-P2 variants, 45 μL 60 μM RTA was added into 50 μL 35 μM GST fusion P2 variants, incubated for 30 min at 4 °C. 20 μL pre-equilibrated glutathione-affinity resin was added into the system and then incubated for another 30 min. Then the beads were washed thoroughly by PBS pH7.4 buffer twice. Finally, the bound protein was eluted from beads with 20 μL 10 mM reduced glutathione. RTA alone was used as control. All the samples were detected by 15% SDS-PAGE.

4.5. Isothermal Titration Calorimetry (ITC)

The binding affinities of RTA toward P2 variants were determined by ITC-200 (Microcal, Northampton, MA, USA). All protein solutions were degassed without stirring and kept on ice before ITC experiments. Titrations were performed at 25 °C in buffer containing 20 mM Tris-Cl, 100 mM NaCl, pH 7.0 by injecting 20 consecutive aliquots (2 μL) of 600 μM P2 variant peptides solution into the ITC cell (200 μL) containing 35 μM RTA solution, using P2 variant peptides titrated into buffer as blank control. The data were collected and analyzed using Origin software (Microcal, Northampton, MA, USA). Thermodynamic parameters were determined using nonlinear least squares fitting, assuming a one-site binding model.

Supplementary Materials: The following are available online at www.mdpi.com/2072-6651/8/10/296/s1, Figure S1: Superposition of the structure of RTA-C9-P2 and RTA-C11-P2, Figure S2: The putative binding model of RTA and C11-P2, Figure S3: Structural superposition of RTA-C6-P2 with previous docking model, Table S1: Crystal parameters and data collection of RTA –C11-P2.

Acknowledgments: We thank Ka-Yee Au, Yinhua Yang and Alice Yuen-Ting Wong for protein expression and crystallization trials. We also thank beamline staff at BL13B1, Biological Crystallography Facility at NSRRC, Hsinchu, Taiwan for their excellent technical support. The work was supported by a One-Off Funding for Research (Project Ref. No. C4045-14G) from the Chinese University of Hong Kong.

Author Contributions: Wei-Wei Shi, Yun-Sang Tang and See-Yuen Sze designed the experiments and performed the experiments; Wei-Wei Shi, Yun-Sang Tang analyzed the data; See-Yuen Sze and Zhen-Ning Zhu contributed to protein expression and purification for in vitro GST-pull down and ITC measurements. Kam-Bo Wong and Pang-Chui Shaw supervised and revised the paper.

Conflicts of Interest: The authors declare no conflict of interest.

References

1. Olsnes, S. The history of ricin, abrin and related toxins. *Toxicon* **2004**, *44*, 361–370. [CrossRef] [PubMed]
2. Endo, Y.; Tsurugi, K. RNA *N*-glycosidase activity of ricin A-chain. Mechanism of action of the toxic lectin ricin on eukaryotic ribosomes. *J. Biol. Chem.* **1987**, *262*, 8128–8130. [PubMed]

3. Correll, C.C.; Munishkin, A.; Chan, Y.L.; Ren, Z.; Wool, I.G.; Steitz, T.A. Crystal structure of the ribosomal RNA domain essential for binding elongation factors. *Proc. Natl. Acad. Sci. USA* **1998**, *95*, 13436–13441. [CrossRef] [PubMed]

4. Lancaster, L.; Lambert, N.J.; Maklan, E.J.; Horan, L.H.; Noller, H.F. The sarcin-ricin loop of 23S rRNA is essential for assembly of the functional core of the 50S ribosomal subunit. *RNA* **2008**, *14*, 1999–2012. [CrossRef] [PubMed]

5. Wang, P.; Tumer, N.E. Pokeweed antiviral protein cleaves double-stranded supercoiled DNA using the same active site required to depurinate rRNA. *Nucleic Acids Res.* **1999**, *27*, 1900–1905. [CrossRef] [PubMed]

6. Endo, Y.; Tsurugi, K. The RNA N-glycosidase activity of ricin A-chain. The characteristics of the enzymatic activity of ricin A-chain with ribosomes and with rRNA. *J. Biol. Chem.* **1988**, *263*, 8735–8739. [PubMed]

7. Vater, C.A.; Bartle, L.M.; Leszyk, J.D.; Lambert, J.M.; Goldmacher, V.S. Ricin A chain can be chemically cross-linked to the mammalian ribosomal proteins L9 and L10e. *J. Biol. Chem.* **1995**, *270*, 12933–12940. [PubMed]

8. Chiou, J.C.; Li, X.P.; Remacha, M.; Ballesta, J.P.; Tumer, N.E. The ribosomal stalk is required for ribosome binding, depurination of the rRNA and cytotoxicity of ricin A chain in *Saccharomyces cerevisiae*. *Mol. Microbiol.* **2008**, *70*, 1441–1452. [CrossRef] [PubMed]

9. Tchorzewski, M. The acidic ribosomal P proteins. *Int. J. Biochem. Cell Biol.* **2002**, *34*, 911–915. [CrossRef]

10. May, K.L.; Li, X.P.; Martinez-Azorin, F.; Ballesta, J.P.; Grela, P.; Tchorzewski, M.; Tumer, N.E. The P1/P2 proteins of the human ribosomal stalk are required for ribosome binding and depurination by ricin in human cells. *FEBS J.* **2012**, *279*, 3925–3936. [CrossRef] [PubMed]

11. Nomura, N.; Honda, T.; Baba, K.; Naganuma, T.; Tanzawa, T.; Arisaka, F.; Noda, M.; Uchiyama, S.; Tanaka, I.; Yao, M.; et al. Archaeal ribosomal stalk protein interacts with translation factors in a nucleotide-independent manner via its conserved C terminus. *Proc. Natl. Acad. Sci. USA* **2012**, *109*, 3748–3753. [CrossRef] [PubMed]

12. Chan, D.S.; Chu, L.O.; Lee, K.M.; Too, P.H.; Ma, K.W.; Sze, K.H.; Zhu, G.; Shaw, P.C.; Wong, K.B. Interaction between trichosanthin, a ribosome-inactivating protein, and the ribosomal stalk protein P2 by chemical shift perturbation and mutagenesis analyses. *Nucleic Acids Res.* **2007**, *35*, 1660–1672. [CrossRef] [PubMed]

13. Chan, S.H.; Hung, F.S.; Chan, D.S.; Shaw, P.C. Trichosanthin interacts with acidic ribosomal proteins P0 and P1 and mitotic checkpoint protein MAD2B. *Eur. J. Biochem.* **2001**, *268*, 2107–2112. [CrossRef] [PubMed]

14. Yang, Y.; Mak, A.N.; Shaw, P.C.; Sze, K.H. Solution structure of an active mutant of maize ribosome-inactivating protein (MOD) and its interaction with the ribosomal stalk protein P2. *J. Mol. Biol.* **2010**, *395*, 897–907. [CrossRef] [PubMed]

15. McCluskey, A.J.; Poon, G.M.; Bolewska-Pedyczak, E.; Srikumar, T.; Jeram, S.M.; Raught, B.; Gariepy, J. The catalytic subunit of Shiga-like toxin 1 interacts with ribosomal stalk proteins and is inhibited by their conserved C-terminal domain. *J. Mol. Biol.* **2008**, *378*, 375–386. [CrossRef] [PubMed]

16. McCluskey, A.J.; Bolewska-Pedyczak, E.; Jarvik, N.; Chen, G.; Sidhu, S.S.; Gariepy, J. Charged and hydrophobic surfaces on the a chain of Shiga-like toxin 1 recognize the C-terminal domain of ribosomal stalk proteins. *PLoS ONE* **2012**, *7*, e31191. [CrossRef] [PubMed]

17. Too, P.H.; Ma, M.K.; Mak, A.N.; Wong, Y.T.; Tung, C.K.; Zhu, G.; Au, S.W.; Wong, K.B.; Shaw, P.C. The C-terminal fragment of the ribosomal P protein complexed to trichosanthin reveals the interaction between the ribosome-inactivating protein and the ribosome. *Nucleic Acids Res.* **2009**, *37*, 602–610. [CrossRef] [PubMed]

18. Lee, K.M.; Yu, C.W.; Chan, D.S.; Chiu, T.Y.; Zhu, G.; Sze, K.H.; Shaw, P.C.; Wong, K.B. Solution structure of the dimerization domain of ribosomal protein P2 provides insights for the structural organization of eukaryotic stalk. *Nucleic Acids Res.* **2010**, *38*, 5206–5216. [CrossRef] [PubMed]

19. Lee, K.M.; Yu, C.W.; Chiu, T.Y.; Sze, K.H.; Shaw, P.C.; Wong, K.B. Solution structure of the dimerization domain of the eukaryotic stalk P1/P2 complex reveals the structural organization of eukaryotic stalk complex. *Nucleic Acids Res.* **2012**, *40*, 3172–3182. [CrossRef] [PubMed]

20. Lee, K.M.; Yusa, K.; Chu, L.O.; Yu, C.W.; Oono, M.; Miyoshi, T.; Ito, K.; Shaw, P.C.; Wong, K.B.; Uchiumi, T. Solution structure of human P1*P2 heterodimer provides insights into the role of eukaryotic stalk in recruiting the ribosome-inactivating protein trichosanthin to the ribosome. *Nucleic Acids Res.* **2013**, *41*, 8776–8787. [CrossRef] [PubMed]

21. Bargis-Surgey, P.; Lavergne, J.P.; Gonzalo, P.; Vard, C.; Filhol-Cochet, O.; Reboud, J.P. Interaction of elongation factor eEF-2 with ribosomal p proteins. *Eur. J. Biochem.* **1999**, *262*, 606–611. [CrossRef] [PubMed]

22. Helgstrand, M.; Mandava, C.S.; Mulder, F.A.; Liljas, A.; Sanyal, S.; Akke, M. The ribosomal stalk binds to translation factors IF2, EF-Tu, EF-G and RF3 via a conserved region of the L12 C-terminal domain. *J. Mol. Biol.* **2007**, *365*, 468–479. [CrossRef] [PubMed]

23. Mochizuki, M.; Kitamyo, M.; Miyoshi, T.; Ito, K.; Uchiumi, T. Analysis of chimeric ribosomal stalk complexes from eukaryotic and bacterial sources: Structural features responsible for specificity of translation factors. *Genes Cells* **2012**, *17*, 273–284. [CrossRef] [PubMed]

24. Choi, A.K.; Wong, E.C.; Lee, K.M.; Wong, K.B. Structures of eukaryotic ribosomal stalk proteins and its complex with trichosanthin, and their implications in recruiting ribosome-inactivating proteins to the ribosomes. *Toxins* **2015**, *7*, 638–647. [CrossRef] [PubMed]

25. Li, X.P.; Kahn, P.C.; Kahn, J.N.; Grela, P.; Tumer, N.E. Arginine residues on the opposite side of the active site stimulate the catalysis of ribosome depurination by ricin A chain by interacting with the P-protein stalk. *J. Biol. Chem.* **2013**, *288*, 30270–30284. [CrossRef] [PubMed]

26. Santos, C.; Ballesta, J.P. The highly conserved protein P0 carboxyl end is essential for ribosome activity only in the absence of proteins P1 and P2. *J. Biol. Chem.* **1995**, *270*, 20608–20614. [CrossRef] [PubMed]

27. Mlsna, D.; Monzingo, A.F.; Katzin, B.J.; Ernst, S.; Robertus, J.D. Structure of recombinant ricin A chain at 2.3 A. *Protein Sci.* **1993**, *2*, 429–435. [CrossRef] [PubMed]

28. Robertus, J.D.; Monzingo, A.F. The structure of ribosome inactivating proteins. *Mini Rev. Med. Chem.* **2004**, *4*, 477–486. [CrossRef] [PubMed]

29. Wong, Y.T.; Ng, Y.M.; Mak, A.N.; Sze, K.H.; Wong, K.B.; Shaw, P.C. Maize ribosome-inactivating protein uses Lys158-Lys161 to interact with ribosomal protein P2 and the strength of interaction is correlated to the biological activities. *PLoS ONE* **2012**, *7*, e49608. [CrossRef] [PubMed]

30. Korennykh, A.V.; Correll, C.C.; Piccirilli, J.A. Evidence for the importance of electrostatics in the function of two distinct families of ribosome inactivating toxins. *RNA* **2007**, *13*, 1391–1396. [CrossRef] [PubMed]

31. Otwinowski, Z.; Minor, W. Processing of X-ray diffraction data collected in oscillation mode. *Macromol. Crystallogr.* **1997**, *276*, 307–326.

32. Collaborative Computational Project, Number 4. The CCP4 suite: Programs for protein crystallography. *Acta Crystallogr. D Biol. Crystallogr.* **1994**, *50*, 760–763. [CrossRef] [PubMed]

33. Murshudov, G.N.; Skubak, P.; Lebedev, A.A.; Pannu, N.S.; Steiner, R.A.; Nicholls, R.A.; Winn, M.D.; Long, F.; Vagin, A.A. Refmac5 for the refinement of macromolecular crystal structures. *Acta Crystallogr. D Biol. Crystallogr.* **2011**, *67*, 355–367. [CrossRef] [PubMed]

34. Emsley, P.; Cowtan, K. Coot: Model-building tools for molecular graphics. *Acta Crystallogr. D Biol. Crystallogr.* **2004**, *60*, 2126–2132. [CrossRef] [PubMed]

35. Davis, I.; Leaver-Fay, A.; Chen, V.; Block, J.; Kapral, G.; Wang, X.; Murray, L.; Arendall, W. Molprobity: All-atom contacts and structure validation for proteins and nucleic acids. *Nucleic Acids Res.* **2007**, *35*, W375–W383. [CrossRef] [PubMed]

36. Laskowski, R.; Macarthur, M.; Moss, D.; Thornton, J. PROCHECK: A program to check the stereochemical quality of protein structures. *J. Appl. Crystallogr.* **1993**, *26*, 283–291. [CrossRef]

37. DeLano, W. *The PyMOL Molecular Graphics System*; De Lano Scientific: San Carlos, CA, USA, 2002.

toxins

MDPI

Review

Microvesicle Involvement in Shiga Toxin-Associated Infection

Annie Villysson, Ashmita Tontanahal and Diana Karpman *

Department of Pediatrics, Clinical Sciences Lund, Lund University, 22184 Lund, Sweden;
annie.villysson@med.lu.se (A.V.); ashmita.tontanahal@med.lu.se (A.T.)
* Correspondence: diana.karpman@med.lu.se; Fax: +46-46-222-0748

Academic Editors: Julien Barbier and Daniel Gillet
Received: 31 October 2017; Accepted: 16 November 2017; Published: 19 November 2017

Abstract: Shiga toxin is the main virulence factor of enterohemorrhagic *Escherichia coli*, a non-invasive pathogen that releases virulence factors in the intestine, causing hemorrhagic colitis and, in severe cases, hemolytic uremic syndrome (HUS). HUS manifests with acute renal failure, hemolytic anemia and thrombocytopenia. Shiga toxin induces endothelial cell damage leading to platelet deposition in thrombi within the microvasculature and the development of thrombotic microangiopathy, mostly affecting the kidney. Red blood cells are destroyed in the occlusive capillary lesions. This review focuses on the importance of microvesicles shed from blood cells and their participation in the prothrombotic lesion, in hemolysis and in the transfer of toxin from the circulation into the kidney. Shiga toxin binds to blood cells and may undergo endocytosis and be released within microvesicles. Microvesicles normally contribute to intracellular communication and remove unwanted components from cells. Many microvesicles are prothrombotic as they are tissue factor- and phosphatidylserine-positive. Shiga toxin induces complement-mediated hemolysis and the release of complement-coated red blood cell-derived microvesicles. Toxin was demonstrated within blood cell-derived microvesicles that transported it to renal cells, where microvesicles were taken up and released their contents. Microvesicles are thereby involved in all cardinal aspects of Shiga toxin-associated HUS, thrombosis, hemolysis and renal failure.

Keywords: Shiga toxin; hemolytic uremic syndrome; enterohemorrhagic *Escherichia coli*; microvesicles; kidney

1. Introduction

Shiga toxin-producing *Escherichia coli* (STEC) or enterohemorrhagic *E. coli* (EHEC) may cause disease in humans manifesting with diarrhea, bloody diarrhea (hemorrhagic colitis) and, in approximately 15% of cases, the severe systemic complication of hemolytic uremic syndrome (HUS) [1]. HUS is characterized by the post-diarrheal acute onset of non-immune hemolytic anemia, thrombocytopenia and renal failure. The most common clinical EHEC isolate is *E. coli* O157:H7 [2], although many other non-O157 serotypes have been described, notably the *E. coli* O104:H4 serotype that caused a huge outbreak in 2011 [3]. EHEC is a non-invasive pathogen [4] that colonizes the intestine where it expresses and also releases virulence factors. Some of these allow adherence to the intestinal mucosa by forming attaching and effacing lesions leading to colonization [5], while flagella are associated with bacterial motility [6]. EHEC interaction with commensal strains and host hormones enhances colonization and virulence by a genetically determined phenomenon known as quorum sensing [7]. The major and unique virulence factor strongly associated with EHEC-induced morbidity is Shiga toxin [8]. In addition, EHEC possesses lipopolysaccharide (LPS) and other factors capable of activating the host response [9]. A prerequisite for the strain to cause systemic and target organ damage, such as renal failure or brain damage [10], is the ability of virulence factors to gain access to the bloodstream and thereby reach target organ cells.

Shiga toxin may be capable of binding to intestine epithelial cells and thereafter translocate [11–13]. The intestinal inflammatory response is multifactorial depending on the interaction between the toxin, other virulence factors, and the host response [9]. Shiga toxin-producing EHEC strains are diarrheogenic. The diarrhea may become bloody leading to hemorrhagic colitis. This form of intestinal injury appears to be specifically associated with Shiga toxin production, as demonstrated in a monkey model of Shigella infection [14]. The massive erosion of the intestinal mucosal lining allows virulence factors released from EHEC to gain access to the circulation. Once within the bloodstream most of the toxin does not circulate in free form [15,16] but rather bound to blood cells such as leukocytes [17] and platelets as well as aggregates between these cells [18]. Red blood cells are also capable of binding the toxin [19,20]. Blood cells are activated by toxin binding and, thereafter, shed microvesicles which are pro-inflammatory, pro-thrombotic [18], and, importantly, transport the toxin to its target organ [21]. This does not exclude other mechanisms of toxin transfer from blood cells to affected cells [22], but has been suggested to be one of the main mechanisms of toxin-induced systemic and targeted organ injury [1].

Microvesicles are a subtype of extracellular vesicles shed directly from the plasma membrane of cells upon activation, stress and apoptosis [23]. Microvesicles can originate from blood cells [24–26] as well as non-circulating organ-specific cells [27,28]. Vesicles may be enriched in components of the parent cells such as proteins, receptors, RNAs (mRNA and miRNA) and lipids, enabling them to interact with cells in their immediate vicinity and at a distance [29]. Vesicle release may also maintain cellular integrity by ridding the cell of harmful substances [30]. Increasing evidence suggests that microvesicles are key players in several diseases, including cancer [31], renal diseases [32], cardiovascular disease [33] and inflammatory diseases [34]. In these diseases, the number of circulating microvesicles is significantly increased, indicating a disruption in physiological processes. In Shiga toxin-associated disease, Shiga toxin-bearing microvesicles have been found in the circulation of EHEC-infected patients as well as within the kidney [21], enabling toxin evasion of the immune system and thereby protection of the toxin from degradation. This review will mainly focus on the functions of microvesicles, in general and in the context of bacterial infections, particularly with respect to Shiga toxin-associated infection.

2. Shiga Toxin

Shiga toxin, encoded by a bacteriophage, is released from bacteria in the gut, most probably during bacterial lysis [35]. Shiga toxin is a ribosomal-inactivating protein. It is an AB_5 toxin composed of two subunits, an A-subunit and a pentameric B-subunit, linked together by non-covalent bonds [36]. The A-subunit accounts for the enzymatic cytotoxic activity whereas the pentameric B-subunit binds to glycosphingolipid receptors mainly the globotriaosylceramide (Gb3) receptor [37,38] and, to a lesser extent, the Gb4 receptor [39]. The density of Gb3 in the cell membrane and its association with lipid rafts affect toxin binding [40].

After Shiga toxin binds to its glycolipid receptor it can be taken up by endocytosis. Various endocytic routes have been described involving formation of membrane microtubular structures mainly in a clathrin-independent manner but also by a clathrin-dependent mechanism [41–44], as recently reviewed [45]. Uptake in intestinal cells by macropinocytosis, in a Gb3-independent manner, has also been reported [46,47]. Once within a cell, Shiga toxin is ultimately destined to reach ribosomes in the cytosol [48]. Shiga toxin is transported in a retrograded manner from early endosomes to the trans-Golgi network and further to the endoplasmic reticulum. Within the endoplasmic reticulum the A subunit is cleaved by furin into the A1 and A2 subunits [49]. From the endoplasmic reticulum, Shiga toxin is transported out to the cytosol, accessing the ribosomes [50].

2.1. Cytotoxicity of Shiga Toxin

The enzymatically active A1 subunit of Shiga toxin exerts a cytotoxic effect by *N*-glycosidic cleavage of a specific adenine base from 28S rRNA [51] leading to inhibition of protein synthesis followed by cell death. Moreover, Shiga toxin activates apoptotic pathways [52,53], most probably by inducing ribosomal damage and further activation of mitogen-activated protein (MAP) kinase pathways, the so-called ribosomal stress response [54,55]. Shiga toxin has been shown to induce intestinal cell apoptosis [56] and renal cell apoptosis in vivo and in vitro [52].

2.2. Inflammatory Effects of Shiga Toxin and Lipopolysaccharide in the Intestine

In addition to its cytotoxic effects, Shiga toxin is capable of activating an inflammatory response in the intestine, in its interaction with blood cells and after binding to target organ cells [9]. These effects occur simultaneously with the cytotoxic effects and are associated with the release of a wide range of pro-inflammatory cytokines and chemokines. In vitro studies have shown that Shiga toxin can trigger neutrophil influx into the intestine by inducing the release of interleukin (IL)-8 and other C-X-C chemokines [57–59]. The interaction with peritoneal macrophages also led to release of pro-inflammatory cytokines such as tumor necrosis factor-alpha (TNF-α) and IL-6 [60]. LPS may also contribute to the inflammatory response in the intestine [60]. Studies in mice have indicated that the initial host response to *E. coli* O157:H7 LPS is essential for bacterial elimination from the gut, thus mice lacking an adequate response to LPS were subject to more severe disease [61,62].

2.3. Interactions between Shiga Toxin and Blood Cells

During HUS, elevated neutrophil counts and decreased platelet counts suggest a worse prognosis [63]. Shiga toxin may circulate bound to neutrophils in vivo [17,64], and aggregates between platelets and neutrophils [18]. Both Shiga toxin and LPS induce the formation of these cell aggregates in which neutrophils are activated. Thus, the toxin may induce neutrophil activation and degranulation [9]. Similarly, the toxin may bind to monocytes [65] and has been detected on monocytes or platelet-monocyte aggregates in patient samples [18]. Shiga toxin induces the release of pro-inflammatory cytokines from monocytes including IL-6, IL-8, TNF-α, IL-1β and RANTES [65,66], as well as the expression of pro-thrombotic tissue factor activating the extrinsic pathway of coagulation leading to thrombin generation and blood clotting [18,67].

Platelets are activated during HUS and deposit in microthrombi. Their aggregation and consumption on injured endothelium leads to lowered platelet counts [68,69]. In the circulation, platelets are degranulated [70] with consequent elevated platelet-derived proteins such as platelet factor-4, β-thromboglobulin and P-selectin [71,72]. Shiga toxin binds to platelets via Gb3 and an alternative glycolipid receptor [73]. LPS binds to platelets via a toll-like receptor (TLR)4-CD62 receptor complex [74] and activates platelets. LPS derived from *E. coli* O157 was particularly potent in this respect [74]. Once activated, platelets further respond to Shiga toxin [75,76], may take up the toxin and exhibit excessive fibrinogen binding, enhancing their thrombotic potential. Platelet activation induced by Shiga toxin may be further exacerbated by toxin-mediated endothelial cell damage, exposing the subendothelium with release of von Willebrand factor, fibrinogen and collagen [68,69], as well as complement deposition. Furthermore, complement deposits on platelet and platelet–leukocyte aggregates in response to stimulation with Shiga toxin and *E. coli* O157 LPS [77]. Studies have shown that complement deposition on platelets initiates thrombin-mediated aggregation [78]. Moreover, once thrombin is formed [79], it can further propagate complement system activation [80,81] enhancing the inflammatory activity at the interface of platelets and the endothelium [82]. Once activated, platelets may contribute to the inflammatory state by the release of potent chemokines, as previously reviewed [9,69].

Patients with HUS exhibit acute hemolysis with fragmented red blood cells [1]. Hemolysis may be caused by the mechanical breakdown of red blood cells within occluded blood vessels, but complement activation on red blood cells may also contribute to this process, since it is known that complement deposition on red blood cells induces hemolysis [83]. Patients with HUS exhibit C3 deposition on red blood cells [19], suggesting that complement is activated on blood cell surfaces. Shiga toxin binds to the Gb3 receptor on red blood cells, also known as the P^k antigen of the P1Pk blood group system [20,84]. In vitro experiments demonstrated that Shiga toxin induced hemolysis by activating complement on human red blood cells in the presence of plasma, an effect abrogated by complement inactivation or by addition of the terminal complement pathway inhibitor eculizumab, directed against C5 [19].

Taken together, Shiga toxin is capable of binding to neutrophils, monocytes, platelets and red blood cells and thus it may be transported on, or within, blood cells in the circulation. Blood cells are activated thus potentiating the inflammatory and thrombotic process occurring during HUS. Furthermore, the toxin may thereby reach its target organs [85], mainly the kidney and the brain, although this mechanism of transfer does not fully clarify how the toxin is released from blood cells and taken up by recipient cells [86].

2.4. Thrombus Formation During HUS

Shiga toxin binds to endothelial and epithelial cells expressing the Gb3 receptor, such as the glomerular endothelium and the tubular epithelium in the human renal cortex [52,87]. Interestingly, cytokine release and exposure to bacterial LPS enhances Gb3 expression and augments toxin binding [88,89]. The mechanisms contributing to thrombus formation are multifactorial involving toxin-mediated endothelial cell damage [90], platelet activation [74,76], the formation of platelet-monocyte aggregates expressing tissue factor [18] as well as the release of inflammatory mediators from monocytes and endothelial cells that further activate platelets [66,91]. In addition, activation of coagulation and impaired fibrinolysis occur before HUS develops [79] promoting thrombus stabilization during HUS, after which fibrinolysis is enhanced [92–94]. Of note, the fibrinolytic system is activated in the murine kidney both in the endothelium and in tubular epithelial cells [95] but Shiga toxin may actually lower production of fibrinolytic parameters by inhibiting protein synthesis [94] and thereby tip the balance towards enhanced thrombus formation.

2.5. Shiga Toxin Induces the Release of Blood Cell-Derived Microvesicles

Binding of Shiga toxin to circulating blood cells may initiate cell activation and the release of blood cell-derived microvesicles, particularly in the presence of LPS [18,77]. During the acute phase of HUS, levels of circulating microvesicles are elevated [18,19,21,96] and Shiga toxin was detected within blood cell-derived microvesicles originating from neutrophils, monocytes, platelets and red blood cells [21]. In the following section, we will describe extracellular vesicles, their formation, shedding and uptake. We will also describe the function of microvesicles in normal cellular interactions and contribution to pathological processes, in particular during infectious diseases. Thereafter, we will focus on their contribution to HUS.

3. Characteristics of Extracellular Vesicles

Extracellular vesicles are membranous particles, released from cells, that are characterized based on their cell of origin and their size (Figure 1). Exosomes (30–100 nm in diameter) are formed by the release of the contents of intracellular endosomal multivesicular bodies containing intraluminal vesicles. Once extruded, these vesicles are termed exosomes [97]. Microvesicles (100–1000 nm) are released from cells by direct budding of the plasma membrane, while apoptotic bodies (1–5 μm) are released by the breakdown of cells during programmed cell death [98]. The latter are distinctly different from other vesicles because they contain larger cellular degradation products, such as organelles. The properties of extracellular vesicles are summarized in Table 1. Detection methods include flow cytometry, nanoparticle tracking and transmission electron microscopy, among others, as well as proteomic analysis of vesicular contents, as reviewed elsewhere [99–102]. Extracellular vesicles can be differentiated based on their mechanism of secretion and based on cellular markers [29,103], often allowing to determine the parent cell from which the vesicles were released. Web-based databases are available in which data regarding vesicle content and properties is summarized (see Vesiclepedia, ExoCarta or EVpedia, online).

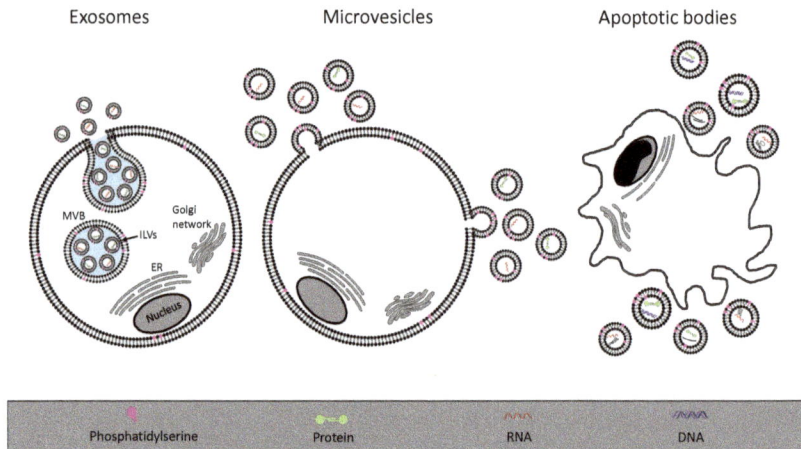

Figure 1. Schematic presentation of the mechanisms of extracellular vesicle release. Exosomes, shown on the left, are derived from multivesicular bodies (MVB) that are late endosomes carrying intraluminal vesicles (ILVs). MVBs fuse with the plasma membrane to release the ILVs that are termed exosomes once shed into the extracellular space. Microvesicles, shown in the middle panel, originate from a direct budding of the plasma membrane. Apoptotic bodies, depicted on the right, are released during apoptosis and cellular degradation. ER: endoplasmic reticulum.

Table 1. Classification of extracellular vesicles.

Extracellular Vesicle	Origin	Size	Content	References
Exosome	Intraluminal vesicle within multivesicular bodies	30–100 nm	mRNA, miRNA, proteins, lipids	[97]
Microvesicle	Shedding from the plasma membrane with cellular content	100–1000 nm	mRNA, miRNA, proteins, lipids	[29]
Apoptotic body	Cellular breakdown and shrinkage during apoptosis	1–5 μm	Organelles, proteins, DNA, RNAs, lipids	[97]

3.1. Exosomes

Intracellular multivesicular bodies are late endosomes containing intraluminal vesicles. When they dock onto the plasma membrane and fuse, the vesicles are released as exosomes into the extracellular space [104]. The formation of intraluminal vesicles and their release as exosomes is mainly regulated by two pathways, either via the endosomal sorting complex required for transport (ESCRT) machinery [105,106] or by an ESCRT-independent process (including tetraspanins and ceramide generation) [107,108]. Exosomes are characteristically different from other vesicle populations due to their endosomal origin with a unique composition of lipids and proteins. Compared to the parent cell, exosomes may be enriched in cholesterol, glycosphingolipids, sphingomyelin and phosphatidylserine [109]. Proteins enriched in exosomes include tetraspanins [110], heat shock proteins, component of the ESCRT family [111] and MHC class I [112]. While these proteins are generally enriched in exosomal populations, other proteins may be specifically enriched in exosomes dependent on the cell of origin [111,113]. Comprehensive reviews regarding exosomal formation and their biological functions have been published elsewhere [103,104,114], and herein we will focus on microvesicles.

3.2. Microvesicles

Microvesicles are shed directly from the plasma membrane together with contents of the parent cell. The content may include proteins (cytokines, chemokines, enzymes, receptors) [32,115], RNAs (mRNAs and non-coding RNAs, particularly miRNAs) [23] and lipids [116]. The content will vary depending on the cell of origin and reflect the biological status of the cell and degree of activation [117]. The structure of microvesicles protects bioactive materials and enables the transfer of such substances from one cell to another as a means of cell-to-cell communication [29,118,119]. Microvesicles can be taken up by neighboring or distant cells and have the potential to phenotypically alter the recipient cell [120,121], as elaborated on below. Another important function of microvesicles is to discard the parent cell of unwanted substances and thus preserve cellular integrity [122–124].

3.2.1. Microvesicle Formation

The underlying mechanism of microvesicle formation involves numerous cellular events leading to local budding of the plasma membrane, followed by a fission event whereby the vesicle is pinched off and released into the extracellular space [125]. Shedding of microvesicles occurs spontaneously in resting cells [23], and is increased during cell activation in response to various stimuli such as hypoxia, oxidative stress, shear stress, pro-inflammatory mediators, cell damage or ligand-binding [126–128]. A common feature during activation is an increase in the cytosolic calcium level, which promotes microvesicle release [129]. Increasing evidence suggests that the inner content as well as the surface of microvesicles are selectively packaged, rather than randomly, by a controlled mechanism of cargo trafficking within microvesicles, as reviewed [121]. The regulatory protein, ADP-ribosylation factor 6, selectively recruits protein cargo onto microvesicles and has been shown to be an important component of this process [130]. The exact mechanisms by which proteins, lipids and RNAs are targeted into microvesicles remain to be elucidated. The process is affected by external stimuli triggering vesicle release, as well as by the cellular environment [126]. The same cell may thus release microvesicles with varied content [131]. Interestingly, microvesicles may incorporate membranous contents at a higher density than the parent cell due to enrichment in lipid rafts [132], thereby enhancing the procoagulant potency of platelet-derived microvesicles [133].

In resting cells, plasma membrane lipids are arranged in an asymmetric pattern whereby the outer leaflet contains phosphatidylcholine and sphingomyelin, and the inner membrane leaflet is enriched with aminophospholipids (phosphatidylserine and phosphatidylethanolamine) [134,135]. Phospholipid asymmetry is controlled by specific enzymes such as flippase (transfers phosphatidylserine to the inner leaflet), floppase (controls outward phospholipid translocation), and enzymes with scramblase activity (transport lipids in a bidirectional Ca^{++}-dependent manner) [136,137]. During cell activation, concurrent with a rise in the intracellular calcium level, floppase and scramblase activities are activated, while flippase is inactivated leading to disruption of lipid asymmetry with exposure of phosphatidylserine on the outer leaflet. The increased calcium level also enhances the activity of cytosolic enzymes such as calpain, involved in cytoskeleton changes that facilitate microvesicle shedding [138]. Microvesicle membranes are characterized by a loss of lipid asymmetry, in comparison to the parent cell, and may expose phosphatidylserine, although this is not a feature of all microvesicles [126]. The mechanism associated with the release of phosphatidylserine-negative microvesicles is poorly understood. A possible mechanism involving cellular proteins able to alter the membrane curvature has been described [139].

3.2.2. Microvesicle Uptake

Uptake of microvesicles by recipient cells can be mediated by various endocytic pathways or fusion with the plasma membrane. Endocytosis may be facilitated by receptor-ligand interactions.

Uptake may involve clathrin-dependent or -independent endocytosis, caveolin-dependent endocytosis, macropinocytosis, phagocytosis or lipid raft-mediated uptake, as reviewed [140]. Fusion of microvesicles with the membrane of the target cell is a highly dynamic process, resembling the process utilized by retroviruses, involving high affinity binding to the target cell, lipid reorganization and restructuring of proteins [141]. The fusion event enables the bioactive vesicular content to be inserted into the cytosol of the target cell [142]. Several factors have been shown to affect fusion; these include an acidic environment, the degree of ligand-receptor binding and the lipid composition of the microvesicle [143].

The uptake of a microvesicle may depend on the presence of receptors and/or proteins on the microvesicle surface and their interaction with counterparts on the recipient cell [140,144]. These interactions may lead to the formation of a microvesicle–cell complex, thereby coating the cell and possibly initiating intracellular signaling events due to receptor-ligand interactions. The microvesicle may be taken up by the cell and release its content into the cell [119]. Ligands specifically enriched on microvesicles such as glycoproteins (integrins, selectins and tissue factor) and phospholipids may favor cell-specific uptake [145–148]. For example, platelet-derived microvesicles transfer tissue factor to monocytes but not to neutrophils [149], demonstrating uptake by a defined cell type.

3.2.3. The role of Microvesicles in Intercellular Communication

The diversity of bioactive materials covering the surface of microvesicles and contained within microvesicles may affect the phenotype of the recipient cell by transfer of nucleic acids, proteins, receptors and lipids. RNAs may be translated in the target cell activating or silencing certain properties [150–152]. The physiological or pathological consequences may result in induction of angiogenesis, thrombosis, altered hemostasis, immune modulation, invasive potential, matrix alteration and tissue regeneration, as recently reviewed [32]. The transfer of functionally active receptors may activate pro-inflammatory or pro-thrombotic signaling pathways as well as proliferative capacity in the recipient cell, as has been exemplified regarding chemokine receptors [153,154], epidermal growth factor receptor [155], kinin B1 receptor [156] and platelet receptors [157]. The transfer of the platelet receptor GPIIb/IIIa to neutrophils will induce neutrophil binding to the endothelium via fibronectin and thereby have a pro-inflammatory effect [158]. Cancer cell-derived microvesicles have been shown to bear epidermal growth factor receptor, that could be transferred to neighboring cells and influence cell morphology and growth capacity [120]. The invasive capacity has been demonstrated in microvesicles derived from cancer cells which are enriched in metalloproteinases capable of breaking down extracellular matrix to promote tumor growth [159].

Microvesicles derived from blood and endothelial cells may induce an inflammatory response. As mentioned above, microvesicles can transfer chemokine receptors [153]. Furthermore, microvesicles from platelets or from endothelial cells can induce monocyte [160] or neutrophil chemotaxis [161]. The recruited monocytes will deposit on endothelial cells [162]. Likewise, platelet-derived microvesicles can recruit CD34-positive hematopoietic cells [157]. The immunomodulatory effects of microvesicles also encompass the transfer of anti-inflammatory mediators (RNAs or proteins), particularly studied in microvesicles derived from stem cells [163] or neutrophils [164], potentially contributing to a beneficial effect during tissue regeneration.

Importantly, microvesicles may possess pro-thrombotic potential. The exposure of phosphatidylserine on the outer leaflet of microvesicles generates a negatively-charged surface promoting binding of prothrombin, factor Va and factor Xa [165]. When scramblase activity is defective, such as in Scott syndrome, the number of phosphatidylserine-positive microvesicles is reduced and coagulation is impaired resulting in a bleeding disorder [166,167]. Phosphatidylserine exposure on monocytic microvesicles was shown to differ depending on the stimulus inducing microvesicle release suggesting that environmental factors can affect the composition of microvesicles and their pro-thrombotic properties [128]. In addition, certain microvesicles, mainly from monocytes and platelets, carry tissue factor on their surface contributing to thrombus formation [18,146,168,169]. Monocyte-derived microvesicles transfer tissue factor to platelets [170]. Red blood cell-derived microvesicles are also thrombogenic and may induce thrombin generation in the absence of tissue factor by activation of the intrinsic coagulation pathway on their surfaces [171]. Similarly, platelet-derived microvesicles containing arachidonic acid metabolized to thromboxane A2 induce platelet aggregation [116].

4. Microvesicles in Infectious Diseases

Infections, and sepsis in particular, are accompanied by a pro-inflammatory and pro-thrombotic host response. The properties of blood cell-derived microvesicles can contribute to these effects and to multi-organ failure. Patients with sepsis have high levels of platelet- and leukocyte-derived microvesicles [172–175]. Blood cell-derived microvesicles were tissue factor-positive in patients with meningococcal sepsis [172] and febrile urinary tract infection [176]. Patients with disseminated intravascular coagulopathy had elevated endothelial and leukocyte microvesicles [177]. An enhanced inflammatory response during sepsis, as reflected by microvesicle levels, may, however, predict a more favorable outcome [178], and circulatory microvesicle levels correlated negatively with the degree of acute kidney injury during sepsis, although this may actually reflect the degree of microvesicle deposition in damaged tissue [174].

When microvesicles released during sepsis were injected into mice they enhanced vascular reactivity and thromboxane A2 production [175]. Likewise, microvesicles derived from rats with peritonitis induced an inflammatory response when injected into healthy rats [179]. Thus, microvesicles may have damaging effects during sepsis [180,181]. Beneficial effects may also prevail as neutrophil-derived microvesicles, that were elevated during bacteremia, exhibited anti-microbial effects [182]. Similar effects were also demonstrated as blood cell-derived microvesicles prevented in vivo spreading of *Streptococcus pyogenes* [183].

Microvesicles May Transfer Infectious Agents or Their Virulence Factors

Bacteria and viruses have developed ways of utilizing extracellular vesicles for the transfer of their antigens and virulence factors. For example, *Mycobacterium tuberculosis* infected macrophages release microvesicles containing *M. tuberculosis* antigen in complex with MHC-II that can induce an antimicrobial T-cell response [184]. HIV budding from cells can utilize components of the ESCRT pathway and yet be shed directly from the cell membrane [185,186] suggesting that viral dissemination may occur via extracellular vesicles [187] possibly utilizing phosphatidylserine receptors [188]. Other viruses, such as herpes viruses, also utilize extracellular vesicles to transfer viral RNAs and proteins from cell to cell [189,190].

Pathogens can thereby exploit the microvesicle transport system for their own benefit to evade host response by covering themselves with the host membrane within a vesicle. The vesicle containing pathogen virulence factors is taken up by recipient cells thereby affecting target cells. Interestingly, pathogens can also release their own vesicles. Protozoan parasites, such as *Trypanosoma cruzi*, release vesicles capable of interacting with host cells [191,192]. Similarly, mycobacterium may release vesicles capable of inducing an inflammatory response in the host [193]. Outer membrane vesicles are released from gram-negative bacteria, such as EHEC, and may transfer toxins, such as Shiga toxin, into intestinal epithelial cells as well as renal or brain endothelial cells [194]. The major difference between pathogen-derived vesicles [194] and host blood cell-derived microvesicles [21], both containing virulence factors, such as Shiga toxin, is that the latter, host-derived extracellular vesicles, will be recognized as "self" and thereby avoid attack by the immune system whereas pathogen-derived vesicles are "non-self" and subject to immune attack.

5. Shiga Toxin-Induced Microvesicles in Laboratory Models

Shiga toxin binds to neutrophils, monocytes, platelets and red blood cells [20,22,64,65,73]. These blood cells are resistant to the cytotoxic effects of the toxin, in part due to minimal protein synthesis in platelets and red blood cells. The cells may instead become activated and release microvesicles [18,19,66,76,77]. In vitro experiments have shown that Shiga toxin induces the release of microvesicles from human monocytes, platelets [18,77] and red blood cells [19]. Co-stimulation with Shiga toxin and *E. coli* O157-LPS enhanced microvesicle release from platelets and leukocytes compared to each stimulant alone [18]. Microvesicles shed from platelets and monocytes carry tissue factor and phosphatidylserine [18], as well as complement C3 and C9 [77]. C5b-9 was also detected on microvesicles from red blood cells which also exposed phosphatidylserine after stimulation with Shiga toxin [19]. The toxin itself may be incorporated in blood cell-derived microvesicles originating from neutrophils, monocytes, platelets and red blood cells [21]. Permeabilization of microvesicles was required to detect Shiga toxin, suggesting that the toxin was mainly localized within the microvesicles and not on their outer membrane [21].

Toxin within microvesicles can be transferred to target organ cells and be taken up after endocytosis of the entire microvesicle [21] (Figure 2). In a mouse model of EHEC infection [62], blood cell-derived microvesicles carrying Shiga toxin were demonstrated to be taken up by renal glomerular endothelial cells and tubular cells [21]. One notable aspect of this finding is that, in contrast to the human glomerulus, mouse glomerular endothelial cells lack the toxin Gb3 receptor [87]. Thus, it is possible that this process is Gb3- or toxin receptor-dependent up until the toxin binds to its receptor on blood cells and is internalized in these cells. After the toxin is shed from the blood cells, within microvesicles, the toxin-containing microvesicles may be taken up even by cells that lack the Gb3 [1].

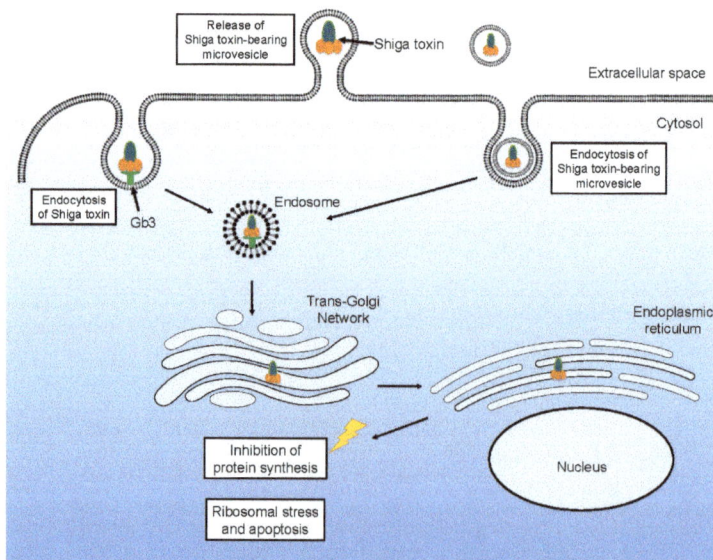

Figure 2. Shiga toxin uptake and intracellular routing. Shiga toxin may be taken up directly after binding to its glycolipid receptor Gb3. Alternatively, the toxin may be endocytosed within a microvesicle. Either way the toxin will undergo retrograde transport via the Golgi to the endoplasmic reticulum and the ribosomes where it may inhibit protein synthesis or induce ribosomal stress and apoptosis.

6. Microvesicles in the Pathogenesis of Hemolytic Uremic Syndrome

During EHEC-associated HUS, only minimal amounts of freely circulating Shiga toxin have been found in patient samples and Shiga toxin is mainly found bound to blood cells [15,18]. In acute phase HUS patients, the number of circulating blood cell-derived microvesicles is elevated [18,19,21,96], decreasing after recovery. Shiga toxin circulates within microvesicles originating from neutrophils, monocytes, platelets and red blood cells [21]. Shiga toxin-containing microvesicles from blood cells were further demonstrated to be taken up by glomerular endothelial cells in a patient sample [21]. Kidney cells, including glomerular endothelial cells, mesangial cells, podocytes and tubular cells are highly sensitive to Shiga toxin [52,195,196]. The effects of Shiga toxin and blood cell-derived microvesicles in the circulation of acute phase HUS patients are described in Figure 3.

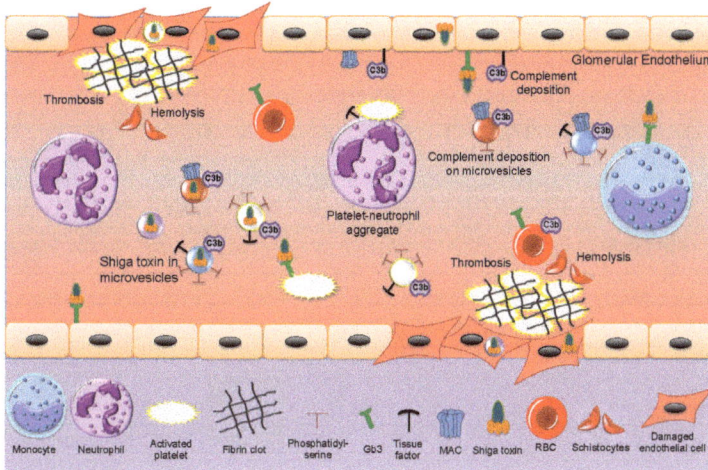

Figure 3. Shiga toxin-induced activation of blood cells and damage to the glomerular endothelium. Within the bloodstream Shiga toxin can bind to monocytes, neutrophils, activated platelets and red blood cells. Aggregates are formed between platelets and leukocytes (depicted in the figure between a neutrophil and a platelet). Microvesicles are shed from blood cells (neutrophils, monocytes, platelets and red blood cells). Some of these contain the toxin. All these microvesicles may expose phosphatidylserine. Microvesicles from platelets and monocytes expose tissue factor. Complement may deposit on platelet, monocyte and red blood cell-derived microvesicles, as well as on the glomerular endothelium. Shiga toxin binds to glomerular endothelial cells via the Gb3 glycolipid receptor or undergoes endocytosis within blood cell-derived microvesicles. Shiga toxin-induced cytotoxicity and complement deposition lead to glomerular endothelial cell damage. Thrombosis is induced by toxin-mediated endothelial cell damage, platelet activation and the pro-thrombotic effects of microvesicles. Hemolysis is induced by mechanical breakdown of red blood cells in occluded capillaries as well as by complement deposition on red blood cells. Fragmented red blood cells are termed schistocytes. Gb3: globotriaosylceramide, MAC: membrane attack complex (C5b-9), RBC: red blood cell.

The main features of HUS are thrombocytopenia due to platelet consumption in microthrombotic lesions on the damaged endothelium, hemolysis with fragmented red blood cells and renal failure. Microvesicles may contribute to and partake in each of these processes. The pro-thrombotic properties of microvesicles, bearing tissue factor and exposing phosphatidylserine on their outer leaflet, as described above, and reviewed [32], may contribute to thrombin generation and platelet activation on damaged microvasculature. Complement deposition on red blood cells will induce hemolysis and the release of complement-coated red blood cell-derived microvesicles during HUS [19] suggesting that hemolysis is induced not only by mechanical fragmentation of red blood cells in occluded blood vessels but also by complement activation. The pro-thrombotic effect may be further enhanced by circulating red blood cell-derived microvesicles capable of activating the intrinsic coagulation pathway [171].

The presence of complement C3 and C9 on platelet and monocyte-derived microvesicles in HUS patients [77] would reflect complement activation on the parent cells and could also contribute to the thrombotic process as complement activation on platelet membranes may promote their activation [197]. During HUS complement is deposited on the injured vascular wall [198], on platelets and microvesicles derived thereof [77], suggesting that complement contributes to the tissue injury during HUS [82]. Indeed, mouse models of EHEC infection or Shiga toxin injection in which the alternative, lectin or terminal complement pathways were blocked, exhibited reduced renal injury [198–200].

Shiga toxin can be transferred from the bloodstream to the kidney within blood cell-derived microvesicles and taken up in glomerular endothelial cells or peritubular capillary endothelial cells [21]. This mechanism could explain how the toxin enters target organ cells after circulating within activated blood cells that release microvesicles. In a mouse model, toxin-positive microvesicles were released from blood cells during the early stages of infection before symptoms develop [21]. They were further taken up by renal cells before clinical signs of disease were evident. Microvesicles were demonstrated to release their content of Shiga toxin within 12 h of entering the cell and the toxin reached ribosomes within 24 h, inhibiting protein synthesis, thus providing evidence that toxin within microvesiscles retains its cytotoxicity. An interesting finding was that not all microvesicles emptied their contents after endocytosis, certain microvesicles transcytosed the recipient cells, migrated through the corresponding glomerular or tubular basement membranes and passed into podocytes or tubular cells, respectively [21]. The capacity of microvesicles to navigate through cells and basement membranes could be secondary to tissue injury during HUS or possibly the toxic content of the microvesicles.

7. Conclusions

During Shiga toxin-associated HUS, microvesicles are released from blood cells capable of transferring the toxin to target organ cells inducing renal cell death and promoting thrombosis. Red blood cell-derived microvesicles are also involved in the hemolytic process. Microvesicles are thereby involved in all aspects of HUS.

Microvesicles transfer an array of nucleic acids, proteins and lipids, and are thus a potent mechanism for intercellular communication, which may have detrimental effects during infection and inflammation, but may also have beneficial effects in tissue regeneration. Furthermore, the shedding of microvesicles carrying unwanted cellular components may maintain cell integrity after a trigger of cell activation. Therefore, future studies should address the importance of microvesicles and possible effects of blocking their release in Shiga toxin-mediated infection and kidney damage.

Acknowledgments: D.K. is supported by The Swedish Research Council (K2013-64X-14008-13-5 and K2015-99X-22877-01-6), The Knut and Alice Wallenberg Foundation (Wallenberg Clinical Scholar 2015.0320), The Torsten Söderberg Foundation, Skåne Centre of Excellence in Health, Crown Princess Lovisa's Society for Child Care, Region Skåne, the IngaBritt and Arne Lundberg Research Foundation and The Konung Gustaf V:s 80-års minnesfond.

Conflicts of Interest: The authors declare no conflict of interest.

References

1. Karpman, D.; Loos, S.; Tati, R.; Arvidsson, I. Haemolytic uraemic syndrome. *J. Intern. Med.* **2017**, *281*, 123–148. [CrossRef] [PubMed]
2. Griffin, P.M.; Ostroff, S.M.; Tauxe, R.V.; Greene, K.D.; Wells, J.G.; Lewis, J.H.; Blake, P.A. Illnesses associated with *Escherichia coli* O157:H7 infections. A broad clinical spectrum. *Ann. Intern. Med.* **1988**, *109*, 705–712. [CrossRef] [PubMed]

3. Frank, C.; Werber, D.; Cramer, J.P.; Askar, M.; Faber, M.; an der Heiden, M.; Bernard, H.; Fruth, A.; Prager, R.; Spode, A.; et al. Epidemic profile of Shiga-toxin-producing *Escherichia coli* O104:H4 outbreak in Germany. *N. Engl. J. Med.* **2011**, *365*, 1771–1780. [CrossRef] [PubMed]

4. McKee, M.L.; O'Brien, A.D. Investigation of enterohemorrhagic *Escherichia coli* o157:H7 adherence characteristics and invasion potential reveals a new attachment pattern shared by intestinal *E. coli*. *Infect. Immun.* **1995**, *63*, 2070–2074. [PubMed]

5. Kaper, J.B. The locus of enterocyte effacement pathogenicity island of Shiga toxin-producing *Escherichia coli* O157:H7 and other attaching and effacing *E. coli*. *Jpn. J. Med. Sci. Biol.* **1998**, *51* (Suppl. 1), S101–S107. [CrossRef] [PubMed]

6. Rogers, T.J.; Paton, J.C.; Wang, H.; Talbot, U.M.; Paton, A.W. Reduced virulence of an fliC mutant of Shiga-toxigenic *Escherichia coli* O113:H21. *Infect. Immun.* **2006**, *74*, 1962–1966. [CrossRef] [PubMed]

7. Pacheco, A.R.; Sperandio, V. Inter-kingdom signaling: Chemical language between bacteria and host. *Curr. Opin. Microbiol.* **2009**, *12*, 192–198. [CrossRef] [PubMed]

8. Karmali, M.A.; Petric, M.; Lim, C.; Fleming, P.C.; Arbus, G.S.; Lior, H. The association between idiopathic hemolytic uremic syndrome and infection by verotoxin-producing *Escherichia coli*. *J. Infect. Dis.* **1985**, *151*, 775–782. [CrossRef] [PubMed]

9. Karpman, D.; Ståhl, A.L. Enterohemorrhagic *Escherichia coli* pathogenesis and the host response. *Microbiol. Spectr.* **2014**, *2*. [CrossRef] [PubMed]

10. Loos, S.; Ahlenstiel, T.; Kranz, B.; Staude, H.; Pape, L.; Hartel, C.; Vester, U.; Buchtala, L.; Benz, K.; Hoppe, B.; et al. An outbreak of Shiga toxin-producing *Escherichia coli* O104:H4 hemolytic uremic syndrome in Germany: Presentation and short-term outcome in children. *Clin. Infect. Dis.* **2012**, *55*, 753–759. [CrossRef] [PubMed]

11. Schuller, S.; Frankel, G.; Phillips, A.D. Interaction of Shiga toxin from *Escherichia coli* with human intestinal epithelial cell lines and explants: Stx2 induces epithelial damage in organ culture. *Cell Microbiol.* **2004**, *6*, 289–301. [CrossRef] [PubMed]

12. Schuller, S.; Heuschkel, R.; Torrente, F.; Kaper, J.B.; Phillips, A.D. Shiga toxin binding in normal and inflamed human intestinal mucosa. *Microbes Infect.* **2007**, *9*, 35–39. [CrossRef] [PubMed]

13. Hurley, B.P.; Thorpe, C.M.; Acheson, D.W. Shiga toxin translocation across intestinal epithelial cells is enhanced by neutrophil transmigration. *Infect. Immun.* **2001**, *69*, 6148–6155. [CrossRef] [PubMed]

14. Fontaine, A.; Arondel, J.; Sansonetti, P.J. Role of Shiga toxin in the pathogenesis of bacillary dysentery, studied by using a tox- mutant of *Shigella dysenteriae* 1. *Infect. Immun.* **1988**, *56*, 3099–3109. [PubMed]

15. Brigotti, M.; Tazzari, P.L.; Ravanelli, E.; Carnicelli, D.; Rocchi, L.; Arfilli, V.; Scavia, G.; Minelli, F.; Ricci, F.; Pagliaro, P.; et al. Clinical relevance of Shiga toxin concentrations in the blood of patients with hemolytic uremic syndrome. *Pediatr. Infect. Dis. J.* **2011**, *30*, 486–490. [CrossRef] [PubMed]

16. He, X.; Quinones, B.; Loo, M.T.; Loos, S.; Scavia, G.; Brigotti, M.; Levtchenko, E.; Monnens, L. Serum Shiga toxin 2 values in patients during acute phase of diarrhoea-associated haemolytic uraemic syndrome. *Acta Paediatr.* **2015**, *104*, e564–e568. [CrossRef] [PubMed]

17. Te Loo, D.M.; van Hinsbergh, V.W.; van den Heuvel, L.P.; Monnens, L.A. Detection of verocytotoxin bound to circulating polymorphonuclear leukocytes of patients with hemolytic uremic syndrome. *J. Am. Soc. Nephrol.* **2001**, *12*, 800–806. [PubMed]

18. Ståhl, A.L.; Sartz, L.; Nelsson, A.; Békássy, Z.D.; Karpman, D. Shiga toxin and lipopolysaccharide induce platelet-leukocyte aggregates and tissue factor release, a thrombotic mechanism in hemolytic uremic syndrome. *PLoS ONE* **2009**, *4*, e6990. [CrossRef] [PubMed]

19. Arvidsson, I.; Ståhl, A.L.; Hedström, M.M.; Kristoffersson, A.C.; Rylander, C.; Westman, J.S.; Storry, J.R.; Olsson, M.L.; Karpman, D. Shiga toxin-induced complement-mediated hemolysis and release of complement-coated red blood cell-derived microvesicles in hemolytic uremic syndrome. *J. Immunol.* **2015**, *194*, 2309–2318. [CrossRef] [PubMed]

20. Bitzan, M.; Richardson, S.; Huang, C.; Boyd, B.; Petric, M.; Karmali, M.A. Evidence that verotoxins (Shiga-like toxins) from *Escherichia coli* bind to p blood group antigens of human erythrocytes in vitro. *Infect. Immun.* **1994**, *62*, 3337–3347. [PubMed]

21. Ståhl, A.L.; Arvidsson, I.; Johansson, K.E.; Chromek, M.; Rebetz, J.; Loos, S.; Kristoffersson, A.C.; Békássy, Z.D.; Mörgelin, M.; Karpman, D. A novel mechanism of bacterial toxin transfer within host blood cell-derived microvesicles. *PLoS Pathog.* **2015**, *11*, e1004619. [CrossRef] [PubMed]

22. Te Loo, D.M.; Monnens, L.A.; van Der Velden, T.J.; Vermeer, M.A.; Preyers, F.; Demacker, P.N.; van Den Heuvel, L.P.; van Hinsbergh, V.W. Binding and transfer of verocytotoxin by polymorphonuclear leukocytes in hemolytic uremic syndrome. *Blood* **2000**, *95*, 3396–3402. [PubMed]

23. Ratajczak, J.; Wysoczynski, M.; Hayek, F.; Janowska-Wieczorek, A.; Ratajczak, M.Z. Membrane-derived microvesicles: Important and underappreciated mediators of cell-to-cell communication. *Leukemia* **2006**, *20*, 1487–1495. [CrossRef] [PubMed]

24. Flaumenhaft, R.; Dilks, J.R.; Richardson, J.; Alden, E.; Patel-Hett, S.R.; Battinelli, E.; Klement, G.L.; Sola-Visner, M.; Italiano, J.E., Jr. Megakaryocyte-derived microparticles: Direct visualization and distinction from platelet-derived microparticles. *Blood* **2009**, *113*, 1112–1121. [CrossRef] [PubMed]

25. Angelillo-Scherrer, A. Leukocyte-derived microparticles in vascular homeostasis. *Circ. Res.* **2012**, *110*, 356–369. [CrossRef] [PubMed]

26. Rubin, O.; Canellini, G.; Delobel, J.; Lion, N.; Tissot, J.D. Red blood cell microparticles: Clinical relevance. *Transfus. Med. Hemother.* **2012**, *39*, 342–347. [CrossRef] [PubMed]

27. Chironi, G.N.; Boulanger, C.M.; Simon, A.; Dignat-George, F.; Freyssinet, J.M.; Tedgui, A. Endothelial microparticles in diseases. *Cell Tissue Res.* **2009**, *335*, 143–151. [CrossRef] [PubMed]

28. Turco, A.E.; Lam, W.; Rule, A.D.; Denic, A.; Lieske, J.C.; Miller, V.M.; Larson, J.J.; Kremers, W.K.; Jayachandran, M. Specific renal parenchymal-derived urinary extracellular vesicles identify age-associated structural changes in living donor kidneys. *J. Extracell. Vesicles* **2016**, *5*, 29642. [CrossRef] [PubMed]

29. Camussi, G.; Deregibus, M.C.; Bruno, S.; Cantaluppi, V.; Biancone, L. Exosomes/microvesicles as a mechanism of cell-to-cell communication. *Kidney Int.* **2010**, *78*, 838–848. [CrossRef] [PubMed]

30. Abid Hussein, M.N.; Boing, A.N.; Sturk, A.; Hau, C.M.; Nieuwland, R. Inhibition of microparticle release triggers endothelial cell apoptosis and detachment. *Thromb. Haemost.* **2007**, *98*, 1096–1107. [CrossRef] [PubMed]

31. Pap, E. The role of microvesicles in malignancies. *Adv. Exp. Med. Biol.* **2011**, *714*, 183–199. [PubMed]

32. Karpman, D.; Ståhl, A.L.; Arvidsson, I. Extracellular vesicles in renal disease. *Nat. Rev. Nephrol.* **2017**, *13*, 545–562. [CrossRef] [PubMed]

33. Lawson, C.; Vicencio, J.M.; Yellon, D.M.; Davidson, S.M. Microvesicles and exosomes: New players in metabolic and cardiovascular disease. *J. Endocrinol.* **2016**, *228*, R57–R71. [CrossRef] [PubMed]

34. Distler, J.H.; Huber, L.C.; Gay, S.; Distler, O.; Pisetsky, D.S. Microparticles as mediators of cellular cross-talk in inflammatory disease. *Autoimmunity* **2006**, *39*, 683–690. [CrossRef] [PubMed]

35. Waldor, M.K.; Friedman, D.I. Phage regulatory circuits and virulence gene expression. *Curr. Opin. Microbiol.* **2005**, *8*, 459–465. [CrossRef] [PubMed]

36. Stein, P.E.; Boodhoo, A.; Tyrrell, G.J.; Brunton, J.L.; Read, R.J. Crystal structure of the cell-binding b oligomer of verotoxin-1 from *E. coli*. *Nature* **1992**, *355*, 748–750. [CrossRef] [PubMed]

37. Lingwood, C.A.; Law, H.; Richardson, S.; Petric, M.; Brunton, J.L.; De Grandis, S.; Karmali, M. Glycolipid binding of purified and recombinant *Escherichia coli* produced verotoxin in vitro. *J. Biol. Chem.* **1987**, *262*, 8834–8839. [PubMed]

38. Lindberg, A.A.; Brown, J.E.; Strömberg, N.; Westling-Ryd, M.; Schultz, J.E.; Karlsson, K.A. Identification of the carbohydrate receptor for Shiga toxin produced by *Shigella dysenteriae* type 1. *J. Biol. Chem.* **1987**, *262*, 1779–1785. [PubMed]

39. Nakajima, H.; Kiyokawa, N.; Katagiri, Y.U.; Taguchi, T.; Suzuki, T.; Sekino, T.; Mimori, K.; Ebata, T.; Saito, M.; Nakao, H.; et al. Kinetic analysis of binding between Shiga toxin and receptor glycolipid gb3cer by surface plasmon resonance. *J. Biol. Chem.* **2001**, *276*, 42915–42922. [CrossRef] [PubMed]

40. Kovbasnjuk, O.; Edidin, M.; Donowitz, M. Role of lipid rafts in Shiga toxin 1 interaction with the apical surface of caco-2 cells. *J. Cell Sci.* **2001**, *114*, 4025–4031. [PubMed]

41. Romer, W.; Berland, L.; Chambon, V.; Gaus, K.; Windschiegl, B.; Tenza, D.; Aly, M.R.; Fraisier, V.; Florent, J.C.; Perrais, D.; et al. Shiga toxin induces tubular membrane invaginations for its uptake into cells. *Nature* **2007**, *450*, 670–675. [CrossRef] [PubMed]

42. Torgersen, M.L.; Lauvrak, S.U.; Sandvig, K. The a-subunit of surface-bound Shiga toxin stimulates clathrin-dependent uptake of the toxin. *FEBS J.* **2005**, *272*, 4103–4113. [CrossRef] [PubMed]

43. Sandvig, K.; Skotland, T.; van Deurs, B.; Klokk, T.I. Retrograde transport of protein toxins through the golgi apparatus. *Histochem. Cell Biol.* **2013**, *140*, 317–326. [CrossRef] [PubMed]

44. Romer, W.; Pontani, L.L.; Sorre, B.; Rentero, C.; Berland, L.; Chambon, V.; Lamaze, C.; Bassereau, P.; Sykes, C.; Gaus, K.; et al. Actin dynamics drive membrane reorganization and scission in clathrin-independent endocytosis. *Cell* **2010**, *140*, 540–553. [CrossRef] [PubMed]

45. Johannes, L.; Parton, R.G.; Bassereau, P.; Mayor, S. Building endocytic pits without clathrin. *Nat. Rev. Mol. Cell Biol.* **2015**, *16*, 311–321. [CrossRef] [PubMed]

46. Malyukova, I.; Murray, K.F.; Zhu, C.; Boedeker, E.; Kane, A.; Patterson, K.; Peterson, J.R.; Donowitz, M.; Kovbasnjuk, O. Macropinocytosis in Shiga toxin 1 uptake by human intestinal epithelial cells and transcellular transcytosis. *Am. J. Physiol. Gastrointest. Liver Physiol.* **2009**, *296*, G78–G92. [CrossRef] [PubMed]

47. Schuller, S. Shiga toxin interaction with human intestinal epithelium. *Toxins (Basel)* **2011**, *3*, 626–639. [CrossRef] [PubMed]

48. Sandvig, K.; Garred, O.; Prydz, K.; Kozlov, J.V.; Hansen, S.H.; van Deurs, B. Retrograde transport of endocytosed Shiga toxin to the endoplasmic reticulum. *Nature* **1992**, *358*, 510–512. [CrossRef] [PubMed]

49. Garred, O.; van Deurs, B.; Sandvig, K. Furin-induced cleavage and activation of Shiga toxin. *J Biol. Chem.* **1995**, *270*, 10817–10821. [CrossRef] [PubMed]

50. Sandvig, K. Shiga toxins. *Toxicon* **2001**, *39*, 1629–1635. [CrossRef]

51. Endo, Y.; Tsurugi, K.; Yutsudo, T.; Takeda, Y.; Ogasawara, T.; Igarashi, K. Site of action of a vero toxin (vt2) from *Escherichia coli* O157:H7 and of Shiga toxin on eukaryotic ribosomes. RNA n-glycosidase activity of the toxins. *Eur. J. Biochem.* **1988**, *171*, 45–50. [CrossRef] [PubMed]

52. Karpman, D.; Håkansson, A.; Perez, M.T.; Isaksson, C.; Carlemalm, E.; Caprioli, A.; Svanborg, C. Apoptosis of renal cortical cells in the hemolytic-uremic syndrome: In vivo and in vitro studies. *Infect. Immun.* **1998**, *66*, 636–644. [PubMed]

53. Burlaka, I.; Liu, X.L.; Rebetz, J.; Arvidsson, I.; Yang, L.; Brismar, H.; Karpman, D.; Aperia, A. Ouabain protects against Shiga toxin-triggered apoptosis by reversing the imbalance between bax and bcl-xl. *J. Am. Soc. Nephrol.* **2013**, *24*, 1413–1423. [CrossRef] [PubMed]

54. Tesh, V.L. Activation of cell stress response pathways by Shiga toxins. *Cell. Microbiol.* **2012**, *14*, 1–9. [CrossRef] [PubMed]

55. Lee, M.S.; Koo, S.; Jeong, D.G.; Tesh, V.L. Shiga toxins as multi-functional proteins: Induction of host cellular stress responses, role in pathogenesis and therapeutic applications. *Toxins (Basel)* **2016**, *8*, 77. [CrossRef] [PubMed]

56. Békássy, Z.D.; Calderon Toledo, C.; Leoj, G.; Kristoffersson, A.; Leopold, S.R.; Perez, M.T.; Karpman, D. Intestinal damage in enterohemorrhagic *Escherichia coli* infection. *Pediatr. Nephrol.* **2011**, *26*, 2059–2071. [CrossRef] [PubMed]

57. Thorpe, C.M.; Hurley, B.P.; Lincicome, L.L.; Jacewicz, M.S.; Keusch, G.T.; Acheson, D.W. Shiga toxins stimulate secretion of interleukin-8 from intestinal epithelial cells. *Infect. Immun.* **1999**, *67*, 5985–5993. [PubMed]

58. Thorpe, C.M.; Smith, W.E.; Hurley, B.P.; Acheson, D.W. Shiga toxins induce, superinduce, and stabilize a variety of c-x-c chemokine mRNAs in intestinal epithelial cells, resulting in increased chemokine expression. *Infect. Immun.* **2001**, *69*, 6140–6147. [CrossRef] [PubMed]

59. Yamasaki, C.; Natori, Y.; Zeng, X.T.; Ohmura, M.; Yamasaki, S.; Takeda, Y.; Natori, Y. Induction of cytokines in a human colon epithelial cell line by Shiga toxin 1 (stx1) and stx2 but not by non-toxic mutant stx1 which lacks n-glycosidase activity. *FEBS Lett.* **1999**, *442*, 231–234. [CrossRef]

60. Tesh, V.L.; Ramegowda, B.; Samuel, J.E. Purified Shiga-like toxins induce expression of proinflammatory cytokines from murine peritoneal macrophages. *Infect. Immun.* **1994**, *62*, 5085–5094. [PubMed]

61. Karpman, D.; Connell, H.; Svensson, M.; Scheutz, F.; Alm, P.; Svanborg, C. The role of lipopolysaccharide and Shiga-like toxin in a mouse model of *Escherichia coli* O157:H7 infection. *J. Infect. Dis.* **1997**, *175*, 611–620. [CrossRef] [PubMed]

62. Calderon Toledo, C.; Rogers, T.J.; Svensson, M.; Tati, R.; Fischer, H.; Svanborg, C.; Karpman, D. Shiga toxin-mediated disease in myd88-deficient mice infected with *Escherichia coli* O157:H7. *Am. J. Pathol.* **2008**, *173*, 1428–1439. [CrossRef] [PubMed]

63. Robson, W.L.; Fick, G.H.; Wilson, P.C. Prognostic factors in typical postdiarrhea hemolytic-uremic syndrome. *Child Nephrol. Urol.* **1988**, *9*, 203–207. [PubMed]

64. Tazzari, P.L.; Ricci, F.; Carnicelli, D.; Caprioli, A.; Tozzi, A.E.; Rizzoni, G.; Conte, R.; Brigotti, M. Flow cytometry detection of Shiga toxins in the blood from children with hemolytic uremic syndrome. *Cytometry B. Clin. Cytom.* **2004**, *61*, 40–44. [CrossRef] [PubMed]

65. Van Setten, P.A.; Monnens, L.A.; Verstraten, R.G.; van den Heuvel, L.P.; van Hinsbergh, V.W. Effects of verocytotoxin-1 on nonadherent human monocytes: Binding characteristics, protein synthesis, and induction of cytokine release. *Blood* **1996**, *88*, 174–183. [PubMed]

66. Guessous, F.; Marcinkiewicz, M.; Polanowska-Grabowska, R.; Keepers, T.R.; Obrig, T.; Gear, A.R. Shiga toxin 2 and lipopolysaccharide cause monocytic thp-1 cells to release factors which activate platelet function. *Thromb. Haemost.* **2005**, *94*, 1019–1027. [CrossRef] [PubMed]

67. Murata, K.; Higuchi, T.; Takada, K.; Oida, K.; Horie, S.; Ishii, H. Verotoxin-1 stimulation of macrophage-like THP-1 cells up-regulates tissue factor expression through activation of c-yes tyrosine kinase: Possible signal transduction in tissue factor up-regulation. *Biochim. Biophys. Acta* **2006**, *1762*, 835–843. [CrossRef] [PubMed]

68. Zoja, C.; Buelli, S.; Morigi, M. Shiga toxin-associated hemolytic uremic syndrome: Pathophysiology of endothelial dysfunction. *Pediatr. Nephrol.* **2010**, *25*, 2231–2240. [CrossRef] [PubMed]

69. Karpman, D.; Manea, M.; Vaziri-Sani, F.; Ståhl, A.L.; Kristoffersson, A.C. Platelet activation in hemolytic uremic syndrome. *Semin. Thromb. Hemost.* **2006**, *32*, 128–145. [CrossRef] [PubMed]

70. Fong, J.S.; Kaplan, B.S. Impairment of platelet aggregation in hemolytic uremic syndrome: Evidence for platelet "exhaustion". *Blood* **1982**, *60*, 564–570. [PubMed]

71. Appiani, A.C.; Edefonti, A.; Bettinelli, A.; Cossu, M.M.; Paracchini, M.L.; Rossi, E. The relationship between plasma levels of the factor viii complex and platelet release products (beta-thromboglobulin and platelet factor 4) in children with the hemolytic-uremic syndrome. *Clin. Nephrol.* **1982**, *17*, 195–199. [PubMed]

72. Katayama, M.; Handa, M.; Araki, Y.; Ambo, H.; Kawai, Y.; Watanabe, K.; Ikeda, Y. Soluble p-selectin is present in normal circulation and its plasma level is elevated in patients with thrombotic thrombocytopenic purpura and haemolytic uraemic syndrome. *Br. J. Haematol.* **1993**, *84*, 702–710. [CrossRef] [PubMed]

73. Cooling, L.L.; Walker, K.E.; Gille, T.; Koerner, T.A. Shiga toxin binds human platelets via globotriaosylceramide (pk antigen) and a novel platelet glycosphingolipid. *Infect. Immun.* **1998**, *66*, 4355–4366. [PubMed]

74. Ståhl, A.L.; Svensson, M.; Mörgelin, M.; Svanborg, C.; Tarr, P.I.; Mooney, J.C.; Watkins, S.L.; Johnson, R.; Karpman, D. Lipopolysaccharide from enterohemorrhagic *Escherichia coli* binds to platelets through TLR4 and CD62 and is detected on circulating platelets in patients with hemolytic uremic syndrome. *Blood* **2006**, *108*, 167–176. [CrossRef] [PubMed]

75. Ghosh, S.A.; Polanowska-Grabowska, R.K.; Fujii, J.; Obrig, T.; Gear, A.R. Shiga toxin binds to activated platelets. *J. Thromb. Haemost.* **2004**, *2*, 499–506. [CrossRef] [PubMed]

76. Karpman, D.; Papadopoulou, D.; Nilsson, K.; Sjögren, A.C.; Mikaelsson, C.; Lethagen, S. Platelet activation by Shiga toxin and circulatory factors as a pathogenetic mechanism in the hemolytic uremic syndrome. *Blood* **2001**, *97*, 3100–3108. [CrossRef] [PubMed]

77. Ståhl, A.L.; Sartz, L.; Karpman, D. Complement activation on platelet-leukocyte complexes and microparticles in enterohemorrhagic *Escherichia coli*-induced hemolytic uremic syndrome. *Blood* **2011**, *117*, 5503–5513. [CrossRef] [PubMed]

78. Polley, M.J.; Nachman, R.L. Human complement in thrombin-mediated platelet function: Uptake of the C5b-9 complex. *J. Exp. Med.* **1979**, *150*, 633–645. [CrossRef] [PubMed]

79. Chandler, W.L.; Jelacic, S.; Boster, D.R.; Ciol, M.A.; Williams, G.D.; Watkins, S.L.; Igarashi, T.; Tarr, P.I. Prothrombotic coagulation abnormalities preceding the hemolytic-uremic syndrome. *N. Engl. J. Med.* **2002**, *346*, 23–32. [CrossRef] [PubMed]

80. Polley, M.J.; Nachman, R. The human complement system in thrombin-mediated platelet function. *J. Exp. Med.* **1978**, *147*, 1713–1726. [CrossRef] [PubMed]

81. Huber-Lang, M.; Sarma, J.V.; Zetoune, F.S.; Rittirsch, D.; Neff, T.A.; McGuire, S.R.; Lambris, J.D.; Warner, R.L.; Flierl, M.A.; Hoesel, L.M.; et al. Generation of C5a in the absence of C3: A new complement activation pathway. *Nat. Med.* **2006**, *12*, 682–687. [CrossRef] [PubMed]

82. Karpman, D.; Tati, R. Complement contributes to the pathogenesis of Shiga toxin-associated hemolytic uremic syndrome. *Kidney Int.* **2016**, *90*, 726–729. [CrossRef] [PubMed]

83. Risitano, A.M.; Marotta, S. Therapeutic complement inhibition in complement-mediated hemolytic anemias: Past, present and future. *Semin. Immunol.* **2016**, *28*, 223–240. [CrossRef] [PubMed]

84. Spitalnik, P.F.; Spitalnik, S.L. The P blood group system: Biochemical, serological, and clinical aspects. *Transfus. Med. Rev.* **1995**, *9*, 110–122. [CrossRef]

85. Karpman, D. Management of Shiga toxin-associated *Escherichia coli*-induced haemolytic uraemic syndrome: Randomized clinical trials are needed. *Nephrol. Dial. Transplant.* **2012**, *27*, 3669–3674. [CrossRef] [PubMed]

86. Geelen, J.M.; van der Velden, T.J.; van den Heuvel, L.P.; Monnens, L.A. Interactions of Shiga-like toxin with human peripheral blood monocytes. *Pediatr. Nephrol.* **2007**, *22*, 1181–1187. [CrossRef] [PubMed]

87. Psotka, M.A.; Obata, F.; Kolling, G.L.; Gross, L.K.; Saleem, M.A.; Satchell, S.C.; Mathieson, P.W.; Obrig, T.G. Shiga toxin 2 targets the murine renal collecting duct epithelium. *Infect. Immun.* **2009**, *77*, 959–969. [CrossRef] [PubMed]

88. Van de Kar, N.C.; Monnens, L.A.; Karmali, M.A.; van Hinsbergh, V.W. Tumor necrosis factor and interleukin-1 induce expression of the verocytotoxin receptor globotriaosylceramide on human endothelial cells: Implications for the pathogenesis of the hemolytic uremic syndrome. *Blood* **1992**, *80*, 2755–2764. [PubMed]

89. Stone, M.K.; Kolling, G.L.; Lindner, M.H.; Obrig, T.G. P38 mitogen-activated protein kinase mediates lipopolysaccharide and tumor necrosis factor alpha induction of Shiga toxin 2 sensitivity in human umbilical vein endothelial cells. *Infect. Immun.* **2008**, *76*, 1115–1121. [CrossRef] [PubMed]

90. Louise, C.B.; Obrig, T.G. Specific interaction of Escherichia coli O157:H7-derived Shiga-like toxin ii with human renal endothelial cells. *J. Infect. Dis.* **1995**, *172*, 1397–1401. [CrossRef] [PubMed]

91. Guessous, F.; Marcinkiewicz, M.; Polanowska-Grabowska, R.; Kongkhum, S.; Heatherly, D.; Obrig, T.; Gear, A.R. Shiga toxin 2 and lipopolysaccharide induce human microvascular endothelial cells to release chemokines and factors that stimulate platelet function. *Infect. Immun.* **2005**, *73*, 8306–8316. [CrossRef] [PubMed]

92. Nevard, C.H.; Jurd, K.M.; Lane, D.A.; Philippou, H.; Haycock, G.B.; Hunt, B.J. Activation of coagulation and fibrinolysis in childhood diarrhoea-associated haemolytic uraemic syndrome. *Thromb. Haemost.* **1997**, *78*, 1450–1455. [PubMed]

93. Van Geet, C.; Proesmans, W.; Arnout, J.; Vermylen, J.; Declerck, P.J. Activation of both coagulation and fibrinolysis in childhood hemolytic uremic syndrome. *Kidney Int.* **1998**, *54*, 1324–1330. [CrossRef] [PubMed]

94. Van de Kar, N.C.; van Hinsbergh, V.W.; Brommer, E.J.; Monnens, L.A. The fibrinolytic system in the hemolytic uremic syndrome: In vivo and in vitro studies. *Pediatr. Res.* **1994**, *36*, 257–264. [CrossRef] [PubMed]

95. Sappino, A.P.; Huarte, J.; Vassalli, J.D.; Belin, D. Sites of synthesis of urokinase and tissue-type plasminogen activators in the murine kidney. *J. Clin. Investig.* **1991**, *87*, 962–970. [CrossRef] [PubMed]

96. Ge, S.; Hertel, B.; Emden, S.H.; Beneke, J.; Menne, J.; Haller, H.; von Vietinghoff, S. Microparticle generation and leucocyte death in Shiga toxin-mediated hus. *Nephrol. Dial. Transplant.* **2012**, *27*, 2768–2775. [CrossRef] [PubMed]

97. Colombo, M.; Raposo, G.; Thery, C. Biogenesis, secretion, and intercellular interactions of exosomes and other extracellular vesicles. *Annu. Rev. Cell Dev. Biol.* **2014**, *30*, 255–289. [CrossRef] [PubMed]

98. Gyorgy, B.; Szabo, T.G.; Pasztoi, M.; Pal, Z.; Misjak, P.; Aradi, B.; Laszlo, V.; Pallinger, E.; Pap, E.; Kittel, A.; et al. Membrane vesicles, current state-of-the-art: Emerging role of extracellular vesicles. *Cell. Mol. Life. Sci.* **2011**, *68*, 2667–2688. [CrossRef] [PubMed]

99. Erdbrügger, U.; Lannigan, J. Analytical challenges of extracellular vesicle detection: A comparison of different techniques. *Cytometry A.* **2016**, *89*, 123–134. [CrossRef] [PubMed]

100. Witwer, K.W.; Buzas, E.I.; Bemis, L.T.; Bora, A.; Lasser, C.; Lotvall, J.; Nolte-'t Hoen, E.N.; Piper, M.G.; Sivaraman, S.; Skog, J.; et al. Standardization of sample collection, isolation and analysis methods in extracellular vesicle research. *J. Extracell. Vesicles* **2013**, *2*. [CrossRef] [PubMed]

101. Choi, D.S.; Kim, D.K.; Kim, Y.K.; Gho, Y.S. Proteomics of extracellular vesicles: Exosomes and ectosomes. *Mass Spectrom. Rev.* **2015**, *34*, 474–490. [CrossRef] [PubMed]

102. Van der Pol, E.; Coumans, F.A.; Grootemaat, A.E.; Gardiner, C.; Sargent, I.L.; Harrison, P.; Sturk, A.; van Leeuwen, T.G.; Nieuwland, R. Particle size distribution of exosomes and microvesicles determined by transmission electron microscopy, flow cytometry, nanoparticle tracking analysis, and resistive pulse sensing. *J. Thromb. Haemost.* **2014**, *12*, 1182–1192. [CrossRef] [PubMed]

103. Thery, C.; Zitvogel, L.; Amigorena, S. Exosomes: Composition, biogenesis and function. *Nat. Rev. Immunol.* **2002**, *2*, 569–579. [PubMed]

104. Kowal, J.; Tkach, M.; Thery, C. Biogenesis and secretion of exosomes. *Curr. Opin. Cell Biol.* **2014**, *29*, 116–125. [CrossRef] [PubMed]

105. Colombo, M.; Moita, C.; van Niel, G.; Kowal, J.; Vigneron, J.; Benaroch, P.; Manel, N.; Moita, L.F.; Thery, C.; Raposo, G. Analysis of escrt functions in exosome biogenesis, composition and secretion highlights the heterogeneity of extracellular vesicles. *J. Cell Sci.* **2013**, *126*, 5553–5565. [CrossRef] [PubMed]

106. Hanson, P.I.; Roth, R.; Lin, Y.; Heuser, J.E. Plasma membrane deformation by circular arrays of ESCRT-III protein filaments. *J. Cell Biol.* **2008**, *180*, 389–402. [CrossRef] [PubMed]

107. Van Niel, G.; Charrin, S.; Simoes, S.; Romao, M.; Rochin, L.; Saftig, P.; Marks, M.S.; Rubinstein, E.; Raposo, G. The tetraspanin CD63 regulates ESCRT-independent and -dependent endosomal sorting during melanogenesis. *Dev. Cell* **2011**, *21*, 708–721. [CrossRef] [PubMed]

108. Trajkovic, K.; Hsu, C.; Chiantia, S.; Rajendran, L.; Wenzel, D.; Wieland, F.; Schwille, P.; Brugger, B.; Simons, M. Ceramide triggers budding of exosome vesicles into multivesicular endosomes. *Science* **2008**, *319*, 1244–1247. [CrossRef] [PubMed]

109. Skotland, T.; Sandvig, K.; Llorente, A. Lipids in exosomes: Current knowledge and the way forward. *Prog. Lipid Res.* **2017**, *66*, 30–41. [CrossRef] [PubMed]

110. Escola, J.M.; Kleijmeer, M.J.; Stoorvogel, W.; Griffith, J.M.; Yoshie, O.; Geuze, H.J. Selective enrichment of tetraspan proteins on the internal vesicles of multivesicular endosomes and on exosomes secreted by human b-lymphocytes. *J. Biol. Chem.* **1998**, *273*, 20121–20127. [CrossRef] [PubMed]

111. Thery, C.; Boussac, M.; Veron, P.; Ricciardi-Castagnoli, P.; Raposo, G.; Garin, J.; Amigorena, S. Proteomic analysis of dendritic cell-derived exosomes: A secreted subcellular compartment distinct from apoptotic vesicles. *J. Immunol.* **2001**, *166*, 7309–7318. [CrossRef] [PubMed]

112. Lynch, S.; Santos, S.G.; Campbell, E.C.; Nimmo, A.M.; Botting, C.; Prescott, A.; Antoniou, A.N.; Powis, S.J. Novel MHC class I structures on exosomes. *J. Immunol.* **2009**, *183*, 1884–1891. [CrossRef] [PubMed]

113. Haraszti, R.A.; Didiot, M.C.; Sapp, E.; Leszyk, J.; Shaffer, S.A.; Rockwell, H.E.; Gao, F.; Narain, N.R.; DiFiglia, M.; Kiebish, M.A.; et al. High-resolution proteomic and lipidomic analysis of exosomes and microvesicles from different cell sources. *J. Extracell. Vesicles* **2016**, *5*, 32570. [CrossRef] [PubMed]

114. Qin, J.; Xu, Q. Functions and application of exosomes. *Acta Pol. Pharm.* **2014**, *71*, 537–543. [PubMed]

115. Little, K.M.; Smalley, D.M.; Harthun, N.L.; Ley, K. The plasma microparticle proteome. *Semin. Thromb. Hemost.* **2010**, *36*, 845–856. [CrossRef] [PubMed]

116. Barry, O.P.; Pratico, D.; Lawson, J.A.; FitzGerald, G.A. Transcellular activation of platelets and endothelial cells by bioactive lipids in platelet microparticles. *J. Clin. Investig.* **1997**, *99*, 2118–2127. [CrossRef] [PubMed]

117. Sapet, C.; Simoncini, S.; Loriod, B.; Puthier, D.; Sampol, J.; Nguyen, C.; Dignat-George, F.; Anfosso, F. Thrombin-induced endothelial microparticle generation: Identification of a novel pathway involving rock-II activation by caspase-2. *Blood* **2006**, *108*, 1868–1876. [CrossRef] [PubMed]

118. Ratajczak, J.; Miekus, K.; Kucia, M.; Zhang, J.; Reca, R.; Dvorak, P.; Ratajczak, M.Z. Embryonic stem cell-derived microvesicles reprogram hematopoietic progenitors: Evidence for horizontal transfer of mRNA and protein delivery. *Leukemia* **2006**, *20*, 847–856. [CrossRef] [PubMed]

119. Cocucci, E.; Racchetti, G.; Meldolesi, J. Shedding microvesicles: Artefacts no more. *Trends Cell Biol.* **2009**, *19*, 43–51. [CrossRef] [PubMed]

120. Al-Nedawi, K.; Meehan, B.; Micallef, J.; Lhotak, V.; May, L.; Guha, A.; Rak, J. Intercellular transfer of the oncogenic receptor EGFRvIII by microvesicles derived from tumour cells. *Nat. Cell Biol.* **2008**, *10*, 619–624. [CrossRef] [PubMed]

121. Tricarico, C.; Clancy, J.; D'Souza-Schorey, C. Biology and biogenesis of shed microvesicles. *Small GTPases* **2016**, *8*, 220–232. [CrossRef] [PubMed]

122. Pomatto, M.A.C.; Gai, C.; Bussolati, B.; Camussi, G. Extracellular vesicles in renal pathophysiology. *Front. Mol. Biosci.* **2017**, *4*, 37. [CrossRef] [PubMed]

123. Iida, K.; Whitlow, M.B.; Nussenzweig, V. Membrane vesiculation protects erythrocytes from destruction by complement. *J. Immunol.* **1991**, *147*, 2638–2642. [PubMed]

124. Willekens, F.L.; Werre, J.M.; Groenen-Dopp, Y.A.; Roerdinkholder-Stoelwinder, B.; de Pauw, B.; Bosman, G.J. Erythrocyte vesiculation: A self-protective mechanism? *Br. J. Haematol.* **2008**, *141*, 549–556. [CrossRef] [PubMed]

125. McMahon, H.T.; Boucrot, E. Membrane curvature at a glance. *J. Cell Sci.* **2015**, *128*, 1065–1070. [CrossRef] [PubMed]

126. Jimenez, J.J.; Jy, W.; Mauro, L.M.; Soderland, C.; Horstman, L.L.; Ahn, Y.S. Endothelial cells release phenotypically and quantitatively distinct microparticles in activation and apoptosis. *Thromb. Res.* **2003**, *109*, 175–180. [CrossRef]

127. Nomura, S.; Tandon, N.N.; Nakamura, T.; Cone, J.; Fukuhara, S.; Kambayashi, J. High-shear-stress-induced activation of platelets and microparticles enhances expression of cell adhesion molecules in THP-1 and endothelial cells. *Atherosclerosis* **2001**, *158*, 277–287. [CrossRef]

128. Bernimoulin, M.; Waters, E.K.; Foy, M.; Steele, B.M.; Sullivan, M.; Falet, H.; Walsh, M.T.; Barteneva, N.; Geng, J.G.; Hartwig, J.H.; et al. Differential stimulation of monocytic cells results in distinct populations of microparticles. *J. Thromb. Haemost.* **2009**, *7*, 1019–1028. [CrossRef] [PubMed]

129. Gonzalez, L.J.; Gibbons, E.; Bailey, R.W.; Fairbourn, J.; Nguyen, T.; Smith, S.K.; Best, K.B.; Nelson, J.; Judd, A.M.; Bell, J.D. The influence of membrane physical properties on microvesicle release in human erythrocytes. *PMC Biophys.* **2009**, *2*, 7. [CrossRef] [PubMed]

130. Muralidharan-Chari, V.; Clancy, J.; Plou, C.; Romao, M.; Chavrier, P.; Raposo, G.; D'Souza-Schorey, C. ARF6-regulated shedding of tumor cell-derived plasma membrane microvesicles. *Curr. Biol.* **2009**, *19*, 1875–1885. [CrossRef] [PubMed]

131. Lai, R.C.; Tan, S.S.; Yeo, R.W.; Choo, A.B.; Reiner, A.T.; Su, Y.; Shen, Y.; Fu, Z.; Alexander, L.; Sze, S.K.; et al. MSC secretes at least 3 EV types each with a unique permutation of membrane lipid, protein and RNA. *J. Extracell. Vesicles.* **2016**, *5*, 29828. [CrossRef] [PubMed]

132. Biro, E.; Akkerman, J.W.; Hoek, F.J.; Gorter, G.; Pronk, L.M.; Sturk, A.; Nieuwland, R. The phospholipid composition and cholesterol content of platelet-derived microparticles: A comparison with platelet membrane fractions. *J. Thromb. Haemost.* **2005**, *3*, 2754–2763. [CrossRef] [PubMed]

133. Sinauridze, E.I.; Kireev, D.A.; Popenko, N.Y.; Pichugin, A.V.; Panteleev, M.A.; Krymskaya, O.V.; Ataullakhanov, F.I. Platelet microparticle membranes have 50- to 100-fold higher specific procoagulant activity than activated platelets. *Thromb. Haemost.* **2007**, *97*, 425–434. [CrossRef] [PubMed]

134. Seigneuret, M.; Zachowski, A.; Hermann, A.; Devaux, P.F. Asymmetric lipid fluidity in human erythrocyte membrane: New spin-label evidence. *Biochemistry* **1984**, *23*, 4271–4275. [CrossRef] [PubMed]

135. Manno, S.; Takakuwa, Y.; Mohandas, N. Identification of a functional role for lipid asymmetry in biological membranes: Phosphatidylserine-skeletal protein interactions modulate membrane stability. *Proc. Natl. Acad. Sci. USA* **2002**, *99*, 1943–1948. [CrossRef] [PubMed]

136. Daleke, D.L. Regulation of transbilayer plasma membrane phospholipid asymmetry. *J. Lipid Res.* **2003**, *44*, 233–242. [CrossRef] [PubMed]

137. Yang, H.; Kim, A.; David, T.; Palmer, D.; Jin, T.; Tien, J.; Huang, F.; Cheng, T.; Coughlin, S.R.; Jan, Y.N.; et al. Tmem16f forms a Ca^{2+}-activated cation channel required for lipid scrambling in platelets during blood coagulation. *Cell* **2012**, *151*, 111–122. [CrossRef] [PubMed]

138. Morel, O.; Jesel, L.; Freyssinet, J.M.; Toti, F. Cellular mechanisms underlying the formation of circulating microparticles. *Arterioscler. Thromb. Vasc. Biol.* **2011**, *31*, 15–26. [CrossRef] [PubMed]

139. Farsad, K.; De Camilli, P. Mechanisms of membrane deformation. *Curr. Opin. Cell Biol.* **2003**, *15*, 372–381. [CrossRef]

140. Mulcahy, L.A.; Pink, R.C.; Carter, D.R. Routes and mechanisms of extracellular vesicle uptake. *J. Extracell. Vesicles* **2014**, *3*. [CrossRef] [PubMed]

141. Podbilewicz, B. Virus and cell fusion mechanisms. *Annu. Rev. Cell Dev. Biol.* **2014**, *30*, 111–139. [CrossRef] [PubMed]

142. Prada, I.; Meldolesi, J. Binding and fusion of extracellular vesicles to the plasma membrane of their cell targets. *Int. J. Mol. Sci.* **2016**, *17*, 1296. [CrossRef] [PubMed]

143. Maas, S.L.; Breakefield, X.O.; Weaver, A.M. Extracellular vesicles: Unique intercellular delivery vehicles. *Trends Cell Biol.* **2017**, *27*, 172–188. [CrossRef] [PubMed]

144. French, K.C.; Antonyak, M.A.; Cerione, R.A. Extracellular vesicle docking at the cellular port: Extracellular vesicle binding and uptake. *Semin. Cell Dev. Biol.* **2017**, *67*, 48–55. [CrossRef] [PubMed]

145. Pluskota, E.; Woody, N.M.; Szpak, D.; Ballantyne, C.M.; Soloviev, D.A.; Simon, D.I.; Plow, E.F. Expression, activation, and function of integrin alphaMbeta2 (Mac-1) on neutrophil-derived microparticles. *Blood* **2008**, *112*, 2327–2335. [CrossRef] [PubMed]

146. Falati, S.; Liu, Q.; Gross, P.; Merrill-Skoloff, G.; Chou, J.; Vandendries, E.; Celi, A.; Croce, K.; Furie, B.C.; Furie, B. Accumulation of tissue factor into developing thrombi in vivo is dependent upon microparticle p-selectin glycoprotein ligand 1 and platelet p-selectin. *J. Exp. Med.* **2003**, *197*, 1585–1598. [CrossRef] [PubMed]

147. Piccin, A.; Murphy, W.G.; Smith, O.P. Circulating microparticles: Pathophysiology and clinical implications. *Blood Rev.* **2007**, *21*, 157–171. [CrossRef] [PubMed]

148. Rossaint, J.; Kuhne, K.; Skupski, J.; Van Aken, H.; Looney, M.R.; Hidalgo, A.; Zarbock, A. Directed transport of neutrophil-derived extracellular vesicles enables platelet-mediated innate immune response. *Nat. Commun.* **2016**, *7*, 13464. [CrossRef] [PubMed]

149. Losche, W.; Scholz, T.; Temmler, U.; Oberle, V.; Claus, R.A. Platelet-derived microvesicles transfer tissue factor to monocytes but not to neutrophils. *Platelets* **2004**, *15*, 109–115. [CrossRef] [PubMed]

150. Deregibus, M.C.; Cantaluppi, V.; Calogero, R.; Lo Iacono, M.; Tetta, C.; Biancone, L.; Bruno, S.; Bussolati, B.; Camussi, G. Endothelial progenitor cell derived microvesicles activate an angiogenic program in endothelial cells by a horizontal transfer of mrna. *Blood* **2007**, *110*, 2440–2448. [CrossRef] [PubMed]

151. Mause, S.F.; Weber, C. Microparticles: Protagonists of a novel communication network for intercellular information exchange. *Circ. Res.* **2010**, *107*, 1047–1057. [CrossRef] [PubMed]

152. Collino, F.; Bruno, S.; Incarnato, D.; Dettori, D.; Neri, F.; Provero, P.; Pomatto, M.; Oliviero, S.; Tetta, C.; Quesenberry, P.J.; et al. AKI recovery induced by mesenchymal stromal cell-derived extracellular vesicles carrying micrornas. *J. Am. Soc. Nephrol.* **2015**, *26*, 2349–2360. [CrossRef] [PubMed]

153. Mack, M.; Kleinschmidt, A.; Bruhl, H.; Klier, C.; Nelson, P.J.; Cihak, J.; Plachy, J.; Stangassinger, M.; Erfle, V.; Schlondorff, D. Transfer of the chemokine receptor CCR5 between cells by membrane-derived microparticles: A mechanism for cellular human immunodeficiency virus 1 infection. *Nat. Med.* **2000**, *6*, 769–775. [CrossRef] [PubMed]

154. Rozmyslowicz, T.; Majka, M.; Kijowski, J.; Murphy, S.L.; Conover, D.O.; Poncz, M.; Ratajczak, J.; Gaulton, G.N.; Ratajczak, M.Z. Platelet- and megakaryocyte-derived microparticles transfer CXCR4 receptor to CXCR4-null cells and make them susceptible to infection by X4-HIV. *AIDS* **2003**, *17*, 33–42. [CrossRef] [PubMed]

155. Al-Nedawi, K.; Meehan, B.; Kerbel, R.S.; Allison, A.C.; Rak, J. Endothelial expression of autocrine VEGF upon the uptake of tumor-derived microvesicles containing oncogenic EGFR. *Proc. Natl. Acad. Sci. USA* **2009**, *106*, 3794–3799. [CrossRef] [PubMed]

156. Kahn, R.; Mossberg, M.; Ståhl, A.L.; Johansson, K.; Lopatko Lindman, I.; Heijl, C.; Segelmark, M.; Mörgelin, M.; Leeb-Lundberg, L.M.; Karpman, D. Microvesicle transfer of kinin B1-receptors is a novel inflammatory mechanism in vasculitis. *Kidney Int.* **2017**, *91*, 96–105. [CrossRef] [PubMed]

157. Baj-Krzyworzeka, M.; Majka, M.; Pratico, D.; Ratajczak, J.; Vilaire, G.; Kijowski, J.; Reca, R.; Janowska-Wieczorek, A.; Ratajczak, M.Z. Platelet-derived microparticles stimulate proliferation, survival, adhesion, and chemotaxis of hematopoietic cells. *Exp. Hematol.* **2002**, *30*, 450–459. [CrossRef]

158. Salanova, B.; Choi, M.; Rolle, S.; Wellner, M.; Luft, F.C.; Kettritz, R. Beta2-integrins and acquired glycoprotein IIb/IIIa (gpIIb/IIIa) receptors cooperate in NF-kappaB activation of human neutrophils. *J. Biol. Chem.* **2007**, *282*, 27960–27969. [CrossRef] [PubMed]

159. Taraboletti, G.; D'Ascenzo, S.; Borsotti, P.; Giavazzi, R.; Pavan, A.; Dolo, V. Shedding of the matrix metalloproteinases mmp-2, mmp-9, and mt1-mmp as membrane vesicle-associated components by endothelial cells. *Am. J. Pathol.* **2002**, *160*, 673–680. [CrossRef]

160. Barry, O.P.; Pratico, D.; Savani, R.C.; FitzGerald, G.A. Modulation of monocyte-endothelial cell interactions by platelet microparticles. *J. Clin. Investig.* **1998**, *102*, 136–144. [CrossRef] [PubMed]

161. Mossberg, M.; Ståhl, A.L.; Kahn, R.; Kristoffersson, A.C.; Tati, R.; Heijl, C.; Segelmark, M.; Leeb-Lundberg, L.M.F.; Karpman, D. C1-inhibitor decreases the release of vasculitis-like chemotactic endothelial microvesicles. *J. Am. Soc. Nephrol.* **2017**, *28*, 2472–2481. [CrossRef] [PubMed]

162. Mause, S.F.; von Hundelshausen, P.; Zernecke, A.; Koenen, R.R.; Weber, C. Platelet microparticles: A transcellular delivery system for rantes promoting monocyte recruitment on endothelium. *Arterioscler. Thromb. Vasc. Biol.* **2005**, *25*, 1512–1518. [CrossRef] [PubMed]

163. Biancone, L.; Bruno, S.; Deregibus, M.C.; Tetta, C.; Camussi, G. Therapeutic potential of mesenchymal stem cell-derived microvesicles. *Nephrol. Dial. Transplant.* **2012**, *27*, 3037–3042. [CrossRef] [PubMed]

164. Gasser, O.; Schifferli, J.A. Activated polymorphonuclear neutrophils disseminate anti-inflammatory microparticles by ectocytosis. *Blood* **2004**, *104*, 2543–2548. [CrossRef] [PubMed]

165. Bevers, E.M.; Williamson, P.L. Getting to the outer leaflet: Physiology of phosphatidylserine exposure at the plasma membrane. *Physiol. Rev.* **2016**, *96*, 605–645. [CrossRef] [PubMed]

166. Sims, P.J.; Wiedmer, T.; Esmon, C.T.; Weiss, H.J.; Shattil, S.J. Assembly of the platelet prothrombinase complex is linked to vesiculation of the platelet plasma membrane. Studies in scott syndrome: An isolated defect in platelet procoagulant activity. *J. Biol. Chem.* **1989**, *264*, 17049–17057. [PubMed]

167. Zwaal, R.F.; Comfurius, P.; Bevers, E.M. Surface exposure of phosphatidylserine in pathological cells. *Cell. Mol. Life Sci.* **2005**, *62*, 971–988. [CrossRef] [PubMed]

168. Ståhl, A.L.; Vaziri-Sani, F.; Heinen, S.; Kristoffersson, A.C.; Gydell, K.H.; Raafat, R.; Gutierrez, A.; Beringer, O.; Zipfel, P.F.; Karpman, D. Factor H dysfunction in patients with atypical hemolytic uremic syndrome contributes to complement deposition on platelets and their activation. *Blood* **2008**, *111*, 5307–5315. [CrossRef] [PubMed]

169. Diamant, M.; Nieuwland, R.; Pablo, R.F.; Sturk, A.; Smit, J.W.; Radder, J.K. Elevated numbers of tissue-factor exposing microparticles correlate with components of the metabolic syndrome in uncomplicated type 2 diabetes mellitus. *Circulation* **2002**, *106*, 2442–2447. [CrossRef] [PubMed]

170. Del Conde, I.; Shrimpton, C.N.; Thiagarajan, P.; Lopez, J.A. Tissue-factor-bearing microvesicles arise from lipid rafts and fuse with activated platelets to initiate coagulation. *Blood* **2005**, *106*, 1604–1611. [CrossRef] [PubMed]

171. Rubin, O.; Delobel, J.; Prudent, M.; Lion, N.; Kohl, K.; Tucker, E.I.; Tissot, J.D.; Angelillo-Scherrer, A. Red blood cell-derived microparticles isolated from blood units initiate and propagate thrombin generation. *Transfusion* **2013**, *53*, 1744–1754. [CrossRef] [PubMed]

172. Nieuwland, R.; Berckmans, R.J.; McGregor, S.; Boing, A.N.; Romijn, F.P.; Westendorp, R.G.; Hack, C.E.; Sturk, A. Cellular origin and procoagulant properties of microparticles in meningococcal sepsis. *Blood* **2000**, *95*, 930–935. [PubMed]

173. Fujimi, S.; Ogura, H.; Tanaka, H.; Koh, T.; Hosotsubo, H.; Nakamori, Y.; Kuwagata, Y.; Shimazu, T.; Sugimoto, H. Activated polymorphonuclear leukocytes enhance production of leukocyte microparticles with increased adhesion molecules in patients with sepsis. *J. Trauma* **2002**, *52*, 443–448. [CrossRef] [PubMed]

174. Tokes-Fuzesi, M.; Woth, G.; Ernyey, B.; Vermes, I.; Muhl, D.; Bogar, L.; Kovacs, G.L. Microparticles and acute renal dysfunction in septic patients. *J. Crit. Care* **2013**, *28*, 141–147. [CrossRef] [PubMed]

175. Mostefai, H.A.; Meziani, F.; Mastronardi, M.L.; Agouni, A.; Heymes, C.; Sargentini, C.; Asfar, P.; Martinez, M.C.; Andriantsitohaina, R. Circulating microparticles from patients with septic shock exert protective role in vascular function. *Am. J. Respir. Crit. Care Med.* **2008**, *178*, 1148–1155. [CrossRef] [PubMed]

176. Woei, A.J.F.J.; van der Starre, W.E.; Tesselaar, M.E.; Garcia Rodriguez, P.; van Nieuwkoop, C.; Bertina, R.M.; van Dissel, J.T.; Osanto, S. Procoagulant tissue factor activity on microparticles is associated with disease severity and bacteremia in febrile urinary tract infections. *Thromb. Res.* **2014**, *133*, 799–803. [CrossRef] [PubMed]

177. Delabranche, X.; Boisrame-Helms, J.; Asfar, P.; Berger, A.; Mootien, Y.; Lavigne, T.; Grunebaum, L.; Lanza, F.; Gachet, C.; Freyssinet, J.M.; et al. Microparticles are new biomarkers of septic shock-induced disseminated intravascular coagulopathy. *Intensive Care Med.* **2013**, *39*, 1695–1703. [CrossRef] [PubMed]

178. Soriano, A.O.; Jy, W.; Chirinos, J.A.; Valdivia, M.A.; Velasquez, H.S.; Jimenez, J.J.; Horstman, L.L.; Kett, D.H.; Schein, R.M.; Ahn, Y.S. Levels of endothelial and platelet microparticles and their interactions with leukocytes negatively correlate with organ dysfunction and predict mortality in severe sepsis. *Crit. Care Med.* **2005**, *33*, 2540–2546. [CrossRef] [PubMed]

179. Mortaza, S.; Martinez, M.C.; Baron-Menguy, C.; Burban, M.; de la Bourdonnaye, M.; Fizanne, L.; Pierrot, M.; Cales, P.; Henrion, D.; Andriantsitohaina, R.; et al. Detrimental hemodynamic and inflammatory effects of microparticles originating from septic rats. *Crit. Care Med.* **2009**, *37*, 2045–2050. [CrossRef] [PubMed]

180. Meziani, F.; Delabranche, X.; Asfar, P.; Toti, F. Bench-to-bedside review: Circulating microparticles—a new player in sepsis? *Crit. Care* **2010**, *14*, 236. [CrossRef] [PubMed]

181. Camussi, G.; Cantaluppi, V.; Deregibus, M.C.; Gatti, E.; Tetta, C. Role of microvesicles in acute kidney injury. *Contrib. Nephrol.* **2011**, *174*, 191–199. [PubMed]

182. Timar, C.I.; Lorincz, A.M.; Csepanyi-Komi, R.; Valyi-Nagy, A.; Nagy, G.; Buzas, E.I.; Ivanyi, Z.; Kittel, A.; Powell, D.W.; McLeish, K.R.; et al. Antibacterial effect of microvesicles released from human neutrophilic granulocytes. *Blood* **2013**, *121*, 510–518. [CrossRef] [PubMed]

183. Oehmcke, S.; Westman, J.; Malmström, J.; Mörgelin, M.; Olin, A.I.; Kreikemeyer, B.; Herwald, H. A novel role for pro-coagulant microvesicles in the early host defense against streptococcus pyogenes. *PLoS Pathog.* **2013**, *9*, e1003529. [CrossRef] [PubMed]

184. Ramachandra, L.; Qu, Y.; Wang, Y.; Lewis, C.J.; Cobb, B.A.; Takatsu, K.; Boom, W.H.; Dubyak, G.R.; Harding, C.V. Mycobacterium tuberculosis synergizes with ATP to induce release of microvesicles and exosomes containing major histocompatibility complex class ii molecules capable of antigen presentation. *Infect. Immun.* **2010**, *78*, 5116–5125. [CrossRef] [PubMed]

185. Bieniasz, P.D. The cell biology of HIV-1 virion genesis. *Cell Host Microbe* **2009**, *5*, 550–558. [CrossRef] [PubMed]

186. Hurley, J.H. ESCRTs are everywhere. *EMBO J.* **2015**, *34*, 2398–2407. [CrossRef] [PubMed]

187. Izquierdo-Useros, N.; Naranjo-Gomez, M.; Archer, J.; Hatch, S.C.; Erkizia, I.; Blanco, J.; Borras, F.E.; Puertas, M.C.; Connor, J.H.; Fernandez-Figueras, M.T.; et al. Capture and transfer of HIV-1 particles by mature dendritic cells converges with the exosome-dissemination pathway. *Blood* **2009**, *113*, 2732–2741. [CrossRef] [PubMed]

188. Sims, B.; Farrow, A.L.; Williams, S.D.; Bansal, A.; Krendelchtchikov, A.; Gu, L.; Matthews, Q.L. Role of tim-4 in exosome-dependent entry of HIV-1 into human immune cells. *Int. J. Nanomed.* **2017**, *12*, 4823–4833. [CrossRef] [PubMed]

189. Meckes, D.G., Jr.; Raab-Traub, N. Microvesicles and viral infection. *J. Virol.* **2011**, *85*, 12844–12854. [CrossRef] [PubMed]

190. Walker, J.D.; Maier, C.L.; Pober, J.S. Cytomegalovirus-infected human endothelial cells can stimulate allogeneic CD4+ memory t cells by releasing antigenic exosomes. *J. Immunol.* **2009**, *182*, 1548–1559. [CrossRef] [PubMed]

191. Garcia-Silva, M.R.; Cabrera-Cabrera, F.; das Neves, R.F.; Souto-Padron, T.; de Souza, W.; Cayota, A. Gene expression changes induced by *Trypanosoma cruzi* shed microvesicles in mammalian host cells: Relevance of trna-derived halves. *Biomed. Res. Int.* **2014**, *2014*, 305239. [CrossRef] [PubMed]

192. Garcia-Silva, M.R.; das Neves, R.F.; Cabrera-Cabrera, F.; Sanguinetti, J.; Medeiros, L.C.; Robello, C.; Naya, H.; Fernandez-Calero, T.; Souto-Padron, T.; de Souza, W.; et al. Extracellular vesicles shed by *Trypanosoma cruzi* are linked to small RNA pathways, life cycle regulation, and susceptibility to infection of mammalian cells. *Parasitol. Res.* **2014**, *113*, 285–304. [CrossRef] [PubMed]

193. Prados-Rosales, R.; Baena, A.; Martinez, L.R.; Luque-Garcia, J.; Kalscheuer, R.; Veeraraghavan, U.; Camara, C.; Nosanchuk, J.D.; Besra, G.S.; Chen, B.; et al. Mycobacteria release active membrane vesicles that modulate immune responses in a TLR2-dependent manner in mice. *J. Clin. Investig.* **2011**, *121*, 1471–1483. [CrossRef] [PubMed]

194. Bielaszewska, M.; Ruter, C.; Bauwens, A.; Greune, L.; Jarosch, K.A.; Steil, D.; Zhang, W.; He, X.; Lloubes, R.; Fruth, A.; et al. Host cell interactions of outer membrane vesicle-associated virulence factors of enterohemorrhagic *Escherichia coli* O157: Intracellular delivery, trafficking and mechanisms of cell injury. *PLoS Pathog.* **2017**, *13*, e1006159. [CrossRef] [PubMed]

195. Simon, M.; Cleary, T.G.; Hernandez, J.D.; Abboud, H.E. Shiga toxin 1 elicits diverse biologic responses in mesangial cells. *Kidney Int.* **1998**, *54*, 1117–1127. [CrossRef] [PubMed]

196. Locatelli, M.; Buelli, S.; Pezzotta, A.; Corna, D.; Perico, L.; Tomasoni, S.; Rottoli, D.; Rizzo, P.; Conti, D.; Thurman, J.M.; et al. Shiga toxin promotes podocyte injury in experimental hemolytic uremic syndrome via activation of the alternative pathway of complement. *J. Am. Soc. Nephrol.* **2014**, *25*, 1786–1798. [CrossRef] [PubMed]

197. Polley, M.J.; Nachman, R.L. Human platelet activation by C3a and C3a des-arg. *J. Exp. Med.* **1983**, *158*, 603–615. [CrossRef] [PubMed]

198. Arvidsson, I.; Rebetz, J.; Loos, S.; Herthelius, M.; Kristoffersson, A.C.; Englund, E.; Chromek, M.; Karpman, D. Early terminal complement blockade and C6 deficiency are protective in enterohemorrhagic *Escherichia coli*-infected mice. *J. Immunol.* **2016**, *197*, 1276–1286. [CrossRef] [PubMed]

199. Morigi, M.; Galbusera, M.; Gastoldi, S.; Locatelli, M.; Buelli, S.; Pezzotta, A.; Pagani, C.; Noris, M.; Gobbi, M.; Stravalaci, M.; et al. Alternative pathway activation of complement by Shiga toxin promotes exuberant C3a formation that triggers microvascular thrombosis. *J. Immunol.* **2011**, *187*, 172–180. [CrossRef] [PubMed]
200. Ozaki, M.; Kang, Y.; Tan, Y.S.; Pavlov, V.I.; Liu, B.; Boyle, D.C.; Kushak, R.I.; Skjoedt, M.O.; Grabowski, E.F.; Taira, Y.; et al. Human mannose-binding lectin inhibitor prevents Shiga toxin-induced renal injury. *Kidney Int.* **2016**, *90*, 774–782. [CrossRef] [PubMed]

toxins

MDPI

Article

Shiga Toxins Induce Apoptosis and ER Stress in Human Retinal Pigment Epithelial Cells

Jun-Young Park [1,5,†], Yu-Jin Jeong [2,4,†], Sung-Kyun Park [2], Sung-Jin Yoon [1], Song Choi [1], Dae Gwin Jeong [2], Su Wol Chung [6], Byung Joo Lee [7], Jeong Hun Kim [7], Vernon L. Tesh [3], Moo-Seung Lee [2,5,*] and Young-Jun Park [1,5,*]

[1]   Metabolic Regulation Research Center, Korea Research Institute of Bioscience and Biotechnology, 125 Gwahak-ro, Daejeon 34141, South Korea; pi1812@kribb.re.kr (J.-Y.P.); sjy04@kribb.re.kr (S.-J.Y.); ssong@kribb.re.kr (S.C.)
[2]   Infectious Disease Research Center, Korea Research Institute of Bioscience and Biotechnology, 125 Gwahak-ro, Daejeon 34141, South Korea; lfllying@kribb.re.kr (Y.-J.J.); skpark@kribb.re.kr (S.-K.P.); dgjeong@kribb.re.kr (D.G.J.)
[3]   Department of Microbial Pathogenesis and Immunology, Texas A&M University Health Science Center, Bryan, TX 77807, USA; tesh@medicine.tamhsc.edu
[4]   Department of Biochemistry, College of Medicine, Konyang University, 158 Gwanjeo-ro, Daejeon 35365, South Korea
[5]   Department of Biomolecular Science, KRIBB School of Bioscience, Korea University of Science and Technology (UST), 127 Gajeong-ro, Yuseong-gu, Daejeon 34113, South Korea
[6]   School of Biological Sciences, College of Natural Sciences, University of Ulsan, 93 Daehak-ro, Ulsan 44610, South Korea; swchung@ulsan.ac.kr
[7]   Fight Against Angiogenesis-Related Blindness Laboratory, Biomedical Research Institute, Seoul National University Hospital, Seoul 03080, South Korea; ozma805@empas.com (B.J.L.); steph25@snu.ac.kr (J.H.K.)
*   Correspondence: msl031000@kribb.re.kr (M.-S.L.); pyj71@kribb.re.kr (Y.-J.P.); Tel.: +82-42-879-8279 (ext. 8279) (M.-S.L.); +82-42-879-4219 (ext.4219) (Y.-J.P.)
†   These authors contributed equally to this work.

Academic Editors: Daniel Gillet and Julien Barbier
Received: 29 May 2017; Accepted: 6 October 2017; Published: 13 October 2017

Abstract: Shiga toxins (Stxs) produced by Shiga toxin-producing bacteria *Shigella dysenteriae* serotype 1 and select serotypes of *Escherichia coli* are the most potent known virulence factors in the pathogenesis of hemorrhagic colitis progressing to potentially fatal systemic complications such as acute renal failure, blindness and neurological abnormalities. Although numerous studies have defined apoptotic responses to Shiga toxin type 1 (Stx1) or Shiga toxin type 2 (Stx2) in a variety of cell types, the potential significance of Stx-induced apoptosis of photoreceptor and pigmented cells of the eye following intoxication is unknown. We explored the use of immortalized human retinal pigment epithelial (RPE) cells as an in vitro model of Stx-induced retinal damage. To the best of our knowledge, this study is the first report that intoxication of RPE cells with Stxs activates both apoptotic cell death signaling and the endoplasmic reticulum (ER) stress response. Using live-cell imaging analysis, fluorescently labeled Stx1 or Stx2 were internalized and routed to the RPE cell endoplasmic reticulum. RPE cells were significantly sensitive to wild type Stxs by 72 h, while the cells survived challenge with enzymatically deficient mutant toxins (Stx1A$^-$ or Stx2A$^-$). Upon exposure to purified Stxs, RPE cells showed activation of a caspase-dependent apoptotic program involving a reduction of mitochondrial transmembrane potential ($\Delta\psi_m$), increased activation of ER stress sensors IRE1, PERK and ATF6, and overexpression CHOP and DR5. Finally, we demonstrated that treatment of RPE cells with Stxs resulted in the activation of c-Jun N-terminal kinase (JNK) and p38 mitogen-activated protein kinase (p38MAPK), suggesting that the ribotoxic stress response may be triggered. Collectively, these data support the involvement of Stx-induced apoptosis in ocular complications of intoxication. The evaluation of apoptotic responses to Stxs by cells isolated from multiple organs may reveal unique functional patterns of the cytotoxic actions of these toxins in the systemic complications that follow ingestion of toxin-producing bacteria.

Keywords: Shiga toxins; Shiga toxin type 1 and 2; Shiga toxin-producing *Escherichia coli*; hemolytic uremic syndrome; signaling pathways; apoptosis; retinal pigment epithelial cells

1. Introduction

Shiga toxins (Stxs) are genetically, structurally, and functionally conserved protein exotoxins produced by several bacterial species, including *Shigella dysenteriae* serotype 1 and Stx-producing *Escherichia coli* (STEC). Following ingestion and adherence of STEC in the intestinal tract, patients may experience bloody diarrhea followed by a complicated and potentially fatal disease course that frequently includes microangiopathic hemolytic anemia, thrombocytopenia and acute renal failure, also known as hemolytic uremic syndrome (HUS), and neurological complications [1]. Stxs are critical virulence determinants in these systemic complications. The neutral glycolipid globotriaosylceramide (Gb3) serves as the toxin receptor on the surface of host cells, and sites of tissue damage often correlate with Gb3 expression [2–5]. Once Stxs are internalized following Gb3 receptor binding, they are trafficked in a retrograde manner into early endosomes, and then through the *trans*-Golgi network and Golgi apparatus to reach the endoplasmic reticulum (ER). Stxs cross the ER membrane to enter the cytosol [6–8]. All Stxs are ribosome-inactivating proteins and possess an AB5 configuration comprised of a monomeric enzymatic A-subunit in non-covalent association with a pentameric ring of identical B-subunit proteins [9,10]. Following retro-translocation across the ER membrane, a processed fragment of the toxin enzymatic A-subunit cleaves a single adenine residue from the 28S rRNA component of the ribosome, leading to host cellular protein synthesis inhibition and induction of apoptosis through ER stress signaling [11–15]. Despite the fact that STEC constitute a significant public health concern, the pathophysiology of systemic complications following ingestion of toxin-producing bacteria is not well understood. We and others have shown that the administration of purified Stxs to mice or baboons reproduced many of the pathophysiologic changes seen patients infected with the toxin-producing bacteria, including intestinal, renal and central nervous system (CNS) damage [16–21]. Histologic studies of HUS cases showed that Stxs target vascular endothelial cells for destruction, leading to blood vessel damage in the intestine, kidneys and CNS [22–25].

A strong association between neurological involvement and impairment of the visual system has been reported in pediatric HUS patients [26]. Recently, a 10-month-old pediatric patient infected with Shiga toxin-producing *Escherichia coli* O104 developed lethargy that necessitated admission to the intensive care unit. The patient presented with severe HUS with retinal and choroidal hemorrhages, as well as ischemic events due to thrombotic microangiopathic lesions. After three months, the infant neurologically had minor physical disabilities and no apparent cognitive disabilities and was discharged from the hospital with complete blindness and severe chronic renal failure [27]. Thus, physicians have become aware of ocular involvement in STEC-mediated HUS because of possible vision-endangering consequences. Retinal pigment epithelium (RPE) found at the base of the retina are just posterior to the photoreceptors, a specialized type of neuron in the retina. Photoreceptors are capable of converting light into signals for vision by stimulating neuronal impulse transmission [28]. Polarized RPE cells are essential for maintaining the proper visual function in the retinal physiology. However, despite recent clinical case reports in which patients present with ocular involvement, there are no precise mechanisms defined by which Stxs contribute to the injury of RPE cells that are closely associated with proper visual function. Thus, we determined whether Stx1- and Stx2-induced apoptosis with toxins induced the ribotoxic and ER stress response signaling using the ARPE-19 human retinal pigment epithelial cell line. In the present study, we first report that receptor Gb3-dependent Stx endocytosis activates the MAPK-mediated ribotoxic stress response and apoptotic and ER stress pathways, triggering caspase-3/7/8 cleavage as well as disrupting the mitochondrial membrane potential in the newly identified toxin-sensitive RPE cell line ARPE-19.

2. Results

2.1. ARPE-19 Cells Are Sensitive to the Cytotoxic Effects of Stx1 and Stx2

Previous studies have indicated that Stxs induce cytotoxic effects in various cell types including monocytic, macrophage-like, and epithelial cell lines [11,29]. To establish the effect of Stxs on ARPE-19 cells, we first investigated the morphologic features of ARPE-19 cells when treated with Stx1 (100 ng/mL), Stx1A$^-$ (100 ng/mL), Stx2 (10 ng/mL), or Stx2A$^-$ (10 ng/mL). ARPE-19 cells presented the typical morphology under control conditions, while Stx1- and Stx2-treated cells exhibited dramatic morphological changes and cytopathic effects at the indicated incubation times. However, both Stx1A$^-$ and Stx2A$^-$ which lack enzymatic activity due to mutations in the A subunit catalytic residue of each toxin, showed similar features to control cells (Figure 1A). The cytotoxic effects of Stxs on ARPE-19 cells were assessed by cell viability measurements following the incubation of cells with Stx1 (100 ng/mL) and Stx2 (10 ng/mL) for 0–72 h. Cell viability rapidly decreased beginning 24 h after incubation with Stxs. In contrast, major changes in cell viability were not detected after 24 h of exposure of ARPE-19 cells to Stxs with mutations in the A subunit (Figure 1B). As shown in Figure S1, a dose- and time-dependent increase of cytotoxicity was observed for all Stxs (Stx1 and Stx2) at the range of concentrations from 1.0 to 400 ng/mL. CD$_{50}$ values of ~100 ng/mL and ~10 ng/mL were estimated for Stx1 and Stx2, respectively. These results indicate that ARPE-19 cells are sensitive to Stx-induced cytotoxicity. This makes RPEs much less sensitive to Stxs compared to Vero or HeLa cells that have CD$_{50s}$ in the pg/mL [15].

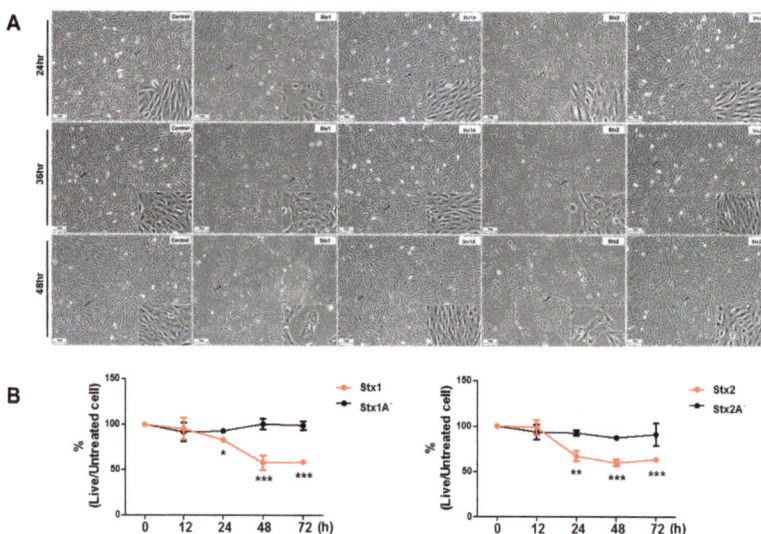

Figure 1. The effects of Shiga toxin treatment on cell death in ARPE-19 cells. (**A**) Stx1 or Stx2 treatment, but not Stx1A$^-$ or Stx2A$^-$ treatment, show dramatic changes in cell morphology. Images were collected from cells incubated with or without Stxs for 24, 36, and 48 h; (**B**) ARPE-19 cells were seeded in 96-well plates at a total cell density of approximately 1.0×10^4 cells/well. Cells were incubated with Stx1 (100 ng/mL), Stx1A$^-$ (100 ng/mL), Stx2 (10 ng/mL) or Stx2A$^-$ (10 ng/mL) for the indicated time points. Cell viability was measured by colorimetric assay using the dye tetrazolium compound [3-(4,5-dimethylthiazol-2-yl)-5-(3-carboxymethoxyphenyl)-2-(4-sulfophenyl)-2*H*-tetrazolium] inert salt (MTS). Data are expressed as % viability compared to untreated control cells at each time point. Results shown are mean ± SEM from two independent experiments using triplicate samples. Asterisks denote statistical significance, Stx1 or Stx2 treatment vs. Stx1A− or Stx2A− treatment, * = $p < 0.05$; ** = $p < 0.01$; *** = $p < 0.001$.

2.2. ARPE-19 Cells Express Membrane-Associated Stx-Receptor Gb3 Expression at the Cell Surface

As reported previously, the binding of Stxs to cell surfaces requires specific glycolipids, with globotriaosylceramide (Gb3) being essential for toxin internalization and subsequent cytotoxicity [5]. Specific cell types in the kidney expressing high Gb3 levels are much more sensitive to the cytotoxic effects of Stxs compared to non-kidney cells with lower Gb3 expression (e.g., human brain endothelial cells) [30–32]. To correlate the cytotoxicity of Stxs with Gb3 expression, we examined Gb3 expression on the surface of ARPE-19 cells. Basal levels of Gb3 expression on the cell surface were measured by Fluorescence- Activated Cell Sorter (FACS) analysis using monoclonal anti-Gb3/CD77 antibodies (Figure 2). Cultured ARPE-19 cells showed expression of Gb3. Levels of Gb3 expression in ARPE-19 cells treated with Stxs at early time points were not significantly altered. In summary, the ARPE-19 cell line expressed Gb3 on the cell surface but did not show changes in Gb3 expression when treated with Stxs for up to 1 h.

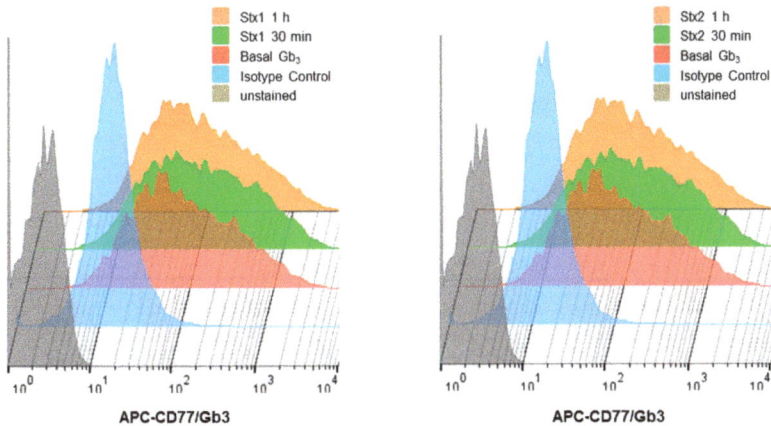

Figure 2. Analysis of Gb3 expression on ARPE-19 cells by flow cytometry. ARPE-19 cells were stained with Alexa Fluor 647-conjugated anti-Gb3/CD77 antibody or an isotype-matched antibody (mouse IgM-Alexa Fluor 647) to determine the background fluorescence at 4 °C for 30 min. or after 30 min or 1 h of stimulation with (left panel) Stx1 (100 ng/mL) or (right panel) Stx2 (10 ng/mL).

2.3. Intracellular Trafficking of Stxs to the ER in ARPE-19 Cells

Previous studies have revealed that Stxs are internalized via Gb3 receptors to the *trans*-Golgi network and lumen of the endoplasmic reticulum (ER), a process is known as retrograde transport [6,33]. To visualize toxin internalization in ARPE-19 cells, we used immunofluorescence microscopy to examine the intracellular trafficking of Alexa Fluor 488-conjugated Stx1 or Alexa Fluor 594-conjugated Stx2 with 4′,6-diamidino-2-phenylindole (DAPI) co-staining. In ARPE-19 cells, we observed that Alexa Fluor 488-conjugated Stx1 and Alexa Fluor 594-conjugated Stx2 fluorescence were significantly increased in ARPE-19 cells over 30 min (Figure 3A). The fluorescence signal was confirmed to coalesce around the nucleus. In the live cell imaging studies, we observed that the perinuclear localization of Alexa Fluor 488-conjugated Stx1 co-localized with an ER specific [6] fluorescent marker, producing yellow fluorescence in merged images (Figure 3B, Movie S1). These experiments indicate that Stxs undergo retrograde intracellular trafficking to the ER in ARPE-19 cells.

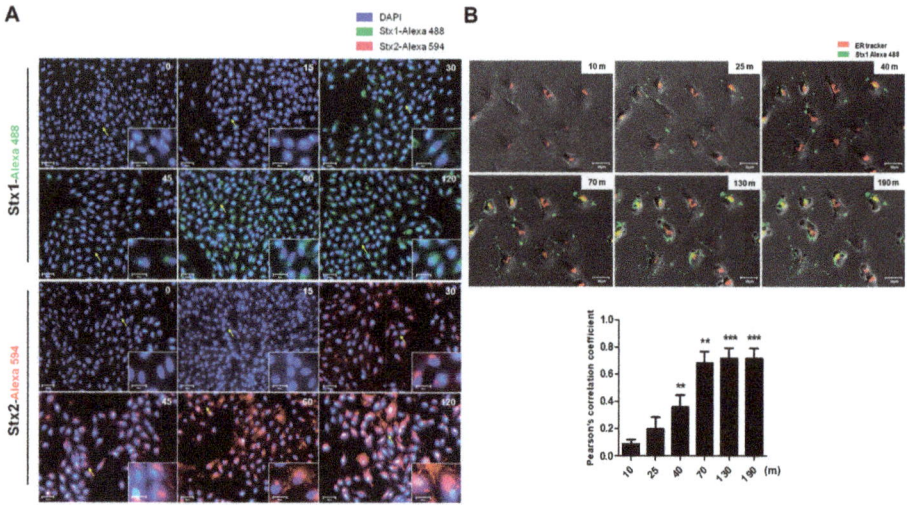

Figure 3. Detection of toxin intracellular trafficking to the ER in ARPE-19 cells treated with Stx1 or Stx2. (**A**) ARPE-19 cells were seeded in 12-well plates at a total cell density of approximately 1.0×10^5 cells/well. Cells were incubated with Alexa Fluor 488-conjugated Stx1 (100 ng/mL) or Alexa Fluor 594-conjugated Stx2 (10 ng/mL) for the time points indicated in the right upper corner of each panel. After washing, cells were fixed and DAPI reaction solution added to stain nuclei. Representative DAPI-positive cells visualized by fluorescence microscopy are shown. In each panel, the arrow depicts cells shown in higher magnification in the insert. Scale bar: 80 μm, insert: 30 μm; (**B**) To detect Stx trafficking to the ER, 1.0×10^5 ARPE-19 cells/well were seeded in 12-well plates. Subsequently, cells were stimulated with complete growth media containing Alexa Fluor 488-conjugated Stx1 (100 ng/mL) with 50 nM ER-tracker (red) live-cell staining dyes for detection of ER. After washing, cells were captured in time-lapse live imaging at the time points indicated in the right upper corner of each panel. Yellow fluorescence indicates co-localization of Stx1 with the ER-specific marker. The scale bars represent 30 μm. The bar graphs show the mean ± SEM of the Pearson's correlation coefficient assessment of the co-localization of Stx1-Alexa Flour 488/ER tracker (lower panel). More than 30 cells were counted per condition ($n = 3$). Asterisks indicate significant differences between treated time at 10 min vs. indicated time points; ** = $p < 0.01$; *** = $p < 0.001$.

2.4. Stx1 and Stx2 Activate Stress-Associated MAPKs in ARPE-19 Cells

In numerous studies, MAPK signaling has been shown to be involved in eliciting responses to various stresses [34]. In addition, the MAPK signaling pathways activated by Stxs trigger the ribotoxic stress response in various cell types [35,36]. We investigated whether Stxs activate MAPK signaling in ARPE-19 cells. Stxs induced the phosphorylation of p38, JNK and Extracellular signal–regulated kinase (ERK) after 1–3 h (Figure 4). Stxs with mutations in the enzymatic A subunits did not induce phosphorylation of p38 or JNK. In contrast, enzymatically inactive Stxs activated ERK phosphorylation. These results show that the internalization and intracellular trafficking of Stxs induce MAPK signaling, and Stx enzymatic activity is required for the activation of p38 and JNK, but not ERK.

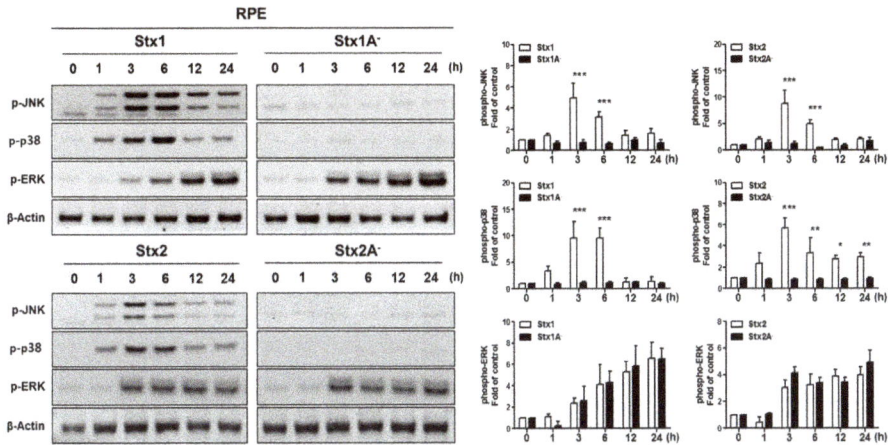

Figure 4. Stx1 and Stx2 activation of stress-associated MAPKs in RPE cells. ARPE-19 cells were stimulated with Stx1 (100 ng/mL), Stx1A$^-$ (100 ng/mL), Stx2 (10 ng/mL) or Stx2A$^-$ (10 ng/mL). After washing, cell lysates were collected at the indicated time points. Phosphorylation of p38MAPK (p-p38), Jun N-terminal kinase (p-JNK), and extracellular signal-regulated kinase (p-ERK) were examined by Western blotting. β-Actin was used as a control for equal protein loading. The results are from one representative experiment of three independent experiments. The data supports the necessity of Stx A$_1$-fragment retro-translocation and action on the ribosomal peptidyl transferase center for activation of p38 and JNK, but not ERK. The bar graphs show the mean ± SEM of fold changes in band densities normalized to β-Actin in comparison to untreated control cell values (right panel). Asterisks indicate significant differences between Stx1 vs. Stx1A$^-$ or Stx 2 vs. Stx2A$^-$; * = $p < 0.05$; ** = $p < 0.01$; *** = $p < 0.001$.

2.5. The ER Stress Response is Induced after Stx1 or Stx2 Exposure in ARPE-19 Cells

Stxs that are trafficked to the ER have the capacity to induce the ER stress response through activation of the ER membrane-bound sensors protein kinase RNA-like endoplasmic reticulum kinase (PERK), inositol-requiring enzyme 1α (IRE1α), and activating transcription factor 6 (ATF6) in several types of cells [7]. If Stx-mediated ER stress is prolonged, it stimulates apoptotic signaling [37]. To assess the induction of ER stress in ARPE-19 cells, the cells were treated for 0 to 12 h with purified Stx1, Stx2, Stx1A$^-$ or Stx2A$^-$. We detected PERK activation that remained elevated up to 12 h after cells were treated with Stx1 or Stx2 (Figure 5A). However, Stx1A$^-$ and Stx2A$^-$ mediated the transient induction of PERK phosphorylation (p-PERK) that dramatically decreased 8 to 12 h after treatment. Consistent with these findings, we showed that Stx1 and Stx2 treatment induced prolonged activation (phosphorylation) of the PERK substrate eukaryotic translation initiation factor-2α (eIF-2α) (Figure 5A). We also observed IRE1α activation (p-IRE1α) that was increased up to 8 h after Stx1, Stx2, Stx1A$^-$ or Stx2A$^-$ treatment. However, when we treated ARPE-19 cells with holotoxins, activation of IRE1α was evident up to 12 h. Treatment of cells with Stx1 or Stx2 toxoids with A subunit mutations showed rapidly decreased IRE1α phosphorylation at 12 h (Figure 5A right panel). In contrast to these findings, we did not detect differences in ATF6 proteolysis between Stx1 or Stx2 and Stx1A$^-$ or Stx2A$^-$ (Figure 5A), suggesting that, following retrograde transport to the ER, individual toxin subunits may be sensed as proteins in an unfolded state in ARPE-19 cells. Immunoglobulin heavy chain binding protein (BiP) is a member of the Hsp70 ATP-dependent molecular chaperone family localized to the ER lumen. BiP is involved in both the translocation of newly synthesized polypeptides and retro-translocation of aberrantly folded proteins for degradation by the proteasome [38]. BiP appears to regulate the ER stress response by dissociation from PERK and ATF-6, and sequestration of the stress sensor IRE1α to prevent oligomerization and activation [39,40]. During ER stress, the dissociation of BiP from the stress sensors may activate transcription factors that markedly up-regulate BiP gene expression [41]. All of the Stx (Stx1, Stx2) or mutant toxin (Stx1A$^-$ and Stx2A$^-$) treatments induced higher expression levels of BiP compared to the results for control cells. Expression of the transcription factor C/EBP homologous protein (CHOP) is regulated by ER stress in response to unfolded proteins in ER [42]. Previous studies showed that Stx treatment of undifferentiated or macrophage-like THP-1 cells induced CHOP mRNA and protein expression [43]. Therefore, we determined whether CHOP mRNA expression is up-regulated in APRE-19 cells stimulated with Stxs (Figure 5B). CHOP mRNA expression dramatically increased 4 h after Stx1 or Stx2 treatment in ARPE-19 cells. Although treatment of APRE-19 cells with Stx1A$^-$ or Stx2A$^-$ mutant toxoids transiently activated the ER stress sensors PERK and IRE1α, the A subunit enzymatic mutants did not induce CHOP mRNA expression. ER-stress induced apoptosis is known to be involved in death receptor 5 (DR5) up-regulation by a CHOP dependent mechanism in some human cancer cells [44]. As shown in Figure 5B, increased levels of DR5 mRNA were detected beginning at 4 h and peaking at 8 h following stimulation with Stx1 or Stx2. However, Stx1A$^-$ or Stx2A$^-$ mutant toxoids did not induce DR5 mRNA expression.

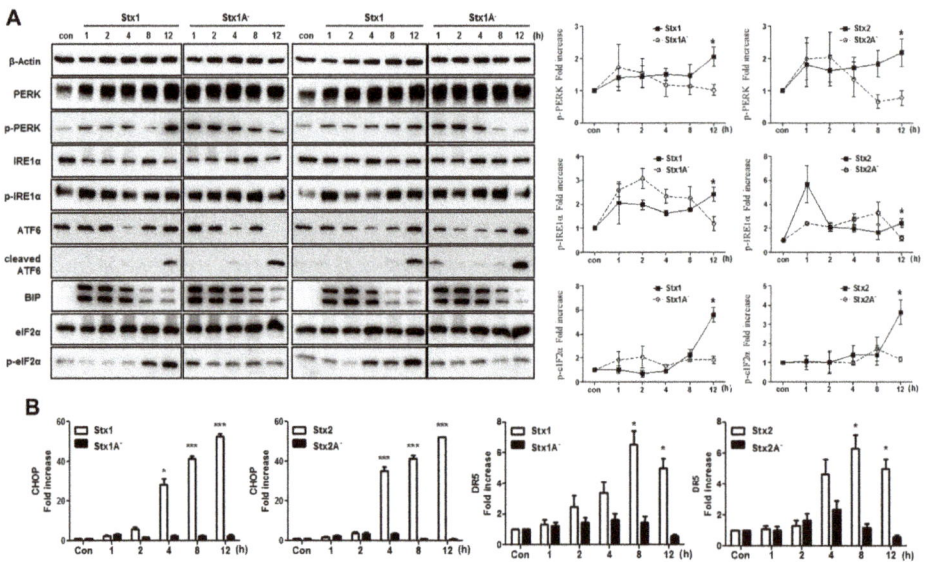

Figure 5. The ER stress response is induced in RPE cells after Stx1 or Stx2 exposure. (**A**) ARPE-19 cells (5.0×10^5 cells/well) were stimulated with Stx1 (100 ng/mL), Stx1A$^-$ (100 ng/mL), Stx2 (10 ng/mL) or Stx2A$^-$ (10 ng/mL). At the indicated times (h), cells were lysed and the presence of activated ER stress sensors and their downstream targets in cellular lysates were determined by Western blotting. Untreated control cells (con) were used to determine baseline protein expression. β-Actin was used as a control for equal protein loading. The line graphs show the mean ± SEM of band densities normalized by the division of β-Actin band densities and compared to untreated control cell values (right panel). Statistical analyses of densitometric scans from at least three independent experiments are shown; (**B**) ARPE-19 cells were treated as outlined above and CHOP and DR5 mRNA expression were measured using RT-PCR and normalized using Glyceraldehyde 3-phosphate dehydrogenase (GAPDH). The data are expressed as mean and SEM increases in (**A** right panel) phospho-PERK (p-PERK), phospho-IRE1α (p-IRE1α), phospho-eIF2α (p-eIF2α), and (**B**) CHOP and DR5 mRNA levels compared to the levels in untreated control cells. Asterisks indicate significant differences between Stx1 and 2 vs. Stx1A$^-$ and 2A$^-$; * = $p < 0.05$; *** = $p < 0.001$.

2.6. Stxs Activate Apoptotic Signaling Pathways in ARPE-19 Cells and Toxin Enzymatic Activity is Required for Apoptosis

Previous studies revealed that Stxs induce apoptotic cell death in human epithelial, endothelial and monocytic THP-1 cell lines [45]. We have shown that the internalization of Stxs in ARPE-19 cells induced stress-related MAPK signaling and ER stress protein activation. ER stress is known to induce apoptosis through p38 and JNK MAPK stress signaling [46]. ER stress-induced apoptosis can activate initiator caspase-8, a process that leads to the loss of mitochondrial membrane potential and activation of downstream effector caspase-3 [47]. Therefore, we explored whether Stx1 or Stx2 induces apoptosis in ARPE-19 cells. To analyze DNA fragmentation as an indicator of apoptotic cell death, we performed the terminal deoxynucleotidyltransferase-mediated 2′-Deoxyuridine, 5′-Triphosphate (dUTP)-biotin nick end labeling (TUNEL) assay. The TUNEL assays demonstrated increased TUNEL-positive APRE-19 cells 24 h after treatment with Stx1 or Stx2 compared to untreated control cells. Increased DNA fragmentation was not detected in Stx1A$^-$ or Stx2A$^-$ treated cells, suggesting that Stx enzymatic activity is required to induce apoptosis (Figure 6A). In order to confirm the occurrence of apoptosis, we also measured the activity of caspase-3/7. Caspase-3/7 activity in ARPE-19 cells increased progressively 12–48 h after treatment with Stx1 or Stx2. In contrast, caspase 3/7 activity in ARPE-19 cells treated with Stx1A$^-$ or Stx2A$^-$ slightly increased compared to control cells (Figure 6B). These results suggest that enzymatically active Stxs induce apoptosis in a caspase-dependent manner in ARPE-19 cells. As shown in Figure 6C, the uncleaved (inactive) form of caspase-8 significantly decreased beginning 3 h after incubation with Stx1 or Stx2. Since activated caspase-8 is known to cleave Bid, which in turn targets mitochondrial membrane perturbation and activates the pro-apoptotic protein Bax [48], we also confirmed that the expression of Bax was increased following the treatment of ARPE-19 cells with Stxs. Cleavage of caspase-3 and Poly (ADP-ribose) polymerase (PARP) was also observed by 24 h of treatment with Stxs. However, these reactions were not induced by treatment with Stx1A$^-$ or Stx2A$^-$ (Figure 6D). Interestingly, the levels of antiapoptotic protein Bcl-2 were decreased by Stx1 or Stx2 treatment in ARPE-19 cells while the levels of Bcl-2 were not altered by Stx1A$^-$ or Stx2A$^-$ (Figure S2). Because the Stxs activated the MAPK signaling in retinal pigment epithelial cell lines (see Figure 4), we performed inhibitor assays using SB203580, SP60125, U0126 to define the effects of MAPK signaling in Stx-induced apoptosis. ARPE-19 cells were treated with MAPK inhibitors and then incubated for 48 h with Stx1 or Stx2. Cell viability measurements indicated that p38 and JNK inhibitors reduced Stx1- and Stx2-induced cell death, but there was no significant difference in cell viability following ERK inhibition (Figure S3A). Also, we identified caspase-3/7 activation by Stx1 or Stx2 in MAPK inhibitor-treated ARPE-19 cells. Stx1/2-induced caspase-3/7 activity was reduced by p38 and JNK inhibitor treatments in ARPE-19 cells. Caspase-3/7 activity was less decreased when treated with ERK inhibitor in comparison to treatment with p38 and JNK inhibitors (Figure S3B). These results suggest that the MAPK signaling contributes to Stx-induced apoptosis. Taken together, these results suggest that Stxs, via a mechanism requiring enzymatic activity, induce apoptosis in a caspase-dependent manner in ARPE-19 cells.

2.7. Stxs Induce Loss of Mitochondrial-Membrane Potential

Mitochondria contain important regulators of apoptotic processes including caspases [49]. Loss of mitochondrial-membrane potential has been shown to contribute to apoptosis [50]. To assess the induction of mitochondrial depolarization in ARPE-19 cells, the cells were treated for 12 h with purified Stx1, Stx2, Stx1A$^-$ or Stx2A$^-$. Normal mitochondria with high $\Delta\psi_m$ accumulate red fluorescent JC-1 aggregates, whereas carbonyl cyanide m-chlorophenyl hydrazone (CCCP)-treated mitochondria with low $\Delta\psi_m$ display the green fluorescent monomeric form of JC-1. After Stx1 or Stx2 treatment, mitochondria showed an increase in monomeric green fluorescence and a decrease in JC-1 aggregated red fluorescence, indicating $\Delta\psi$ depolarization (Figure 7). These responses were not induced in Stx1A$^-$ or Stx2A$^-$ treated ARPE-19 cells (Figure 7). These findings demonstrate that Stx1- and Stx2-induced mitochondrial membrane depolarization is dependent on enzymatic activity.

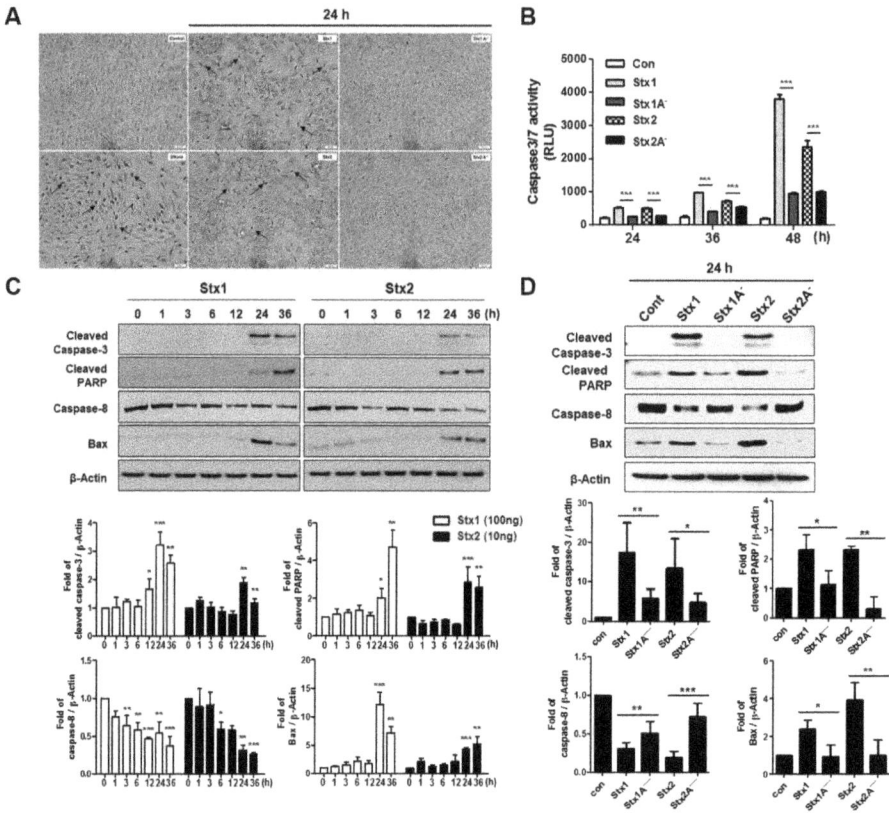

Figure 6. Stxs activate apoptotic signaling pathways in RPE cells. (**A**) The TUNEL assay was performed to detect apoptotic APRE-19 cells after treating with Stx1, Stx2, Stx1A⁻ or Stx2A⁻ for the indicated times. Cells staining dark brown are TUNEL positive and are indicated by arrows. Scale bars = 50 μm; (**B**) Analysis of caspase-3/7 activity. ARPE-19 cells were incubated with Stx1 (100 ng/mL), Stx1A⁻ (100 ng/mL), Stx2 (10 ng/mL) and Stx2A⁻ (10 ng/mL) for the indicated time points, and caspase 3/7 activity was measured using the Caspase-Glo 3/7 Assay; (**C,D**) ARPE-19 cells (1.0×10^5 cells/well) were treated with Stx1 (100 ng/mL), Stx1A⁻ (100 ng/mL), Stx2 (10 ng/mL) and Stx2A⁻ (10 ng/mL) for the indicated time points. Protein samples were prepared and analyzed by Western blotting using anti-cleaved caspase-3, anti-caspase-8, anti-cleaved PARP, anti-Bax and anti-β-Actin antibodies. β-Actin was used as a control for equal protein loading. The results are from one representative experiment of three independent experiments. The bar graphs show the mean ± SEM of fold changes in band densities normalized by division of β-Actin band densities and compared to untreated control cell values (lower panels). Asterisks indicate significant differences between control cell values vs. Stxs-treated cells (panel **C**) or Stx 1 vs. Stx1A⁻ and Stx2 vs. Stx2A⁻ (panel **D**); * = $p < 0.05$; ** = $p < 0.01$; *** = $p < 0.001$.

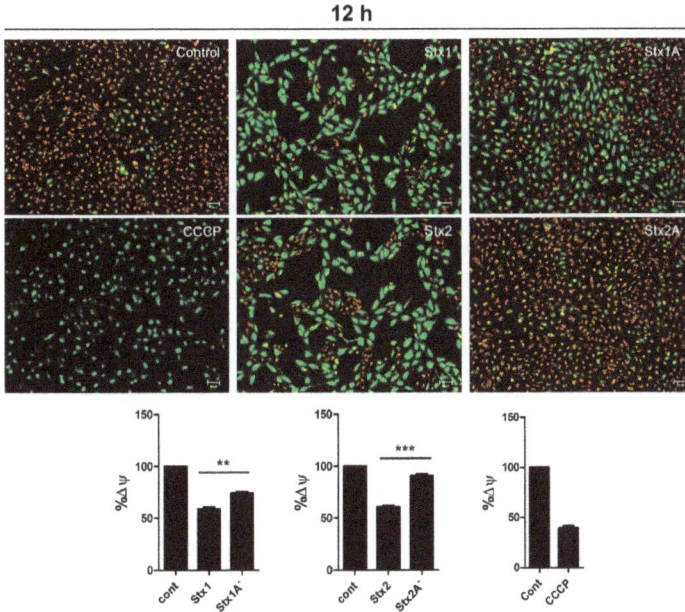

Figure 7. Stxs induced loss of mitochondrial membrane potential in RPE cells. Mitochondrial membrane potential in Stx-treated ARPE-19 cells was evaluated using JC-1 staining to monitor alterations of mitochondrial membrane potential. Red fluorescence indicates the accumulation of JC-1 aggregates in normal mitochondrial membranes. Green fluorescence indicates JC-1 monomers and membrane depolarization. ARPE-19 cells (1.0×10^5 cells/well) were treated with Stx1 (100 ng/mL), Stx1A$^-$ (100 ng/mL), Stx2 (10 ng/mL) and Stx2A$^-$ (10 ng/mL) for 12 h. For the positive control, CCCP was added to 50 µM final concentration and cells were incubated at 37 °C for 5 min. The scale bars represent 200 µm. The quantitative analysis of JC-1 staining was measured by a microplate reader with 520 nm for emission of green fluorescence, and 590 nm for emission of red fluorescence (lower panel). The bar graphs show the mean ± SEM of percentages of cells with division of green fluorescence into red fluorescence normalized using untreated control cell values. Asterisks indicate significant differences between Stx1 vs. 1A$^-$ or Stx2 vs. 2A$^-$; ** = $p < 0.01$; *** = $p < 0.001$.

3. Discussion

The role of Stxs in the diarrheal phase of disease and the myriad of systemic complications that may follow ingestion of toxin-producing bacteria remain to be fully characterized. In particular, the mechanisms by which the toxins mediate ocular disease are incompletely understood. A primary aim of this study was to explore retinal pigment epithelial cell responses to Stxs. In previous studies, we and others presented data to support the notion that Stxs are capable of inducing apoptosis (programmed cell death). Apoptosis is a well-characterized form of cell death that may follow activation of intracellular signaling pathways in response to a variety of cell stressors, including ER stress, in many different types of cells [15,51]. Besides the cell death signals triggered by Stxs through the ER stress response, the toxins as multifunctional proteins have also been shown to activate the ribotoxic stress response, the induction of autophagy and the elicitation of proinflammatory responses including the assembly of the NLRP3 inflammasome [52]. The capacity of Stxs to induce apoptotic signaling may play an essential role in pathogenesis including intestinal damage as well as extra intestinal complications such as HUS and the vascular damage implicated in the pathogenesis of CNS dysfunction [53]. For example, in HUS, extensive damage to glomerular endothelial cells and renal tubular epithelial cells was observed with pyknotic nuclei and sloughing of cells into the tubule lumina.

A high coincidence between central nervous system disorders, which are associated with a higher rate of death, and cortical visual dysfunction has been reported in pediatric HUS patients [26], suggesting that ocular involvement may indicate a severe form of HUS. Previous ocular findings in HUS patients, including a 10-month-old infant with stools positive for STEC, consisted of retinal, choroidal and vitreal hemorrhages, as well as ischemic signs due to thrombotic microangiopathic lesions along with increased platelet consumption after initial vascular endothelial cell damage caused by Stxs [27,54,55]. Retinal pigment epithelial cells may be the targets of Stxs in disease progression that may follow the prodromal diarrheal disease. In the present study, we demonstrate that the ARPE-19 retinal epithelial cell line is sensitive to the cytotoxic action of Stx1 or Stx2, with estimated CD_{50}s of 100 ng/mL and 10 ng/mL, respectively (Figure S1). Our results demonstrating cytotoxicity by utilizing light microscopy and MTS-based cell viability assays (Figure 1) suggested to us that the toxin receptor Gb3 should be expressed on the surface of ARPE-19 cells. In agreement with previous studies using various toxin-sensitive cell types [30], FACS analysis examining Gb3 expression on ARPE-19 cells showed high membrane expression of the Stx receptor in vitro (Figure 2). Purified fluorescently-labeled recombinant Stx1 and Stx2 proteins have been extensively employed to characterize toxin intracellular trafficking to different host cellular compartments [30,56]. As shown in Figure 3, Stx1 and Stx2 were internalized and routed to the ER in ARPE-19 cells, suggesting that rapid retrograde transport to the ER occurs with maximal fluorescence detected 40–190 min after the toxin exposure. This pattern of rapid intracellular toxin transport to the ER correlates with findings using other highly sensitive cell lines [57].

Smith et al. [29] associated the apoptotic cell death pathway with signaling through the ribotoxic stress response activated by exposure to Stx1 in the human epithelial cell line Hct8. Functionally active Stx1 holotoxin, but not an enzymatic activity deficient Stx1 mutant, triggered caspase-3 (ultimate executioner caspase) activation and nuclear fragmentation. Hct8 cells were partially protected from Stx1-induced apoptosis, with decreased caspase-3 activation and DNA fragmentation, when the cells were treated with a p38 MAPK inhibitor prior to exposure to the toxin [29]. The work of Iordanov et al. [14] suggested that ribosomal inactivating proteins that act on the ribosomal peptidyl transferase center (includes Stxs) activate stress-activated protein kinase cascades to initiate proinflammatory and proapoptotic signaling following the alteration of ribosomes. Accordingly, our findings summarized in Figure 4 specifically link Stx binding, internalization and retrograde transport to the ER with stress-activated protein kinase activation (ribotoxic stress response) in ARPE-19 cells. Interestingly, activation of p38 and JNK MAPKs requires toxin enzymatic activity, as toxoids with mutations that eliminate the ribosomal depurination reaction, fail to initiate signaling pathways leading to the phosphorylation of p38 and JNK. In contrast, ERK is activated in the absence of enzymatic activity, suggesting that intracellular routing to the ER or the ER membrane retrotranslocation process itself may be sufficient to trigger ERK signaling in APRE-19 cells. The comprehensive signaling cascades associating alterations in ribosomal function with MAPK activation in retinal epithelial cells exposed to Stxs remain to be fully characterized. There may be some cell-type specific responses among the multiple signaling pathways for the activation of the ribotoxic stress response. For example, using Hct8 and Vero cells, Jandhyala et al. demonstrated that DHP-2 (a pharmacological inhibitor of the upstream MAP3K ZAK) blocked Stx2-mediated activation of JNK and p38 MAPK, partially protected cells from apoptosis and partially reduced active caspase-3 levels without altering protein synthesis inhibition caused by the toxin [58].

In the ER, following synthesis by ribosomes, nascent polypeptides are folded correctly and the processed proteins transported to other locations in the cells or secreted. When overloading of the ER lumen with misfolded proteins occurs, it may lead to the induction of the ER stress response, also referred to as the unfolded protein response (UPR), to cope with the stress. Lee et al. used the toxin-sensitive monocytic THP-1 cell line to first report that, following binding to the membrane glycolipid Gb3 and transport to the ER, Stx1 was capable of inducing ER stress and activating the ER stress sensors PERK, IRE1 and ATF6 [11]. Based on these earlier findings, we initiated studies to characterize the ER stress response in RPE to define the ocular involvement of Stxs in visual impairment that may follow the ingestion of toxin-producing bacteria. We show here that Stx1 and Stx2 activate the ER stress sensors PERK, IRE1 and ATF-6 in ARPE-19 cells, along with enhanced expression of key downstream substrates of ER sensors BiP, CHOP (also called GADD153) and DR5 (TRAIL-R2) (Figure 5). In this regard, we speculate that the retrograde routing and enzymatic activity of functional Stxs may induce apoptosis via initiation of the ER stress response, leading to the rapid activation of the caspase3/7/8 cascade and the CHOP signaling pathways in retinal pigment epithelial cells. DR5 expression is regulated by CHOP [44], and we have previously demonstrated that Stx1 treatment of monocytic and macrophage-like THP-1 cells up-regulate the expression of CHOP and DR5 [11,37]. The precise role of TRAIL and DR5 in retinal pigment epithelial cells undergoing apoptosis induced by Stxs requires further scrutiny.

Apoptotic nuclei localized to tubular epithelial cells and glomerular endothelial cells were detected after TUNEL staining of renal cortical tissues in in vivo and in vitro studies [59]. Moreover, using renal biopsies from seven HUS patients, Te Loo et al. [60] subjected the tissue samples to dual staining with TUNEL and SC-35 (a dye to label for RNA synthesis and splicing factor to avoid non-specific TUNEL positive staining), and found that 80% of apoptotic cells were observed in tubules and 20% in glomeruli. The TUNEL analyses presented here suggest that Stx1- or Stx2-induced cell death is mediated by apoptosis, as evidenced by a marked increase of TUNEL$^+$ cells following the intoxication of ARPE-19 cells. The induction of apoptosis in Stx-treated ARPE-19 cells was further characterized by caspase-3/7 activity measurement. Based on our observations, the enzymatic deficient variants (Stx1A$^-$ and Stx2A$^-$) could not activate either caspase-3 or caspase-8 processing whereas their active counterparts did activate these caspases. (Figure 6A,B). Rapid cleavage of pro-caspase-3 or pro-caspase-8, beginning 4–8 h after intoxication, was reported to occur coincidentally with the decreased expression of anti-apoptotic regulator Bcl-2 and the increased expression of pro-apoptotic factor Bax in Stx1-treated THP-1 cells [11,37]. We reasoned that if the intrinsic pathway was important in Stx-induced retinal pigment epithelial cell apoptosis, then ARPE-19 cells should have an increased loss of mitochondrial membrane potential ($\Delta\psi_m$), and we observed that this indeed was the case. $\Delta\psi_m$ was significantly disrupted when ARPE-19 cells were incubated with Stx1 or Stx2 holotoxin. However $\Delta\psi_m$ was not affected when the cells were exposed to Stx1A$^-$ and Stx2A$^-$ (Figure 7), supporting a requirement for toxin enzymatic activity to induce apoptosis in ARPE-19 cells. Similarly, we and other workers have previously shown that several toxin-sensitive cell types share functional apoptotic signaling mechanisms triggering major caspase activation and mitochondrial membrane depolarization after exposure to Stx1 or Stx2 [43,61,62].

In conclusion, our data support the model of Stx1 or Stx2-induced apoptosis and ER stress pathways in human retinal pigment epithelial cells (Figure 8), suggesting the ocular involvement of Stxs in the pathogenesis of disease leading to visual impairment or blindness. Although there has been reported a high coincidence between neurological and visual system abnormalities, further experimentation will be required to clarify the pathophysiological role of Stxs in causing ocular disease. Funduscopy capable of observing the retinal microvasculature revealed retinal hemorrhages, attenuation of retinal venules, and extensive retinal detachment in the eyes of a patient infected with Shiga toxin-producing *Escherichia coli* O104 [9]. These clinical findings suggest that Stxs may be transmitted to the retinal epithelium via damaged ocular blood vessels. Alternatively, Stxs may disrupt the blood–brain barrier (BBB) due to thrombotic microangiopathic lesions formed after initial vascular endothelial injury. Landoni et al. [63] observed that Stx1 affected the permeability of the brain endothelium to influence the BBB once the toxin reached the brain parenchyma, contributing to the development of neuropathological symptoms observed in HUS. The defined location of Gb3 in the mammalian CNS, and the role of apoptosis in neuropathogenesis are controversial [41,64–67]. Unlike Stxs-mediated renal apoptosis, programmed neuronal cell death induced by Stxs has not been extensively characterized, although rabbit neurons appear to be susceptible to Stx2-induced apoptosis via focal proinflammatory response [68]. The identification of intermediate signaling molecules in Stx-induced retinal apoptosis may provide new insights into discovering therapeutic targets to disrupt the intoxicated cell death, thereby alleviating the retinal injury caused by the toxin. However, in human retinal pigment epithelium, many mechanistic aspects of the apoptotic signaling pathways initiated by Stxs remain to be explored. Moreover, further assessments to clarify the apoptosis signaling mechanisms activated by Stxs in in vivo retinal systems are needed.

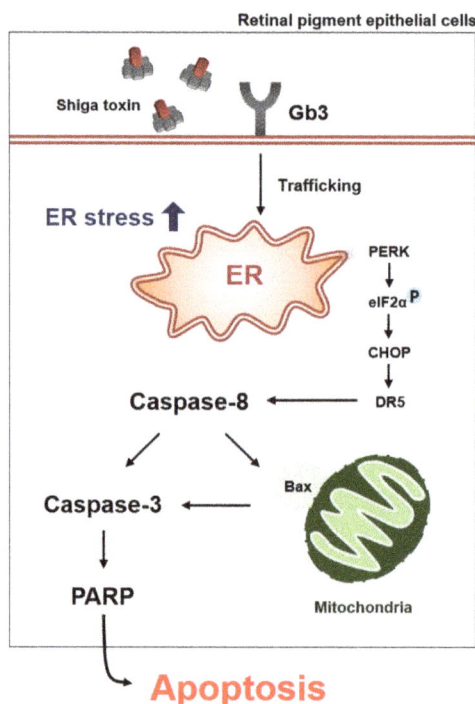

Figure 8. Proposed model of RPE apoptosis triggered by Stxs.

4. Materials and Methods

4.1. Antibodies and Reagents

Rabbit monoclonal antibodies specific for human phospho-p38, phospho-p42/44(ERK), phospho-SAPK/JNK, cleaved caspase-3, cleaved caspase-8, cleaved caspase-9, cleaved PARP, Bax, Bcl-2, BiP, PERK, IRE1α, eIF-2α, phospho-eIF2α and mouse monoclonal anti-β-actin antibodies were purchased from Cell Signaling Technology (Danvers, MA, USA). Rabbit monoclonal antibodies specific for human ATF6 were purchased from Santa Cruz Biotechnology (Santa Cruz, CA, USA). Rabbit monoclonal antibodies specific for human phospho-IRE1α and phospho-PERK were purchased from Abcam (Cambridge, MA, USA). For mitochondrial membrane potential measurement, the Mitochondrial Membrane Potential Assay kit (Cell Signaling Technology, Danvers, MA, USA) using the JC-1 fluorescence was obtained to detect alterations in transmembrane potential.

4.2. Toxins

Stx1 was prepared as previously described [16]. Briefly, Stx1 was purified from cell lysates prepared from *E. coli* DH5α (pCKS112), a recombinant strain containing a plasmid encoding the *stx1* operon, by sequential ion exchange and immunoaffinity chromatography. The purity of toxin preparations was assessed by sodium dodecyl sulfate-polyacylamide gel electrophoresis (SDS-PAGE) with silver staining and western blotting analysis. Toxin preparations were passed through ActiClean

Etox columns (Sterogene Bioseparations, Carlsbad, CA, USA) to remove trace endotoxin contaminants and were determined to contain <0.1 ng of endotoxin per mL as determined by the *Limulus* amoebocyte lysate assay (Associates of Cape Cod, East Falmouth, MA, USA). Purified Stx1 holotoxin containing a double mutation (E167Q and R170L) in the A subunit (Stx1A$^-$) which dramatically reduces enzymatic activity was a kind gift from Dr. Shinji Yamasaki, Osaka Prefecture University, Osaka, Japan [69]. Recombinant purified Stx2, Stx2A$^-$ toxoid (Y77S/E167Q/R170L) were obtained from the NIAID, NIH Biodefense and Emerging Infections Research Repository (BEI Resources, Manassas, VA, USA).

4.3. Cytotoxicity Assay

ARPE-19 cells (5.0×10^4 cells/well) were seeded in 96-well microtiter plates prior to treatment with Stx1 (100 ng/mL), Stx2 (10 ng/mL), Stx1A$^-$ (100 ng/mL) or Stx2A$^-$ (10 ng/mL). Cytotoxicity was determined by colorimetric assay [70] using the tetrazolium compound [3-(4,5-dimethylthiazol-2-yl)-5-(3-carboxymethoxyphenyl)-2-(4-sulfophenyl)-2*H*-tetrazolium] inert salt; (MTS, Promega, Madison, WI, USA). MTS (50 µL/5000 cells) were added to each well and incubation continued for 2 h at 37 °C in 5% CO_2. Optical density was recorded with an automated microtiter plate reader at an absorbance of 490 nm (Microplate Reader with SoftMax; Molecular Devices, Sunnyvale, CA, USA). The percentage of cell death was determined using the following equation: percentage of cell death = [(average OD_{490} of treated cells − average OD_{490} of control cells) ÷ average OD_{490} of control cells] × 100. A background absorbance at 630 nm measured with untreated cells was subtracted from each sample reading. The reference wavelength 630 nm was used to subtract the background contributed by excess cell debris and other nonspecific absorbance.

4.4. Flow Cytometric Analysis of Gb3 (CD77) Membrane Expression

ARPE-19 human retinal pigment epithelial cells were seeded at approximately 2.0×10^5 cells/well into 12-well plates and cultured in Dulbecco's Modified Eagle Medium/Nutrient Mixture F-12 (DMEM/F12) media containing 10% FBS and supplemented with 5.0 µg/mL streptomycin and 5 U/mL penicillin. Cells were maintained at 37 °C in a humidified 5% CO_2 atmosphere. After 24 h, ARPE-19 cells were mock-treated (basal level) or treated with Stx1 (100 ng/mL) or Stx2 (10 ng/mL) for 30 min or 1 h. After washing with cold PBS three times, ARPE-19 cells were detached with trypsin-EDTA and collected by centrifugation at $780 \times g$ for 5 min. Then cells were stained with Alexa Fluor 647-conjugated anti-human Gb3/CD77 monoclonal antibody (BD Biosciences, San Jose, CA, USA) or Alexa Fluor 647-conjugated mouse IgM (isotype control) at 4 °C for 30 min in the dark and analyzed by flow cytometry using a BD FACSCanto^{TM} II cytometer (BD Bioscience, San Jose, CA, USA).

4.5. Intracellular Trafficking Assay

Intracellular trafficking of Stxs into RPE cells was determined using purified Stx1 conjugated to fluorescent tags (Alexa488 and Alexa594, respectively). Purified Stx1 or Stx2 were labeled with Alexa Fluor-488 or -594 dyes (Molecular Probes, Inc., Invitrogen, Eugene, OR, USA) as described in the manufacturer's protocol. Briefly, ARPE-19 cells (1.0×10^5 cells/well) were seeded overnight into 12-well plates. The cells were washed twice with PBS and incubated in complete DMEM/F12 growth media supplemented with 0.5% FBS before further staining for 30 min at 37 °C in the presence of 5% CO_2. Subsequently, cells were treated with Alexa Fluor 488-conjugated Stx1 (100 ng/mL) or Alexa Fluor 594-conjugated Stx2 (10 ng/mL) for the indicated time points. Cells were washed in PBS and fixed in 4% paraformaldehyde in PBS for 10 min at room temperature and then cell nuclei were labeled by DAPI staining. Fluorescently conjugated Stx1 or Stx2 was observed by fluorescence inverted microscopy. To detect Stx trafficking to the ER, ARPE-19 cells were treated with Alexa Fluor 488-conjugated Stx1 (100 ng/mL) with an endoplasmic reticulum-specific dye (50 nM ER-Tracker Live Cell Staining dyes; Molecular Probes, Inc., Eugene, OR, USA). Using time-lapse live-cell imaging microscopy (EVOS® FL Auto Cell imaging System, Thermo Fisher Scientific, Waltham, MA, USA), ARPE-19 cells were extensively imaged over the next 190 min. Binding competition between the anti-Gb3 antibody and the Stxs at the cell surface was observed, which prevents the use of the antibody to observe the intracellular colocalization of the receptor Gb3 and the toxins. The images shown are representative of at least two independent experiments. All data within each experiment were collected at identical settings.

4.6. Western Blotting

The cells were stimulated with Stxs, harvested, and lysed in a CHAPS buffer supplemented with protease and phosphatase inhibitors (GenDEPOT, Barker, TX, USA). Equal amounts of proteins (70–100 µg/lane) were separated by 8% or 12% Tris-glycine SDS–PAGE and transferred to nitrocellulose membranes. Membranes were blocked with 5% non-fat milk prepared with TBST (20 mM Tris (pH 7.6), 137 mM NaCl, 0.1% Tween 20). Membranes were incubated with primary antibodies at 4 °C for 24 h. After washing, the membranes were incubated with horseradish peroxidase-labeled secondary antibodies for 2 h at room temperature. Bands were visualized using the Western Lightning Chemiluminescence System (Atto Co., Tokyo, Japan). Data were obtained from at least three independent experiments. The integrated density was measured using the ImageJ software (NIH, Bethesda, Rockville, ML, USA).

4.7. Real-Time Quantitative RT-PCR

ARPE-19 human retinal pigment epithelial cells were seeded at approximately 2.0×10^5 cells/well into 12-well plates and treated with Stx1 (100 ng/mL), Stx1A$^-$ (100 ng/mL), Stx2 (10 ng/mL) or Stx2A$^-$ (10 ng/mL). At the indicated times, total RNA was extracted using a RNeasy mini kit (Qiagen, Hilden, Germany) and cDNA was synthesized using ReverTra Ace-α-$^®$ reverse transcriptase (Toyobo Life Sciences, Osaka, Japan). Real-time PCR was performed using the ViiA7 real-time PCR system (Life Technologies Corporation, Carlsbad, CA, USA) and Thunderbird SYBR qPCR Mix (Toyobo Life Sciences). The following primers were employed in these reactions: Human CHOP mRNA 5′-TAGGGGACATGTGTGAGCATGA-3′ (sense) and 5′-TCACGGCAAAGAGA TCGGAGA-3′ (anti-sense); Human DR5 mRNA 5′-AAGACCCTTGTGCTCGTTGT-3′ (sense) and 5′-GGAGCTAGGTCTTGTTGGGT-3′ (anti-sense); Human GAPDH mRNA 5′-GCACCGTCAA GGCTGAGAAC-3′ (sense) and 5′-TGGTGAAGACGCCAGTGGA-3′ (anti-sense). The relative mRNA levels for all genes were normalized to the levels of GAPDH.

4.8. TUNEL Assay

For terminal deoxynucleotidyltransferase-mediated dUTP-biotin nick end labeling (TUNEL) analysis, ARPE-19 cells were grown on glass coverslips. Cells were treated with Stx1, Stx1A$^-$, Stx2 and Stx2A$^-$ for 12 and 24 h. After fixing with 4% paraformaldehyde for 10 min, the coverslips were washed twice with 1× PBS for 5 min, and permeabilized with 0.2% Triton X-100 in PBS for 5 min at room temperature and then analyzed using a DeadEnd Colorimetric TUNEL Analysis Kit (Promega, Madison, WI, USA).

4.9. Caspase 3/7 Activity

Apoptosis induction of ARPE-19 cells was measured via caspase activity. Cells were seeded at approximately 5.0×10^5 cells/well in 6-well plates, cultured overnight and then treated with Stx1 (100 ng/mL), Stx1A$^-$ (100 ng/mL), Stx2 (10 ng/mL), or Stx2A$^-$ (10 ng/mL) in DMEM/F12 containing 0.5% FBS for the indicated times. The cells were washed three times with PBS and lysed with Lysis buffer containing protease inhibitors. Caspase activity was assessed using a luminescent assay (Caspase-Glo 3/7 Assay, Promega Corp., Madison, WI, USA) according to the manufacturer's directions.

4.10. Mitochondrial Membrane Potential Assay

The lipophilic, cationic dye JC-1 (5,5′,6,6′-tetrachloro-1,1′,3,3′-tetraethylbenzimidazolcarbocyanine iodide) [71] fluorescence associated with mitochondrial membranes was detected using immunofluorescence microscopy. Red fluorescence indicated healthy cells and green fluorescence indicated cells with mitochondrial membrane dysfunction. ARPE-19 cells were seeded and treated with Stxs for 12 h. After treatment, cells were incubated with JC-1 to produce a final concentration of 2.0 μM for 30 min at 37 °C in the presence of 5% CO_2. The plates were washed once with warm PBS. The quantitative analysis of the JC-1 staining was measured by a microplate reader with 514 nm for excitation and 529 nm for emission of green (monomer form) fluorescence, and 585 nm for excitation and 590 nm for emission of red (aggregate form) fluorescence. The percentage change of the mitochondrial transmembrane potential was determined using the following equation: percentage of $\Delta\psi_m$ = (average $OD_{514-529}$ of treated cells ÷ average $OD_{585-590}$ of control cells) × 100.

4.11. Quantitative Analysis

Quantitative analysis of the images was carried out using ImageJ. Pearson's correlation coefficients of multiple sets of images were quantified using the 'Colocalization' tool in the ImageJ. The values were between 0 and 1; a value of 1 meant complete co-localization, while a value of 0 meant no co-localization.

4.12. Statistical Analysis

The differences among the mean values of the different groups were assessed, and all data are expressed as mean ± SEM. All of the statistical calculations were performed by one-way ANOVA followed by the Tukey post test using GraphPad Prism version 5.00. (GraphPad Software, Inc., La Jolla, CA, USA). Values of * $p < 0.05$ were considered statistically significant. (* = $p < 0.05$; ** = $p < 0.01$; *** = $p < 0.001$).

Supplementary Materials: The following are available online at www.mdpi.com/2072-6651/9/10/319/s1, Figure S1: Dose- and time-dependent cytotoxic effects of Shiga toxin in ARPE-19 cells, Movie S1: Live imaging movie of intracellular trafficking to the ER in RPE cells treated with Stx1, Figure S2: In ARPE-19 cells, protein levels of pro-survival factor Bcl-2 were significantly decreased by Stx1 or Stx2, Figure S3: Stx1 and Stx2-induced cell death and caspase3/7 activity were dependent on MAPK signaling in ARPE-19 cells.

Acknowledgments: This work was supported by Korea Research Institute of Bioscience and Biotechnology (KRIBB) Research Initiative Program, the National Research Foundation (NRF) funded by the Korean government (MSIP) (NRF-2015M3A9E6028953) and by Basic Research Program through the National Research Foundation of Korea (NRF) funded by the Ministry of Science, ICT & Future Planning (grant number: 2017R1C1B1005137). This study was also supported by the Bio & Medical Technology Development Program of the National Research Foundation, and MSIP (NRF-2015M3A9E6028949 to Je.H.K.).

Author Contributions: J.-Y.P., Y.-J.J., S.-K.P., S.-J.Y., S.C., D.G.J., S.W.C., and B.J.L. performed the experiments; J.H.K. and V.L.T. analyzed the data; M.-S.L., and Y.-J.P. conceived and designed the experiments; J.-Y.P., Y.-J.J., M.-S.L. and Y.-J.P. wrote the paper.

Conflicts of Interest: The authors declare no conflict of interest.

References

1. Tarr, P.I.; Gordon, C.A.; Chandler, W.L. Shiga-toxin-producing *Escherichia coli* and haemolytic uraemic syndrome. *Lancet* **2005**, *365*, 1073–1086. [CrossRef]
2. Jacewicz, M.; Clausen, H.; Nudelman, E.; Donohue-Rolfe, A.; Keusch, G.T. Pathogenesis of shigella diarrhea. Xi. Isolation of a shigella toxin-binding glycolipid from rabbit jejunum and HeLa cells and its identification as globotriaosylceramide. *J. Exp. Med.* **1986**, *163*, 1391–1404. [CrossRef] [PubMed]
3. Lindberg, A.A.; Brown, J.E.; Stromberg, N.; Westling-Ryd, M.; Schultz, J.E.; Karlsson, K.A. Identification of the carbohydrate receptor for Shiga toxin produced by *Shigella dysenteriae* type 1. *J. Biol. Chem.* **1987**, *262*, 1779–1785. [PubMed]
4. Lingwood, C.A.; Law, H.; Richardson, S.; Petric, M.; Brunton, J.L.; De Grandis, S.; Karmali, M. Glycolipid binding of purified and recombinant *Escherichia coli* produced verotoxin in vitro. *J. Biol. Chem.* **1987**, *262*, 8834–8839. [PubMed]
5. Lingwood, C.A.; Binnington, B.; Manis, A.; Branch, D.R. Globotriaosyl ceramide receptor function—Where membrane structure and pathology intersect. *FEBS Lett.* **2010**, *584*, 1879–1886. [CrossRef] [PubMed]

6. Sandvig, K.; Garred, O.; Prydz, K.; Kozlov, J.V.; Hansen, S.H.; van Deurs, B. Retrograde transport of endocytosed Shiga toxin to the endoplasmic reticulum. *Nature* **1992**, *358*, 510–512. [CrossRef] [PubMed]
7. Lee, M.S.; Cherla, R.P.; Tesh, V.L. Shiga toxins: Intracellular trafficking to the ER leading to activation of host cell stress responses. *Toxins* **2010**, *2*, 1515–1535. [CrossRef] [PubMed]
8. Sandvig, K.; Bergan, J.; Kavaliauskiene, S.; Skotland, T. Lipid requirements for entry of protein toxins into cells. *Prog. Lipid Res.* **2014**, *54*, 1–13. [CrossRef] [PubMed]
9. Fraser, M.E.; Chernaia, M.M.; Kozlov, Y.V.; James, M.N. Crystal structure of the holotoxin from *Shigella dysenteriae* at 2.5 Å resolution. *Nat. Struct. Biol.* **1994**, *1*, 59–64. [CrossRef] [PubMed]
10. Fraser, M.E.; Fujinaga, M.; Cherney, M.M.; Melton-Celsa, A.R.; Twiddy, E.M.; O'Brien, A.D.; James, M.N. Structure of Shiga toxin type 2 (Stx2) from *Escherichia coli* O157:H7. *J. Biol. Chem.* **2004**, *279*, 27511–27517. [CrossRef] [PubMed]
11. Lee, S.Y.; Lee, M.S.; Cherla, R.P.; Tesh, V.L. Shiga toxin 1 induces apoptosis through the endoplasmic reticulum stress response in human monocytic cells. *Cell. Microbiol.* **2008**, *10*, 770–780. [CrossRef] [PubMed]
12. Endo, Y.; Tsurugi, K.; Yutsudo, T.; Takeda, Y.; Ogasawara, T.; Igarashi, K. Site of action of a Vero toxin (VT2) from *Escherichia coli* O157:H7 and of Shiga toxin on eukaryotic ribosomes. RNA *N*-glycosidase activity of the toxins. *Eur. J. Biochem.* **1988**, *171*, 45–50. [CrossRef] [PubMed]
13. Saxena, S.K.; O'Brien, A.D.; Ackerman, E.J. Shiga toxin, Shiga-like toxin II variant, and ricin are all single-site RNA *N*-glycosidases of 28S rRNA when microinjected into *Xenopus* oocytes. *J. Biol. Chem.* **1989**, *264*, 596–601. [PubMed]
14. Iordanov, M.S.; Pribnow, D.; Magun, J.L.; Dinh, T.H.; Pearson, J.A.; Chen, S.L.; Magun, B.E. Ribotoxic stress response: Activation of the stress-activated protein kinase JNK1 by inhibitors of the peptidyl transferase reaction and by sequence-specific RNA damage to the alpha-sarcin/ricin loop in the 28S rRNA. *Mol. Cell. Biol.* **1997**, *17*, 3373–3381. [CrossRef] [PubMed]
15. Tesh, V.L. Induction of apoptosis by Shiga toxins. *Future Microbiol.* **2010**, *5*, 431–453. [CrossRef] [PubMed]
16. Tesh, V.L.; Burris, J.A.; Owens, J.W.; Gordon, V.M.; Wadolkowski, E.A.; O'Brien, A.D.; Samuel, J.E. Comparison of the relative toxicities of Shiga-like toxins type I and type II for mice. *Infect. Immun.* **1993**, *61*, 3392–3402. [PubMed]
17. Rutjes, N.W.; Binnington, B.A.; Smith, C.R.; Maloney, M.D.; Lingwood, C.A. Differential tissue targeting and pathogenesis of Verotoxins 1 and 2 in the mouse animal model. *Kidney Int.* **2002**, *62*, 832–845. [CrossRef] [PubMed]
18. Sauter, K.A.; Melton-Celsa, A.R.; Larkin, K.; Troxell, M.L.; O'Brien, A.D.; Magun, B.E. Mouse model of hemolytic-uremic syndrome caused by endotoxin-free Shiga toxin 2 (Stx2) and protection from lethal outcome by anti-Stx2 antibody. *Infect. Immun.* **2008**, *76*, 4469–4478. [CrossRef] [PubMed]
19. Mohawk, K.L.; O'Brien, A.D. Mouse models of *Escherichia coli* O157:H7 infection and Shiga toxin injection. *J. Biomed. Biotechnol.* **2011**, *2011*. [CrossRef] [PubMed]
20. Taylor, F.B., Jr.; Tesh, V.L.; DeBault, L.; Li, A.; Chang, A.C.; Kosanke, S.D.; Pysher, T.J.; Siegler, R.L. Characterization of the baboon responses to Shiga-like toxin: Descriptive study of a new primate model of toxic responses to Stx-1. *Am. J. Pathol.* **1999**, *154*, 1285–1299. [CrossRef]
21. Stearns-Kurosawa, D.J.; Oh, S.Y.; Cherla, R.P.; Lee, M.S.; Tesh, V.L.; Papin, J.; Henderson, J.; Kurosawa, S. Distinct renal pathology and a chemotactic phenotype after enterohemorrhagic *Escherichia coli* Shiga toxins in non-human primate models of hemolytic uremic syndrome. *Am. J. Pathol.* **2013**, *182*, 1227–1238. [CrossRef] [PubMed]
22. Murray, K.F.; Patterson, K. *Escherichia coli* O157:H7-induced hemolytic-uremic syndrome: Histopathologic changes in the colon over time. *Pediatr. Dev. Pathol.* **2000**, *3*, 232–239. [CrossRef] [PubMed]
23. Richardson, S.E.; Karmali, M.A.; Becker, L.E.; Smith, C.R. The histopathology of the hemolytic uremic syndrome associated with verocytotoxin-producing *Escherichia coli* infections. *Hum. Pathol.* **1988**, *19*, 1102–1108. [CrossRef]
24. Tzipori, S.; Chow, C.W.; Powell, H.R. Cerebral infection with *Escherichia coli* O157:H7 in humans and gnotobiotic piglets. *J. Clin. Pathol.* **1988**, *41*, 1099–1103. [CrossRef] [PubMed]
25. Theobald, I.; Kuwertz-Broking, E.; Schiborr, M.; Heindel, W. Central nervous system involvement in hemolytic uremic syndrome (HUS)—A retrospective analysis of cerebral CT and MRI studies. *Clin. Nephrol.* **2001**, *56*, S3–S8. [PubMed]

26. Sheth, K.J.; Swick, H.M.; Haworth, N. Neurological involvement in hemolytic-uremic syndrome. *Ann. Neurol.* **1986**, *19*, 90–93. [CrossRef] [PubMed]

27. Loudon, S.E.; Dorresteijn, E.M.; Catsman-Berrevoets, C.E.; Verdijk, R.M.; Simonsz, H.J.; Jansen, A.J. Blinded by Shiga toxin-producing O104 *Escherichia coli* and hemolytic uremic syndrome. *J. Pediatr.* **2014**, *165*, 410–410.e1. [CrossRef] [PubMed]

28. Bhutto, I.; Lutty, G. Understanding age-related macular between the photoreceptdegeneration (AMD): Relationships or/retinal pigment epithelium/Bruch's membrane/choriocapillaris complex. *Mol. Asp. Med.* **2012**, *33*, 295–317. [CrossRef] [PubMed]

29. Smith, W.E.; Kane, A.V.; Campbell, S.T.; Acheson, D.W.; Cochran, B.H.; Thorpe, C.M. Shiga toxin 1 triggers a ribotoxic stress response leading to p38 and JNK activation and induction of apoptosis in intestinal epithelial cells. *Infect. Immun.* **2003**, *71*, 1497–1504. [CrossRef] [PubMed]

30. Lentz, E.K.; Leyva-Illades, D.; Lee, M.S.; Cherla, R.P.; Tesh, V.L. Differential response of the human renal proximal tubular epithelial cell line HK-2 to Shiga toxin types 1 and 2. *Infect. Immun.* **2011**, *79*, 3527–3540. [CrossRef] [PubMed]

31. Hughes, A.K.; Ergonul, Z.; Stricklett, P.K.; Kohan, D.E. Molecular basis for high renal cell sensitivity to the cytotoxic effects of shigatoxin-1: Upregulation of globotriaosylceramide expression. *J. Am. Soc. Nephrol.* **2002**, *13*, 2239–2245. [CrossRef] [PubMed]

32. Ramegowda, B.; Samuel, J.E.; Tesh, V.L. Interaction of Shiga toxins with human brain microvascular endothelial cells: Cytokines as sensitizing agents. *J. Infect. Dis.* **1999**, *180*, 1205–1213. [CrossRef] [PubMed]

33. Sandvig, K.; Grimmer, S.; Lauvrak, S.U.; Torgersen, M.L.; Skretting, G.; van Deurs, B.; Iversen, T.G. Pathways followed by ricin and Shiga toxin into cells. *Histochem. Cell Biol.* **2002**, *117*, 131–141. [CrossRef] [PubMed]

34. Johnson, G.L.; Lapadat, R. Mitogen-activated protein kinase pathways mediated by ERK, JNK, and p38 protein kinases. *Science* **2002**, *298*, 1911–1912. [CrossRef] [PubMed]

35. Tesh, V.L. Activation of cell stress response pathways by Shiga toxins. *Cell. Microbiol.* **2012**, *14*, 1–9. [CrossRef] [PubMed]

36. Lee, M.S.; Kwon, H.; Lee, E.Y.; Kim, D.J.; Park, J.H.; Tesh, V.L.; Oh, T.K.; Kim, M.H. Shiga toxins activate the NLRP3 inflammasome pathway to promote both production of the proinflammatory cytokine interleukin-1β and apoptotic cell death. *Infect. Immun.* **2016**, *84*, 172–186. [CrossRef] [PubMed]

37. Lee, M.S.; Cherla, R.P.; Leyva-Illades, D.; Tesh, V.L. Bcl-2 regulates the onset of Shiga toxin 1-induced apoptosis in THP-1 cells. *Infect. Immun.* **2009**, *77*, 5233–5244. [CrossRef] [PubMed]

38. Otero, J.H.; Lizak, B.; Hendershot, L.M. Life and death of a BiP substrate. *Semin. Cell Dev. Biol.* **2010**, *21*, 472–478. [CrossRef] [PubMed]

39. Sano, R.; Reed, J.C. ER stress-induced cell death mechanisms. *Biochim. Biophys. Acta* **2013**, *1833*, 3460–3470. [CrossRef] [PubMed]

40. Pincus, D.; Chevalier, M.W.; Aragon, T.; van Anken, E.; Vidal, S.E.; El-Samad, H.; Walter, P. BiP binding to the ER-stress sensor IRE1 tunes the homeostatic behavior of the unfolded protein response. *PLoS Biol.* **2010**, *8*, e1000415. [CrossRef] [PubMed]

41. Yoshida, H.; Haze, K.; Yanagi, H.; Yura, T.; Mori, K. Identification of the cis-acting endoplasmic reticulum stress response element responsible for transcriptional induction of mammalian glucose-regulated proteins. Involvement of basic leucine zipper transcription factors. *J. Biol. Chem.* **1998**, *273*, 33741–33749. [CrossRef] [PubMed]

42. Wang, X.Z.; Lawson, B.; Brewer, J.W.; Zinszner, H.; Sanjay, A.; Mi, L.J.; Boorstein, R.; Kreibich, G.; Hendershot, L.M.; Ron, D. Signals from the stressed endoplasmic reticulum induce C/EBP-homologous protein (CHOP/GADD153). *Mol. Cell. Biol.* **1996**, *16*, 4273–4280. [CrossRef] [PubMed]

43. Lee, M.S.; Cherla, R.P.; Lentz, E.K.; Leyva-Illades, D.; Tesh, V.L. Signaling through C/EBP homologous protein and death receptor 5 and calpain activation differentially regulate THP-1 cell maturation-dependent apoptosis induced by Shiga toxin type 1. *Infect. Immun.* **2010**, *78*, 3378–3391. [CrossRef] [PubMed]

44. Yamaguchi, H.; Wang, H.G. CHOP is involved in endoplasmic reticulum stress-induced apoptosis by enhancing DR5 expression in human carcinoma cells. *J. Biol. Chem.* **2004**, *279*, 45495–45502. [CrossRef] [PubMed]

45. Cherla, R.P.; Lee, S.Y.; Tesh, V.L. Shiga toxins and apoptosis. *FEMS Microbiol. Lett.* **2003**, *228*, 159–166. [CrossRef]

46. Nishitoh, H.; Matsuzawa, A.; Tobiume, K.; Saegusa, K.; Takeda, K.; Inoue, K.; Hori, S.; Kakizuka, A.; Ichijo, H. ASK1 is essential for endoplasmic reticulum stress-induced neuronal cell death triggered by expanded polyglutamine repeats. *Genes Dev.* **2002**, *16*, 1345–1355. [CrossRef] [PubMed]

47. Chandra, D.; Choy, G.; Deng, X.; Bhatia, B.; Daniel, P.; Tang, D.G. Association of active caspase 8 with the mitochondrial membrane during apoptosis: Potential roles in cleaving BAP31 and caspase 3 and mediating mitochondrion-endoplasmic reticulum cross talk in etoposide-induced cell death. *Mol. Cell. Biol.* **2004**, *24*, 6592–6607. [CrossRef] [PubMed]

48. Li, H.; Zhu, H.; Xu, C.J.; Yuan, J. Cleavage of bid by caspase 8 mediates the mitochondrial damage in the Fas pathway of apoptosis. *Cell* **1998**, *94*, 491–501. [CrossRef]

49. Wang, C.; Youle, R.J. The role of mitochondria in apoptosis. *Annu. Rev. Genet.* **2009**, *43*, 95–118. [CrossRef] [PubMed]

50. Green, D.R.; Reed, J.C. Mitochondria and apoptosis. *Science* **1998**, *281*, 1309–1312. [CrossRef] [PubMed]

51. Tesh, V.L. The induction of apoptosis by Shiga toxins and ricin. *Curr. Top. Microbiol. Immunol.* **2012**, *357*, 137–178. [PubMed]

52. Lee, M.S.; Koo, S.; Jeong, D.G.; Tesh, V.L. Shiga toxins as multi-functional proteins: Induction of host cellular stress responses, role in pathogenesis and therapeutic applications. *Toxins* **2016**, *8*, 77. [CrossRef] [PubMed]

53. Mele, C.; Remuzzi, G.; Noris, M. Hemolytic uremic syndrome. *Semin. Immunopathol.* **2014**, *36*, 399–420. [CrossRef] [PubMed]

54. Sturm, V.; Menke, M.N.; Landau, K.; Laube, G.F.; Neuhaus, T.J. Ocular involvement in paediatric haemolytic uraemic syndrome. *Acta Ophthalmol.* **2010**, *88*, 804–807. [CrossRef] [PubMed]

55. Siegler, R.L.; Brewer, E.D.; Swartz, M. Ocular involvement in hemolytic-uremic syndrome. *J. Pediatr.* **1988**, *112*, 594–597. [CrossRef]

56. Falguières, T.; Mallard, F.; Baron, C.; Hanau, D.; Lingwood, C.; Goud, B.; Salamero, J.; Johannes, L. Targeting of Shiga toxin B-subunit to retrograde transport route in association with detergent-resistant membranes. *Mol. Biol. Cell* **2001**, *12*, 2453–2468. [CrossRef] [PubMed]

57. Sandvig, K.; Torgersen, M.L.; Engedal, N.; Skotland, T.; Iversen, T.G. Protein toxins from plants and bacteria: Probes for intracellular transport and tools in medicine. *FEBS Lett.* **2010**, *584*, 2626–2634. [CrossRef] [PubMed]

58. Jandhyala, D.M.; Ahluwalia, A.; Obrig, T.; Thorpe, C.M. ZAK: A MAP3Kinase that transduces Shiga toxin- and ricin-induced proinflammatory cytokine expression. *Cell. Microbiol.* **2008**, *10*, 1468–1477. [CrossRef] [PubMed]

59. Karpman, D.; Hakansson, A.; Perez, M.T.; Isaksson, C.; Carlemalm, E.; Caprioli, A.; Svanborg, C. Apoptosis of renal cortical cells in the hemolytic-uremic syndrome: In vivo and in vitro studies. *Infect. Immun.* **1998**, *66*, 636–644. [PubMed]

60. Te Loo, D.M.; Monnens, L.A.; van Der Velden, T.J.; Vermeer, M.A.; Preyers, F.; Demacker, P.N.; van Den Heuvel, L.P.; van Hinsbergh, V.W. Binding and transfer of verocytotoxin by polymorphonuclear leukocytes in hemolytic uremic syndrome. *Blood* **2000**, *95*, 3396–3402. [PubMed]

61. Lee, S.Y.; Cherla, R.P.; Tesh, V.L. Simultaneous induction of apoptotic and survival signaling pathways in macrophage-like THP-1 cells by Shiga toxin 1. *Infect. Immun.* **2007**, *75*, 1291–1302. [CrossRef] [PubMed]

62. Fujii, J.; Wood, K.; Matsuda, F.; Carneiro-Filho, B.A.; Schlegel, K.H.; Yutsudo, T.; Binnington-Boyd, B.; Lingwood, C.A.; Obata, F.; Kim, K.S.; et al. Shiga toxin 2 causes apoptosis in human brain microvascular endothelial cells via C/EBP homologous protein. *Infect. Immun.* **2008**, *76*, 3679–3689. [CrossRef] [PubMed]

63. Landoni, V.I.; Schierloh, P.; de Campos Nebel, M.; Fernandez, G.C.; Calatayud, C.; Lapponi, M.J.; Isturiz, M.A. Shiga toxin 1 induces on lipopolysaccharide-treated astrocytes the release of tumor necrosis factor-alpha that alter brain-like endothelium integrity. *PLoS Pathog.* **2012**, *8*, e1002632. [CrossRef] [PubMed]

64. Richardson, S.E.; Rotman, T.A.; Jay, V.; Smith, C.R.; Becker, L.E.; Petric, M.; Olivieri, N.F.; Karmali, M.A. Experimental verocytotoxemia in rabbits. *Infect. Immun.* **1992**, *60*, 4154–4167. [PubMed]

65. Zoja, C.; Corna, D.; Farina, C.; Sacchi, G.; Lingwood, C.; Doyle, M.P.; Padhye, V.V.; Abbate, M.; Remuzzi, G. Verotoxin glycolipid receptors determine the localization of microangiopathic process in rabbits given verotoxin-1. *J. Lab. Clin. Med.* **1992**, *120*, 229–238. [PubMed]

66. Ren, J.; Utsunomiya, I.; Taguchi, K.; Ariga, T.; Tai, T.; Ihara, Y.; Miyatake, T. Localization of verotoxin receptors in nervous system. *Brain Res.* **1999**, *825*, 183–188. [CrossRef]

67. Obata, F.; Tohyama, K.; Bonev, A.D.; Kolling, G.L.; Keepers, T.R.; Gross, L.K.; Nelson, M.T.; Sato, S.; Obrig, T.G. Shiga toxin 2 affects the central nervous system through receptor globotriaosylceramide localized to neurons. *J. Infect. Dis.* **2008**, *198*, 1398–1406. [CrossRef] [PubMed]

68. Takahashi, K.; Funata, N.; Ikuta, F.; Sato, S. Neuronal apoptosis and inflammatory responses in the central nervous system of a rabbit treated with Shiga toxin-2. *J. Neuroinflamm.* **2008**, *5*, 11. [CrossRef] [PubMed]

69. Ohmura, M.; Yamasaki, S.; Kurazono, H.; Kashiwagi, K.; Igarashi, K.; Takeda, Y. Characterization of non-toxic mutant toxins of Vero toxin 1 that were constructed by replacing amino acids in the A subunit. *Microb. Pathog.* **1993**, *15*, 169–176. [CrossRef] [PubMed]

70. Cory, A.H.; Owen, T.C.; Barltrop, J.A.; Cory, J.G. Use of an aqueous soluble tetrazolium/formazan assay for cell growth assays in culture. *Cancer Commun.* **1991**, *3*, 207–212. [PubMed]

71. Salvioli, S.; Ardizzoni, A.; Franceschi, C.; Cossarizza, A. JC-1, but not DiOC6(3) or rhodamine 123, is a reliable fluorescent probe to assess delta psi changes in intact cells: Implications for studies on mitochondrial functionality during apoptosis. *FEBS Lett.* **1997**, *411*, 77–82. [CrossRef]

toxins

MDPI

Article

Abrin Toxicity and Bioavailability after Temperature and pH Treatment

Christina C. Tam, Thomas D. Henderson II, Larry H. Stanker, Xiaohua He and Luisa W. Cheng *

Foodborne Toxin Detection and Prevention Research Unit, Western Regional Research Center,
Agricultural Research Services, United States Department of Agriculture, 800 Buchanan Street,
Albany, CA 94710, USA; christina.tam@ars.usda.gov (C.C.T.); thomas.henderson@ars.usda.gov (T.D.H.II);
larry.stanker@ars.usda.gov (L.H.S.); Xiaohua.he@ars.usda.gov (X.H.)
* Correspondence: luisa.cheng@ars.usda.gov; Tel.: +1-510-559-6337; Fax: +1-510-559-5880

Academic Editors: Julien Barbier and Daniel Gillet
Received: 1 September 2017; Accepted: 10 October 2017; Published: 13 October 2017

Abstract: Abrin, one of most potent toxins known to man, is derived from the rosary pea (jequirity pea), *Abrus precatorius* and is a potential bioterror weapon. The temperature and pH stability of abrin was evaluated with an in vitro cell free translation (CFT) assay, a Vero cell culture cytotoxicity assay, and an in vivo mouse bioassay. pH treatment of abrin had no detrimental effect on its stability and toxicity as seen either in vitro or in vivo. Abrin exposure to increasing temperatures did not completely abrogate protein translation. In both the cell culture cytotoxicity model and the mouse bioassay, abrin's toxic effects were completely abrogated if the toxin was exposed to temperatures of 74 °C or higher. In the cell culture model, 63 °C-treated abrin had a 30% reduction in cytotoxicity which was validated in the in vivo mouse bioassay with all mice dying but with a slight time-to-death delay as compared to the non-treated abrin control. Since temperature inactivation did not affect abrin's ability to inhibit protein synthesis (A-chain), we hypothesize that high temperature treatment affected abrin's ability to bind to cellular receptors (affecting B-chain). Our results confirm the absolute need to validate in vitro cytotoxicity assays with in vivo mouse bioassays.

Keywords: abrin; *Abrus precatorius*; mouse bioassay; food safety; temperature stability; pH stability

1. Introduction

Abrus precatorius, native to India as well as tropical/subtropical regions, was originally introduced to Florida and the Caribbean [1,2] but has established itself in many parts of the southeastern United States [3]. The seeds of *A. precatorius*, also known as jequirity, rosary pea, and Crab's Eye, contain one of the most potent toxins in the world, abrin. These seeds have traditionally been made into bracelets and necklaces for use as jewelry as well as rosary beads.

Abrin and ricin are both members of the Type II family of ribosome-inactivating proteins (RIP) that inhibit eukaryotic protein synthesis leading to apoptosis and cell death. Abrin consists of an A-chain of ≈30 kDa and a B-chain of ≈33 kDa linked together by a single disulfide bond. Abrin A-chain is an N-glycosidase that cleaves the C–N bond of adenine at position 4324 on the 28S ribosomal RNA thereby preventing ribosomes from binding to elongation factor (EF) 1 and 2, leading to inhibition of protein synthesis and eventual cell death [2–6]. Abrin B-chain is a lectin that binds to cell surface carbohydrate receptors that facilitate receptor-mediated endocytosis of the A-B toxin. After internalization and reduction of the disulfide bond, the A-chain interacts with ribosomes in the cytosol thereby causing the inhibition of protein synthesis.

Abrin isolated from *Abrus precatorius* seeds has many different potential isoforms [7–9]. Three to four isoforms have been isolated by different laboratories and reported to have different toxin activities or median lethal doses (LD_{50}) [7,9–14]. Our abrin toxin preparations also contained a 120 kDa heterotetrameric species, agglutinin (APA-1). APA-1 consists of two A-chains and two B-chains that are stabilized through hydrophilic and hydrophobic interactions [15]. APA-1 has reduced toxicity as compared to abrin [7,16,17].

Abrin intoxication can occur via the gastrointestinal, inhalational, or cutaneous routes. Accidental abrin intoxication cases have been primarily attributed to children having ingested *A. precatorius* seeds, usually with no serious effects if the integrity of the seed was not compromised [15]. Since there is no antidote for abrin poisoning, the only treatment available is supportive care to mitigate the effects of the toxin. In the case of abrin intoxication via ingestion, the current treatment consists of induction of emesis, gastric lavage, activated charcoal, and whole bowel irrigation [3].

The LD_{50} of abrin for humans has been reported to be from 10 to 1000 µg/kg via oral ingestion and 3.3 µg/kg if injected [18]. In mice, abrin has been shown to be 31.4 times more lethal than ricin at 0.7 µg/kg vs. 22 µg/kg when given intravenously [19]. Since abrin is so much more toxic than ricin with no remedy for intoxication, one can imagine the devastation that would ensue if our nation's food supply were to be contaminated deliberatively in a terrorist attack. To help mitigate this potential disaster, we would need to evaluate the effectiveness of the current food safety processing practices in inactivating abrin toxicity.

Previous work on the thermal and pH stability of protein contaminants such as abrin have primarily used calorimetry, intrinsic fluorescence, ELISA, and cell culture cytotoxicity assays to measure the activity of abrin [20–22]. The results of these studies suggest that abrin is extremely heat stable and pH tolerant, and that food matrices such as dairy negatively affect the biological activity of the toxin [22]. However, these studies do not assess the actual in vivo impact of these inactivation strategies in a mouse bioassay to mimic a true intoxication scenario. In our current study, we will determine the thermal and pH stability of abrin via three different assays: (1) an in vitro cell free translation assay to measure active abrin A-chain activity; (2) an in vitro Vero cell cytotoxicity assay; and (3) a validation of the in vitro results with a mouse bioassay which will mimic a real-life intoxication model and evaluate the ability of the toxin to be absorbed and cause cytotoxicity.

2. Results

2.1. Abrin Toxin is Heterogeneous

Work in many laboratories has established that the isolation of abrin from *Abrus precatorius* seeds can yield three to four isoforms [7–9]. These isoforms can vary slightly at the amino acid level and have different toxin activity levels. Interestingly, these seed preparations also contain a related 120 kDa heterotetrameric species, agglutinin (APA-1). APA-1 consists of two A-chains and two B-chains that are stabilized through hydrophilic and hydrophobic interactions [15]. Agglutinin was shown to have reduced toxicity as compared to abrin [7,16,17].

Due to these reports, we sought to determine the components of our "abrin" toxin stock before further experiments in order to characterize its stability and bioavailability. Abrin with or without the reducing agent, DTT, along with the individual abrin A-and B-chain controls were loaded at a concentration of 100 ng per well and separated by SDS-PAGE electrophoresis. This gel was subsequently silver-stained and the components in each lane were analyzed. As seen in Figure 1, abrin without DTT had one predominant molecular weight species ≈ 55–60 kDa (holotoxin). A higher molecular weight species ≈ 120 kDa (agglutinin) and two smaller molecular weight species ≈ 25–35 kDa (individual B- and A-chains) were also present in this extract, but in much smaller amounts than the abrin holotoxin. In the presence of DTT, abrin is mostly reduced to the individual A- and B-chains. The trace amounts of the two higher molecular weight species found in this sample could be from residual non-reduced abrin holotoxin and/or agglutinin which was not fully reduced to individual A- and B-chains. The abrin A-chains seem to have at least two different molecular weight species, indicating different A-chain isomers. The B-chain control has one predominant band ≈ 30–35 kDa whereas the reduced abrin sample indicates heterogeneity of B-chains.

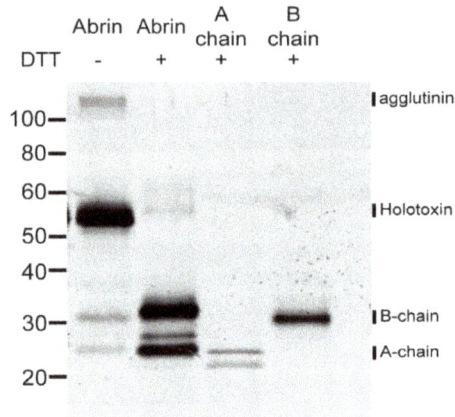

Figure 1. SDS-PAGE analysis of heterogeneous toxin complexes and subunits. An amount of 100 ng per lane of sample (abrin, abrin A-chain, abrin B-chain) treated or not treated with 0.05 M DTT was loaded onto a NuPAGE 4–12% Bis-Tris gel and subjected to SDS-PAGE electrophoresis. The gel was silver stained with the SilverXpress kit. In the absence of the reducing agent, abrin predominantly consists of the holotoxin with small amounts of agglutinin, A-chain, and B-chain. Once reduced with DTT, abrin is almost all reduced to the smaller individual A- and B-chains. The abrin A-chain control sample with DTT has two predominant A-chain species. The reduced abrin B-chain control has one predominant species that is of higher molecular weight than the individual A-chain control.

2.2. Cell Free Translation Assay Measurements of Temperature Effects on Abrin Stability

Abrin has been reported to be very stable and able to withstand extreme heat and cold [15,20–22]. Similarly, other RIPs such as the Type I RIP saporin have been reported to be highly stable [23,24]. Since most known abrin poisonings are accidental, studies to understand how abrin is affected by food processing protocols in the case of deliberate poisoning is of critical importance. Work on other plant lectins have shown that they are very heat-stable and are able to withstand denaturation at 80 °C [25]. The food industry treats foods at certain temperatures as a way to inactivate pathogens and toxins. We wondered if these same temperature inactivation protocols would have any applicable effects in inactivating abrin.

To assay for active toxin, we took advantage of the fact that abrin, like ricin, are both plant toxins that are ribosome-inactivating proteins. We modified an established cell free translation (CFT) assay used for both ricin and Shiga toxin (Stx2) [26,27] to detect active abrin. In Figure 2A, a representative abrin toxin standard curve shows that increasing concentrations of toxin correlate with increasing translation inhibition (decreasing ability to translate the reporter luciferase mRNA, therefore less luminescence measured). At concentrations above 125 ng/mL of toxin, the CFT assay is unable to respond to changes in toxin concentration since translation is completely inhibited. Therefore, we decided to validate all in vitro toxin cell free translation activity assays using a single concentration of 100 ng/mL of toxin because this dose shows high translation inhibition but allows for any detrimental effects on toxin function to be noticed.

A

B

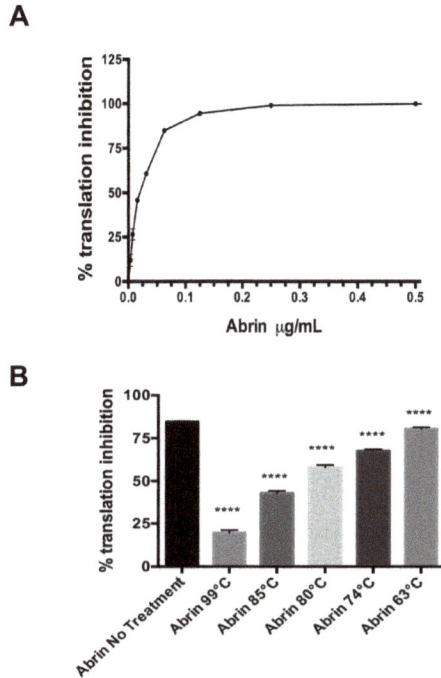

Figure 2. Translation inhibition by the plant toxin Abrin was abrogated by exposure to high temperature. (**A**) A representative standard curve shows increasing translation inhibition until saturation as one increased the abrin concentration in the in vitro cell free translation (CFT) assay run in parallel with the experimental samples shown in (**B**). Values represent means of triplicate samples ± SD. (**B**) Cell free translation assay using a single concentration of toxin (100 ng/mL) for all the different conditions. Increasing the temperature that abrin is exposed to decreases the translational inhibition seen in the CFT assay. Values represent means of triplicate samples ± SD. Statistical significance was determined by two-tailed unpaired Student's *t*-test, (****) $p < 0.0001$. Two independent experiments in triplicate were performed and one representative data set is presented.

The abrin toxin was exposed to increasing temperatures (63 °C, 74 °C, 80 °C, 85 °C, 99 °C) for 3 min in a thermocycler. Since food manufacturing processes vary tremendously depending on the actual food product and the pathogen and/or toxin chosen as the target to be inactivated, these temperatures and times were chosen based upon previous work in the laboratory for ricin (unpublished) as well as guidelines by the U.S. Food & Drug Administration to inactivate 90% of the pathogens and toxins present in sample [28–30]. These conditions certainly may not reflect the actual food processing conditions. The CFT assay shows that abrin is thermally stable since exposure to the highest temperature of 99 °C for 3 min did not completely abrogate its ability to inhibit luciferase mRNA translation. Samples treated at 99 °C had ~19.55 ± 0.61% translation inhibition while abrin without treatment was at ~84.5 ± 0.2% ($p < 0.0001$). When abrin was exposed to 85 °C for 3 min, toxin inhibition of translation activity decreased ≈ 50% to ~42.8 ± 0.6% ($p < 0.0001$) as compared to the untreated toxin control. Exposure of abrin to 63 °C for 3 min, though statistically significant, had the least impact in the CFT assay at ~80.43 ± 0.36% as compared to abrin with no treatment ($p < 0.0001$).

2.3. Cytotoxic Effects of Temperature-Treated Abrin on Vero Cells

Since abrin showed a temperature-dependent decrease in translation inhibition in the cell free translation assay, we decided to evaluate the cytotoxic effect of the toxins in the well-known Vero (African green monkey epithelial) cell culture model that has been used extensively in toxin research [31,32]. In the Vero cell assay, the cytotoxic effects of abrin were completely abrogated if the toxin was treated with temperatures $\geq 74\,^\circ$C for 3 min, similar to the cells with Dulbeco's Modified Eagle Medium (DMEM) alone (Table 1, $p < 0.0001$). These results were completely different from the results shown in Figure 2B with the CFT, which measures the catalytic function of the toxin A-chain. The 63 $^\circ$C-treated abrin, though still toxic to cells, had a statistically significant $\approx 30\%$ reduction in cytotoxicity compared with the untreated abrin in the cell culture assay, but showed a decrease of only 4% of catalytic abrin A-chain activity in the translation inhibition CFT assay (Figure 2B).

Table 1. In vitro Vero cell cytotoxicity assay of temperature-treated toxins.

Treatment	Relative Cytotoxicity (%)
DMEM	0
Abrin	100
99 $^\circ$C	0
85 $^\circ$C	0
80 $^\circ$C	0
74 $^\circ$C	0
63 $^\circ$C	72 ± 4

Vero cell cytotoxicity after treatment with abrin, either treated or not with increasing temperatures. Abrin was used at 5 ng/mL for these assays. Values represent means of six samples $\pm$ SD. Statistical significance was determined by two-tailed unpaired Student's *t*-test, $p < 0.0001$ for all conditions compared with abrin non-treated. Two independent experiments were performed and one representative data set is presented.

2.4. Mouse Bioassay Determination of Temperature Effects on Abrin Stability and Toxicity

The Vero cell cytotoxicity results suggest that substantial amounts of abrin remain active after thermal exposure at $\leq 63\,^\circ$C (Table 1). We wanted to validate these cell culture cytotoxicity results with the in vivo mouse bioassay. From the Vero cell assay, one can expect intoxication and death in the mouse bioassay for abrin treated at $\leq 63\,^\circ$C. Since the mean lethal dose (LD_{50}) varies widely due to the route of intoxication as well as the activity of the toxin extracts, one must establish an LD_{50} and mouse bioassay for individual toxin batch preparations. The mouse tail vein injection (iv) model was used to assess the toxicity of abrin as well as temperature exposed toxin solutions. Groups of mice were injected iv with seven different dosages of abrin starting with the highest concentration at 2 μg/mouse decreasing two-fold in phosphate buffer + 0.2% gelatin (Figure 3A). Mice were monitored for at least 8 days and standard time-to-death (median survival) vs. toxin dose curve was plotted. The time-to-death curve plotted for abrin is a log (inhibitor) vs. response (three parameter) curve rather than a linear regression model ($R^2 = 0.8257$). Intoxication with doses above 0.25 μg/mouse caused the rapid death of all members of the group within the first 48 h. Mice injected with abrin at 0.125 μg/mouse almost all succumbed to death by 72 h whereas lower doses did not cause death and/or show intoxication. An average LD_{50} value of 3.3 μg/kg for our abrin stock using the iv route of infection was derived from the LD_{50} values calculated using the Weil method and the Reed and Muench method [33,34]. This value is higher than the 0.7 μg/kg iv reported in the literature [35]. An earlier study using an abrin toxin stock from the same supplier has determined a LD_{50} value of > 1 mg/kg for oral intoxication and 2–20 μg/kg for iv treated mice [15]. Our LD_{50} for the iv infection model of 3.3 μg/kg is similar to that of the previous study.

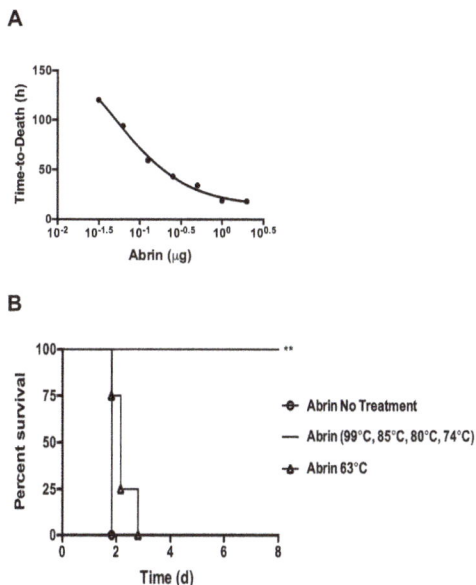

Figure 3. Temperature effects on abrin toxicity in the intravenous mouse bioassay. (**A**) A series of known abrin samples were injected into mice tail vein injection (iv) to derive time-to-death standard curves and LD_{50} values. Data was combined from two independent experiments from seven dosage levels consisting of a total of $n = 6$ to 8 mice per dose. The data was plotted using the log (inhibitor) vs. response (three parameter) curve on GraphPad Prism 6. $R^2 = 0.8257$; (**B**) Temperature treated toxins were administered to mice iv at a lethal dose of 1 µg per mouse ($n = 4$ mice per experimental condition). Mice given abrin treated at 74 °C or higher all survived as compared to the untreated controls (**, $p = 0.0082$). The time-to-death delay seen from mice given the abrin treated at 63 °C was not statistically significant ($p = 0.1580$). Two independent experiments were performed and one representative set of survival curves is shown. Survival curves were plotted for each condition and the log-rank (Mantel–Cox) test was used to evaluate statistical significance on GraphPad Prism 6.

To determine the effectiveness of abrin inactivation by thermal treatment in vivo, treated and non-treated toxins were injected iv into groups of mice ($n = 4$ per experimental condition) at 1 µg/mouse (equal to 13 x mouse iv LD_{50}). Survival curves were calculated and the results are shown in Figure 3B. Mice that were given abrin inactivated at 74 °C or higher survived compared to the abrin no treatment control (abrin no treatment median survival $\approx$ 1.83 days vs. abrin $\geq$ 74 °C median survival $\approx$ end of experiment, log-rank (Mantel–Cox) $p = 0.0082$) in agreement with the Vero cell cytotoxicity assay (Table 1). Mice given abrin treated at 63 °C all died similarly to the abrin no heat treatment control but with a slight time-to-death delay that was not statistically significant (abrin no treatment median survival was $\approx$ 1.83 days vs. abrin 63 °C median survival at $\approx$ 2.167 days, log-rank (Mantel–Cox) $p = 0.1580$). Abrin treated at 63 °C similarly was still able to cause a significant amount of cytotoxicity in the Vero cell cytotoxicity assay but was reduced in comparison to the non-treated abrin control (Table 1). This reduction in cytotoxicity is reflected in the slight time-to-death delay seen in the in vivo mouse bioassay.

2.5. pH Treatment of Abrin Has No Detrimental Effect on Its Function

One of the main routes for abrin intoxication is via the gastrointestinal route. After oral ingestion of the toxin, abrin must travel through the digestive system, encountering a steadily decreasing pH gradient. How the toxin survives this change to an increasing acidic milieu and what effects this may have on toxin function are of critical importance. In addition, different food matrices may present a range of pH conditions that may affect toxin function. To determine the effect of pH on toxicity, abrin was subjected to different pH levels in the phosphate buffer gelatin matrix (pH 2.0–pH 9.0) for 1 h at room temperature. Gelatin was used because of the tendency of low amounts of abrin to bind to tube walls. After exposure, the toxin mixes were neutralized with phosphate buffer gelatin to pH 7.0 before addition to the CFT assay, Vero cell cytotoxicity assay, or mouse bioassay. In Figure 4A, no decrease in toxicity was observed between abrin pH 7.0 or any of the other pH samples in the cell free translation assay. pH-treated abrin was still able to inhibit luciferase mRNA translation (% translation inhibition $\cong$ 99.6–99.9%). Additionally, pH-treated toxin did not have significant deficient cytotoxic effects on Vero cells (Table S1). For the mouse bioassay, mice in all the experimental groups died rapidly within 2.25 days (n = 5 mice per experimental condition). The median survival for the various pH-treated abrin were pH 9.0 $\approx$ 1.92 days, pH 5.0–9.0 $\approx$ 1.83 days, pH 4.0 $\approx$ 1.94 days, and pH 2.0 $\approx$ 1.81 days but they were not statistically significant compared to pH 7.0-treated abrin. However, there was a statistical difference between abrin pH 7.0 and abrin pH 3.0 (abrin pH 7.0 median survival $\approx$ 1.83 days vs. abrin pH 3.0 median survival $\approx$ 1.81, log-rank (Mantel–Cox) p = 0.0027).

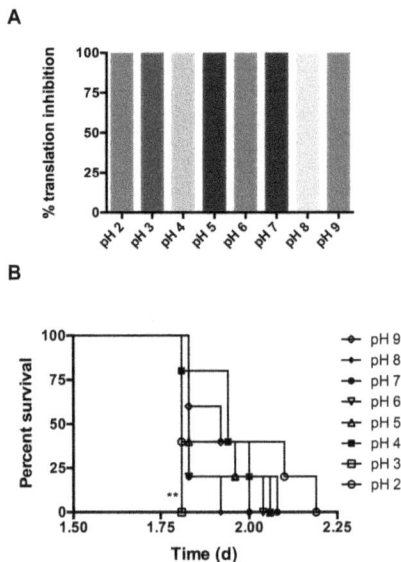

Figure 4. pH treatment of abrin has no detrimental effect on toxin activity. (**A**) A single concentration of 100 ng/mL abrin was used in cell free translation assays for all pH conditions. There is no significant difference between the various pH-exposed toxins. Values represent means of triplicate samples $\pm$ SD. Statistical significance was determined by two-tailed unpaired Student's t-test. The data represents a single experiment; (**B**) pH-treated toxins were administered to mice iv at a lethal dose of 1 µg per mouse (n = 5 mice per experimental condition) for one experiment. There was no detrimental effect on abrin's ability to cause intoxication and subsequent death. Survival curves were plotted for mice for each condition and the log-rank (Mantel–Cox) test was used to evaluate statistical significance on GraphPad Prism 6. The only statistical significant decrease in toxicity was seen in abrin treated at pH 3.0 which shortened the time-to-death compared to abrin treated at pH 7.0 (** p = 0.0027).

3. Discussion

In this study, we wanted to determine the most effective parameters to use against abrin should bioterror attacks occur on our food supply. We proceeded to evaluate the effectiveness of thermal inactivation against abrin. To measure the successful inactivation of abrin toxin activity, we utilized three different assays: two in vitro and one in vivo. As shown in Figure 2B, the in vitro cell free translation assay showed that there was a correlation between increasing temperature exposure and decreasing toxin activity. However, treatment, even at the highest temperature of 99 °C for 3 min, still gave a statistically significant translation inhibition activity of ≈20% which was approximately a 4-fold decrease from toxin that was not exposed to any thermal inactivation conditions. The heat-stability of abrin in this study is similar to previous published data that suggest that batch pasteurization at 85 °C for 30 min is required to completely inactivate ricin and abrin activity in milk when measured by ELISA [22]. When these toxins were tested for toxicity in the Vero cell model, high temperature inactivation (74 °C or higher) completely abrogated cell toxicity while 70% of the Vero cells were killed after intoxication with 63 °C-treated toxin (Table 1). The in vivo mouse bioassay validated the defect in Vero cell cytotoxicity treated with abrin exposed to ≥74 °C as well as the slight decrease in toxicity (i.e., slight time-to-death delay) with the 63 °C-treated abrin (Table 1, Figure 4B). Since we know that the in vitro CFT assay measures the catalytic ability of the A-chain to bind to ribosomes to inhibit protein synthesis, and that the cell culture cytotoxicity and mouse bioassay models account for receptor binding and enzymatic activity, these in vivo results suggest that higher temperatures did not affect A-chain catalytic activity. The decrease in toxicity thus could be due to the disruption of the B-chain's ability to bind to the proper cell surface receptors to cause internalization of the toxin and/or cause the toxin complex to be more susceptible to degradation in vivo. To determine if proteolysis/degradation of the holotoxin or individual subunits was causing the decrease in toxicity, the toxins were examined via silver staining in the absence or presence of a reducing agent. In Figure S1, we show a similar protein profile present in the extracts after thermal or pH treatment within each silver stained gel. This indicates that there is still a significant amount of toxin left to cause cytotoxicity. Therefore, protein degradation does not seem to be the cause of the toxicity differences (Table 1, Figures 3B and 4B). Future work is needed to validate our hypothesis that the thermal effect that we see of 63 °C-treated abrin is potentially due to abrin conformational changes that could affect their binding to carbohydrate receptors using antibodies that specifically recognize native lectin chains [36] and/or binding measurements to the receptors themselves.

In addition to evaluating the thermal stability of abrin, we decided to investigate the stability of the toxin at different pH. Abrin can be found in foods of different pH and toxin ingestion requires passage through increasing acidic pH in the digestive system. In addition, low pH is needed for A-chain translocation to the cytosol to bind to and affect ribosomes. Abrin toxin activity investigated from pH 2.0 to pH 9.0. pH treatment did not significantly affect the ability of abrin to inhibit the translation seen in the CFT assay (Figure 4A) or the Vero cell cytotoxicity (Table S1), nor did it inhibit its ability to cause death in the mouse bioassay (Figure 4B).

Our research reaffirms the need for in vivo mouse bioassays to validate in vitro experimental assays, as seen in the thermal stability study. Our study on abrin thermal stability suggests that the food safety procedures utilizing thermal inactivation of pathogens and toxins must be set to ≥74 °C to be truly effective. Any food safety procedures that comprise a pH only treatment may not be sufficient for abrin inactivation. Since we treated the toxin for 1 h at room temperature at the different pHs and then neutralized the solution, one must wonder if there may be differences if one increases the incubation time and/or does not neutralize the pH. Additionally, our studies were performed using a phosphate buffer gelatin and not any food matrices. Future studies should take this into account to test the effectiveness of these procedures in different food matrices, i.e., milk, beef, and eggs.

4. Materials and Methods

4.1. Materials

Abrin toxin (mixed isomers, Cat# ABR-1) was obtained from Toxin Technology, Inc. (Sarasota, FL, USA) as 1 mg lyophilized powder before resuspension with 1 mL of 1× phosphate buffered saline (PBS) pH 7.2 to give a 1 mg/mL stock stored at 4 °C. Abrin A-chain (BEI Resources, NR-43945),

Abrin B-chain (BEI Resources, NR-43946) were supplied as 1 mg/mL stock stored at −20 °C. The following components for the cell free translation assay were bought from Promega (Madison, WI, USA): nuclease-treated rabbit reticulocyte lysate (L4960), complete amino acid mixture (1 mM, N2111), nuclease-free water (P1193), luciferase mRNA (1 mg/mL, L4561), and the Bright-Glo Luciferase Assay System (10 mL, E2610). The NuPAGE 4–12% Bis-Tris gels and SilverXpress kit (LC6100) were supplied from Invitrogen (Carlsbad, CA, USA).

4.2. Temperature Stability Assays

Abrin was diluted in 1× PBS pH 7.2 + 0.2% phosphate buffer gelatin to 10 µg/mL. Aliquots of these toxins were treated at various temperatures (63 °C, 74 °C, 80 °C, 85 °C, and 99 °C) for 3 min in a thermocycler. The toxin samples per condition were split into two for further analysis. One half of the samples were used for the cell free translation assay at a single concentration of 100 ng/mL. Statistical significance was calculated using the two-tailed unpaired Student's *t*-test between the non-heat-treated abrin sample with each of the other heat-treated toxin samples. The other half of the heat-treated and non-heat-treated toxin samples were administered to mice iv at a lethal dose of 1 µg per mouse ($n = 4$ mice per experimental condition). These conditions were tested in two independent experiments for a total of $n = 8$ mice per experimental condition. Survival curves were generated for each condition and compared for statistical significance using GraphPad Prism 6 for each independent experiment. The log-rank (Mantel–Cox) test was used to determine statistical significance between the survival curves with *p*-values < 0.05 considered significant.

4.3. pH Stability Assays

Abrin was diluted to 20 µg/mL in 1× PBS + 0.2% phosphate buffer gelatin at different pH (pH 2.0–pH 9.0) for 1 h at room temperature. After this incubation, the pH-treated toxin samples were neutralized with 1× PBS + 0.2% phosphate buffer gelatin 1:1 (*v*/*v*) to generate a stock of 10 µg/mL at pH 7.0 before further analysis. One half of the samples was used for the cell free translation assay at a single concentration of 100 ng/mL. Statistical significance was calculated using the two-tailed unpaired Student's *t*-test between the pH 7.0-treated abrin sample with each of the other different pH-treated toxin samples. The rest of the toxin samples were administered to mice iv at a lethal dose of 1 µg per mouse ($n = 5$ mice per experimental condition) in one experiment. Survival curves were generated for each condition and compared for statistical significance using GraphPad Prism 6 for each independent experiment. The log-rank (Mantel–Cox) test was used to determine statistical significance between the survival curves with *p*-values < 0.05 considered significant.

4.4. SDS-PAGE Electrophoresis for Silver Stain

Abrin, abrin A-chain, and abrin B-chain at the concentration of 100 ng per well in the presence or absence of a reducing agent (0.05 M DTT) were separated by sodium dodecyl sulfate-polyacrylamide gel electrophoresis with NuPAGE 4–12% Bis-Tris gels (Invitrogen) followed by silver staining using the SilverXpress kit according to the manufacturer's instructions (Invitrogen). The silver stained gel image was acquired using the ChemiDoc MP imaging system (BIO-RAD, Hercules, CA, USA). Molecular weight standards were purchased from Invitrogen. Abrin untreated or treated with either increasing temperatures or pH in 1× PBS with 0.2% phosphate buffer gelatin were loaded at 100 ng per well as above. Electrophoresis and silver-staining procedures were followed as stated.

4.5. Abrin In Vitro Cell Free Translation Assay

The in vitro cell free translation assay for abrin was modified from the protocols used to detect ricin and Shiga toxin (Stx2) activity in regards to the concentration of toxins evaluated [26,27]. Briefly, serial dilutions of abrin (for standard curve) or a single concentration from 100 ng/mL (for heat-treated or pH-treated toxin experiments) were added to the translational lysate mixture: nuclease-treated rabbit reticulocyte lysate, complete amino acid mixture, RNasin, nuclease-free water, and luciferase mRNA at ratio of (v/v) [35:1:1:36:2]. The ratio of toxin to translational lysate mixture was 1:5. Fifteen microliters of lysate mixture was added to 3 μL of each sample and gently mixed. A quick spin was used to concentrate the entire reaction mixture to the bottom of the tube. The reaction mixtures were incubated at 30 °C with shaking at 80 rpm for 90 min. Each reaction mixture was then aliquoted into triplicate wells containing 5 μL of the translation reaction in a 96-well black plate. Translational efficiency was measured by the addition of 100 μL of the Bright-Glo Luciferase Assay System and luminescence measured as counts per second (cps) was read on a Victor 3 (Perkin-Elmer, Shelton, CT, USA). The negative control (full translation efficiency) consisted of the translational lysate mixture with 1× PBS with 0.2% phosphate buffer gelatin. Toxin samples were diluted with 1× PBS with phosphate buffer gelatin to a final concentration of 0.2% before any experiments. Toxin activity was calculated as percentage (%) translation inhibition due to inhibition of translation [(negative control cps − toxin sample cps)/negative control cps] × 100. All values shown represent the mean ± standard deviation (SD) of triplicate samples measured in a representative experiment. Standard curves with serial dilutions of toxin were always performed in parallel to experimental conditions for each independent experiment to account for potential variations due to changes in toxin activity and other conditions. Statistical significance was calculated using the two-tailed unpaired Student's t-test from GraphPad Prism 6 with p-values < 0.05 considered significant.

4.6. Vero Cell Cytotoxicity Assay

Vero cells (African green monkey kidney epithelial cells) were cultured in Dulbeco's Modified Eagle Medium high glucose + 10% fetal bovine serum and incubated in a humidified incubator (37 °C, 5% CO_2). Cells were trypsinized and adjusted to 5×10^4 cells/mL, seeded into black-sided, clear-bottom 96-well tissue culture plates at 100 μL/well and allowed to adhere overnight (18 h) at 37 °C. The media was removed and 100 μL of fresh DMEM—containing DMEM, untreated abrin, or the toxins inactivated at different temperatures—was then added. Plates were then incubated at 37 °C for 2 h, the media was removed and more fresh media was added. The cells were then incubated for 48 h at 37 °C. Luminescence was measured as follows: 100 μL of CellTiter-Glo (Promega, G7570) at 1:5 in PBS was added to each well and the plates shaken for 2 min in order to lyse the cells. Upon another 10-min incubation at room temperature, luminescence was measured using a Victor 3 plate reader (lid was removed from the plate for a better signal). The percentage (%) cytotoxicity for each well of the toxin treated samples was calculated as follows: [(average DMEM negative control cps − experimental cps)/average DMEM negative control cps] × 100. The average % cytotoxicity was calculated for all the conditions. The non-treated or pH 7.0-treated abrin was set as 100% relative cytotoxicity. To obtain the relative cytotoxicity percentages (%) for all the other conditions, we calculated [average % cytotoxicity condition/average % cytotoxicity of positive control non-treated abrin or pH 7.0 abrin] × 100.

4.7. Abrin Intravenous Route (iv) Mouse Bioassay

Randomly assigned female Swiss-Webster mice in groups of two to four mice were challenged intravenously (iv) with abrin toxin resuspended in 1× PBS with 0.2% phosphate buffer gelatin per dosage level. Seven dosage levels were tested per experiment and two independent experiments were performed. Previous work from our laboratory did not find any differences in the murine strain's susceptibility to ricin. Additionally, previous studies have used this outbred strain for both ricin and abrin [37,38]. Between the two independent experiments, n = 6–8 mice were tested per dosage level. Animals were monitored for at least 8 days for signs of intoxication or death. Abrin iv LD_{50} values were calculated using both the Weil method and the Reed and Muench method [33,34]. All procedures involving animals were reviewed and approved by the Institutional Animal Care and Use Committee of the United States Department of Agriculture, Western Regional Research Center. Animal use protocols for abrin mouse bioassays (Protocol # 16-2) were approved by the Western Regional Research Center Institutional Animal Care and Use Committee (WRRC- IACUC) on 2 September 2016.

Supplementary Materials: The following are available online at www.mdpi.com/2072-6651/9/10/320/s1, Table S1: The cytotoxic effect on Vero cells by pH-treated abrin, Figure S1: SDS-PAGE analysis of temperature (**A**) and pH-treated (**B**) toxin complexes and subunits.

Acknowledgments: This work was funded by the United States Department of Agriculture, Agricultural Research Service, National Program project NP108, CRIS 5325-42000-049-00D. LHS was also funded by interagency agreement IAA # 60-2030-5-004 with Department of Homeland Security.

Author Contributions: L.W.C. and C.C.T. conceived and designed the experiments; T.D.H.II, L.W.C., and C.C.T. performed the experiments; L.W.C., X.H. and C.C.T. analyzed the data; L.H.S. contributed reagents; C.C.T., L.H.S. and L.W.C. wrote the paper.

Conflicts of Interest: The authors declare no conflict of interest.

References

1. Felder, E.; Mossbrugger, I.; Lange, M.; Wolfel, R. Simultaneous detection of ricin and abrin DNA by real-time pcr (qpcr). *Toxins* **2012**, *4*, 633–642. [CrossRef] [PubMed]

2. Zhou, H.; Zhou, B.; Ma, H.; Carney, C.; Janda, K.D. Selection and characterization of human monoclonal antibodies against abrin by phage display. *Bioorg. Med. Chem. Lett.* **2007**, *17*, 5690–5692. [CrossRef] [PubMed]

3. Reedman, L.; Shih, R.D.; Hung, O. Survival after an intentional ingestion of crushed abrus seeds. *West. J. Emerg. Med.* **2008**, *9*, 157–159. [PubMed]

4. Cheng, J.; Lu, T.H.; Liu, C.L.; Lin, J.Y. A biophysical elucidation for less toxicity of agglutinin than abrin-A from the seeds of abrus precatorius in consequence of crystal structure. *J. Biomed. Sci.* **2010**, *17*, 34. [CrossRef] [PubMed]

5. Narayanan, S.; Surolia, A.; Karande, A.A. Ribosome-inactivating protein and apoptosis: Abrin causes cell death via mitochondrial pathway in jurkat cells. *Biochem. J.* **2004**, *377*, 233–240. [CrossRef] [PubMed]

6. Narayanan, S.; Surendranath, K.; Bora, N.; Surolia, A.; Karande, A.A. Ribosome inactivating proteins and apoptosis. *FEBS Lett.* **2005**, *579*, 1324–1331. [CrossRef] [PubMed]

7. Hegde, R.; Maiti, T.K.; Podder, S.K. Purification and characterization of three toxins and two agglutinins from abrus precatorius seed by using lactamyl-sepharose affinity chromatography. *Anal. Biochem.* **1991**, *194*, 101–109. [CrossRef]

8. Qing, L.T.; Qu, X.L. Cloning, expression of the abrin-a a-chain in escherichia coli and measurement of the biological activities in vitro. *Sheng Wu Hua Xue Yu Sheng Wu Wu Li Xue Bao* **2002**, *34*, 405–410. [PubMed]

9. Lin, J.Y.; Lee, T.C.; Hu, S.T.; Tung, T.C. Isolation of four isotoxic proteins and one agglutinin from jequiriti bean (abrus precatorius). *Toxicon* **1981**, *19*, 41–51. [PubMed]

10. Olsnes, S.; Saltvedt, E.; Pihl, A. Isolation and comparison of galactose-binding lectins from abrus precatorius and ricinus communis. *J. Biol. Chem.* **1974**, *249*, 803–810. [PubMed]

11. Wei, C.H.; Hartman, F.C.; Pfuderer, P.; Yang, W.K. Purification and characterization of two major toxic proteins from seeds of abrus precatorius. *J. Biol. Chem.* **1974**, *249*, 3061–3067. [PubMed]

12. Roy, J.; Som, S.; Sen, A. Isolation, purification, and some properties of a lectin and abrin from abrus precatorius linn. *Arch. Biochem. Biophys.* **1976**, *174*, 359–361. [CrossRef]

13. Lin, J.Y.; Lee, T.C.; Tung, T.C. Isolation of antitumor proteins abrin-A and abrin-B from abrus precatorius. *Int. J. Pept. Protein Res.* **1978**, *12*, 311–317. [CrossRef] [PubMed]

14. Herrmann, M.S.; Behnke, W.D. A characterization of abrin A from the seeds of the abrus precatorius plant. *Biochim. Biophys. Acta* **1981**, *667*, 397–410. [CrossRef]

15. Garber, E.A. Toxicity and detection of ricin and abrin in beverages. *J. Food Prot.* **2008**, *71*, 1875–1883. [CrossRef] [PubMed]

16. Stirpe, F.; Barbieri, L.; Battelli, M.G.; Soria, M.; Lappi, D.A. Ribosome-inactivating proteins from plants: Present status and future prospects. *Biotechnology* **1992**, *10*, 405–412. [CrossRef] [PubMed]

17. Liu, C.L.; Tsai, C.C.; Lin, S.C.; Wang, L.I.; Hsu, C.I.; Hwang, M.J.; Lin, J.Y. Primary structure and function analysis of the abrus precatorius agglutinin a chain by site-directed mutagenesis. Pro(199) of amphiphilic alpha-helix H impairs protein synthesis inhibitory activity. *J. Biol. Chem.* **2000**, *275*, 1897–1901. [CrossRef] [PubMed]

18. Johnson, R.C.; Zhou, Y.; Jain, R.; Lemire, S.W.; Fox, S.; Sabourin, P.; Barr, J.R. Quantification of l-abrine in human and rat urine: A biomarker for the toxin abrin. *J. Anal. Toxicol.* **2009**, *33*, 77–84. [CrossRef] [PubMed]

19. Gill, D.M. Bacterial toxins: A table of lethal amounts. *Microbiol. Rev.* **1982**, *46*, 86–94. [PubMed]

20. Krupakar, J.; Swaminathan, C.P.; Das, P.K.; Surolia, A.; Podder, S.K. Calorimetric studies on the stability of the ribosome-inactivating protein abrin ii: Effects of pH and ligand binding. *Biochem. J.* **1999**, *338 Pt 2*, 273–279. [CrossRef] [PubMed]

21. Tolleson, W.H.; Jackson, L.S.; Triplett, O.A.; Aluri, B.; Cappozzo, J.; Banaszewski, K.; Chang, C.W.; Nguyen, K.T. Chemical inactivation of protein toxins on food contact surfaces. *J. Agric. Food Chem.* **2012**, *60*, 6627–6640. [CrossRef] [PubMed]

22. Jackson, L.S.; Triplett, O.A.; Tolleson, W.H. Influence of yogurt fermentation and refrigerated storage on the stability of protein toxin contaminants. *Food Chem. Toxicol.* **2015**, *80*, 101–107. [CrossRef] [PubMed]

23. Santanche, S.; Bellelli, A.; Brunori, M. The unusual stability of saporin, a candidate for the synthesis of immunotoxins. *Biochem. Biophys. Res. Commun.* **1997**, *234*, 129–132. [CrossRef] [PubMed]

24. Manosroi, J.; von Kleist, S.; Manosroi, A.; Grunert, F. Thermo-stability and antitumor activity on colon cancer cell lines of monoclonal anti-cea antibody-saporin immunotoxin. *J. Korean Med. Sci.* **1992**, *7*, 128–135. [CrossRef] [PubMed]

25. Biswas, S.; Kayastha, A.M. Thermal stability of phaseolus vulgaris leucoagglutinin: A differential scanning calorimetry study. *J. Biochem. Mol. Biol.* **2002**, *35*, 472–475. [CrossRef] [PubMed]

26. He, X.; Lu, S.; Cheng, L.W.; Rasooly, R.; Carter, J.M. Effect of food matrices on the biological activity of ricin. *J. Food Prot.* **2008**, *71*, 2053–2058. [CrossRef] [PubMed]

27. He, X.; Quinones, B.; Carter, J.M.; Mandrell, R.E. Validation of a cell-free translation assay for detecting shiga toxin 2 in bacterial culture. *J. Agric. Food Chem.* **2009**, *57*, 5084–5088. [CrossRef] [PubMed]

28. Reyes-De-Corcuera, J.I.; Goodrich-Schneider, R.M.; Barringer, S.A.; Landeros-Urbina, M.A. Processing of fruit and vegetable beverages. In *Food Processing: Principles and Applications*, 2nd ed.; John Wiley & Sons: Hoboken, NJ, USA, 2014.

29. Peng, J.; Tang, J.; Barrett, D.M.; Sablani, S.S.; Anderson, N.; Powers, J.R. Thermal pasteurization of vegetables: Critical factors for process design and effects on quality. *Crit. Rev. Food Sci. Nutr.* **2015**, *57*, 2970–2995. [CrossRef] [PubMed]

30. U.S. Food & Drug Adminstration. Available online: https://www.fda.gov/Food/GuidanceRegulation/default.htm (accessed on 6 October 2017).

31. Sandvig, K.; Olsnes, S. Entry of the toxic proteins abrin, modeccin, ricin, and diphtheria toxin into cells. II. Effect of pH, metabolic inhibitors, and ionophores and evidence for toxin penetration from endocytotic vesicles. *J. Biol. Chem.* **1982**, *257*, 7504–7513. [PubMed]

32. Pauly, D.; Worbs, S.; Kirchner, S.; Shatohina, O.; Dorner, M.B.; Dorner, B.G. Real-time cytotoxicity assay for rapid and sensitive detection of ricin from complex matrices. *PLoS ONE* **2012**, *7*, e35360. [CrossRef] [PubMed]

33. Weil, C.S.; Smyth, H.F., Jr.; Nale, T.W. Quest for a suspected industrial carcinogen. *AMA Arch. Ind. Hyg. Occup. Med.* **1952**, *5*, 535–547. [PubMed]

34. Reed, L.J.; Muench, H. A simple method of estimating fifty percent endpoints. *Am. J. Hyg.* **1938**, *27*, 493–497.

35. Fodstad, O.; Johannessen, J.V.; Schjerven, L.; Pihl, A. Toxicity of abrin and ricin in mice and dogs. *J. Toxicol. Environ. Health* **1979**, *5*, 1073–1084. [CrossRef] [PubMed]

36. Goldman, E.R.; Anderson, G.P.; Zabetakis, D.; Walper, S.; Liu, J.L.; Bernstein, R.; Calm, A.; Carney, J.P.; O'Brien, T.W.; Walker, J.L.; et al. Llama-derived single domain antibodies specific for abrus agglutinin. *Toxins* **2011**, *3*, 1405–1419. [CrossRef] [PubMed]

37. Niyogi, S.K.; Rieders, F. Toxicity studies with fractions from abrus precatorius seed kernels. *Toxicon* **1969**, *7*, 211–216. [CrossRef]

38. Herrera, C.; Tremblay, J.M.; Shoemaker, C.B.; Mantis, N.J. Mechanisms of ricin toxin neutralization revealed through engineered homodimeric and heterodimeric camelid antibodies. *J. Biol. Chem.* **2015**, *290*, 27880–27889. [CrossRef] [PubMed]

Article

A Monoclonal–Monoclonal Antibody Based Capture ELISA for Abrin

Christina C. Tam [1], Luisa W. Cheng [1], Xiaohua He [1], Paul Merrill [1], David Hodge [2] and Larry H. Stanker [1,*

[1]   Foodborne Toxin Detection and Prevention Research Unit, Western Regional Research Center, Agricultural Research Services, United States Department of Agriculture, 800 Buchanan Street, Albany, CA 94710, USA; christina.tam@ars.usda.gov (C.C.T.); luisa.cheng@ars.usda.gov (L.W.C.); xiaohua.he@ars.usda.gov (X.H.); paul.merrill@ars.usda.gov (P.M.)

[2]   United States Department of Homeland Security, Washington, DC 20528, USA; david.hodge@hq.dhs.gov

*   Correspondence: larry.stanker@ars.usda.gov; Tel.: +1-510-559-5984; Fax: +1-510-559-5880

Academic Editors: Julien Barbier and Daniel Gillet
Received: 1 September 2017; Accepted: 13 October 2017; Published: 18 October 2017

Abstract: Abrin, one of the most highly potent toxins in the world, is derived from the plant, *Abrus precatorius*. Because of its high toxicity, it poses potential bioterror risks. Therefore, a need exists for new reagents and technologies that would be able to rapidly detect abrin contamination as well as lead to new therapeutics. We report here a group of abrin-specific monoclonal antibodies (mAbs) that recognize abrin A-chain, intact A–B chain toxin, and agglutinin by Western blot. Additionally, these mAbs were evaluated for their ability to serve as capture antibodies for a sandwich (capture) ELISA. All possible capture–detector pairs were evaluated and the best antibody pair identified and optimized for a capture ELISA. The capture ELISA based on this capture–detector mAb pair had a limit of detection (L.O.D) of ≈1 ng/mL measured using three independent experiments. The assay did not reveal any false positives with extracts containing other potential ribosome-inactivating proteins (RIPs). Thus, this new capture ELISA uses mAbs for both capture and detection; has no cross-reactivity against other plant RIPs; and has a sensitivity comparable to other reported capture ELISAs using polyclonal antibodies as either capture or detector.

Keywords: abrin; *abrus precatorius*; ELISA; monoclonal antibodies; ribosome-inactivating protein

---

1. Introduction

Abrin and ricin are both members of the Type II family of ribosome-inactivating proteins (RIP) that inhibit eukaryotic protein synthesis leading to apoptosis and cell death. Abrin is found in jequirity seeds or rosary peas of the plant, *Abrus precatorius* L., and native to tropical/subtropical regions of the world. Abrin like ricin is considered a Select Agent due to its potential use in bio warfare. Abrin is a heterodimeric protein toxin consisting of an A-chain and a B-chain linked together with a single disulfide. The catalytic active A-chain is an *N*-glycosidase that deadenylates the rat 28S ribosomal RNA at position 4324 [1] thereby preventing ribosomes from binding to elongation factor (EF) 1 and 2 leading to inhibition of protein synthesis and eventual cell death [2–7]. Abrin B-chain is a lectin that binds to cell surface carbohydrate receptors facilitating receptor-mediated endocytosis of the A–B toxin.

Abrin isolated from *Abrus precatorius* seeds is a heterogeneous mixture [7–9]. Multiple isoforms have been reported, and each form has different toxicity [8–14]. In addition to these toxin isoforms, the presence of a 120 kDa heterotetrameric species, referred to as the agglutinin (APA-1), also can be isolated from seed preparations. APA-1 consists of two A-chains and two B-chains that are stabilized through hydrophilic and hydrophobic interactions [15]. Though APA-1 and abrin toxin isoforms have very similar sequences, APA-1 has reduced toxicity as compared to abrin [8,16,17].

Abrin intoxication can occur via multiple routes that affect its toxicity. The routes of intoxication are gastrointestinal, inhalational, or cutaneous. Abrin intoxication cases have been attributed to the accidental ingestion of *A. precatorius* seeds by children. Since there is no antidote for abrin poisoning, the only available treatment is supportive care to mitigate the effects of the toxin. In the case of abrin intoxication via the ingestion, the current treatment consists of inducing emesis, gastric lavage, activated charcoal, and whole bowel irrigation [3]. The LD_{50} of abrin for humans has been reported to range from 10–1000 µg/kg via oral ingestion and 3.3 µg/kg if injected [18]. In mice, abrin has been shown to be 31.4 times more lethal than ricin with an LD of 0.7 µg/kg vs. 22 µg/kg when given intravenously [19].

Multiple technologies (electrospray MS, portable sensor, etc.) have been developed to detect L-abrine (*N*-methyltryptophan) [18,20,21] as a biomarker for abrin exposure. ELISA, ECL, and lateral flow technologies have been reported to detect abrin. Earlier work reported that polyclonal–monoclonal antibody, as well as polyclonal–polyclonal based capture ELISAs and ECL assays have limits of detection (LODs) ranging from 0.1–0.5 ng/mL in buffer [15]. In the same study, the authors report that the ECL platform had LODs from 0.1–0.5 ng/mL in different food matrices whereas the ELISAs had higher LODs (0.5–10 ng/mL). Other studies detecting abrin in food using monoclonal antibodies as capture with polyclonal antibodies as detector reported an LOD of ≈0.7 ng/mL in PBS or PBSTM [22]. A monoclonal capture with a polyclonal detector ELISA was reported to have a LOD in buffer of ≈0.5 µg/L (0.5 ng/mL) [23]. We speculated that a capture ELISA based on monoclonal antibodies for both capture and detection could be developed to have sensitivities comparable to those seen in literature. Monoclonal antibodies have defined epitopes and they represent a more consistent source for a critical reagent in comparison to polyclonal antibodies. Furthermore, they could potentially be used as therapeutics to neutralize toxins in the future.

To help mitigate a potential bioterrorist event using abrin, new reagents and technologies that would rapidly detect this toxin are highly desirable. In this study, we characterized a group of monoclonal antibodies that led to the development of a monoclonal-monoclonal based capture ELISA for abrin detection with a LOD comparable to those reported in literature and commercially available.

2. Results

2.1. Characterization of Anti-Abrin Monoclonal Antibodies

Anti-abrin mAbs from ten hybridoma cell lines were grown as previously described [24] and purified from the ascites fluid by affinity chromatography on Protein-G Sepharose. The antibodies referred to as LS02ABx, LS03ABx, LS04ABx, LS05ABx, LS06ABx, LS07ABx, LS08ABx, LS10ABx, LS11ABx, and LS13ABx. Seven of the mAbs have IgG1 kappa heavy chains with kappa light chains, while the rest were either IgG2a/IgG2b kappa and one had a lambda light chain (Table 1). Each of the antibodies was successfully labeled with biotin, with an average of 2–4 biotin molecules per IgG molecule (Table 1).

Table 1. Isotypes and biotin loading of anti-abrin monoclonal antibodies.

Monoclonals	Isotype	Biotin/IgG
LS02ABx	IgG1, kappa	4.0 ± 0.17
LS03ABx	IgG1, kappa	3.9 ± 0.10
LS04ABx	IgG2a, kappa	3.8 ± 0.20
LS05ABx	IgG2b, kappa	2.5 ± 0.50
LS06ABx	IgG1, kappa	1.5 ± 0.5
LS07ABx	IgG1, kappa	3.7 ± 0.3
LS08ABx	IgG2a, lambda	4.0 ± 0.16
LS10ABx	IgG1, kappa	2.5 ± 0.5
LS11ABx	IgG1, kappa	2.0 ± 0.01
LS13ABx	IgG1, kappa	1.8 ± 0.20

2.2. Silver Stain and Western Blots

In an effort to determine if these mAbs bound the A- or B-chain of the toxin each mAb was used to probe Western blots. Electrophoresis of our abrin and ricin stocks followed by silver staining revealed multiple bands (Figure 1A). Ricin consists of the holotoxin, the agglutinin, and the individual A and B subunits when electrophoresed without DTT. Similarly, non-reduced abrin consists primarily of the holotoxin (*), some agglutinin ([1]), A-chain, and B-chain. Addition of DTT to the abrin material resulted in a significant reduction of the agglutinin and the holotoxin but an increase of the A and B subunits. The abrin A-chain control sample (without reducing agent) has two predominant A-chain isomers. The non-reduced abrin B-chain control shows the expected B-chain as well as a higher molecular weight species that is most likely the holotoxin. The same samples were subjected to immunoblotting with each of the 10 LSABx monoclonal antibodies to determine if there was cross-reactivity with ricin as well as to determine which abrin chain, A or B, the mAbs bound. Typical results are shown in Figure 1B for mAbs LS04ABx and LS13ABx but similar results were obtained for each of the mAbs (data not shown). All were observed to bind strongly to abrin holotoxin and A-chain. Binding to the agglutinin is weak but present on the immunoblots. None of the mAbs recognize the abrin B-chain or any form of ricin.

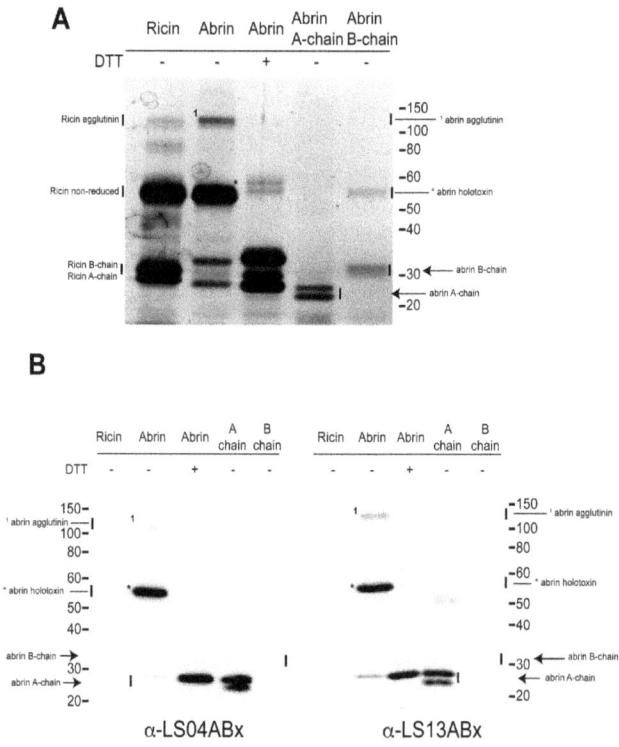

Figure 1. SDS-PAGE analysis of toxin complexes and subunits. 500 ng per lane of sample (ricin, abrin, abrin A-chain, abrin B-chain) treated or not treated with 0.05 M DTT was loaded onto a NuPAGE 4–12% Bis-Tris gel and subjected to SDS-PAGE electrophoresis. Gels were processed for silver staining or Western blotting. (**A**) Silver staining with the SilverXpress. (**B**) Representative Western blots probed with mAbs LS04ABx and LS13ABx. Similar exposure times and images were obtained for the remaining mAbs (data not shown). *—Abrin holotoxin non-reduced; [1]—agglutinin, and corresponding arrows denote A- and B-chains.

2.3. Monoclonal Antibody Binding to Abrin Toxin and Individual A and B-Chain Subunits

The Western blots indicated that these antibodies were directed against the A-chain, we used a direct ELISA to confirm this specificity. Abrin toxin, A-chain, and B-chain were used to coat 96-well plates and binding of biotin-labeled anti-abrin mAbs was measured. The results of these experiments are summarized in Figure 2. With the exception of mAb LS04ABx, which bound poorly to abrin immobilized on plastic ELISA plates, the Abs strongly bound abrin toxin (black columns) and abrin A-chain (grey columns) compared to no toxin control (open grey columns), p-values < 0.05. In contrast, greatly reduced or no binding to B-chain (hashed columns) was observed as compared to the abrin toxin or abrin-A-chain. The weak binding (* p < 0.05) observed in ELISA by some of the mAbs to the B-chain standard may reflect the observation in the above experiments (Figure 1A) suggesting some A-chain contamination of the B-chain preparation. B-chain binding is at least an order of magnitude less than the signal seen with either abrin or abrin A-chain after the signal of the control is subtracted (Figure 2).

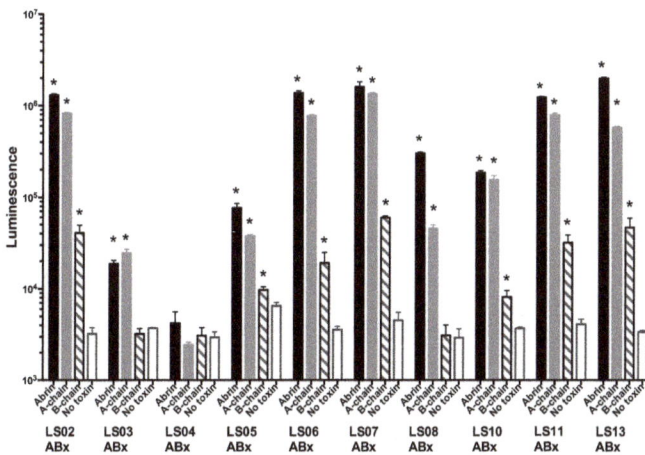

Figure 2. LSABx monoclonal antibodies recognize the abrin A-chain in a direct ELISA. Samples of 1 µg/mL of abrin, A-chain, B-chain, or no toxin were assayed for each of the 10 LSABx-biotinylated antibodies at a concentration of 1 µg/mL. Luminescence was measured from triplicate wells; error bars equal ± one standard deviation (SD). Statistical significance was determined by two-tailed unpaired Student's t-test comparing each column with the no toxin control for each set of mAbs, (*) p-values < 0.05 are considered significant. Abrin toxin (black columns); A-chain (grey columns); B-chain (hashed columns); and no toxin (open grey columns).

2.4. Monoclonal–Monoclonal Capture ELISA for Abrin

The 10 anti-Abrin mAbs were evaluated for their ability to function as either capture or detector antibodies in a capture (sandwich) ELISA format. These results expressed as the lower limit of detection (LOD) for each possible antibody combination are summarized in Table 2. They are expressed as the value in ng/mL that gave a signal above the average signal of the no-toxin control plus three SD. Clearly the majority of antibodies failed to bind toxin in solution, to capture the toxin. Of the 100 possible combinations only 4 capture–detector pairs resulted in LODs ≤10 ng/mL. These capture-detection pairs were LS04ABx-LS11ABx, LS04ABx-LS13ABx, LS06ABx-LS08ABx, and LS08ABx-LS11ABx. Since these represent preliminary experiments not using the most sensitive substrate, they were further evaluated.

Table 2. Observed LODs (ng/mL) for all possible combinations of anti-Abrin mAbs.

					Detectorm Abs					
	LS02ABx	LS03ABx	LS04ABx	LS05ABx	LS06ABx	LS07ABx	LS08ABx	LS10ABx	LS11ABx	LS13ABx
Capture mAbs										
LS02ABx	-	200	200	-	-	-	50	-	-	-
LS03ABx	100	-	-	-	200	200	-	-	-	200
LS04ABx	-	-	-	200	-	-	-	-	5	10
LS05ABx	-	-	-	-	-	-	-	-	-	-
LS06ABx	-	-	-	-	-	75	10	100	-	-
LS07ABx	-	-	-	-	-	75	40	-	-	-
LS08ABx	100	40	-	100	-	100	-	75	5	50
LS10ABx	-	-	-	-	-	-	-	-	-	-
LS11ABx	-	-	-	-	-	-	30	-	-	-
LS13ABx	100	200	200	-	-	-	100	-	-	-

The minus sign indicates limits of detection (LODs) greater than 300 ng/mL.

These four capture–detector mAb pairs were further optimized by adjusting antibody concentrations and use of SuperSignal Femto Max Sensitivity substrate (data not shown). Results using mAb LS04ABx as a capture with biotin-labeled mAb LS13ABx as a detector are shown in a representative ELISA curve in Figure 3 representing one independent experiment consisting of three replicates. The assay has a good dynamic range with starting concentrations of 300 ng/mL with sequential two-fold dilutions of the toxin in buffer. Each point on the curve was statistically significant compared with the no toxin control. The limit of detection was determined by calculating the average of the zero toxin controls plus 3 times the standard deviation. In Figure 3, the LOD for this assay was 1.1 ng/mL. An additional two independent experiments each with replicates ($n = 6$) was performed on different days with different buffers and toxin stocks (Supplemental Figure S1). The LOD's from these experiments were $\approx$1 ng/mL and 0.8 ng/mL. The overall LOD calculated from the three independent experiments was $\approx$1 ng/mL $\pm$ 0.15.

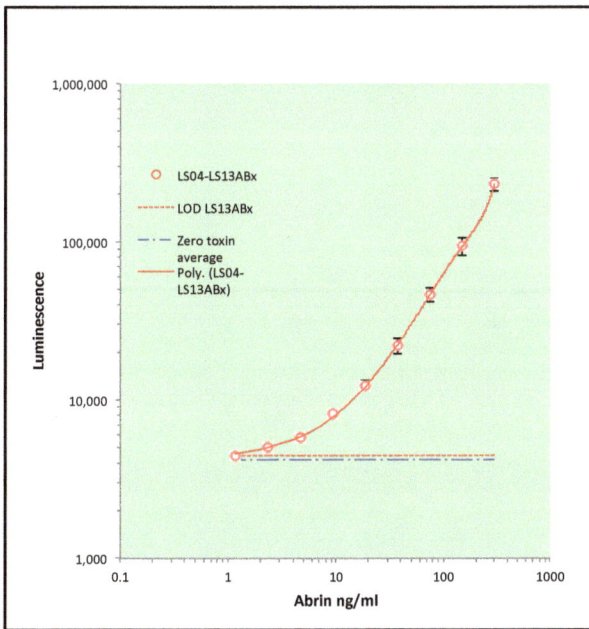

Figure 3. Capture ELISA optimized for detection of Abrin. Mab LS04ABx used as the capture reagent at 2 μg/mL. Detector antibody LS13ABx was used at 1 μg/mL. Representative capture ELISA result showing one out of three independent experiments (see Figure S1 for the remaining experiments). Points represent the average of a triplicate ($n = 3$), error bars = standard deviation. Solid lines show 4-Parameter curve fitting. Red dashed line represents the lowest detection level cut offs calculated as the average of the zero plus three standard deviations. Blue broken line represents the average value of the zero toxin control. Statistical significance was determined by two-tailed unpaired Student's t-test, $p < 0.05$ considered statistically significant for all points compared to the average zero toxin value.

2.5. Specificity of Abrin Capture ELISA against Other Plant Ribosome-Inactivating Proteins (RIPs)

Ribosome-inactivating proteins from both Type I and Type II families are prevalent in the plant kingdom. One must take into account whether there may be cross-reactivity between abrin with these other RIPs resulting in 'false' positives in an assay. Earlier studies had tested the specificity of a commercially available lateral flow device against 35 near neighbor crude extracts [25]. The authors reported that the extracts with the exception of *Abrus laevigatus* tested negative at a concentration of

10 µg/mL. Abrin and the identical extracts used above in [24] plus an additional 10 extracts were tested here. These extracts were tested in triplicate at 3 µg/mL with the optimized capture ELISA shown in Figure 3. All 46 near neighbor extracts tested here were negative (Table 3) including *Abrus laevigatus*. Only abrin from the *Abrus precatorius* extract was positive in the assay.

Table 3. Results of Abrin capture ELISA with near neighbor extracts.

Near Neighbours	Capture ELISA
Abrus laevigatus E. Mey.	-
Abrus precatorius L. +	+
Abrus schimperi subsp. *Africanus* (Vatke) Verdc.	-
Acalypha rhomboidea Raf.	-
Acalypha rhomboidea 11837	-
Adriana quadripartite (Labill.) Gaudich.	-
Adriana quadripartita 76851	-
Bryonia dioica Jacq.	-
Canavalia gladiate (Jacq.) DC.	-
Canavalia rosea (Sw.) DC	-
Canavalia madagascariensis J.D. Sauer	-
Cinnamomum camphora (L.) J.Presl.	-
Cucurbita moschata Duchesne	-
Dianthus caryophyllus Linnaeus	-
Fatsia japonica (Thunb.) Decne & Planch.	-
Fatsia japonica 39610	-
Galactia striata (Jacq.) Urban	-
Galactia wrightii A. Gray	-
Iris hollandica Bf.	-
Jubernardia globifera (Benth.) Troupin	-
Jubernardia globifera 76851	-
Luffa acutangula (L.) Roxb.	-
Luffa cylindrical (aegyptica) Bergquist, 1995	-
Lychnis chalcedonica (L.) E.H.L. Krause	-
Macaranga grandifolia (Blanco) Merr.	-
Macaranga grandifolia 3403	-
Mallotus nudiflorus (L.) Kulju & Welzen	-
Mallotus philippensis (Lam.) Müll. Arg.	-
Manihot escuelenta Crantz.	-
Manihot escuelenta 13562	-
Mercurialis annua L.	-
Mercurialis annua 279720	-
Momordica charantia L.	-
Phytolacca Americana L.	-
Phytolacca americana 39161	-
Phytolacca dioica L.	-
Plukenetia volubilis 72130	-
Plukenetia volubilis L.	-
Sambucus ebulus L.	-
Sambucus nigra L.	-
Saponaria officinalis L.	-
Saponaria officinalis 391915	-
Senna occidentalis (L.) Link	-
Trewia nudiflora 85123	-
Trichosanthes kirilowi Maxim.	-
Viscum album L.	-
Viscum album 22397	-

- = Zero activity at 3 µg/mL of sample in capture ELISA. + = Activity at 3 µg/mL of sample in capture ELISA.

2.6. MAb Neutralization of Abrin Cytotoxicity in Cells

Selected mAbs were premixed with toxin in order to determine if they could neutralize the toxin. In these experiments abrin was used at 5 ng/mL and mixtures of toxin and antibody contained the mAb at 50 μg/mL. These data, are summarized in Table 4, and demonstrate that administration of the antibody alone had no toxic effect on the Vero cell growth. Further, premixing toxin with mAbs LS02ABx, LS03ABx, LS04ABx, LS07ABx, LS08ABx, LS10ABx, or LS11ABx had only minimal effects on abrin cytotoxicity. In contrast, mixing mAb LS13ABx with toxin showed a relative toxicity of 54%. Mixing three mAbs, LS02 + 07 + 13ABx resulted in an even greater decrease in relative cytotoxicity, to 41% as compared with exposure to abrin by itself. As a control for antibody specificity, ricin cytotoxicity was observed not to be inhibited by the same antibodies used in the Vero cell assay (Table S1).

Table 4. Neutralization of abrin cytotoxicity with monoclonal antibodies. Values represent means of two samples ± SD. Statistical significance was determined by two-tailed unpaired Student's *t*-test, $p < 0.05$ for all conditions compared to no toxin (DMEM) except for LS07ABx, LS02ABx, and LS02 + 07 + 13.

Treatment	Relative Toxicity (%)
DMEM	0
Abrin	100
LS07ABx	4 + 3
LS13ABx	4 ± 1
LS02ABx	4 ± 3
LS02 + 07 + 13	4 ± 3
LS04Abx + Abrin	94 ± 1
LS07Abx + Abrin	69 ± 2
LS08Abx + Abrin	88 ± 1
LS10Abx + Abrin	94 ± 1
LS13Abx + Abrin	54 ± 1
LS11Abx + Abrin	77 ± 1
LS03Abx + Abrin	96 ± 0
LS02Ax + Abrin	66 ± 2
LS02 + 07 + 13 + Abrin	41 ± 2

3. Discussion

Due to the highly toxic nature of abrin resulting in its classification as a select agent we evaluated a set of mAbs to abrin. Both polyclonal- and polyclonal/monoclonal-based capture immunoassays have been described [15,22,23,26]. In addition, a sensitive lateral flow immunoassay (LFA) incorporating up-converting phosphor technology (UPT) and laser interrogation has recently been described [27]. The authors developed the LFA with up converting phosphor-conjugated antibodies that would required a specially built reader that excites the lanthanide-doped crystals by infrared light and emits visible light. The abrin-UPT-LFA had detection limits as low as 0.1 ng/mL. A monoclonal capture ELISA with an LOD of 7.8 ng/mL [27] and use of aptamers for abrin detection in a colorimetric assay with an LOD of 0.05 nM (3.1 ng/mL) [28] also have been reported. Immunoassay performance fundamentally depends on the performance of the antibodies used as well the assay format. Thus it is important to have a large selection of well-characterized antibodies available for assay development. However, it is important that monoclonal-based capture immunoassays have similar LOD's as compared to those reported in literature. Previous works on detection technologies for abrin have been dependent on polyclonal and monoclonal antibodies for both capture and detector antibodies [15,22,23]. Monoclonal-monoclonal based ELISA would not be subjected to the variability inherent in assays based on polyclonal antisera, the assay should be more consistent over time and readily transported to different capture immunoassay formats.

We report here the results of our analysis of ten anti-Abrin mAbs. All demonstrated specific binding to abrin A-chain and to holotoxin on Western blots. This was further validated by our observations that the antibodies bound the A-chain but not the B-chain in ELISA using purified A- and B-chain. Since the antibodies specifically bound the A-chain and the holotoxin, the binding epitope is not obscured in the A–B holotoxin. Antibody binding to the agglutinin also was observed. Additionally, there was no cross-reactivity with any form of ricin (Figure 1). The best capture mAb was LS04ABx, which poorly bound plate-immobilized abrin by ELISA. These observations suggest that LS04ABx recognizes a complex or conformational epitope that is masked or modified when toxin is absorbed onto a plastic surface. We have observed this before in our studies to generate capture antibodies to botulinum neurotoxin [29]. Perhaps this is a result of both toxins being A-B toxins.

Capture–detector pairs were evaluated for their ability to perform in a capture ELISA (Table 2). Of all the possible pair combinations, 4 capture–detector pairs (LS04ABx-LS11ABx; LS04ABx-LS13ABx; LS06ABx-LS08ABx; and LS08ABx-LS11ABx) were found to have LODs low enough to be useful. These assays were further optimized with the aim of reducing their LODs ($\leq$10 ng/mL). One capture–detector pair with LS04ABx as capture and LS13ABx as detector had a LOD $\approx$ 1 ng/mL (Figure 3). When this capture ELISA was evaluated for potential cross-reactivity against other plant RIPS, no false positives were observed (Table 3). The cytotoxicity studies, while preliminary, suggest that mAbLS13ABx can neutralize abrin toxin specifically but that mixing multiple antibodies improved neutralization (Table 4, Table S1). These results are supported by other studies that demonstrated mixtures of anti-botulism mAbs were more effective than single mAbs [30–33].

Therefore, we report the development of a new monoclonal-monoclonal based capture ELISA that has high sensitivity with a LOD $\approx$ 1 ng/mL $\pm$0.15. This capture ELISA detects the holotoxin as well as the A-chain. The lack of false positives when 46 abrin/ricin near neighbor RIPs suggests that the assay is highly specific for abrin. These new monoclonal antibodies could be developed further in other detection assays using different technologies to increase the sensitivity of these reagents i.e., up-converting phosphor technology or used in conjunction with other abrin monoclonal/polyclonal antibodies. Furthermore, the neutralization potential of these mAbs should be tested in vivo with the mouse bioassay to validate whether or not these are truly neutralization antibodies and if so, determine the temporal effectiveness of the antibody-based protection towards abrin intoxication.

4. Materials and Methods

4.1. Reagents

Pure ricin (RCA-II, Cat. #L1090) was obtained from Vector Laboratories (Burlingame, CA, USA) as a 5 mg/mL solution and stored at 4 °C. Abrin toxin (mixed isomers, Cat# ABR-1) was purchased from Toxin Technologies, Inc. (Sarasota, FL, USA) as a lyophilized material at 1 mg/vial. The contents of the vial were dissolved by addition of 1 mL of 1x phosphate buffered saline (PBS) pH 7.4 to give a 1 mg/mL stock stored at 4 °C. Abrin A-chain (BEI Resources, NR-43945) and Abrin B-chain (BEI Resources, NR-43946) were supplied as 1 mg/mL stocks stored at −20 °C. Hybridoma cell lines secreting anti-abrin monoclonal antibody (mAb) were obtained from The Russian Research Center for Molecular Diagnostics and Therapy (Moscow) and are referred to as LS02ABx, LS03ABx, LS04ABx, LS05ABx, LS06ABx, LS07ABx, LS08ABx, LS10ABx, LS11ABX, and LS13ABx. Bovine serum albumin (BSA); goat anti-mouse immunoglobulin G (whole molecule) conjugated to horseradish peroxidase (IgG–peroxidase) #A-4416; polyoxyethylene sorbitan monolaurate (Tween-20); Protein-G conjugated Sepharose #P-32196; 0.02 M TRIS-buffered saline (TBS) #T-5912; 0.9% NaCl, pH 7.4; and 0.01 M phosphate buffered saline (PBS), 0.138 M NaCl, 0.0027 M KCl, pH 7.4 #P-3813; were purchased from Sigma Chemical Co. (St. Louis, MO, USA). HRP conjugated to Streptavidin #SNN2204 was obtained from Invitrogen Inc. (Carlsbad, CA, USA). Black, Maxisorp 96-well Nunc microtiter plates were obtained from PGC Scientific (Gaithersburg, MD, USA), and SuperSignal Pico and Femto Max Sensitivity substrates was purchased from Pierce Inc. (Rockford, IL, USA). Carnation non-fat dry milk (NFDM) was obtained locally. The NuPAGE 4–12% Bis-Tris gels and SilverXpress kit (LC6100) were supplied from Invitrogen. Luminescence was measured using a Perkin-Elmer Victor-3 microplate reader. Data were exported to Microsoft Excel for further analysis.

4.2. Monoclonal Antibody Procedure and Purification

Hybridoma cells were grown using our standard hybridoma media in a 4% CO_2 at 37 °C as previously described [23]. Small amounts (usually less than 10 mL) of ascites fluids were obtained (Covance Research Products, Inc., Denver, PA, USA). Antibodies were purified from the ascites fluid by affinity chromatography on Protein-G Sepharose. Bound antibody was eluted with 0.1 M glycine–HCl, pH 2.7 and then dialyzed against PBS. Protein concentrations were determined with a BCA-kit (Pierce) using the microplate method suggested by the manufacturer. Antibodies were conjugated with biotin using EZ-Link Sulfo-NHS-LC-Biotin (Pierce) as described using a 50-fold molar excess of biotin reagent. Biotin conjugation was evaluated using the Pierce Biotin Quantitation kit according to the manufacturer's protocol (Pierce, 28005) (based on Beer Lambert Law: $A_\lambda = \varepsilon_\lambda bC$). The calculations used were : (1) mmol protein per mL = [protein concentration (mg/mL)/MW of protein (mg/mmol)] = *Calc#1*; (2) $\Delta A500$ = (A500 H\A) − (A500H\A\B) = *Calc#2*; mmol biotin/mL reaction mixture = $[\Delta A500/(34,000 \times b)]$ = *Calc#3*; and mmol biotin in original sample/mmol protein in original sample = [(*Calc#3*) × 10 × dilution factor]/*Calc#1*.

Antibody isotype was determined by ELISA using toxin-coated microtiter plates and horseradish peroxidase-conjugated, isotype-specific anti-bodies (SouthernBiotech, Birmingham, AL, USA) and with IsoStrip mouse monoclonal antibody isotyping kit #11 493 027 001 (Roach Diagnostics, Indianapolis, IN, USA) according to the manufacturers.

4.3. ELISA Methods

<u>Direct ELISA's</u> were as previously described [23] using black microtiter plates coated with 100 µL per well of a 1 µg/mL solution of toxin in 0.05 M sodium carbonate buffer, pH 9.6. Biotin-labeled anti-abrin mAbs at 1 µg/mL were used. SuperSignal Femto Max Sensitivity substrate (Pierce, Rockford, IL, USA) was used according to manufacturer's instructions. The plates were incubated for 3 min at room temperature and luminescent counts recorded using a Victor-3 microplate reader (PerkinElmer Inc., Waltham, MA, USA).

<u>Capture ELISA's</u> for routine evaluation of antibody pairs were as described [23]. Black microtiter plates were coated with 50 µL/well of a 2 µg/mL solution of anti-abrin mAb in carbonate buffer overnight at 4 °C. Plates were washed with TBST and remaining reactive sites blocked with 3% NFDM-TBST for 1 h at 37 °C. Plates were then washed and toxin added starting at 300 ng/mL followed by a 2-fold dilution series. Zero toxin wells served as control. Biotin-labeled anti-abrin mAb at 5 µg/mL in 3% NFDM-TBST was added and the plates incubated for 1 h at 37 °C. HRP-conjugated Streptavidin (1:20,000 dilution in TBST) was added and the plates incubated for 1 h at 37 °C. The plates were given a final wash and SuperSignal Pico Extended Duration substrate was added. Luminescence measured at 3 min in a Perkin-Elmer Victor-3 microplate reader. The optimized capture ELISA protocol used 50 µL/well of a 2 µg/mL solution of LS04ABx to coat microtiter plates and biotin labeled detector antibody LS13ABx was used at 1 µg/mL. SuperSignal Femto Max Sensitivity substrate was used and plates were washed 12 times following the streptavidin step in order to lower non-specific signal.

4.4. SDS-PAGE Electrophoresis for Silver Stain and Western Blotting

Ricin, abrin, abrin A-chain, abrin B-chain at a concentration of 500 ng per well in the presence or absence of reducing agent (0.05 M DTT) were separated by sodium dodecyl sulfate (SDS)-polyacrylamide gel electrophoresis (PAGE) with NuPAGE 4–12% Bis-Tris gels (Invitrogen) followed by either Silver stain or Western blotting. One gel was subsequently silver stained using the SilverXpress kit according to the manufacturer's instructions (Invitrogen). For the remaining gels, the resolved proteins were transferred to a PVDF membrane (Invitrogen iBlot). The membranes were blocked in 5% milk-Tris-buffered saline-0.1% Tween 20 buffer then probed with each of the

10 monoclonal LSABx antibodies at a concentration of 5 μg/mL in blocking solution followed by secondary antibody (goat anti-mouse horseradish peroxidase (HRP)-conjugated; 1:3000, Cell Signaling Technology). The blot was incubated in Pierce ECL Western Blotting Substrate solution (Thermo Scientific). Protein bands from peroxidase activities to chemiluminescent substrates were developed and detected using the ChemiDoc MP imaging system (BIO-RAD, Hercules, CA, USA). Molecular weight standards were purchased from Invitrogen.

4.5. Capture ELISA with Abrin/Ricin near Neighbor Extracts

To evaluate the specificity of the monoclonal abrin antibodies against other plant ribosome-inactivating proteins (RIPs), extracts from 46 abrin and ricin near neighbors were obtained from Julie R. Avila (Lawrence Livermore National Laboratory, Livermore, CA, USA) as described [24]. Crude extracts were at either 13.2 μg/mL or 20 μg/mL. The most sensitive capture ELISA used SuperSignal Femto Max Sensitivity substrate and plates were washed 12 times following the streptavidin step in order to lower non-specific signal.

4.6. Vero Cell Cytotoxicity Assay

Vero cells were cultured in Dulbecco's Modified Eagle Medium (DMEM) high glucose +10% fetal bovine serum and incubated in a humidified incubator (37 °C, 5% CO_2). Cells were trypsinized and adjusted to 0.5×10^5 cells/mL, seeded into black-sided, clear-bottom 96-well tissue culture plates at 100 μL/well and allowed to adhered overnight (18 h) at 37 °C. The media was then removed, and 100 μL of fresh DMEM containing abrin or ricin with and without mAb dilutions (pre-incubated 1 h at 37 °C) was then added. Plates were then incubated at 37 °C for 2 h, the media was removed and fresh media added. The cells were then incubated for 48 h at 37 °C. Cytotoxicity was measured as follows: 100 μL of CellTiter-Glo (Promega, Madison, WI, USA, G7570) at 1:5 in PBS was added to each well and the plates shaken for 2 min to lyse the cells. After 10 min at room temperature, luminescence was measured using a Victor-3 plate reader (lid was removed from the plate for a better signal). Statistical significance was calculated using the two-tailed unpaired Student's *t*-test with *p*-values < 0.05 considered significant.

Supplementary Materials: The following are available online at www.mdpi.com/2072-6651/9/10/328/s1, Table S1: Ricin cytotoxicity is not neutralized with LSABx monoclonal antibodies; Supplemental Figure S1: Two independent experiments for capture ELISA.

Acknowledgments: This work was funded by the United States Department of Agriculture, Agricultural Research Service, National Program project NP108, CRIS 5325-42000-049-00D. L.H.S. was also funded by interagency agreement IAA # 60-2030-5-004 with Department of Homeland Security.

Author Contributions: L.W.C., C.C.T., X.H., and L.H.S. conceived and designed the experiments; C.C.T., P.M., X.H., and L.H.S. performed the experiments; C.C.T., L.W.C., X.H., and L.H.S. analyzed the data; D.H. contributed reagents; C.C.T., L.H.S., and L.W.C. wrote the paper.

Conflicts of Interest: The authors declare no conflict of interests.

References

1. Endo, Y.; Tsurugi, K. Mechanism of action of ricin and related toxic lectins on eukaryotic ribosomes. *Nucleic Acids Symp. Ser.* **1986**, *17*, 187–190.
2. Cheng, J.; Lu, T.H.; Liu, C.L.; Lin, J.Y. A biophysical elucidation for less toxicity of agglutinin than abrin-a from the seeds of abrus precatorius in consequence of crystal structure. *J. Biomed. Sci.* **2010**, *17*, 34. [CrossRef] [PubMed]
3. Zhou, H.; Zhou, B.; Ma, H.; Carney, C.; Janda, K.D. Selection and characterization of human monoclonal antibodies against abrin by phage display. *Bioorg. Med. Chem. Lett.* **2007**, *17*, 5690–5692. [CrossRef] [PubMed]
4. Reedman, L.; Shih, R.D.; Hung, O. Survival after an intentional ingestion of crushed abrus seeds. *West. J. Emerg. Med.* **2008**, *9*, 157–159. [PubMed]

5. Narayanan, S.; Surolia, A.; Karande, A.A. Ribosome-inactivating protein and apoptosis: Abrin causes cell death via mitochondrial pathway in jurkat cells. *Biochem. J.* **2004**, *377*, 233–240. [CrossRef] [PubMed]
6. Narayanan, S.; Surendranath, K.; Bora, N.; Surolia, A.; Karande, A.A. Ribosome inactivating proteins and apoptosis. *FEBS Lett.* **2005**, *579*, 1324–1331. [CrossRef] [PubMed]
7. Qing, L.T.; Qu, X.L. Cloning, expression of the abrin-a a-chain in escherichia coli and measurement of the biological activities in vitro. *Sheng Wu Hua Xue Yu Sheng Wu Wu Li Xue Bao* **2002**, *34*, 405–410. [PubMed]
8. Hegde, R.; Maiti, T.K.; Podder, S.K. Purification and characterization of three toxins and two agglutinins from abrus precatorius seed by using lactamyl-sepharose affinity chromatography. *Anal. Biochem.* **1991**, *194*, 101–109. [CrossRef]
9. Lin, J.Y.; Lee, T.C.; Hu, S.T.; Tung, T.C. Isolation of four isotoxic proteins and one agglutinin from jequiriti bean (abrus precatorius). *Toxicon* **1981**, *19*, 41–51. [PubMed]
10. Olsnes, S.; Saltvedt, E.; Pihl, A. Isolation and comparison of galactose-binding lectins from abrus precatorius and ricinus communis. *J. Biol. Chem.* **1974**, *249*, 803–810. [PubMed]
11. Wei, C.H.; Hartman, F.C.; Pfuderer, P.; Yang, W.K. Purification and characterization of two major toxic proteins from seeds of abrus precatorius. *J. Biol. Chem.* **1974**, *249*, 3061–3067. [PubMed]
12. Roy, J.; Som, S.; Sen, A. Isolation, purification, and some properties of a lectin and abrin from abrus precatorius linn. *Arch. Biochem. Biophys.* **1976**, *174*, 359–361. [CrossRef]
13. Lin, J.Y.; Lee, T.C.; Tung, T.C. Isolation of antitumor proteins abrin-a and abrin-b from abrus precatorius. *Int. J. Pept. Protein Res.* **1978**, *12*, 311–317. [CrossRef] [PubMed]
14. Herrmann, M.S.; Behnke, W.D. A characterization of abrin a from the seeds of the abrus precatorius plant. *Biochim. Biophys. Acta* **1981**, *667*, 397–410. [CrossRef]
15. Garber, E.A. Toxicity and detection of ricin and abrin in beverages. *J. Food Prot.* **2008**, *71*, 1875–1883. [CrossRef] [PubMed]
16. Stirpe, F.; Barbieri, L.; Battelli, M.G.; Soria, M.; Lappi, D.A. Ribosome-inactivating proteins from plants: Present status and future prospects. *Biotechnology* **1992**, *10*, 405–412. [CrossRef] [PubMed]
17. Liu, C.L.; Tsai, C.C.; Lin, S.C.; Wang, L.I.; Hsu, C.I.; Hwang, M.J.; Lin, J.Y. Primary structure and function analysis of the abrus precatorius agglutinin a chain by site-directed mutagenesis. Pro(199) of amphiphilic alpha-helix h impairs protein synthesis inhibitory activity. *J. Biol. Chem.* **2000**, *275*, 1897–1901. [CrossRef] [PubMed]
18. Johnson, R.C.; Zhou, Y.; Jain, R.; Lemire, S.W.; Fox, S.; Sabourin, P.; Barr, J.R. Quantification of l-abrine in human and rat urine: A biomarker for the toxin abrin. *J. Anal. Toxicol.* **2009**, *33*, 77–84. [CrossRef] [PubMed]
19. Gill, D.M. Bacterial toxins: A table of lethal amounts. *Microbiol. Rev.* **1982**, *46*, 86–94. [PubMed]
20. Wooten, J.V.; Pittman, C.T.; Blake, T.A.; Thomas, J.D.; Devlin, J.J.; Higgerson, R.A.; Johnson, R.C. A case of abrin toxin poisoning, confirmed via quantitation of L-abrine (*n*-methyl-L-tryptophan) biomarker. *J. Med. Toxicol.* **2014**, *10*, 392–394. [CrossRef] [PubMed]
21. Cho, H.; Jaworski, J. A portable and chromogenic enzyme-based sensor for detection of abrin poisoning. *Biosens. Bioelectron.* **2014**, *54*, 667–673. [CrossRef] [PubMed]
22. Garber, E.A.; Venkateswaran, K.V.; O'Brien, T.W. Simultaneous multiplex detection and confirmation of the proteinaceous toxins abrin, ricin, botulinum toxins, and staphylococcus enterotoxins a, b, and c in food. *J. Agric. Food Chem.* **2010**, *58*, 6600–6607. [CrossRef] [PubMed]
23. Zhou, Y.; Tian, X.L.; Li, Y.S.; Pan, F.G.; Zhang, Y.Y.; Zhang, J.H.; Wang, X.R.; Ren, H.L.; Lu, S.Y.; Li, Z.H.; et al. Development of a monoclonal antibody-based sandwich-type enzyme-linked immunosorbent assay (elisa) for detection of abrin in food samples. *Food Chem.* **2012**, *135*, 2661–2665. [CrossRef] [PubMed]
24. Stanker, L.H.; Merrill, P.; Scotcher, M.C.; Cheng, L.W. Development and partial characterization of high-affinity monoclonal antibodies for botulinum toxin type a and their use in analysis of milk by sandwich elisa. *J. Immunol. Methods* **2008**, *336*, 1–8. [CrossRef] [PubMed]
25. Ramage, J.G.; Prentice, K.W.; Morse, S.A.; Carter, A.J.; Datta, S.; Drumgoole, R.; Gargis, S.R.; Griffin-Thomas, L.; Hastings, R.; Masri, H.P.; et al. Comprehensive laboratory evaluation of a specific lateral flow assay for the presumptive identification of abrin in suspicious white powders and environmental samples. *Biosecur. Bioterror.* **2014**, *12*, 49–62. [CrossRef] [PubMed]
26. Xu, C.; Li, X.; Liu, G.; Xu, C.; Xia, C.; Wu, L.; Zhang, H.; Yang, W. Development of elisa and colloidal gold-pab conjugate-based immunochromatographic assay for detection of abrin-a. *Monoclon. Antib. Immunodiagn. Immunother.* **2015**, *34*, 341–345. [CrossRef] [PubMed]

27. Liu, X.; Zhao, Y.; Sun, C.; Wang, X.; Wang, X.; Zhang, P.; Qiu, J.; Yang, R.; Zhou, L. Rapid detection of abrin in foods with an up-converting phosphor technology-based lateral flow assay. *Sci. Rep.* **2016**, *6*, 34926. [CrossRef] [PubMed]

28. Li, X.B.; Yang, W.; Zhang, Y.; Zhang, Z.G.; Kong, T.; Li, D.N.; Tang, J.J.; Liu, L.; Liu, G.W.; Wang, Z. Preparation and identification of monoclonal antibody against abrin-a. *J. Agric. Food Chem.* **2011**, *59*, 9796–9799. [CrossRef] [PubMed]

29. Hu, J.; Ni, P.; Dai, H.; Sun, Y.; Wang, Y.; Jiang, S.; Li, Z. Aptamer-based colorimetric biosensing of abrin using catalytic gold nanoparticles. *Analyst* **2015**, *140*, 3581–3586. [CrossRef] [PubMed]

30. Scotcher, M.C.; Cheng, L.W.; Stanker, L.H. Detection of botulinum neurotoxin serotype b at sub mouse ld(50) levels by a sandwich immunoassay and its application to toxin detection in milk. *PLoS ONE* **2010**, *5*, e11047. [CrossRef] [PubMed]

31. Nowakowski, A.; Wang, C.; Powers, D.B.; Amersdorfer, P.; Smith, T.J.; Montgomery, V.A.; Sheridan, R.; Blake, R.; Smith, L.A.; Marks, J.D. Potent neutralization of botulinum neurotoxin by recombinant oligoclonal antibody. *Proc. Natl. Acad. Sci. USA* **2002**, *99*, 11346–11350. [CrossRef] [PubMed]

32. Fan, Y.F.; Barash, J.R.; Lou, J.L.; Conrad, F.; Marks, J.D.; Arnon, S.S. Immunological characterization and neutralizing ability of monoclonal antibodies directed against botulinum neurotoxin type h. *J. Infect. Dis.* **2016**, *213*, 1606–1614. [CrossRef] [PubMed]

33. Fan, Y.; Garcia-Rodriguez, C.; Lou, J.; Wen, W.; Conrad, F.; Zhai, W.; Smith, T.J.; Smith, L.A.; Marks, J.D. A three monoclonal antibody combination potently neutralizes multiple botulinum neurotoxin serotype f subtypes. *PLoS ONE* **2017**, *12*, e0174187. [CrossRef] [PubMed]

toxins

MDPI

Review

Treatments for Pulmonary Ricin Intoxication: Current Aspects and Future Prospects

Yoav Gal [1,*], Ohad Mazor [2], Reut Falach [1], Anita Sapoznikov [1], Chanoch Kronman [1,*] and Tamar Sabo [1]

1   Department of Biochemistry and Molecular Genetics, Israel Institute for Biological Research, Ness-Ziona 76100, Israel; reutf@iibr.gov.il (R.F.); anitas@iibr.gov.il (A.S.); tamars@iibr.gov.il (T.S.)
2   Department of Infectious Diseases, Israel Institute for Biological Research, Ness-Ziona 76100, Israel; ohadm@iibr.gov.il
*   Correspondence: yoavg@iibr.gov.il (Y.G.); chanochk@iibr.gov.il (C.K.); Tel.: +972-8-9381479 (Y.G.); +972-8-9381522 (C.K.)

Academic Editors: Julien Barbier and Daniel Gillet
Received: 6 September 2017; Accepted: 29 September 2017; Published: 3 October 2017

Abstract: Ricin, a plant-derived toxin originating from the seeds of *Ricinus communis* (castor beans), is one of the most lethal toxins known, particularly if inhaled. Ricin is considered a potential biological threat agent due to its high availability and ease of production. The clinical manifestation of pulmonary ricin intoxication in animal models is closely related to acute respiratory distress syndrome (ARDS), which involves pulmonary proinflammatory cytokine upregulation, massive neutrophil infiltration and severe edema. Currently, the only post-exposure measure that is effective against pulmonary ricinosis at clinically relevant time-points following intoxication in pre-clinical studies is passive immunization with anti-ricin neutralizing antibodies. The efficacy of this antitoxin treatment depends on antibody affinity and the time of treatment initiation within a limited therapeutic time window. Small-molecule compounds that interfere directly with the toxin or inhibit its intracellular trafficking may also be beneficial against ricinosis. Another approach relies on the co-administration of antitoxin antibodies with immunomodulatory drugs, thereby neutralizing the toxin while attenuating lung injury. Immunomodulators and other pharmacological-based treatment options should be tailored according to the particular pathogenesis pathways of pulmonary ricinosis. This review focuses on the current treatment options for pulmonary ricin intoxication using anti-ricin antibodies, disease-modifying countermeasures, anti-ricin small molecules and their various combinations.

Keywords: ricin; pulmonary intoxication; countermeasures; antitoxins; disease-modifying agents; anti-ricin small molecules

1. Introduction

Ricin toxin, derived from the castor bean plant *Ricinus communis,* is a highly toxic protein that belongs to the type 2 ribosome-inactivating proteins (RIP) family [1]. Ricin binds to galactose residues at the cell surface via its lectinic B subunit (RTB) and then internalizes and traffics to the endoplasmic reticulum (ER), where the two subunits are reduced [2]. The catalytically active A subunit (RTA) translocates into the cytoplasm, where it depurinates a conserved adenine residue located in the 28S ribosomal RNA of the 60S subunit, thus leading to irreversible inhibition of protein synthesis and ultimately to cell death. Ricin is classified as a Category B agent by the U.S. Centers for Disease Control and Prevention (CDC) and is considered a potential bioterror agent mainly due to its high availability and ease of preparation.

Despite many efforts invested over the past decades, no clinically approved treatment against ricin poisoning has been established. To date, the only post-exposure measure found to be effective against pulmonary ricinosis in pre-clinical studies is passive immunization with anti-ricin neutralizing antibodies [3–6]. Recent data suggests that small-molecule compounds that either interfere with or inhibit the intracellular trafficking of ricin may also be beneficial against ricinosis [7–10]. Another approach has recently shown that the co-administration of antitoxin antibodies with immunomodulatory drugs enables toxin neutralization while decreasing lung injury severity, thus improving treatment outcomes [11,12].

In this review, we survey current treatment options for pulmonary ricin intoxication using anti-ricin antibodies, disease-modifying countermeasures, anti-ricin small molecule drugs or combinations of drugs-antitoxin. The rationale for screening additional drug candidates for combinatorial treatment is discussed, and suggestions for drugs that might be incorporated into future post-exposure therapy regimens are provided.

2. Ricin-Induced Cytotoxicity and Pathophysiology of Pulmonary Ricinosis

A better understanding of the mechanisms underlying both the cellular and physiological effects following pulmonary ricin poisoning are expected to assist in the development of novel treatment modalities against this type of exposure. Manipulation of these pathological processes will hopefully provide tools for clinical interventions that will attenuate lung damage and enhance therapeutic outcomes. In the following section, the current knowledge of the pathological and biochemical changes that occur following ricin intoxication is summarized.

2.1. Pathogenesis

Data regarding the pathological changes that occur following pulmonary ricin intoxication are mostly available from experiments performed in rodents, non-human primates and swine. The overall clinical picture is that the injury is mostly confined to the lungs and that the intoxicated animals suffer from marked interstitial pneumonia associated with massive neutrophil infiltration, perivascular and alveolar edema, fibrin deposition, hemorrhage and diffuse massive airway epithelial necrosis involving all lung lobes [13–17]. Eventually, flooding of the lungs leads to respiratory insufficiency and death. Recently, a swine model for pulmonary ricin intoxication was established by our group, allowing us to further assess the physiological and pathological changes that occur over time. It was found that the clinical manifestations comply with the accepted diagnostic criteria for acute respiratory distress syndrome (ARDS). As in other tested animal models, the pattern of local pro-inflammatory cytokine storming preceding massive neutrophil infiltration and increased vascular hyper-permeability was demonstrated [18].

2.2. Biomarkers and Cellular Stress Pathways

While the final outcome of ricin activity within the cell is the cessation of protein synthesis, it also leads to the activation of several cellular signaling pathways that in turn may activate multi-organ responses (Figure 1). These signals are further exacerbated by the inflammatory response and damage processes induced by the host, resulting in a "vicious cycle" of damage propagation.

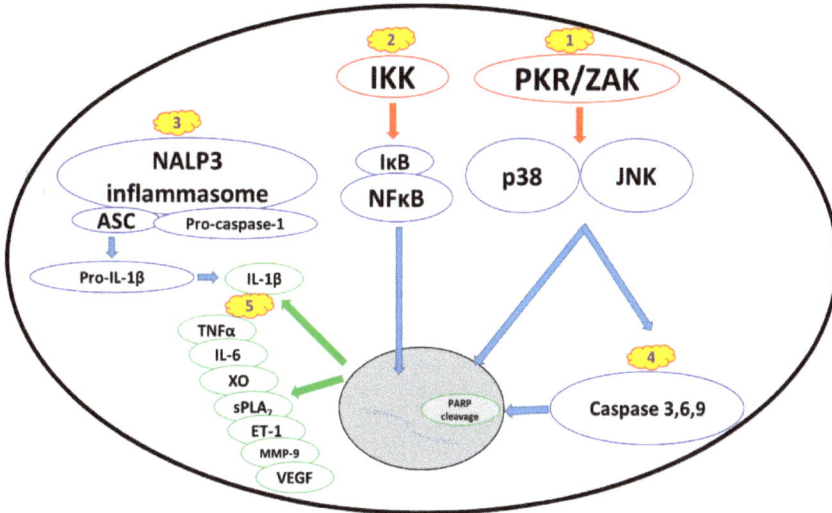

Figure 1. Ricin-induced activation of cell signaling pathways and downstream formation of damage mediators. (**1**) The ribotoxic stress response characterized by MAP3K (PKR and ZAK) activation of MAPK (p38 and JNK) signaling; (**2**) The nuclear factor kappa B pathway, which is activated upon IκK-induced IκB phosphorylation and degradation; (**3**) NALP3 inflammasome-mediated IL-1β activation; (**4**) Apoptotic cell death attributed to pro-apoptotic caspase activation; (**5**) Proinflammatory cytokines and damage mediators released upon activation of the various signal transduction pathways activated by ricin.

2.2.1. Ribotoxic Stress Response

Partial or complete ricin-mediated inactivation of ribosomes leads to the activation of a proinflammatory signaling pathway termed the "ribotoxic stress response." Ricin can trigger the activation of JNK and p38 [19–21], which in turn increase the production of proinflammatory cytokines and apoptosis-mediated cell death [8,20,22–24]. The claim that MAP3K ZAK serves as an upstream activator of the ribotoxic stress response [23] is supported by the inability of ricin to activate p38 and JNK in ZAK-knockout macrophages in vitro and the lower ricin-induced pathology score following oral exposure in ZAK-knockout mice [25]. Another critical upstream mediator of the ribotoxic stress response is ribosome-associated RNA-dependent protein kinase (PKR) (Figure 1) [26], which also induces the activation of JNK and p38, as well as other signaling factors [27–29].

2.2.2. Nuclear Factor Kappa B Pathway

An additional central downstream signaling pathway in the ricin-elicited stress response is the activation of nuclear factor kappa B (NFκB) (Figure 1). NFκB translocates to the nucleus upon IκB kinase (IκK)-induced IκB degradation and transactivates proinflammatory genes [30–32]. Activation of this signaling pathway is associated with many types of sterile lung injuries [33–36] and specifically with pulmonary ricinosis, where nuclear localization of NFκB was detected in mice intratracheally exposed to ricin [37]. Ricin was also shown to activate NFκB in pulmonary epithelial cell cultures [38].

2.2.3. Proinflammatory Cytokines and Damage Mediators

As mentioned above, one of the hallmarks of pulmonary exposure to ricin is the activation of a massive inflammatory response in the lungs. The NALP3 inflammasome (also known as the NLRP3 inflammasome) promotes the cleavage of pro-IL-1β to active IL-1β by caspase-1 (Figure 1) and has a major impact on neutrophil infiltration and exacerbation of the overall inflammatory-mediated damage. It has previously been demonstrated [39] that inflammation— in particular pulmonary neutrophil infiltration and ensuing edema formation—is initiated by macrophage-dependent IL-1β signaling in mice exposed intratracheally to ricin and that ricin-induced IL-1β secretion from macrophages is dependent on NALP3 inflammasome activity. This primary event of IL-1β production is critical for the development of lung injury because the depletion of this cytokine significantly attenuates inflammation, as well as neutrophil pulmonary infiltration.

In addition to the early macrophage-dependent production of IL-1β in the lung tissue, various other pro-inflammatory cytokines and damage mediators (Figure 1) have been detected. Our laboratory has previously demonstrated an early and transient secretion of TNFα into the broncho-alveolar fluid (BALF) of pulmonary ricin-intoxicated mice [11]. TNFα is a major mediator of neutrophil-dependent vascular hyperpermeability [40], which plays a key role in lung pathologies, including ARDS [41].

Other cytokines and chemokines have also been detected in the BALF of ricin-intoxicated mice. For example, a rapid and significant rise in IL-6 levels was discerned as early as 6 h post-exposure [11]. This cytokine, identified as an early biomarker of acute lung injury and a predictive marker of morbidity and mortality, acts as a major proinflammatory mediator for the induction of an acute-phase response leading to a wide range of effects, including leukocyte recruitment and activation [42–45].

In addition to the proinflammatory cytokines described above, diverse damage mediators, such as secretory phospholipase A2 (sPLA$_2$), vascular endothelial growth factor (VEGF), matrix-metalloproteinase-9 (MMP-9) and xanthine oxidase (XO), were detected in the BALF of mice following intranasal ricin intoxication [11]. The levels of the lipolytic enzyme sPLA$_2$, a potent mediator of inflammation via hydrolysis and degradation of surfactant phospholipids [46–48], are significantly elevated following ricin intoxication. In rodent acute lung injury models, sPLA$_2$ was found to promote neutrophil infiltration and edema formation [49,50].

An altered lung fluid balance, leading to increased permeability pulmonary edema, is a major pathophysiological characteristic of intranasal ricin intoxication. VEGF, which promotes vascular permeability and interstitial edema [51], is significantly increased following ricin intoxication. Consistent with this finding, the levels of the serum-resident enzyme cholinesterase, as well as total protein, are increased significantly in the BALF of intoxicated mice, indicating that the blood-lung barrier is severely impaired. Increased Evans Blue dye extravasation, as a marker for increased vascular permeability, has also been reported [12,39].

Consistent with these findings, the levels of the gelatinolytic enzyme MMP-9 rapidly increase after ricin intoxication, displaying a peak level equivalent to a >100-fold increase as soon as 24 h post-exposure. MMP-9 plays an important role in lung injuries [52] and correlates with alveolar-capillary permeability [53].

Increased levels of XO, an enzyme associated with oxidative damage and endothelial dysfunction-mediated edema formation [54], were measured after ricin intoxication and displayed a three-fold increase over control levels at 48 h post-exposure, followed by a ~10-fold increase over the next 24 h. The expression of XO, which results in localized formation of reactive oxygen species (ROS), was found to correlate with the severity of lung damage [55,56]. Ricin-induced oxidative stress has been reported in vivo following systemic intoxication of mice [57–60], and it has been suggested that NALP3 activation promotes ROS generation, which indirectly activates the inflammasome [61]. Hence, XO-derived ROS may also enhance inflammation via this route. Furthermore, ROS have been reported to mediate ricin-induced apoptotic cell death [62], a pathological process that will be further discussed below.

We have recently observed a significant increase in vasoconstrictor peptide endothelin-1 (ET-1) in the BALF of ricin-intoxicated pigs [18]. ET-1, the most abundant isoform of the endothelin peptide family, is produced by a variety of cells, including the airway epithelium and alveolar epithelial cells [63–65], and is known to be released in response to various pathological states [66,67]. Elevated levels of circulating ET-1 are considered a marker for endothelial dysfunction [68] and also correlate with increased pulmonary water contents [69]. Recent studies have suggested that ET-1 not only induces edema accumulation but also prevents edema resolution by impairing alveolar fluid clearance [64,70].

2.2.4. Apoptosis and Changes in Cell Morphology

The induction of an apoptotic response following ricin intoxication (Figure 1) has been demonstrated in several cell types, including epithelial cells [16], endothelial cells [71–73] and macrophages [22,74,75]. This effect is associated with caspase-3 and PARP cleavage, which might be counteracted by the anti-apoptotic protein Bcl-2 [76,77]. It has also been suggested that ricin-induced apoptosis is mediated by p38 activation and is associated with TNFα production [22]. A single in vitro study demonstrated that the redistribution of intracellular zinc ions occurs early during ricin-induced apoptosis and that the exogenous addition of zinc ions may reduce apoptosis by inhibiting caspase 3, 6 and 9 activation, without affecting protein synthesis inhibition [24]. In addition to apoptotic events *per se*, ricin stimulates very rapid and dramatic morphological changes in primary endothelial cells, including the rounding of cells and formation of intercellular gaps, resulting in the passage of molecules through cell monolayers in vitro. These changes precede protein synthesis inhibition and may explain the vascular leak syndrome, which is associated with systemic ricin intoxication [78]. The enhanced sensitivity of endothelial cells may also explain ricin-induced vascular hyper-permeability in the setting of pulmonary exposure, similar to the vascular leak syndrome developed following systemic exposure.

3. Countermeasures for Ricin Intoxication

To date, there are no clinically approved post-exposure medical countermeasures against ricin intoxication. Pre-clinical studies indicate that anti-ricin small molecules may confer protection against pulmonary ricinosis, but only if administered before or shortly after intoxication. To obtain significant surviving ratios at clinically relevant time points following intoxication, the only effective countermeasure, to the best of our knowledge, is anti-ricin antibody-based therapy ("antitoxin"). The combination of antitoxin with small molecules that are anti-ricin targeted, or with compounds that attenuate pathological outcomes ("disease-modifying agents") may improve protection in comparison to antitoxin treatment alone.

The next part of this review summarizes the current knowledge regarding the following potential countermeasures for ricin intoxication: (i) antitoxins; (ii) disease-modifying agents and (iii) small molecules. While some of the reviewed studies use systemic models of ricin intoxication, it is reasonable to assume that these therapies would also be beneficial, to some extent, during the course of pulmonary ricinosis.

3.1. Antitoxins

Over the years, various polyclonal and monoclonal ricin-neutralizing antibodies exhibiting a range of protection efficiencies have been described as post-exposure measures. The ricin molecule participates in many protein-carbohydrate and protein-protein interactions during the intoxication process. Accordingly, antibodies that effectively block the different binding steps have been found. For example, antibodies directed against the B-subunit of the toxin molecule can block the attachment of the toxin to the cell surface and thus inhibit its ability to enter the cell [4,79]. Antibodies directed against other epitopes located either on the A- or the B- subunit of ricin can interfere with its ability to interact with proteins involved in the retrograde transport of RTA into the cytosol [4,79,80]. Antibodies directed against the A-subunit of the toxin inhibit the catalytic activity of ricin in vitro [10,80–82] and are also found effective in vivo [5,6]; however, because these antibodies may dissociate from RTA during the retrograde transport process, it is highly possible that other mechanisms are responsible for the neutralizing activity of these antibodies.

From a clinical perspective, an extended therapeutic time window may be required for efficient treatment, because under various scenarios, therapeutic intervention may be implemented only after the passage of a considerable span of time following exposure. Taking into account the existing assays for the reliable and sensitive detection of ricin in a variety of samples, it can be reasoned that the trigger to treat, namely, the identification of ricin as the cause of intoxication, will be in the range of 24–48 h post-exposure. Unfortunately, the protection efficiencies of most of the reported antibodies decline sharply if they are not applied within several hours of exposure [11,83,84]. Table 1 enlists the monoclonal antibodies that were shown to elicit post-exposure protection. Importantly, extensive efforts are being made to improve the efficacy of the antibody [5] and to shorten the identification time in clinical samples. In this respect, we recently developed a unique method to detect active ricin in clinical samples. This method enables identification of the toxin in samples from pulmonary-intoxicated mice at early time points such as 3 h following intoxication [85]. In another set of experiments, we also demonstrated as a proof-of-concept that a combined treatment of anti-ricin antibodies with diverse drugs ("add-on therapy") improved the survival outcome of intoxicated animals when treatment was initiated at 24 h post-exposure (described in detail below). In the following sections, other appealing possibilities of drugs that may be synergistically combined with anti-ricin antibodies and improve the treatment outcome are discussed.

Table 1. Monoclonal antibodies shown to be protective against ricin when administered post-exposure.

Antibody Name	Antibody Type	Target	Reference
RAC18	murine; chimeric		[86]
PB10	chimeric		[87]
RA36	murine		[88]
43RCA-G1	humanized	RTA	[89]
GD12	murine; chimeric		[90]
MH1	chimeric		[6]
MH36	chimeric		[6]
JB4	chimeric		[87]
RB34	murine		[88]
RB37	murine		[88]
D9	murine; humanized	RTB	[91,92]
MH2	chimeric		[6]
MH73	chimeric		[6]
MH75	chimeric		[6]
MH77	chimeric		[6]

3.2. Disease-Modifying Countermeasures

At later times following pulmonary ricin exposure, the pathophysiological condition of the intoxicated animals may have deteriorated due to the concomitant activation of several stress pathways, which exacerbate the pathological outcome. Accordingly, it can be hypothesized that by mitigating the activation of these pathways, a more favorable outcome will be attained. Such disease-modifying countermeasures can target the stress pathways described above, i.e., proinflammatory cytokines, pathologic damage mediators, inflammasome activation, stress-activated signaling pathways, apoptosis and many others (Table 2).

Table 2. Summary of disease-modifying countermeasures.

Pathway	Target	Inhibitors
Proinflammatory cytokines	IL-1β	anakinra, immunomodulators
	TNFα	anti-TNFα agents, immunomodulators
	IL-6	tocilizumab, immunomodulators
Damage mediators	XO	allopurinol, febuxostat, antioxidants
	sPLA2	Mepacrine
	ET-1	bosentan, tezosentan
	MMP-9	Doxycycline
	VEGF	bevacizumab, aflibercept

Table 2. *Cont.*

Pathway	Target	Inhibitors
NFκB pathway	NFκB	NFκB inhibitors, 'Compound A'
	IKK	IκK inhibitors, auranofin, BMS-345541
MAP3K	PKR	2-AP, C16, imoxine, PKRi
	ZAK	sorafenib, nilotinib, DHP-2
MAPK	p38	PW66, UM101, p38 inhibitors
	JNK	PW66, SP600125, JNK inhibitors
NALP3 inflammasome	NALP3 inflammasome	MCC950, parthenolide, glyburide, BHB, isoliquiritigenen
	IL-1β	Anakinra
Apoptosis	Apoptosis	antioxidants, zinc, apoptosis inhibitors
	caspases 3, 6, 7, 9	PW69, bithionol

3.2.1. Attenuation of Proinflammatory Cytokines and Damage Mediators

As mentioned above, the proinflammatory cytokines IL-1β, TNFα and IL-6 are elevated in the lungs of all animal models in which they are tested. Clinically approved drugs are frequently used to target these cytokines in many inflammatory-related pathologies [93]. Therefore, it is highly reasonable to apply anti-cytokine treatment in the course of pulmonary ricinosis. IL-1β and TNFα are early-formed cytokines; therefore anti-IL-1β or anti-TNFα drugs should be administered as soon as possible following intoxication, even if a co-administrated antitoxin is administered at a later time point. Supporting this notion, the clinically approved IL-1R antagonist (IL-1Ra) anakinra, a competitive inhibitor interfering with binding of IL-1α and IL-1β to their related receptors, attenuated lung injury in mice intratracheally intoxicated with ricin [39]. Consequently, this treatment induced a significant attenuation of cytokine storming and neutrophil recruitment, an improved histological score and extension of the mean time to death of ricin-intoxicated mice [39].

Inhibitors of other proinflammatory cytokines, such as TNFα and IL-6, have not yet been tested in ricin-intoxicated animals. Several anti-TNFα drugs are available clinically and can be chosen according to the species or effect requested (chimeric vs. humanized anti-TNFα, among others). Regarding IL-6, tocilizumab is a clinically approved anti-IL-6 drug, and additional anti-IL-6 agents are under clinical development [94], yet the proper animal model should be chosen carefully because tocilizumab does not cross-react with murine-IL-6 [95]. While most studies suggest that there is a positive correlation between the severity of the pulmonary damage and IL-6 levels following ricin intoxication, it should be noted that in other models of pulmonary damage, IL-6 attenuates lung injury [96,97].

Inhibitors of the enzymes involved in lung tissue degradation following ricin intoxication, namely sPLA$_2$, MMP-9 and XO, which are markers of lipolytic, proteolytic, and oxidant activities, respectively, should also be tested. Although there are no approved specific inhibitors for MMP-9, doxycycline, as well as other tetracyclines were shown to directly interact with this enzyme, thus inhibiting its activity [98]. Several inhibitors of sPLA$_2$ and XO are clinically available for diverse pathological indications. For example, mepacrine, an anti-malarial drug [99], is a potent inhibitor of sPLA$_2$, demonstrating anti-inflammatory activities in lung pathologies. Specifically, mepacrine attenuated pulmonary vascular leakage following the intratracheal instillation of IL-1 [49,100], which, as stated above, is an important early mediator of murine pulmonary ricinosis [39]. The clinically approved XO inhibitor allopurinol significantly attenuated edema and improved the histological score in ventilator-induced lung injury [101]. In the setting of bleomycin-induced lung injury in mice, allopurinol reduced both pulmonary neutrophil infiltration and IL-1β levels [102]. Febuxostat, a non-purinic XO inhibitor, protected rats from LPS-induced lung injury, as reflected by decreased oxidative stress markers and reduced TNFα levels [103]. Furthermore, febuxostat significantly attenuated pulmonary neutrophil infiltration in acid-induced acute lung injury in mice [104].

XO is a representative marker of oxidative stress-induced damage. However, experimental evidence supports the notion that oxidants and oxidative stress are strongly associated with acute lung injury, having many more potential sources of ROS. These ROS may lead to direct

DNA damage, lipid peroxidation, protein oxidation and proinflammatory gene upregulation [105]. There are many antioxidants used clinically, including N-acetylcysteine (NAC)—a widely used mucus-dissolving over-the-counter medication—which was shown to suppress the release of IL-1β from bone marrow-derived macrophages incubated with ricin [21]. Epigallocatechin gallate (EGCG), an antioxidant found in green tea, diminished ricin-induced cytotoxicity in cell cultures [106,107]. Several antioxidants, such as butylated hydroxyanisole (BHA) and vitamin E succinate, were also shown to provide protective effects in vivo against systemic ricin intoxication [58]. Although the pathological outcomes of systemic ricin exposure are considerably different from those following pulmonary exposure, it is worthwhile to evaluate the above-mentioned antioxidants in a pulmonary ricinosis model because the efficacy of these compounds have also been demonstrated in various types of lung pathologies. For example, vitamin E effectively reduced the oxidative burst and neutrophil pulmonary infiltration in endotoxin-induced lung injury in mice [108]. EGCG attenuated LPS-induced lung injury in mice by reducing neutrophil accumulation, edema formation and pulmonary damage severity [109]. In another study, EGCG was found to reduce seawater aspiration-induced acute lung injury in rats via the regulation of inflammatory cytokines [110].

The elevated levels of VEGF following intranasal ricin intoxication [11] are closely related to pulmonary vascular hyperpermeability and edema formation [51,111], the ultimate cause of respiratory failure and death following pulmonary ricinosis. There are various clinically approved drugs antagonizing the effect of VEGF, i.e., bevacizumab (Avastin) [112] and aflibercept [113], which can be used in the setting of pulmonary ricinosis. Yet, it should be noted that the presence of VEGF in the alveolar space could be protective against diverse settings of murine lung injury [114]; therefore, an anti-VEGF regimen should be assessed carefully. The elevated levels of ET-1 measured in pig BALF following intratracheal ricin intoxication [18] may also be associated with edema formation and progression. Several ET-1 antagonists are clinically available, for example, bosentan [115] and tezosentan [116], which should be evaluated for their efficacy in pulmonary ricin intoxication.

In addition to ET-1 and VEGF, many other pathological effectors involved in the process of edema can also be targeted using a large repertoire of clinically approved and preclinically tested drugs. For example, iloprost improves endothelial barrier function in a murine model of LPS-induced lung injury [117]. Enhancing vascular endothelial barrier integrity with the Tie2-agonist Vasculotide, a patented molecule under experimental investigation, improved survival in murine models of influenza even when administered as late as 72 h following infection [118]. β-agonists, e.g., salbutamol, which is frequently used for the treatment of asthma and other pulmonary pathologies, may accelerate the clearance of extravascular lung water and promote anti-inflammatory effects [119–121].

An additional disease-modifying approach is to use wide-range anti-inflammatory drugs or immunomodulators for the treatment of pulmonary ricinosis. In that respect, we have previously shown that doxycycline [11] and ciprofloxacin [12], antibacterial agents repurposed as immunomodulators, significantly attenuated various proinflammatory markers, such as IL-1β, IL-6, XO, neutrophil lung count and edema markers. These highly effective drugs are widely available and are not expensive, rendering them attractive as anti-ricin therapies. Supporting the notion that the coverage of as many damage mediators in parallel would achieve a better impact, it was demonstrated in several non-ricin-mediated lung pathologies that a combinational treatment of two drugs obtained a better outcome than each drug alone. For example, a combinatory treatment with vitamin E and the corticosteroid dexamethasone resulted in a sharper decrease in BALF neutrophil content and less severe oxidative damage following endotoxin-induced lung injury than each drug alone [108]. Similarly, the co-administration of NAC with steroids and deferoxamine improved the outcome of chlorine-induced [122] or LPS-induced [123] acute lung injury in mice and rats.

3.2.2. Ribotoxic Stress Response Inhibitors

It has previously been demonstrated that the inhibition of p38 protects cells from ricin-induced effects [22]; however, p38 inhibitors are often toxic and therefore exhibit very limited success in clinical trials. Recently, using computer-aided drug design, UM101, a novel inhibitor of p38α, the major isoform responsible for the proinflammatory effects of this MAPK, was evaluated [124]. This compound was more potent than the non-selective p38 inhibitor SB203580 in stabilizing endothelial barrier function and reducing inflammation in the course of LPS-induced murine acute lung injury. In a high-throughput cell-based assay, one of the few identified compounds, PW66, was found to diminish the ricin-induced ribotoxic stress response by interfering with the activation of p38 and JNK as well as inhibiting TNFα secretion [8]. The status of current JNK inhibitors was recently reviewed by Cicenas [125], and it is worth mentioning that several JNK inhibitors, particularly SP600125, exhibit protective effects against acute lung injury in vivo [126–129].

The MAP3Ks ZAK and PKR, which are upstream activators of p38 and JNK, may also serve as targets for the treatment of pulmonary ricinosis. Indeed, sorafenib and nilotinib, which are clinically approved inhibitors with a high affinity for ZAK, decreased the ricin-induced ribotoxic stress response (activation of both p38 and JNK) in macrophages [21]. The ZAK inhibitor DHP-2 decreased the p38 and JNK-induced ribotoxic stress response in epithelial cells, increased cell viability and further decreased caspase-3 activation and proinflammatory gene transcription following incubation with ricin [23]. Small molecule inhibitors of PKR were also identified [130] and were found to be effective both in vitro and in vivo. For example, 2-AP and C16 suppressed the production of the proinflammatory cytokine IL-8 in monocytes [131], whereas PKRi was effective in a mouse model of long-term memory [132]. In addition, the PKR inhibitor imoxine, as well as 2-AP, improved glucose hemostasis and performed anti-inflammatory activities in an obese diabetic murine model [133].

3.2.3. NFκB Inhibitors

Activation of the NFκB signaling pathway induces the transactivation of proinflammatory genes and plays a major role in ricin-mediated pathogenesis. A screen of 2800 clinically approved drugs and bioactive compounds was conducted to identify novel NFκB inhibitors [134], of which 19 exhibited very high potency, including bortezomib, daunorubicin, digitoxin, emetine, sorafenib, sunitinib, tioconazole and zafirlukast. Additional NFκB-inhibiting drug candidates, including IκB kinase inhibitors, are discussed elsewhere [135,136]. Importantly, some NFκB inhibitors were found to be effective in animal models of pulmonary inflammation, such as Bayer's 'Compound A,' which demonstrated broad anti-inflammatory activity in mice and rats following intraperitoneal administration of LPS [137]. In addition, using the BMS-345541 IκK inhibitor, the levels of lung-activated NFκB, proinflammatory cytokine levels, neutrophil influx and edema formation were all reduced in LPS-challenged mice [138].

3.2.4. NALP3 Inflammasome Inhibitors

In addition to the direct induction of the NALP3 Inflammasome by ricin [21,39], the extracellular release of endogenous molecules such as uric acid [102], ATP [139] and neutrophil-derived extracellular histones [140], may stimulate inflammasome activation following ricin-mediated cell necrosis. This highly efficient activation of the NALP3 inflammasome by both ricin and the proinflammatory mediators released by dying cells may be a target for a potential therapeutic intervention. Several NALP3 inhibitors, including parthenolide, glyburide (a clinically approved anti-diabetic drug), 5-chloro-2-methoxy-N-[2-(4-sulfamoylphenyl) ethyl]benzamide and isoliquiritigenen (a chemical compound found in licorice) [141–144], are currently under investigation. The small molecule MCC950 was recently suggested as a therapy for NALP3-associated syndrome, exhibiting a specific and highly potent inhibitory activity that resulted in reduced IL-1β production in mice [145]. Likewise, the ketone metabolite β-hydroxybutyrate (BHB) reduced NALP3 inflammasome-induced IL-1β and IL-18 production [146]. Additional inflammasome inhibitors have been extensively reviewed by Shao [147]. These include clinically approved non-selective inhibitors (interferons-α/β), autophagy-enhancing agents (resveratrol, arglabin and HU-308, which is a representative cannabinoid receptor 2 agonists), and microRNAs (microRNA-223).

3.2.5. Compounds Counteracting Apoptosis and Cell Morphology Changes

Ricin-induced apoptosis may be mediated by various stimuli, such as p38, TNFα [22], ROS [62] and zinc redistribution [24]. Accordingly, there are many options to reduce apoptosis, including the above-mentioned drugs that target these stimuli. In this respect, zinc deficiency aggravates several types of lung injury via NFκB activation, VEGF overexpression and alveolar epithelial/macrophage cell dysfunction [148–150]. The administration of zinc ions alleviates several types of lung injury in rodents [151,152] by decreasing the levels of XO, oxidative stress markers, lung neutrophil recruitment and NFκB activation. Importantly, a link between zinc, caspase-3 and adherent junction-related cell-to-cell contact was demonstrated. Zinc deprivation in the presence of pro-apoptotic stimuli accelerated caspase-3 activation and E-cadherin and β-cathenin proteolysis and induced an increase in paracellular leak across epithelial cell monolayers. Zinc supplementation, but not caspase inhibition, protected the lung epithelial barrier [153]. Consequently, it might be hypothesized that zinc also inhibits changes in cell morphology that might be associated with the vascular leak syndrome [78].

Reducing the progression of apoptotic-mediated cell death may also be achieved by interfering with the apoptotic machinery using clinically approved drugs or compounds that are under clinical investigation [154,155]. For example, the compound PW69 was demonstrated as a ricin-induced caspase 3/7-mediated apoptosis suppressor [8]. Excitingly, a recent work revealed that the antiparasitic drug bithionol potently inhibited caspases- 3/7, 6, and 9 and exhibited pronounced protection against ricin-induced cell death [156].

3.3. Anti-ricin Small Molecules

Anti-ricin antibodies are expected to be non-active when administered after the toxin has already entered the cell, while small molecules that can penetrate the cells might be effective at that time point. Over recent years, an extensive search was held for small molecule inhibitors of ricin. High-content screens have revealed attractive, potentially effective compounds that may be used therapeutically; however, all are still under preliminary investigation. The mechanisms of action of these small molecules are diverse, targeting different cellular pathways of ricin, e.g., membrane binding, intracellular trafficking and the active site (Figure 2 and Table 3). Importantly, there are some limitations and prerequisites for such small-molecule-drug candidates, including safety issues concerning their clinical use. Additionally, they should preferably be water soluble and well absorbed from the GI tract if taken orally. Drug candidates aimed at targeting the active site or inhibiting intracellular trafficking are required to penetrate cell membranes as well.

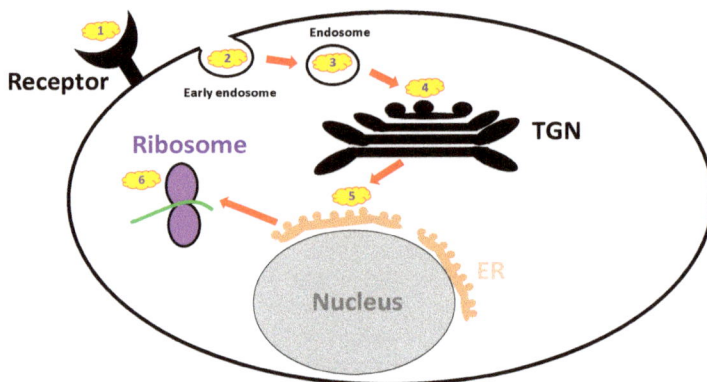

Figure 2. Cellular targets for anti-ricin small molecule compound-based treatment. (**1**) Receptor mimicry; (**2**) Blockers of endocytosis; (**3–5**) Retrograde trafficking blockers; (**6**) Active-site inhibitors. TGN: trans-Golgi network; ER: endoplasmic reticulum.

Table 3. Small molecule anti-ricin inhibitors-mechanisms and targets.

Mechanism	Cell Target	Inhibitors
Receptor mimicry	RTB	Derivatives of glycosphingolipids, lactose and galactose
Endocytosis blockers	Early endosome Endosome	NaN3, cytochalasin D, colchicine
Trafficking blockers	TGN ER	Retro-2, DA2MT, atorvastatin, brefeldin A, mansonone D benzyl alcohol, 3'-Azido-3'-Deoxythimidine
Reductive activation inhibitors	PDI, TrxR, TMX, glutathione disulfide oxidoreductase	auranofin, bacitracin
Active site and RTA inhibitors	Ribosomes	purine- pterin- and pyrimidine-based inhibitors, 4-fluorophenyl methyl 2-(furan-2-yl)quinolone-4-carboxylate, difluoromethylornithine, aptamers, RIP-α-sarcin/ricin loop interface blockers, baicalin

It is a great advantage if the drugs are already clinically approved for other indications (repurposed drugs) and also commonly available and not expensive. A list of candidate anti-ricin small molecule drugs categorized by their mode of actions is provided below.

3.3.1. Receptor Mimicry (RTB Binders)

Asialofetuin (ASF), which contains 12 terminal galactose residues per molecule [157,158] and binds ricin 1000 times better than monovalent galactose [159], is an excellent in vitro scavenger of ricin. However, it is rapidly cleared from the circulation and therefore cannot be used clinically. Extensive work has been performed to obtain better receptor mimicry. Glycosphingolipid (GSL)-, lactose-, and galactose-based derivatives (Table 3) were found to be potentially good candidates for this manner. The gangliosides GM1 and GM3 protect cells from ricin-induced intoxication [160], while the synthetic GSL analogues beta-lactosyl-ceramide, beta-d-galactosyl ceramide, asialo-GM1 and serum albumin-based neoglycoconjugates were shown to be selective and potent ricin binders in vitro [161,162]. Lactose-based glycopolymers were found to be effective for capturing ricin in a cell-free system, as well as for inhibiting cell binding [163,164]. In a different experimental setting, a synthetic galactose-based surfactant efficiently sequestered ricin from aqueous solution, but due to its water-insolubility, it must be formulized prior to its application for ricin intoxication therapy [165]. A galactose-based biantennary oligosaccharide effectively bound to ricin in a cell-free system [166], whereas a chemically modified glycoprotein containing triantennary N-linked oligosaccharides reduced the cytotoxicity of ricin more than 1000-fold in cultured cells [167]. Additional studies were performed with the closely related protein *Ricinus communis* agglutinin (RCA), demonstrating potent binding to Galβ1-4GlcNAc, with specificity for highly branched glycans containing this structure [168]. EGCG, a potent antioxidant possessing anti-inflammatory properties [109,110], was also suggested to interfere with the binding of RTB to lactose-conjugated sepharose [107].

Although all of these molecules effectively antagonize ricin in vitro or in cell free systems, to our knowledge, there are no data available regarding the in vivo efficacy of anti-ricin receptor mimetic-based small molecules.

3.3.2. Endocytosis Blockers

Research conducted decades ago revealed that the co-incubation of an inhibitor of glycolysis (2-deoxyglucose) and an uncoupler of oxidative phosphorylation (sodium azide, NaN3) potently inhibits ricin endocytosis and protects cells against intoxication, indicating that endocytosis is a critical step in ricin cellular entry [169]. Later work demonstrated that cytochalasin D and the clinically approved drug colchicine selectively inhibit the endocytic uptake of ricin from non-clathrin-coated areas of cell membranes. Furthermore, colchicine reduces the catalytic activity of ricin (protein synthesis arrest) in cell culture [170].

3.3.3. Trafficking Blockers

After internalization into the cells, ricin is transported from early endosomes to the ER via the Golgi apparatus, an entrance pathway termed the "retrograde trafficking route." Several molecules were found to block ricin translocation to the cytosol, e.g., brefeldin A (BFA) [171], 3'-azido-3'-deoxythimidine [172] and mansonone-D [173]. BFA, a fungal antibiotic, which inhibits anterograde vesicular transport by disrupting the Golgi apparatus, is considered to be the first small molecule identified that protects cells from ricin [171]. However, whereas BFA protects cells from the cytotoxicity induced by ricin, it may under some circumstances enhance ricin toxicity in other cell lines [174,175]. In addition, it was recently demonstrated that benzyl alcohol, which is widely used as a food and medical preservative, inhibits ricin membrane trafficking between endosomes and the trans-Golgi network, thus providing protection against ricin-induced cytotoxicity [176].

In the past decade, several high-throughput screens were conducted, including a high-content screen of ~3000 compounds that identified several small molecule candidates that interfered in vitro with the retrograde translocation of ricin or stabilized RTA in the ER [177]. With these screens, the greatest progress in the field of ricin trafficking blockers was recently achieved. Small molecules that selectively block retrograde trafficking at the early endosome/trans Golgi network interface were identified. These highly selective, non-toxic molecules were efficient against pulmonary ricinosis in mice, especially Retro-2 administered prophylactically. This molecule was found to be highly potent, exhibiting bioactivity in the nanomolar range [178]. In a different experimental setting, characterization of a common pharmacophore of retrograde trafficking inhibitors, such as Retro-2 and its achiral analog DA2MT, offered new insights into lead compound identification and optimization for ricin and other RIP antidote development [179]. Additional inhibitors of cellular trafficking are discussed elsewhere [180], and some of the molecules may be potentially effective if proven safe when used against ricin intoxication.

In addition to the trafficking inhibitors mentioned above, Bassik et al. demonstrated that ricin trafficking to the ER was effectively blocked in vitro upon hydroxymethylglutharyl (HMG)–CoA reductase inhibition with atorvastatin, a popular cholesterol-lowering drug [181].

3.3.4. Reductive Activation Inhibitors

A reduction-dependent disassociation of the RTA-RTB inter-subunit disulfide bond is required for the intracellular activation of ricin, namely, the translocation of RTA from the ER to its target site, the cytosol. Several enzymes responsible for this process have been identified, e.g., protein disulfide isomerase (PDI), thioredoxin reductase [182], glutathione disulfide oxidoreductase [183] and TMX, a transmembrane thioredoxin-related protein member of the PDI family [184]. Among these enzymes, thioredoxin reductase and PDI may be inhibited by the clinically approved drugs, auranofin (used therapeutically for rheumatoid arthritis, [185]) and the antibacterial agent bacitracin [186], respectively. Indeed, auranofin significantly inhibits ricin-mediated cytotoxicity [182].

3.3.5. Active Site and RTA Inhibitors

The search for RTA active site inhibitors began decades ago and included purine-based inhibitors and pterin-like and single ring pyrimidine-based derivatives, as extensively discussed elsewhere [10]. It should be mentioned, however, that many of these compounds are extremely insoluble and/or toxic to cells. According to a single publication in the literature, the clinically approved polyamine difluoromethylornithine, a positively charged compound, which was proposed to bind ribosomes and RNA, prolonged the time to death in mice intraperitoneally intoxicated with ricin, likely by blocking the active site of the toxin [58]. In addition, in a recent screening of ~80,000 compounds in vitro, 20 molecules with significant anti-ricin activity were identified, one of which (4-fluorophenyl methyl 2-(furan-2-yl)quinolone-4-carboxylate) exhibited significant therapeutic activity [9]. A computer modeling identified this compound as a ricin active site blocker. More specific active site blockers targeting the ribosome inactivating protein (RIP)-α-sarcin/ricin loop (SRL) interface were found to be effective, conferring up to 20% cell protection against ricin at nanomolar concentrations [187]. In addition, an RTA-ligand RNA aptamer-based approach demonstrated partial protection in a cell-based cytotoxicity assay [188], supporting the potential use of anti-RTA aptamers as ricin inhibitors.

The identification of new classes of RTA inhibitors by virtual screening was also conducted. Two compounds (out of 50,000) showed modest to strong ricin inhibition in a cell-based assay, which was, however, accompanied by some cytotoxicity [189]. In a recent study [190], baicalin, extracted from the plant *Scutellaria baicalensis* and used as a Chinese medical herb, reduced ricin-mediated cytotoxicity in vitro and conferred significant post-exposure protection in mice intraperitoneally exposed to ricin. Baicalin is an RTA inhibitor with a novel mechanism of action. Rather than occupying the active site, it induces toxin oligomerization upon extensive hydrogen bond networking formation with RTA. The potential protective effects of baicalin in pulmonary ricinosis remain to be evaluated.

3.4. Multiple Pathway-Interfering Anti-Ricin Agents

As mentioned earlier, auranofin inhibits the reductive activation of ricin by thioredoxin reductase [182], yet it may also decrease NFκB activation via IκK inhibition [191]. Notably, auranofin should be cautiously used since thioredoxin reductase inhibition is often associated with a concomitant increase in ROS formation [192]. Baicalin, a direct anti-ricin agent that forms oligomers through interactions with the RTA, may also function as an antioxidant [193,194] and an anti-inflammatory agent, as reflected by attenuation of LPS-induced lung injury in rats and mice [195–197]. In addition, resveratrol, an antioxidant [198] that may also act as a NALP3 inhibitor through autophagy enhancement [147], may have a protective effect against ricinosis. The actin depolymerizing agent cytochalasin D, which was reported to inhibit the uptake of ricin [170], may also inhibit NALP3-dependent IL-1β production [199,200]. Amifostine, a potent antioxidant [201], inhibited p53-dependent apoptosis by reducing apoptosis-related gene transcription [202,203]. Colchicine, a selective blocker of the cellular endocytic uptake of ricin [170], is also a microtubule-depolymerizing agent [204] that impairs the mobility and phagocytic activity of neutrophils. In particular colchicine was reported to attenuate phosgene-induced lung injury and to improve mouse survival by reducing neutrophil influx into the lungs even when given 30 min post exposure [205]. Atorvastatin, an HMG–CoA reductase demonstrated to operate as a retrograde trafficking blocker of ricin [181], as well as other HMG–CoA reductase inhibitors, were reported to attenuate non-ricin-mediated acute lung injury parameters, as reflected by reduced levels of proinflammatory cytokines and markers of edema [206–208].

3.5. Drug-Drug Combinational Treatment

Due to the multiplicity of ricin-induced damage pathways and effectors, an improved therapeutic outcome may be achieved upon combinational treatment of several drugs. Pathogenesis characterization of any treatment should assist tailoring the optimum additive countermeasure for efficient damage coverage. Clearly, there are many combinational treatment options, and therefore, the use of concomitant medication should be established wisely. For example, ZAK inhibition attenuates p38 and JNK activation in cell culture without any influence on NALP3 activation [21]. Therefore, if relevant to in vivo settings, ZAK inhibition should be combined with NALP3 inhibitors or anakinra following pulmonary ricin intoxication. As mentioned above, auranofin is a good candidate for ricin intoxication therapy due to its dual inhibition activity on both NFκB signaling and thioredoxin reductase. However, as mentioned earlier auranofin sensitizes cells to oxidative stress-mediated damage [192]. Accordingly, the co-administration of an antioxidative compound could improve treatment outcomes. NFκB knockdown in vitro [38] decreases the levels of several proinflammatory mediators but not IL-6. Hence, combined NFκB inhibition with anti-IL-6 medication is worth attempting. In the same experiment, incubating cells with anti-TNFα did not have any influence on NFκB activation, implying a beneficial effect of NFκB/TNFα combined inhibition, pending the relevance of this finding to pulmonary ricin intoxication in vivo. Miller et al. reported that some NFκB inhibitors promote caspase 3/7 activation [134]; therefore, targeting these effectors in parallel to NFκB inhibition may be advantageous. In addition to the aforementioned classical drug-drug combination, inhibition of these stress-related processes with the co-administration of anti-ricin small molecules, i.e., trafficking blockers may be useful.

3.6. Antitoxin-Drug Combinational Treatment

Administration of anti-ricin antibodies is less effective if applied at late time points following intoxication [83,84]. However, it was previously shown that the co-administration of antitoxin together with various anti-inflammatory drugs could significantly expand the therapeutic time window [11,12]. In the first study, the survival rates of mice treated with anti-ricin antibodies at 24 h post-pulmonary intoxication were considerably improved by the co-administration of doxycycline [11], an antibacterial agent with broad anti-inflammatory activity [209–213]. Doxycycline promoted a significant reduction of proinflammatory cytokines and damage mediators (IL-1β, IL-6, XO, VEGF and pulmonary ChE) in intoxicated mouse BALFs. Because doxycycline treatment *per se* did not confer protection, an "add on" effect obtained following antitoxin-doxycycline combination treatment was monitored. Indeed, doxycycline or antitoxin administration as monotherapies did not decrease the levels of BALF MMP-9; however, a combined drug-antitoxin treatment resulted in a significant reduction of this mediator [11]. The second efficient combinational drug-antitoxin-based therapy was achieved using ciprofloxacin as the co-administered drug. Ciprofloxacin is an antibacterial agent that also possesses potent immunomodulatory properties, a feature that is mainly associated with decreased synthesis of proinflammatory cytokines [214]. Co-administration of ciprofloxacin with antitoxin dramatically improved survival through effective attenuation of neutrophil infiltration and edema. These findings illustrated that ciprofloxacin led to a significantly decreased proinflammatory cytokine response.

Thus, in BALFs of ciprofloxacin-treated ricin-intoxicated mice, IL-6 levels decreased by ~90%, while a significant increase in levels of the anti-inflammatory cytokine IL-10 was observed. Pulmonary levels of the damage markers XO, protein, ChE and Evans Blue dye extravasation were also significantly attenuated upon ciprofloxacin treatment [12]. It appears that ricin, by virtue of being highly inflammatory, persistently stimulates acute inflammatory responses with which doxycycline or ciprofloxacin alone cannot cope. Antitoxin-based treatment is therefore required to halt any further propagation of proinflammatory signals by virtue of toxin neutralization, while the combined drugs exert their beneficial effects by dampening the inflammation-related assaults that have already developed.

Indeed, although treatment with the drugs alone reduced inflammation-related factors to a considerable extent, the values remained higher than in control mice. It appears, therefore, that keeping the inflammation at bay promotes the expansion of the therapeutic time window for antitoxin intervention. Supporting this argument, the time to death was delayed in IL-1-knockout mice compared with naïve mice following intratracheal ricin intoxication [39]. It is important to mention, that we have also examined the combination of antitoxin with dexamethasone, a highly potent corticosteroid, as it may be assumed that patients exhibiting an unknown inflammatory syndrome will be treated with these agents long before the specific cause is known. Surprisingly, although steroids possess broad anti-inflammatory activities and seem ideal for use in suppressing the ricin-induced inflammatory pathways, co-administration of dexamethasone did not confer any improvement in surviving ratios, unless given before, or shortly after the onset of intoxication. Under these circumstances we believe that non-steroidal-based immunomodulators are better candidates for combinational antitoxin-drug treatment. Nevertheless, we do believe that the usage of steroids in ricin-intoxicated victims, whether the treatment is given early, or at late times following exposure, should be beneficial. This argument is supported by the fact that treating mice with dexamethasone in combination with the antitoxin-doxycycline regime, conferred enhanced protection as compared to antitoxin-doxycycline alone [11].

Taken together, drug-antitoxin concomitant medication is emerging as the best approach for the treatment of pulmonary ricinosis, and the drugs and small-molecules that were reviewed above may help to improve this therapy.

4. Summary and Future Prospects

To date, the only applicable countermeasure for pulmonary ricinosis at clinically relevant time points following exposure is antitoxin administration. Small molecules inhibiting the intracellular trafficking of ricin (e.g., Retro-2) or interacting with RTA (baicalin) were shown to be efficient in vivo only when administered shortly after intoxication (the latter was not tested in pulmonary ricin exposure). Drug-antitoxin based therapy improves treatment outcome, however, the timing of each pharmacological intervention should be carefully chosen. For example, treatment with anti-IL-1β or anti-TNFα should be optimal when administered at early times following intoxication, while immunomodulators and stress-activated pathways inhibitors may be beneficial also at later time points. Anti-ricin small molecules should be administered as soon as possible following intoxication.

To improve treatment outcomes, it is highly reasonable to simultaneously target several stress pathways, as well as cellular components related to the toxicity of ricin (binding, trafficking, catalytic activity etc.), as long as the additive toxicity of this combinational treatment does not pose a problem. Even small molecules that were not demonstrated to confer protection in vivo may be beneficial when included in an anti-ricin drug cocktail.

In addition to the countermeasures discussed in this review, several other therapeutic strategies should be considered following pulmonary ricinosis. Because this pathology complies with the clinical criteria for ARDS, pharmacologic treatments tested for this syndrome should be evaluated. As it is suggested that treatment against pulmonary ricin intoxication includes an antitoxin, drugs which were tested for ARDS and demonstrated only partial efficacy in clinical trials should be reconsidered as components of a combined drug-antitoxin treatment modality. The relevant pharmacological options include corticosteroids (as mentioned above), vasodilators, decreased alveolar surface tension, thromboxane synthase and 5-lipoxygenase inhibitors, antioxidants, immunonutrition, increased clearance of alveolar edema, enhanced repair of the alveolar epithelium, anticoagulants, hematopoietic colony-stimulating factors, and prevention of fibrosis, all of which have been extensively reviewed elsewhere [215–218].

Targeting the cytokine storm is also an attractive strategy to manipulate the overwhelming inflammation process that develops in the lungs following pulmonary ricin intoxication. Pharmacological strategies for this purpose were exhaustively discussed by D'Elia et al. and include stimulation of the cholinergic anti-inflammatory pathway, e.g., by GTS-21, which is a selective α7-acetylcholine (ACh) nicotinic receptor agonist found in clinical trials, and CNI-1495, an α7ACh agonist under pre-clinical testing. Prostaglandin-, cyclooxygenase- and platelet-activating factor-inhibitors, as well as chemokine manipulation, may also attenuate cytokine storm. The active resolution of tissue damage by resolvins, lipoxins and protectins is also relevant [219]. In particular, the activation of the cholinergic anti-inflammatory pathway following nicotine administration two hours after systemic ricin exposure significantly delayed and reduced mortality in mice [220], although the relevance to pulmonary ricinosis remains to be elucidated.

Neutrophils are considered a major hallmark of pulmonary ricinosis. Aggressive or prolonged neutrophil responses result in deleterious inflammatory conditions and tissue destruction. Potential new drug candidates to control of neutrophil activity were reviewed by Burgos et al. [135]. It is of great importance to follow the progression in this field and consider using agents counteracting neutrophil activity in the course of pulmonary ricinosis. Targeting p38 may be of a specific interest because it is involved in both neutrophil migration and chemotaxis in vivo [221] and in the ribotoxic stress response, as mentioned above. Comprehensive understanding of intravascular danger signals (e.g., formyl-peptide signals released from necrotic cells) that guide neutrophils to the site of sterile injury should be harnessed in an effort to attenuate neutrophil-derived injury [222].

Pharmacological interventions against systemic capillary leak syndrome were reviewed by Druey et al. [223]. Several clinically used drugs were offered to alleviate this syndrome, and these should be tested for their beneficial effects in pulmonary ricinosis.

Immune selective anti-inflammatory derivatives (ImSAIDs), based on salivary gland-derived peptides, should also be considered for evaluation in pulmonary ricin exposure. The tripeptide ImSAID Phe-Glu-Gly (FEG) attenuated both systemic [224] and pulmonary [225,226] inflammation. FEG significantly ameliorated endotoxin-induced lung injury in both a prophylactic and therapeutic manner in rats [227].

In addition, regenerative therapies should also be applied for the treatment of lung pathology following ricin intoxication. These therapies include the aforementioned resolving/lipoxin/protectin-based therapy, as well as keratinocyte growth factor (KGF)-mediated enhanced repair of the alveolar epithelium [216,228,229]. Cell therapy using a mesenchymal stromal cell-based approach, reviewed by Johnson et al., can also be performed, depending on its clinical progress [230].

Small molecules that interfere with the cell trafficking of ricin were comprehensively discussed in this review. Although considerable progress was achieved following the development of Retro-2, additional work should be performed to improve the effectiveness of trafficking blockers. Bassik et al. [181] used a mammalian genetic approach to reveal pathways underlying ricin susceptibility. Many intracellular factors, familiar as well as unexpected, were found to be involved in the toxicity of ricin. Knockdown of several of these factors was strongly protective, while knockdown of others, increased the sensitivity to ricin. Profound sensitization to ricin was found upon depletion of coat protein I (COPI) components, which are normally involved in retrograde transport. This is presumably due to the upregulation of compensatory alternative pathways or to the fact that COPI normally functions in transport steps that divert ricin from ER. Additionally, in contrast to all depleted ribosomal components, which sensitize cells to ricin, the knockdown of the ribosomal proteins RPS25 and ILF2/3, whom interact with RPS25 conferred ricin resistance. The depletion of the two poorly characterized genes *WDR11* and *C17orf75* sensitized cells to ricin. Specifically, *WDR11* was suggested to participate in autophagy-mediated ricin degradation. The clinical aspects of this work have not been elucidated, but it is reasonable to assume that targeting novel proteins that may be related to ricin susceptibility should be utilized in the future to develop novel anti-ricin therapeutic strategies.

In conclusion, anti-ricin post-exposure treatment should include the following: antitoxin, as an obligatory component, combined with (i) a disease modifying countermeasure (i.e., anti-inflammatory/immunomodulator agent); and/or (ii) a small molecule anti-ricin inhibitor (i.e., cell trafficking blocker). Pending clinical progression, the treatment at later stages may include regenerative therapies, gene- and cell-based treatments, and any other beneficial therapy that remains to be discovered. Optimum treatment should be tailored based on thorough pathological studies, specifically focusing on damage mediators that were not effectively attenuated following a treatment of choice.

Conflicts of Interest: The authors declare no conflicts of interest.

References

1. Olsnes, S.; Kozlov, J.V. Ricin. *Toxicon* **2001**, *39*, 1723–1728. [CrossRef]
2. Endo, Y.; Mitsui, K.; Motizuki, M.; Tsurugi, K. The mechanism of action of ricin and related toxic lectins on eukaryotic ribosomes. The site and the characteristics of the modification in 28 s ribosomal rna caused by the toxins. *J. Biol. Chem.* **1987**, *262*, 5908–5912. [PubMed]
3. Colombatti, M.; Johnson, V.G.; Skopicki, H.A.; Fendley, B.; Lewis, M.S.; Youle, R.J. Identification and characterization of a monoclonal antibody recognizing a galactose-binding domain of the toxin ricin. *J. Immunol.* **1987**, *138*, 3339–3344. [PubMed]
4. Maddaloni, M.; Cooke, C.; Wilkinson, R.; Stout, A.V.; Eng, L.; Pincus, S.H. Immunological characteristics associated with the protective efficacy of antibodies to ricin. *J. Immunol.* **2004**, *172*, 6221–6228. [CrossRef] [PubMed]

5. Noy-Porat, T.; Alcalay, R.; Epstein, E.; Sabo, T.; Kronman, C.; Mazor, O. Extended therapeutic window for post-exposure treatment of ricin intoxication conferred by the use of high-affinity antibodies. *Toxicon* **2017**, *127*, 100–105. [CrossRef] [PubMed]

6. Noy-Porat, T.; Rosenfeld, R.; Ariel, N.; Epstein, E.; Alcalay, R.; Zvi, A.; Kronman, C.; Ordentlich, A.; Mazor, O. Isolation of anti-ricin protective antibodies exhibiting high affinity from immunized non-human primates. *Toxins* **2016**, *8*, 64. [CrossRef] [PubMed]

7. Gupta, N.; Noel, R.; Goudet, A.; Hinsinger, K.; Michau, A.; Pons, V.; Abdelkafi, H.; Secher, T.; Shima, A.; Shtanko, O.; et al. Inhibitors of retrograde trafficking active against ricin and shiga toxins also protect cells from several viruses, leishmania and chlamydiales. *Chemico-Biol. Interact.* **2017**, *267*, 96–103. [CrossRef] [PubMed]

8. Wahome, P.G.; Ahlawat, S.; Mantis, N.J. Identification of small molecules that suppress ricin-induced stress-activated signaling pathways. *PLoS ONE* **2012**, *7*, e49075. [CrossRef] [PubMed]

9. Wahome, P.G.; Bai, Y.; Neal, L.M.; Robertus, J.D.; Mantis, N.J. Identification of small-molecule inhibitors of ricin and shiga toxin using a cell-based high-throughput screen. *Toxicon* **2010**, *56*, 313–323. [CrossRef] [PubMed]

10. Wahome, P.G.; Robertus, J.D.; Mantis, N.J. Small-molecule inhibitors of ricin and shiga toxins. *Curr. Top. Microbiol. Immunol.* **2012**, *357*, 179–207. [PubMed]

11. Gal, Y.; Mazor, O.; Alcalay, R.; Seliger, N.; Aftalion, M.; Sapoznikov, A.; Falach, R.; Kronman, C.; Sabo, T. Antibody/doxycycline combined therapy for pulmonary ricinosis: Attenuation of inflammation improves survival of ricin-intoxicated mice. *Toxicol. Rep.* **2014**, *1*, 496–504. [CrossRef] [PubMed]

12. Gal, Y.; Sapoznikov, A.; Falach, R.; Ehrlich, S.; Aftalion, M.; Sabo, T.; Kronman, C. Potent antiedematous and protective effects of ciprofloxacin in pulmonary ricinosis. *Antimicrob. Agents Chemother.* **2016**, *60*, 7153–7158. [PubMed]

13. DaSilva, L.; Cote, D.; Roy, C.; Martinez, M.; Duniho, S.; Pitt, M.L.; Downey, T.; Dertzbaugh, M. Pulmonary gene expression profiling of inhaled ricin. *Toxicon* **2003**, *41*, 813–822. [CrossRef]

14. Poli, M.A.; Rivera, V.R.; Pitt, M.L.; Vogel, P. Aerosolized specific antibody protects mice from lung injury associated with aerosolized ricin exposure. *Toxicon* **1996**, *34*, 1037–1044. [CrossRef]

15. Roy, C.J.; Hale, M.; Hartings, J.M.; Pitt, L.; Duniho, S. Impact of inhalation exposure modality and particle size on the respiratory deposition of ricin in balb/c mice. *Inhal. Toxicol.* **2003**, *15*, 619–638. [CrossRef] [PubMed]

16. Sapoznikov, A.; Falach, R.; Mazor, O.; Alcalay, R.; Gal, Y.; Seliger, N.; Sabo, T.; Kronman, C. Diverse profiles of ricin-cell interactions in the lung following intranasal exposure to ricin. *Toxins* **2015**, *7*, 4817–4831. [CrossRef] [PubMed]

17. Wilhelmsen, C.L.; Pitt, M.L. Lesions of acute inhaled lethal ricin intoxication in rhesus monkeys. *Vet. Pathol.* **1996**, *33*, 296–302. [CrossRef] [PubMed]

18. Katalan, S.; Falach, R.; Rosner, A.; Goldvaser, M.; Brosh-Nissimov, T.; Dvir, A.; Mizrachi, A.; Goren, O.; Cohen, B.; Gal, Y.; et al. A novel swine model of ricin-induced acute respiratory distress syndrome. *Dis. Model. Mech.* **2017**, *10*, 173–183. [CrossRef] [PubMed]

19. Iordanov, M.S.; Pribnow, D.; Magun, J.L.; Dinh, T.H.; Pearson, J.A.; Chen, S.L.; Magun, B.E. Ribotoxic stress response: Activation of the stress-activated protein kinase jnk1 by inhibitors of the peptidyl transferase reaction and by sequence-specific rna damage to the alpha-sarcin/ricin loop in the 28s rrna. *Mol. Cell. Biol.* **1997**, *17*, 3373–3381. [CrossRef] [PubMed]

20. Jetzt, A.E.; Cheng, J.S.; Tumer, N.E.; Cohick, W.S. Ricin a-chain requires c-jun n-terminal kinase to induce apoptosis in nontransformed epithelial cells. *Int. J. Biochem. Cell Biol.* **2009**, *41*, 2503–2510. [CrossRef] [PubMed]

21. Lindauer, M.; Wong, J.; Magun, B. Ricin toxin activates the nalp3 inflammasome. *Toxins* **2010**, *2*, 1500–1514. [CrossRef] [PubMed]

22. Higuchi, S.; Tamura, T.; Oda, T. Cross-talk between the pathways leading to the induction of apoptosis and the secretion of tumor necrosis factor-alpha in ricin-treated raw 264.7 cells. *J. Biochem.* **2003**, *134*, 927–933. [CrossRef] [PubMed]

23. Jandhyala, D.M.; Ahluwalia, A.; Obrig, T.; Thorpe, C.M. Zak: A map3kinase that transduces shiga toxin- and ricin-induced proinflammatory cytokine expression. *Cell. Microbiol.* **2008**, *10*, 1468–1477. [CrossRef] [PubMed]

24. Tamura, T.; Sadakata, N.; Oda, T.; Muramatsu, T. Role of zinc ions in ricin-induced apoptosis in u937 cells. *Toxicol. Lett.* **2002**, *132*, 141–151. [CrossRef]

25. Jandhyala, D.M.; Wong, J.; Mantis, N.J.; Magun, B.E.; Leong, J.M.; Thorpe, C.M. A novel zak knockout mouse with a defective ribotoxic stress response. *Toxins* **2016**, *8*, 259. [CrossRef] [PubMed]

26. Zhou, H.R.; He, K.; Landgraf, J.; Pan, X.; Pestka, J.J. Direct activation of ribosome-associated double-stranded rna-dependent protein kinase (pkr) by deoxynivalenol, anisomycin and ricin: A new model for ribotoxic stress response induction. *Toxins* **2014**, *6*, 3406–3425. [CrossRef] [PubMed]

27. Chu, W.M.; Ostertag, D.; Li, Z.W.; Chang, L.; Chen, Y.; Hu, Y.; Williams, B.; Perrault, J.; Karin, M. Jnk2 and ikkbeta are required for activating the innate response to viral infection. *Immunity* **1999**, *11*, 721–731. [CrossRef]

28. Koromilas, A.E.; Roy, S.; Barber, G.N.; Katze, M.G.; Sonenberg, N. Malignant transformation by a mutant of the ifn-inducible dsrna-dependent protein kinase. *Science* **1992**, *257*, 1685–1689. [CrossRef] [PubMed]

29. Lengyel, P. Tumor-suppressor genes: News about the interferon connection. *Proc. Natl. Acad. Sci. USA* **1993**, *90*, 5893–5895. [CrossRef] [PubMed]

30. Hoffmann, A.; Natoli, G.; Ghosh, G. Transcriptional regulation via the nf-kappab signaling module. *Oncogene* **2006**, *25*, 6706–6716. [CrossRef] [PubMed]

31. Karin, M. Nuclear factor-kappab in cancer development and progression. *Nature* **2006**, *441*, 431–436. [CrossRef] [PubMed]

32. Liu, S.F.; Malik, A.B. Nf-kappa b activation as a pathological mechanism of septic shock and inflammation. *Am. J. Physiol. Lung Cell. Mol. Physiol.* **2006**, *290*, L622–L645. [CrossRef] [PubMed]

33. Blackwell, T.S.; Holden, E.P.; Blackwell, T.R.; DeLarco, J.E.; Christman, J.W. Cytokine-induced neutrophil chemoattractant mediates neutrophilic alveolitis in rats: Association with nuclear factor kappa b activation. *Am. J. Respir. Cell Mol. Biol.* **1994**, *11*, 464–472. [CrossRef] [PubMed]

34. Haddad, E.B.; Salmon, M.; Koto, H.; Barnes, P.J.; Adcock, I.; Chung, K.F. Ozone induction of cytokine-induced neutrophil chemoattractant (cinc) and nuclear factor-kappa b in rat lung: Inhibition by corticosteroids. *FEBS Lett.* **1996**, *379*, 265–268. [CrossRef]

35. Sacks, M.; Gordon, J.; Bylander, J.; Porter, D.; Shi, X.L.; Castranova, V.; Kaczmarczyk, W.; Van Dyke, K.; Reasor, M.J. Silica-induced pulmonary inflammation in rats: Activation of nf-kappa b and its suppression by dexamethasone. *Biochem. Biophys. Res. Commun.* **1998**, *253*, 181–184. [CrossRef] [PubMed]

36. Sadikot, R.T.; Jansen, E.D.; Blackwell, T.R.; Zoia, O.; Yull, F.; Christman, J.W.; Blackwell, T.S. High-dose dexamethasone accentuates nuclear factor-kappa b activation in endotoxin-treated mice. *Am. J. Respir. Crit. Care Med.* **2001**, *164*, 873–878. [CrossRef] [PubMed]

37. Wong, J.; Korcheva, V.; Jacoby, D.B.; Magun, B. Intrapulmonary delivery of ricin at high dosage triggers a systemic inflammatory response and glomerular damage. *Am. J. Pathol.* **2007**, *170*, 1497–1510. [CrossRef] [PubMed]

38. Wong, J.; Korcheva, V.; Jacoby, D.B.; Magun, B.E. Proinflammatory responses of human airway cells to ricin involve stress-activated protein kinases and nf-kappab. *Am. J. Physiol. Lung Cell. Mol. Physiol.* **2007**, *293*, L1385–L1394. [CrossRef] [PubMed]

39. Lindauer, M.L.; Wong, J.; Iwakura, Y.; Magun, B.E. Pulmonary inflammation triggered by ricin toxin requires macrophages and il-1 signaling. *J. Immunol.* **2009**, *183*, 1419–1426. [CrossRef] [PubMed]

40. Finsterbusch, M.; Voisin, M.B.; Beyrau, M.; Williams, T.J.; Nourshargh, S. Neutrophils recruited by chemoattractants in vivo induce microvascular plasma protein leakage through secretion of tnf. *J. Exp. Med.* **2014**, *211*, 1307–1314. [CrossRef] [PubMed]

41. Mukhopadhyay, S.; Hoidal, J.R.; Mukherjee, T.K. Role of tnfalpha in pulmonary pathophysiology. *Respir. Res.* **2006**, *7*, 125. [CrossRef] [PubMed]

42. Fattori, E.; Cappelletti, M.; Costa, P.; Sellitto, C.; Cantoni, L.; Carelli, M.; Faggioni, R.; Fantuzzi, G.; Ghezzi, P.; Poli, V. Defective inflammatory response in interleukin 6-deficient mice. *J. Exp. Med.* **1994**, *180*, 1243–1250. [CrossRef] [PubMed]

43. Parsons, P.E.; Eisner, M.D.; Thompson, B.T.; Matthay, M.A.; Ancukiewicz, M.; Bernard, G.R.; Wheeler, A.P. Lower tidal volume ventilation and plasma cytokine markers of inflammation in patients with acute lung injury. *Crit. Care Med.* **2005**, *33*, 1–6, discussion 230–232. [CrossRef] [PubMed]

44. Ranieri, V.M.; Suter, P.M.; Tortorella, C.; De Tullio, R.; Dayer, J.M.; Brienza, A.; Bruno, F.; Slutsky, A.S. Effect of mechanical ventilation on inflammatory mediators in patients with acute respiratory distress syndrome: A randomized controlled trial. *JAMA* **1999**, *282*, 54–61. [CrossRef] [PubMed]

45. Remick, D.G.; Bolgos, G.; Copeland, S.; Siddiqui, J. Role of interleukin-6 in mortality from and physiologic response to sepsis. *Infect. Immun.* **2005**, *73*, 2751–2757. [CrossRef] [PubMed]

46. Arbibe, L.; Koumanov, K.; Vial, D.; Rougeot, C.; Faure, G.; Havet, N.; Longacre, S.; Vargaftig, B.B.; Bereziat, G.; Voelker, D.R.; et al. Generation of lyso-phospholipids from surfactant in acute lung injury is mediated by type-ii phospholipase a2 and inhibited by a direct surfactant protein a-phospholipase a2 protein interaction. *J. Clin. Investig.* **1998**, *102*, 1152–1160. [CrossRef] [PubMed]

47. Holm, B.A.; Keicher, L.; Liu, M.Y.; Sokolowski, J.; Enhorning, G. Inhibition of pulmonary surfactant function by phospholipases. *J. Appl. Physiol.* **1991**, *71*, 317–321. [PubMed]

48. Touqui, L.; Wu, Y.Z. Interaction of secreted phospholipase a2 and pulmonary surfactant and its pathophysiological relevance in acute respiratory distress syndrome. *Acta Pharmacol. Sin.* **2003**, *24*, 1292–1296. [PubMed]

49. Lee, Y.M.; Hybertson, B.M.; Terada, L.S.; Repine, A.J.; Cho, H.G.; Repine, J.E. Mepacrine decreases lung leak in rats given interleukin-1 intratracheally. *Am. J. Respir. Crit. Care Med.* **1997**, *155*, 1624–1628. [CrossRef] [PubMed]

50. Munoz, N.M.; Meliton, A.Y.; Meliton, L.N.; Dudek, S.M.; Leff, A.R. Secretory group v phospholipase a2 regulates acute lung injury and neutrophilic inflammation caused by lps in mice. *Am. J. Physiol. Lung Cell. Mol. Physiol.* **2009**, *296*, L879–L887. [CrossRef] [PubMed]

51. Kaner, R.J.; Ladetto, J.V.; Singh, R.; Fukuda, N.; Matthay, M.A.; Crystal, R.G. Lung overexpression of the vascular endothelial growth factor gene induces pulmonary edema. *Am. J. Respir. Cell Mol. Biol.* **2000**, *22*, 657–664. [CrossRef] [PubMed]

52. Carney, D.E.; McCann, U.G.; Schiller, H.J.; Gatto, L.A.; Steinberg, J.; Picone, A.L.; Nieman, G.F. Metalloproteinase inhibition prevents acute respiratory distress syndrome. *J. Surg. Res.* **2001**, *99*, 245–252. [CrossRef] [PubMed]

53. Soccal, P.M.; Gasche, Y.; Pache, J.C.; Schneuwly, O.; Slosman, D.O.; Morel, D.R.; Spiliopoulos, A.; Suter, P.M.; Nicod, L.P. Matrix metalloproteinases correlate with alveolar-capillary permeability alteration in lung ischemia-reperfusion injury. *Transplantation* **2000**, *70*, 998–1005. [CrossRef] [PubMed]

54. Grosso, M.A.; Brown, J.M.; Viders, D.E.; Mulvin, D.W.; Banerjee, A.; Velasco, S.E.; Repine, J.E.; Harken, A.H. Xanthine oxidase-derived oxygen radicals induce pulmonary edema via direct endothelial cell injury. *J. Surg. Res.* **1989**, *46*, 355–360. [CrossRef]

55. Komaki, Y.; Sugiura, H.; Koarai, A.; Tomaki, M.; Ogawa, H.; Akita, T.; Hattori, T.; Ichinose, M. Cytokine-mediated xanthine oxidase upregulation in chronic obstructive pulmonary disease's airways. *Pulm. Pharmacol. Ther.* **2005**, *18*, 297–302. [CrossRef] [PubMed]

56. Wright, R.M.; Ginger, L.A.; Kosila, N.; Elkins, N.D.; Essary, B.; McManaman, J.L.; Repine, J.E. Mononuclear phagocyte xanthine oxidoreductase contributes to cytokine-induced acute lung injury. *Am. J. Respir. Cell Mol. Biol.* **2004**, *30*, 479–490. [CrossRef] [PubMed]

57. Kumar, O.; Sugendran, K.; Vijayaraghavan, R. Oxidative stress associated hepatic and renal toxicity induced by ricin in mice. *Toxicon* **2003**, *41*, 333–338. [CrossRef]

58. Muldoon, D.F.; Bagchi, D.; Hassoun, E.A.; Stohs, S.J. The modulating effects of tumor necrosis factor alpha antibody on ricin-induced oxidative stress in mice. *J. Biochem. Toxicol.* **1994**, *9*, 311–318. [CrossRef] [PubMed]

59. Muldoon, D.F.; Hassoun, E.A.; Stohs, S.J. Ricin-induced hepatic lipid peroxidation, glutathione depletion, and DNA single-strand breaks in mice. *Toxicon* **1992**, *30*, 977–984. [CrossRef]

60. Muldoon, D.F.; Hassoun, E.A.; Stohs, S.J. Role of iron in ricin-induced lipid peroxidation and superoxide production. *Res. Commun. Mol. Pathol. Pharmacol.* **1996**, *92*, 107–118. [PubMed]

61. Zhou, R.; Tardivel, A.; Thorens, B.; Choi, I.; Tschopp, J. Thioredoxin-interacting protein links oxidative stress to inflammasome activation. *Nat. Immunol.* **2010**, *11*, 136–140. [CrossRef] [PubMed]

62. Oda, T.; Iwaoka, J.; Komatsu, N.; Muramatsu, T. Involvement of n-acetylcysteine-sensitive pathways in ricin-induced apoptotic cell death in u937 cells. *Biosci. Biotechnol. Biochem.* **1999**, *63*, 341–348. [CrossRef] [PubMed]

63. Blouquit, S.; Sari, A.; Lombet, A.; D'Herbomez, M.; Naline, E.; Matran, R.; Chinet, T. Effects of endothelin-1 on epithelial ion transport in human airways. *Am. J. Respir. Cell Mol. Biol.* **2003**, *29*, 245–251. [CrossRef] [PubMed]

64. Comellas, A.P.; Briva, A.; Dada, L.A.; Butti, M.L.; Trejo, H.E.; Yshii, C.; Azzam, Z.S.; Litvan, J.; Chen, J.; Lecuona, E.; et al. Endothelin-1 impairs alveolar epithelial function via endothelial etb receptor. *Am. J. Respir. Crit. Care Med.* **2009**, *179*, 113–122. [CrossRef] [PubMed]

65. Jain, R.; Shaul, P.W.; Borok, Z.; Willis, B.C. Endothelin-1 induces alveolar epithelial-mesenchymal transition through endothelin type a receptor-mediated production of tgf-beta1. *Am. J. Respir. Cell Mol. Biol.* **2007**, *37*, 38–47. [CrossRef] [PubMed]

66. Frank, J.A.; Briot, R.; Lee, J.W.; Ishizaka, A.; Uchida, T.; Matthay, M.A. Physiological and biochemical markers of alveolar epithelial barrier dysfunction in perfused human lungs. *Am. J. Physiol. Lung Cell. Mol. Physiol.* **2007**, *293*, L52–L59. [CrossRef] [PubMed]

67. Marasciulo, F.L.; Montagnani, M.; Potenza, M.A. Endothelin-1: The yin and yang on vascular function. *Curr. Med. Chem.* **2006**, *13*, 1655–1665. [CrossRef] [PubMed]

68. Nakano, Y.; Tasaka, S.; Saito, F.; Yamada, W.; Shiraishi, Y.; Ogawa, Y.; Koh, H.; Hasegawa, N.; Fujishima, S.; Hashimoto, S.; et al. Endothelin-1 level in epithelial lining fluid of patients with acute respiratory distress syndrome. *Respirology* **2007**, *12*, 740–743. [CrossRef] [PubMed]

69. Kuzkov, V.V.; Kirov, M.Y.; Sovershaev, M.A.; Kuklin, V.N.; Suborov, E.V.; Waerhaug, K.; Bjertnaes, L.J. Extravascular lung water determined with single transpulmonary thermodilution correlates with the severity of sepsis-induced acute lung injury. *Crit. Care Med.* **2006**, *34*, 1647–1653. [CrossRef] [PubMed]

70. Berger, M.M.; Rozendal, C.S.; Schieber, C.; Dehler, M.; Zugel, S.; Bardenheuer, H.J.; Bartsch, P.; Mairbaurl, H. The effect of endothelin-1 on alveolar fluid clearance and pulmonary edema formation in the rat. *Anesthesia Analg.* **2009**, *108*, 225–231. [CrossRef] [PubMed]

71. Brown, R.F.; White, D.E. Ultrastructure of rat lung following inhalation of ricin aerosol. *Int. J. Exp. Pathol.* **1997**, *78*, 267–276. [CrossRef] [PubMed]

72. Hughes, J.N.; Lindsay, C.D.; Griffiths, G.D. Morphology of ricin and abrin exposed endothelial cells is consistent with apoptotic cell death. *Hum. Exp. Toxicol.* **1996**, *15*, 443–451. [CrossRef] [PubMed]

73. Lindstrom, A.L.; Erlandsen, S.L.; Kersey, J.H.; Pennell, C.A. An in vitro model for toxin-mediated vascular leak syndrome: Ricin toxin a chain increases the permeability of human endothelial cell monolayers. *Blood* **1997**, *90*, 2323–2334. [PubMed]

74. Gonzalez, T.V.; Farrant, S.A.; Mantis, N.J. Ricin induces il-8 secretion from human monocyte/macrophages by activating the p38 map kinase pathway. *Mol. Immunol.* **2006**, *43*, 1920–1923. [CrossRef] [PubMed]

75. Hassoun, E.; Wang, X. Ricin-induced toxicity in the macrophage j744a.1 cells: The role of tnf-alpha and the modulation effects of tnf-alpha polyclonal antibody. *J. Biochem. Mol. Toxicol.* **2000**, *14*, 95–101. [CrossRef]

76. Hu, R.; Zhai, Q.; Liu, W.; Liu, X. An insight into the mechanism of cytotoxicity of ricin to hepatoma cell: Roles of bcl-2 family proteins, caspases, ca(2+)-dependent proteases and protein kinase c. *J. Cell. Biochem.* **2001**, *81*, 583–593. [CrossRef] [PubMed]

77. Wu, Y.H.; Shih, S.F.; Lin, J.Y. Ricin triggers apoptotic morphological changes through caspase-3 cleavage of bat3. *J. Biol. Chem.* **2004**, *279*, 19264–19275. [CrossRef] [PubMed]

78. Soler-Rodriguez, A.M.; Ghetie, M.A.; Oppenheimer-Marks, N.; Uhr, J.W.; Vitetta, E.S. Ricin a-chain and ricin a-chain immunotoxins rapidly damage human endothelial cells: Implications for vascular leak syndrome. *Exp. Cell Res.* **1993**, *206*, 227–234. [CrossRef] [PubMed]

79. Yermakova, A.; Mantis, N.J. Protective immunity to ricin toxin conferred by antibodies against the toxin's binding subunit (rtb). *Vaccine* **2011**, *29*, 7925–7935. [CrossRef] [PubMed]

80. Neal, L.M.; O'Hara, J.; Brey, R.N., 3rd; Mantis, N.J. A monoclonal immunoglobulin g antibody directed against an immunodominant linear epitope on the ricin a chain confers systemic and mucosal immunity to ricin. *Infect. Immun.* **2010**, *78*, 552–561. [CrossRef] [PubMed]

81. Li, X.P.; Chiou, J.C.; Remacha, M.; Ballesta, J.P.; Tumer, N.E. A two-step binding model proposed for the electrostatic interactions of ricin a chain with ribosomes. *Biochemistry* **2009**, *48*, 3853–3863. [CrossRef] [PubMed]

82. O'Hara, J.M.; Yermakova, A.; Mantis, N.J. Immunity to ricin: Fundamental insights into toxin-antibody interactions. In *Ricin and Shiga Toxins*; Springer: Berlin, Germany, 2011; Volume 357, pp. 209–241.

83. Griffiths, G.D.; Phillips, G.J.; Holley, J. Inhalation toxicology of ricin preparations: Animal models, prophylactic and therapeutic approaches to protection. *Inhal. Toxicol.* **2007**, *19*, 873–887. [CrossRef] [PubMed]

84. Pratt, T.S.; Pincus, S.H.; Hale, M.L.; Moreira, A.L.; Roy, C.J.; Tchou-Wong, K.M. Oropharyngeal aspiration of ricin as a lung challenge model for evaluation of the therapeutic index of antibodies against ricin a-chain for post-exposure treatment. *Exp. Lung Res.* **2007**, *33*, 459–481. [CrossRef] [PubMed]

85. Israeli, O.; Falach, R.; Sapoznikov, A.; Gal, Y.; Shifman, O.; Ehrlich, S.; Aftalion, M.; Beth-Din, A.; Kronman, C.; Sabo, T. Determination of ricin intoxication in biological samples by monitoring depurinated 28s rrna in a unique reverse transcription-ligase-polymerase chain reaction assay. *Forensic Toxicol.* **2017**, in press. [CrossRef]

86. Pincus, S.H.; Das, A.; Song, K.; Maresh, G.A.; Corti, M.; Berry, J. Role of fc in antibody-mediated protection from ricin toxin. *Toxins* **2014**, *6*, 1512–1525. [CrossRef] [PubMed]

87. Sully, E.K.; Whaley, K.J.; Bohorova, N.; Bohorov, O.; Goodman, C.; Kim, D.H.; Pauly, M.H.; Velasco, J.; Hiatt, E.; Morton, J.; et al. Chimeric plantibody passively protects mice against aerosolized ricin challenge. *Clin. Vaccine Immunol. CVI* **2014**, *21*, 777–782. [CrossRef] [PubMed]

88. Prigent, J.; Panigai, L.; Lamourette, P.; Sauvaire, D.; Devilliers, K.; Plaisance, M.; Volland, H.; Creminon, C.; Simon, S. Neutralising antibodies against ricin toxin. *PLoS ONE* **2011**, *6*, e20166. [CrossRef] [PubMed]

89. Respaud, R.; Marchand, D.; Pelat, T.; Tchou-Wong, K.M.; Roy, C.J.; Parent, C.; Cabrera, M.; Guillemain, J.; Mac Loughlin, R.; Levacher, E.; et al. Development of a drug delivery system for efficient alveolar delivery of a neutralizing monoclonal antibody to treat pulmonary intoxication to ricin. *J. Control. Release* **2016**, *234*, 21–32. [CrossRef] [PubMed]

90. O'Hara, J.M.; Whaley, K.; Pauly, M.; Zeitlin, L.; Mantis, N.J. Plant-based expression of a partially humanized neutralizing monoclonal igg directed against an immunodominant epitope on the ricin toxin a subunit. *Vaccine* **2012**, *30*, 1239–1243. [CrossRef] [PubMed]

91. Hu, W.G.; Yin, J.; Chau, D.; Hu, C.C.; Lillico, D.; Yu, J.; Negrych, L.M.; Cherwonogrodzky, J.W. Conformation-dependent high-affinity potent ricin-neutralizing monoclonal antibodies. *BioMed Res. Int.* **2013**, *2013*, 471346. [CrossRef] [PubMed]

92. Hu, W.G.; Yin, J.; Chau, D.; Negrych, L.M.; Cherwonogrodzky, J.W. Humanization and characterization of an anti-ricin neutralization monoclonal antibody. *PLoS ONE* **2012**, *7*, e45595. [CrossRef] [PubMed]

93. Rider, P.; Carmi, Y.; Cohen, I. Biologics for targeting inflammatory cytokines, clinical uses, and limitations. *Int. J. Cell Biol.* **2016**, *2016*, 9259646. [CrossRef] [PubMed]

94. Jones, S.A.; Scheller, J.; Rose-John, S. Therapeutic strategies for the clinical blockade of il-6/gp130 signaling. *J. Clin. Investig.* **2011**, *121*, 3375–3383. [CrossRef] [PubMed]

95. Ueda, O.; Tateishi, H.; Higuchi, Y.; Fujii, E.; Kato, A.; Kawase, Y.; Wada, N.A.; Tachibe, T.; Kakefuda, M.; Goto, C.; et al. Novel genetically-humanized mouse model established to evaluate efficacy of therapeutic agents to human interleukin-6 receptor. *Sci. Rep.* **2013**, *3*, 1196. [CrossRef] [PubMed]

96. Bhargava, R.; Janssen, W.; Altmann, C.; Andres-Hernando, A.; Okamura, K.; Vandivier, R.W.; Ahuja, N.; Faubel, S. Intratracheal il-6 protects against lung inflammation in direct, but not indirect, causes of acute lung injury in mice. *PLoS ONE* **2013**, *8*, e61405. [CrossRef] [PubMed]

97. Wolters, P.J.; Wray, C.; Sutherland, R.E.; Kim, S.S.; Koff, J.; Mao, Y.; Frank, J.A. Neutrophil-derived il-6 limits alveolar barrier disruption in experimental ventilator-induced lung injury. *J. Immunol.* **2009**, *182*, 8056–8062. [CrossRef] [PubMed]

98. Leite, L.M.; Carvalho, A.G.; Ferreira, P.L.; Pessoa, I.X.; Goncalves, D.O.; Lopes Ade, A.; Goes, J.G.; Alves, V.C.; Leal, L.K.; Brito, G.A.; et al. Anti-inflammatory properties of doxycycline and minocycline in experimental models: An in vivo and in vitro comparative study. *Inflammopharmacology* **2011**, *19*, 99–110. [CrossRef] [PubMed]

99. Zidovetzki, R.; Sherman, I.W.; Prudhomme, J.; Crawford, J. Inhibition of plasmodium falciparum lysophospholipase by anti-malarial drugs and sulphydryl reagents. *Parasitology* **1994**, *108 Pt 3*, 249–255. [CrossRef] [PubMed]

100. Ram, A.; Mabalirajan, U.; Singh, S.K.; Singh, V.P.; Ghosh, B. Mepacrine alleviates airway hyperresponsiveness and airway inflammation in a mouse model of asthma. *Int. Immunopharmacol.* **2008**, *8*, 893–899. [CrossRef] [PubMed]

101. Kuipers, M.T.; Aslami, H.; Vlaar, A.P.; Juffermans, N.P.; Tuip-de Boer, A.M.; Hegeman, M.A.; Jongsma, G.; Roelofs, J.J.; van der Poll, T.; Schultz, M.J.; et al. Pre-treatment with allopurinol or uricase attenuates barrier dysfunction but not inflammation during murine ventilator-induced lung injury. *PLoS ONE* **2012**, *7*, e50559. [CrossRef] [PubMed]

102. Gasse, P.; Riteau, N.; Charron, S.; Girre, S.; Fick, L.; Petrilli, V.; Tschopp, J.; Lagente, V.; Quesniaux, V.F.; Ryffel, B.; et al. Uric acid is a danger signal activating nalp3 inflammasome in lung injury inflammation and fibrosis. *Am. J. Respir. Crit. Care Med.* **2009**, *179*, 903–913. [CrossRef] [PubMed]

103. Fahmi, A.N.; Shehatou, G.S.; Shebl, A.M.; Salem, H.A. Febuxostat protects rats against lipopolysaccharide-induced lung inflammation in a dose-dependent manner. *Naunyn Schmiedebergs Arch. Pharmacol.* **2016**, *389*, 269–278. [CrossRef] [PubMed]

104. Kataoka, H.; Yang, K.; Rock, K.L. The xanthine oxidase inhibitor febuxostat reduces tissue uric acid content and inhibits injury-induced inflammation in the liver and lung. *Eur. J. Pharmacol.* **2015**, *746*, 174–179. [CrossRef] [PubMed]

105. Chow, C.W.; Herrera Abreu, M.T.; Suzuki, T.; Downey, G.P. Oxidative stress and acute lung injury. *Am. J. Respir. Cell Mol. Biol.* **2003**, *29*, 427–431. [CrossRef] [PubMed]

106. Cherubin, P.; Garcia, M.C.; Curtis, D.; Britt, C.B.; Craft, J.W., Jr.; Burress, H.; Berndt, C.; Reddy, S.; Guyette, J.; Zheng, T.; et al. Inhibition of cholera toxin and other ab toxins by polyphenolic compounds. *PLoS ONE* **2016**, *11*, e0166477. [CrossRef] [PubMed]

107. Dyer, P.D.; Kotha, A.K.; Gollings, A.S.; Shorter, S.A.; Shepherd, T.R.; Pettit, M.W.; Alexander, B.D.; Getti, G.T.; El-Daher, S.; Baillie, L.; et al. An in vitro evaluation of epigallocatechin gallate (egcg) as a biocompatible inhibitor of ricin toxin. *Biochim. Biophys. Acta* **2016**, *1860*, 1541–1550. [CrossRef] [PubMed]

108. Rocksen, D.; Ekstrand-Hammarstrom, B.; Johansson, L.; Bucht, A. Vitamin e reduces transendothelial migration of neutrophils and prevents lung injury in endotoxin-induced airway inflammation. *Am. J. Respir. Cell Mol. Biol.* **2003**, *28*, 199–207. [CrossRef] [PubMed]

109. Bae, H.B.; Li, M.; Kim, J.P.; Kim, S.J.; Jeong, C.W.; Lee, H.G.; Kim, W.M.; Kim, H.S.; Kwak, S.H. The effect of epigallocatechin gallate on lipopolysaccharide-induced acute lung injury in a murine model. *Inflammation* **2010**, *33*, 82–91. [CrossRef] [PubMed]

110. Liu, W.; Dong, M.; Bo, L.; Li, C.; Liu, Q.; Li, Y.; Ma, L.; Xie, Y.; Fu, E.; Mu, D.; et al. Epigallocatechin-3-gallate ameliorates seawater aspiration-induced acute lung injury via regulating inflammatory cytokines and inhibiting jak/stat1 pathway in rats. *Mediat. Inflamm.* **2014**, *2014*, 612593. [CrossRef] [PubMed]

111. Thickett, D.R.; Armstrong, L.; Christie, S.J.; Millar, A.B. Vascular endothelial growth factor may contribute to increased vascular permeability in acute respiratory distress syndrome. *Am. J. Respir. Crit. Care Med.* **2001**, *164*, 1601–1605. [CrossRef] [PubMed]

112. Ferrara, N.; Gerber, H.P.; LeCouter, J. The biology of vegf and its receptors. *Nat. Med.* **2003**, *9*, 669–676. [CrossRef] [PubMed]

113. Holash, J.; Davis, S.; Papadopoulos, N.; Croll, S.D.; Ho, L.; Russell, M.; Boland, P.; Leidich, R.; Hylton, D.; Burova, E.; et al. Vegf-trap: A vegf blocker with potent antitumor effects. *Proc. Natl. Acad. Sci. USA* **2002**, *99*, 11393–11398. [CrossRef] [PubMed]

114. Corne, J.; Chupp, G.; Lee, C.G.; Homer, R.J.; Zhu, Z.; Chen, Q.; Ma, B.; Du, Y.; Roux, F.; McArdle, J.; et al. Il-13 stimulates vascular endothelial cell growth factor and protects against hyperoxic acute lung injury. *J. Clin. Investig.* **2000**, *106*, 783–791. [CrossRef] [PubMed]

115. Channick, R.N.; Sitbon, O.; Barst, R.J.; Manes, A.; Rubin, L.J. Endothelin receptor antagonists in pulmonary arterial hypertension. *J. Am. Coll. Cardiol.* **2004**, *43*, 62S–67S. [CrossRef] [PubMed]

116. Kuklin, V.N.; Kirov, M.Y.; Evgenov, O.V.; Sovershaev, M.A.; Sjoberg, J.; Kirova, S.S.; Bjertnaes, L.J. Novel endothelin receptor antagonist attenuates endotoxin-induced lung injury in sheep. *Crit. Care Med.* **2004**, *32*, 766–773. [CrossRef] [PubMed]

117. Birukova, A.A.; Wu, T.; Tian, Y.; Meliton, A.; Sarich, N.; Tian, X.; Leff, A.; Birukov, K.G. Iloprost improves endothelial barrier function in lipopolysaccharide-induced lung injury. *Eur. Respir. J.* **2013**, *41*, 165–176. [CrossRef] [PubMed]

118. Sugiyama, M.G.; Armstrong, S.M.; Wang, C.; Hwang, D.; Leong-Poi, H.; Advani, A.; Advani, S.; Zhang, H.; Szaszi, K.; Tabuchi, A.; et al. The tie2-agonist vasculotide rescues mice from influenza virus infection. *Sci. Rep.* **2015**, *5*, 11030. [CrossRef] [PubMed]

119. Maris, N.A.; de Vos, A.F.; Dessing, M.C.; Spek, C.A.; Lutter, R.; Jansen, H.M.; van der Zee, J.S.; Bresser, P.; van der Poll, T. Antiinflammatory effects of salmeterol after inhalation of lipopolysaccharide by healthy volunteers. *Am. J. Respir. Crit. Care Med.* **2005**, *172*, 878–884. [CrossRef] [PubMed]

120. Perkins, G.D.; McAuley, D.F.; Richter, A.; Thickett, D.R.; Gao, F. Bench-to-bedside review: Beta2-agonists and the acute respiratory distress syndrome. *Crit. Care* **2004**, *8*, 25–32. [CrossRef] [PubMed]

121. Perkins, G.D.; McAuley, D.F.; Thickett, D.R.; Gao, F. The beta-agonist lung injury trial (balti): A randomized placebo-controlled clinical trial. *Am. J. Respir. Crit. Care Med.* **2006**, *173*, 281–287. [CrossRef] [PubMed]

122. Wigenstam, E.; Koch, B.; Bucht, A.; Jonasson, S. N-acetyl cysteine improves the effects of corticosteroids in a mouse model of chlorine-induced acute lung injury. *Toxicology* **2015**, *328*, 40–47. [CrossRef] [PubMed]

123. Ritter, C.; da Cunha, A.A.; Echer, I.C.; Andrades, M.; Reinke, A.; Lucchiari, N.; Rocha, J.; Streck, E.L.; Menna-Barreto, S.; Moreira, J.C.; et al. Effects of n-acetylcysteine plus deferoxamine in lipopolysaccharide-induced acute lung injury in the rat. *Crit. Care Med.* **2006**, *34*, 471–477. [CrossRef] [PubMed]

124. Shah, N.G.; Tulapurkar, M.E.; Ramarathnam, A.; Brophy, A.; Martinez, R., 3rd; Hom, K.; Hodges, T.; Samadani, R.; Singh, I.S.; MacKerell, A.D., Jr.; et al. Novel noncatalytic substrate-selective p38alpha-specific mapk inhibitors with endothelial-stabilizing and anti-inflammatory activity. *J. Immunol.* **2017**, *198*, 3296–3306. [CrossRef] [PubMed]

125. Cicenas, J. Jnk inhibitors: Is there a future? *MAP Kinase* **2015**, *4*, 31–37. [CrossRef]

126. Ishii, M.; Suzuki, Y.; Takeshita, K.; Miyao, N.; Kudo, H.; Hiraoka, R.; Nishio, K.; Sato, N.; Naoki, K.; Aoki, T.; et al. Inhibition of c-jun nh2-terminal kinase activity improves ischemia/reperfusion injury in rat lungs. *J. Immunol.* **2004**, *172*, 2569–2577. [CrossRef] [PubMed]

127. Lee, H.S.; Kim, H.J.; Moon, C.S.; Chong, Y.H.; Kang, J.L. Inhibition of c-jun nh2-terminal kinase or extracellular signal-regulated kinase improves lung injury. *Respir. Res.* **2004**, *5*, 23. [CrossRef] [PubMed]

128. Shen, H.; Wu, N.; Wang, Y.; Han, X.; Zheng, Q.; Cai, X.; Zhang, H.; Zhao, M. Jnk inhibitor sp600125 attenuates paraquat-induced acute lung injury: An in vivo and in vitro study. *Inflammation* **2017**, *40*, 1319–1330. [CrossRef] [PubMed]

129. Zheng, Y.; Zhang, M.; Zhao, Y.; Chen, J.; Li, B.; Cai, W. Jnk inhibitor sp600125 protects against lipopolysaccharide-induced acute lung injury via upregulation of claudin-4. *Exp. Ther. Med.* **2014**, *8*, 153–158. [CrossRef] [PubMed]

130. Jammi, N.V.; Whitby, L.R.; Beal, P.A. Small molecule inhibitors of the rna-dependent protein kinase. *Biochem. Biophys. Res. Commun.* **2003**, *308*, 50–57. [CrossRef]

131. Gray, J.S.; Bae, H.K.; Li, J.C.; Lau, A.S.; Pestka, J.J. Double-stranded rna-activated protein kinase mediates induction of interleukin-8 expression by deoxynivalenol, shiga toxin 1, and ricin in monocytes. *Toxicol. Sci.* **2008**, *105*, 322–330. [CrossRef] [PubMed]

132. Zhu, P.J.; Huang, W.; Kalikulov, D.; Yoo, J.W.; Placzek, A.N.; Stoica, L.; Zhou, H.; Bell, J.C.; Friedlander, M.J.; Krnjevic, K.; et al. Suppression of pkr promotes network excitability and enhanced cognition by interferon-gamma-mediated disinhibition. *Cell* **2011**, *147*, 1384–1396. [CrossRef] [PubMed]

133. Nakamura, T.; Arduini, A.; Baccaro, B.; Furuhashi, M.; Hotamisligil, G.S. Small-molecule inhibitors of pkr improve glucose homeostasis in obese diabetic mice. *Diabetes* **2014**, *63*, 526–534. [CrossRef] [PubMed]

134. Miller, S.C.; Huang, R.; Sakamuru, S.; Shukla, S.J.; Attene-Ramos, M.S.; Shinn, P.; Van Leer, D.; Leister, W.; Austin, C.P.; Xia, M. Identification of known drugs that act as inhibitors of nf-kappab signaling and their mechanism of action. *Biochem. Pharmacol.* **2010**, *79*, 1272–1280. [CrossRef] [PubMed]

135. Burgos, R.A.; Hidalgo, M.A.; Figueroa, C.D.; Conejeros, I.; Hancke, J.L. New potential targets to modulate neutrophil function in inflammation. *Mini Rev. Med. Chem.* **2009**, *9*, 153–168. [CrossRef] [PubMed]

136. Gamble, C.; McIntosh, K.; Scott, R.; Ho, K.H.; Plevin, R.; Paul, A. Inhibitory kappa b kinases as targets for pharmacological regulation. *Br. J. Pharmacol.* **2012**, *165*, 802–819. [CrossRef] [PubMed]

137. Ziegelbauer, K.; Gantner, F.; Lukacs, N.W.; Berlin, A.; Fuchikami, K.; Niki, T.; Sakai, K.; Inbe, H.; Takeshita, K.; Ishimori, M.; et al. A selective novel low-molecular-weight inhibitor of ikappab kinase-beta (ikk-beta) prevents pulmonary inflammation and shows broad anti-inflammatory activity. *Br. J. Pharmacol.* **2005**, *145*, 178–192. [CrossRef] [PubMed]

138. Everhart, M.B.; Han, W.; Sherrill, T.P.; Arutiunov, M.; Polosukhin, V.V.; Burke, J.R.; Sadikot, R.T.; Christman, J.W.; Yull, F.E.; Blackwell, T.S. Duration and intensity of nf-kappab activity determine the severity of endotoxin-induced acute lung injury. *J. Immunol.* **2006**, *176*, 4995–5005. [CrossRef] [PubMed]

139. Riteau, N.; Gasse, P.; Fauconnier, L.; Gombault, A.; Couegnat, M.; Fick, L.; Kanellopoulos, J.; Quesniaux, V.F.; Marchand-Adam, S.; Crestani, B.; et al. Extracellular atp is a danger signal activating p2x7 receptor in lung inflammation and fibrosis. *Am. J. Respir. Crit. Care Med.* **2010**, *182*, 774–783. [CrossRef] [PubMed]

140. Grailer, J.J.; Canning, B.A.; Kalbitz, M.; Haggadone, M.D.; Dhond, R.M.; Andjelkovic, A.V.; Zetoune, F.S.; Ward, P.A. Critical role for the nlrp3 inflammasome during acute lung injury. *J. Immunol.* **2014**, *192*, 5974–5983. [CrossRef] [PubMed]

141. Honda, H.; Nagai, Y.; Matsunaga, T.; Okamoto, N.; Watanabe, Y.; Tsuneyama, K.; Hayashi, H.; Fujii, I.; Ikutani, M.; Hirai, Y.; et al. Isoliquiritigenin is a potent inhibitor of nlrp3 inflammasome activation and diet-induced adipose tissue inflammation. *J. Leukoc. Biol.* **2014**, *96*, 1087–1100. [CrossRef] [PubMed]

142. Juliana, C.; Fernandes-Alnemri, T.; Wu, J.; Datta, P.; Solorzano, L.; Yu, J.W.; Meng, R.; Quong, A.A.; Latz, E.; Scott, C.P.; et al. Anti-inflammatory compounds parthenolide and bay 11-7082 are direct inhibitors of the inflammasome. *J. Biol. Chem.* **2010**, *285*, 9792–9802. [CrossRef] [PubMed]

143. Lamkanfi, M.; Mueller, J.L.; Vitari, A.C.; Misaghi, S.; Fedorova, A.; Deshayes, K.; Lee, W.P.; Hoffman, H.M.; Dixit, V.M. Glyburide inhibits the cryopyrin/nalp3 inflammasome. *J. Cell Biol.* **2009**, *187*, 61–70. [CrossRef] [PubMed]

144. Marchetti, C.; Chojnacki, J.; Toldo, S.; Mezzaroma, E.; Tranchida, N.; Rose, S.W.; Federici, M.; Van Tassell, B.W.; Zhang, S.; Abbate, A. A novel pharmacologic inhibitor of the nlrp3 inflammasome limits myocardial injury after ischemia-reperfusion in the mouse. *J. Cardiovasc. Pharmacol.* **2014**, *63*, 316–322. [CrossRef] [PubMed]

145. Coll, R.C.; Robertson, A.A.; Chae, J.J.; Higgins, S.C.; Munoz-Planillo, R.; Inserra, M.C.; Vetter, I.; Dungan, L.S.; Monks, B.G.; Stutz, A.; et al. A small-molecule inhibitor of the nlrp3 inflammasome for the treatment of inflammatory diseases. *Nat. Med.* **2015**, *21*, 248–255. [CrossRef] [PubMed]

146. Youm, Y.H.; Nguyen, K.Y.; Grant, R.W.; Goldberg, E.L.; Bodogai, M.; Kim, D.; D'Agostino, D.; Planavsky, N.; Lupfer, C.; Kanneganti, T.D.; et al. The ketone metabolite beta-hydroxybutyrate blocks nlrp3 inflammasome-mediated inflammatory disease. *Nat. Med.* **2015**, *21*, 263–269. [PubMed]

147. Shao, B.Z.; Xu, Z.Q.; Han, B.Z.; Su, D.F.; Liu, C. Nlrp3 inflammasome and its inhibitors: A review. *Front. Pharmacol.* **2015**, *6*, 262. [CrossRef] [PubMed]

148. Chen, X.; Bian, J.; Ge, Y. Zinc-deficient diet aggravates ventilation-induced lung injury in rats. *J. Biomed. Res.* **2012**, *26*, 59–65. [CrossRef]

149. Gomez, N.N.; Davicino, R.C.; Biaggio, V.S.; Bianco, G.A.; Alvarez, S.M.; Fischer, P.; Masnatta, L.; Rabinovich, G.A.; Gimenez, M.S. Overexpression of inducible nitric oxide synthase and cyclooxygenase-2 in rat zinc-deficient lung: Involvement of a nf-kappab dependent pathway. *Nitric Oxide* **2006**, *14*, 30–38. [CrossRef] [PubMed]

150. Joshi, P.C.; Mehta, A.; Jabber, W.S.; Fan, X.; Guidot, D.M. Zinc deficiency mediates alcohol-induced alveolar epithelial and macrophage dysfunction in rats. *Am. J. Respir. Cell Mol. Biol.* **2009**, *41*, 207–216. [CrossRef] [PubMed]

151. Bao, S.; Liu, M.J.; Lee, B.; Besecker, B.; Lai, J.P.; Guttridge, D.C.; Knoell, D.L. Zinc modulates the innate immune response in vivo to polymicrobial sepsis through regulation of nf-kappab. *Am. J. Physiol. Lung Cell. Mol. Physiol.* **2010**, *298*, L744–L754. [CrossRef] [PubMed]

152. Turut, H.; Kurutas, E.B.; Bulbuloglu, E.; Yasim, A.; Ozkaya, M.; Onder, A.; Imrek, S.S. Zinc aspartate alleviates lung injury induced by intestinal ischemia-reperfusion in rats. *J. Surg. Res.* **2009**, *151*, 62–67. [CrossRef] [PubMed]

153. Bao, S.; Knoell, D.L. Zinc modulates cytokine-induced lung epithelial cell barrier permeability. *Am. J. Physiol. Lung Cell. Mol. Physiol.* **2006**, *291*, L1132–L1141. [CrossRef] [PubMed]

154. Bouchier-Hayes, L.; Lartigue, L.; Newmeyer, D.D. Mitochondria: Pharmacological manipulation of cell death. *J. Clin. Investig.* **2005**, *115*, 2640–2647. [CrossRef] [PubMed]

155. Green, D.R.; Kroemer, G. Pharmacological manipulation of cell death: Clinical applications in sight? *J. Clin. Investig.* **2005**, *115*, 2610–2617. [CrossRef] [PubMed]

156. Leonardi, W.; Zilbermintz, L.; Cheng, L.W.; Zozaya, J.; Tran, S.H.; Elliott, J.H.; Polukhina, K.; Manasherob, R.; Li, A.; Chi, X.; et al. Bithionol blocks pathogenicity of bacterial toxins, ricin, and zika virus. *Sci. Rep.* **2016**, *6*, 34475. [CrossRef] [PubMed]

157. Dawson, R.M.; Paddle, B.M.; Alderton, M.R. Characterization of the asialofetuin microtitre plate-binding assay for evaluating inhibitors of ricin lectin activity. *J. Appl. Toxicol. JAT* **1999**, *19*, 307–312. [CrossRef]

158. Spiro, R.G.; Bhoyroo, V.D. Structure of the o-glycosidically linked carbohydrate units of fetuin. *J. Biol. Chem.* **1974**, *249*, 5704–5717. [PubMed]

159. Baenziger, J.U.; Fiete, D. Structural determinants of ricinus communis agglutinin and toxin specificity for oligosaccharides. *J. Biol. Chem.* **1979**, *254*, 9795–9799. [PubMed]

160. Tonevitsky, A.G.; Zhukova, O.S.; Mirimanova, N.V.; Omelyanenko, V.G.; Timofeeva, N.V.; Bergelson, L.D. Effect of gangliosides on binding, internalization and cytotoxic activity of ricin. *FEBS Lett.* **1990**, *264*, 249–252. [CrossRef]

161. Blome, M.C.; Schengrund, C.L. Multivalent binding of ricin to bovine serum albumin-based neoglycoconjugates. *Toxicon* **2008**, *51*, 1214–1224. [CrossRef] [PubMed]

162. Uzawa, H.; Ohga, K.; Shinozaki, Y.; Ohsawa, I.; Nagatsuka, T.; Seto, Y.; Nishida, Y. A novel sugar-probe biosensor for the deadly plant proteinous toxin, ricin. *Biosens. Bioelectron.* **2008**, *24*, 929–933. [CrossRef] [PubMed]

163. Nagatsuka, T.; Uzawa, H.; Ohsawa, I.; Seto, Y.; Nishida, Y. Use of lactose against the deadly biological toxin ricin. *ACS Appl. Mater. Interfaces* **2010**, *2*, 1081–1085. [CrossRef] [PubMed]

164. Nagatsuka, T.; Uzawa, H.; Sato, K.; Ohsawa, I.; Seto, Y.; Nishida, Y. Glycotechnology for decontamination of biological agents: A model study using ricin and biotin-tagged synthetic glycopolymers. *ACS Appl. Mater. Interfaces* **2012**, *4*, 832–837. [CrossRef] [PubMed]

165. Dawson, R.M.; Alderton, M.R.; Wells, D.; Hartley, P.G. Monovalent and polyvalent carbohydrate inhibitors of ricin binding to a model of the cell-surface receptor. *J. Appl. Toxicol. JAT* **2006**, *26*, 247–252. [CrossRef] [PubMed]

166. Ganguly, D.; Mukhopadhyay, C. Extended binding site of ricin b lectin for oligosaccharide recognition. *Biopolymers* **2007**, *86*, 311–320. [CrossRef] [PubMed]

167. Lambert, J.M.; McIntyre, G.; Gauthier, M.N.; Zullo, D.; Rao, V.; Steeves, R.M.; Goldmacher, V.S.; Blattler, W.A. The galactose-binding sites of the cytotoxic lectin ricin can be chemically blocked in high yield with reactive ligands prepared by chemical modification of glycopeptides containing triantennary n-linked oligosaccharides. *Biochemistry* **1991**, *30*, 3234–3247. [CrossRef] [PubMed]

168. Itakura, Y.; Nakamura-Tsuruta, S.; Kominami, J.; Sharon, N.; Kasai, K.; Hirabayashi, J. Systematic comparison of oligosaccharide specificity of ricinus communis agglutinin i and erythrina lectins: A search by frontal affinity chromatography. *J. Biochem.* **2007**, *142*, 459–469. [CrossRef] [PubMed]

169. Sandvig, K.; Olsnes, S. Entry of the toxic proteins abrin, modeccin, ricin, and diphtheria toxin into cells. Ii. Effect of ph, metabolic inhibitors, and ionophores and evidence for toxin penetration from endocytotic vesicles. *J. Biol. Chem.* **1982**, *257*, 7504–7513. [PubMed]

170. Sandvig, K.; van Deurs, B. Selective modulation of the endocytic uptake of ricin and fluid phase markers without alteration in transferrin endocytosis. *J. Biol. Chem.* **1990**, *265*, 6382–6388. [PubMed]

171. Yoshida, T.; Chen, C.C.; Zhang, M.S.; Wu, H.C. Disruption of the golgi apparatus by brefeldin a inhibits the cytotoxicity of ricin, modeccin, and pseudomonas toxin. *Exp. Cell Res.* **1991**, *192*, 389–395. [CrossRef]

172. Wellner, R.B.; Pless, D.D.; Thompson, W.L. Characterization of 3′-azido-3′-deoxythymidine inhibition of ricin and pseudomonas exotoxin a toxicity in cho and vero cells. *J. Cell. Physiol.* **1994**, *159*, 495–505. [CrossRef] [PubMed]

173. Nambiar, M.P.; Murugesan, R.; Wu, H.C. Inhibition of the cytotoxicity of protein toxins by a novel plant metabolite, mansonone-d. *J. Cell. Physiol.* **1998**, *176*, 40–49. [CrossRef]

174. Okimoto, T.; Seguchi, T.; Ono, M.; Nakayama, Y.; Funatsu, G.; Fujiwara, T.; Ikehara, Y.; Kuwano, M. Brefeldin a protects ricin-induced cytotoxicity in human cancer kb cell line, but not in its resistant counterpart with altered golgi structures. *Cell Struct. Funct.* **1993**, *18*, 241–251. [CrossRef] [PubMed]

175. Sandvig, K.; Prydz, K.; Hansen, S.H.; van Deurs, B. Ricin transport in brefeldin a-treated cells: Correlation between golgi structure and toxic effect. *J. Cell Biol.* **1991**, *115*, 971–981. [CrossRef] [PubMed]

176. Simm, R.; Kvalvaag, A.S.; van Deurs, B.; Lindback, T.; Sandvig, K. Benzyl alcohol induces a reversible fragmentation of the golgi apparatus and inhibits membrane trafficking between endosomes and the trans-golgi network. *Exp. Cell Res.* **2017**, *357*, 67–78. [CrossRef] [PubMed]

177. Redmann, V.; Gardner, T.; Lau, Z.; Morohashi, K.; Felsenfeld, D.; Tortorella, D. Novel class of potential therapeutics that target ricin retrograde translocation. *Toxins* **2013**, *6*, 33–53. [CrossRef] [PubMed]

178. Stechmann, B.; Bai, S.K.; Gobbo, E.; Lopez, R.; Merer, G.; Pinchard, S.; Panigai, L.; Tenza, D.; Raposo, G.; Beaumelle, B.; et al. Inhibition of retrograde transport protects mice from lethal ricin challenge. *Cell* **2010**, *141*, 231–242. [CrossRef] [PubMed]

179. Yu, S.; Park, J.G.; Kahn, J.N.; Tumer, N.E.; Pang, Y.P. Common pharmacophore of structurally distinct small-molecule inhibitors of intracellular retrograde trafficking of ribosome inactivating proteins. *Sci. Rep.* **2013**, *3*, 3397. [CrossRef] [PubMed]

180. Barbier, J.; Bouclier, C.; Johannes, L.; Gillet, D. Inhibitors of the cellular trafficking of ricin. *Toxins* **2012**, *4*, 15–27. [CrossRef] [PubMed]

181. Bassik, M.C.; Kampmann, M.; Lebbink, R.J.; Wang, S.; Hein, M.Y.; Poser, I.; Weibezahn, J.; Horlbeck, M.A.; Chen, S.; Mann, M.; et al. A systematic mammalian genetic interaction map reveals pathways underlying ricin susceptibility. *Cell* **2013**, *152*, 909–922. [CrossRef] [PubMed]

182. Bellisola, G.; Fracasso, G.; Ippoliti, R.; Menestrina, G.; Rosen, A.; Solda, S.; Udali, S.; Tomazzolli, R.; Tridente, G.; Colombatti, M. Reductive activation of ricin and ricin a-chain immunotoxins by protein disulfide isomerase and thioredoxin reductase. *Biochem. Pharmacol.* **2004**, *67*, 1721–1731. [CrossRef] [PubMed]

183. Barbieri, L.; Battelli, M.G.; Stirpe, F. Reduction of ricin and other plant toxins by thiol:Protein disulfide oxidoreductases. *Arch. Biochem. Biophys.* **1982**, *216*, 380–383. [CrossRef]

184. Pasetto, M.; Barison, E.; Castagna, M.; Della Cristina, P.; Anselmi, C.; Colombatti, M. Reductive activation of type 2 ribosome-inactivating proteins is promoted by transmembrane thioredoxin-related protein. *J. Biol. Chem.* **2012**, *287*, 7367–7373. [CrossRef] [PubMed]

185. Kean, W.F.; Hart, L.; Buchanan, W.W. Auranofin. *Br. J. Rheumatol.* **1997**, *36*, 560–572. [CrossRef] [PubMed]

186. Dickerhof, N.; Kleffmann, T.; Jack, R.; McCormick, S. Bacitracin inhibits the reductive activity of protein disulfide isomerase by disulfide bond formation with free cysteines in the substrate-binding domain. *FEBS J.* **2011**, *278*, 2034–2043. [CrossRef] [PubMed]

187. Pang, Y.P.; Park, J.G.; Wang, S.; Vummenthala, A.; Mishra, R.K.; McLaughlin, J.E.; Di, R.; Kahn, J.N.; Tumer, N.E.; Janosi, L.; et al. Small-molecule inhibitor leads of ribosome-inactivating proteins developed using the doorstop approach. *PLoS ONE* **2011**, *6*, e17883. [CrossRef] [PubMed]

188. Fan, S.; Wu, F.; Martiniuk, F.; Hale, M.L.; Ellington, A.D.; Tchou-Wong, K.M. Protective effects of anti-ricin a-chain rna aptamer against ricin toxicity. *World J. Gastroenterol.* **2008**, *14*, 6360–6365. [CrossRef] [PubMed]

189. Bai, Y.; Watt, B.; Wahome, P.G.; Mantis, N.J.; Robertus, J.D. Identification of new classes of ricin toxin inhibitors by virtual screening. *Toxicon* **2010**, *56*, 526–534. [CrossRef] [PubMed]

190. Dong, J.; Zhang, Y.; Chen, Y.; Niu, X.; Zhang, Y.; Li, R.; Yang, C.; Wang, Q.; Li, X.; Deng, X. Baicalin inhibits the lethality of ricin in mice by inducing protein oligomerization. *J. Biol. Chem.* **2015**, *290*, 12899–12907. [CrossRef] [PubMed]

191. Jeon, K.I.; Jeong, J.Y.; Jue, D.M. Thiol-reactive metal compounds inhibit nf-kappa b activation by blocking i kappa b kinase. *J. Immunol.* **2000**, *164*, 5981–5989. [CrossRef] [PubMed]

192. Martinez-Gonzalez, J.J.; Guevara-Flores, A.; Rendon, J.L.; del Arenal, I.P. Auranofin-induced oxidative stress causes redistribution of the glutathione pool in taenia crassiceps cysticerci. *Mol. Biochem. Parasitol.* **2015**, *201*, 16–25. [CrossRef] [PubMed]

193. Kang, K.A.; Zhang, R.; Piao, M.J.; Chae, S.; Kim, H.S.; Park, J.H.; Jung, K.S.; Hyun, J.W. Baicalein inhibits oxidative stress-induced cellular damage via antioxidant effects. *Toxicol. Ind. Health* **2012**, *28*, 412–421. [CrossRef] [PubMed]

194. Shieh, D.E.; Liu, L.T.; Lin, C.C. Antioxidant and free radical scavenging effects of baicalein, baicalin and wogonin. *Anticancer Res.* **2000**, *20*, 2861–2865. [PubMed]

195. Deng, J.; Wang, D.X.; Liang, A.L.; Tang, J.; Xiang, D.K. Effects of baicalin on alveolar fluid clearance and alpha-enac expression in rats with lps-induced acute lung injury. *Can. J. Physiol. Pharmacol.* **2017**, *95*, 122–128. [CrossRef] [PubMed]

196. Ding, X.M.; Pan, L.; Wang, Y.; Xu, Q.Z. Baicalin exerts protective effects against lipopolysaccharide-induced acute lung injury by regulating the crosstalk between the cx3cl1-cx3cr1 axis and nf-kappab pathway in cx3cl1-knockout mice. *Int. J. Mol. Med.* **2016**, *37*, 703–715. [CrossRef] [PubMed]

197. Huang, K.L.; Chen, C.S.; Hsu, C.W.; Li, M.H.; Chang, H.; Tsai, S.H.; Chu, S.J. Therapeutic effects of baicalin on lipopolysaccharide-induced acute lung injury in rats. *Am. J. Chin. Med.* **2008**, *36*, 301–311. [CrossRef] [PubMed]

198. Das, D.K.; Sato, M.; Ray, P.S.; Maulik, G.; Engelman, R.M.; Bertelli, A.A.; Bertelli, A. Cardioprotection of red wine: Role of polyphenolic antioxidants. *Drugs Exp. Clin. Res.* **1999**, *25*, 115–120. [PubMed]

199. Cassel, S.L.; Eisenbarth, S.C.; Iyer, S.S.; Sadler, J.J.; Colegio, O.R.; Tephly, L.A.; Carter, A.B.; Rothman, P.B.; Flavell, R.A.; Sutterwala, F.S. The nalp3 inflammasome is essential for the development of silicosis. *Proc. Natl. Acad. Sci. USA* **2008**, *105*, 9035–9040. [CrossRef] [PubMed]

200. Duewell, P.; Kono, H.; Rayner, K.J.; Sirois, C.M.; Vladimer, G.; Bauernfeind, F.G.; Abela, G.S.; Franchi, L.; Nunez, G.; Schnurr, M.; et al. Nlrp3 inflammasomes are required for atherogenesis and activated by cholesterol crystals. *Nature* **2010**, *464*, 1357–1361. [CrossRef] [PubMed]

201. Dorr, R.T. Radioprotectants: Pharmacology and clinical applications of amifostine. *Semin. Radiat. Oncol.* **1998**, *8*, 10–13. [PubMed]

202. Acosta, J.C.; Richard, C.; Delgado, M.D.; Horita, M.; Rizzo, M.G.; Fernandez-Luna, J.L.; Leon, J. Amifostine impairs p53-mediated apoptosis of human myeloid leukemia cells. *Mol. Cancer Ther.* **2003**, *2*, 893–900. [PubMed]

203. Lee, E.J.; Gerhold, M.; Palmer, M.W.; Christen, R.D. P53 protein regulates the effects of amifostine on apoptosis, cell cycle progression, and cytoprotection. *Br. J. Cancer* **2003**, *88*, 754–759. [CrossRef] [PubMed]

204. Bergen, L.G.; Borisy, G.G. Tubulin-colchicine complex (tc) inhibits microtubule depolymerization by a capping reaction exerted preferentially at the minus end. *J. Cell. Biochem.* **1986**, *30*, 11–18. [CrossRef] [PubMed]

205. Ghio, A.J.; Kennedy, T.P.; Hatch, G.E.; Tepper, J.S. Reduction of neutrophil influx diminishes lung injury and mortality following phosgene inhalation. *J. Appl. Physiol.* **1991**, *71*, 657–665. [PubMed]

206. Choudhury, S.; Kandasamy, K.; Maruti, B.S.; Addison, M.P.; Kasa, J.K.; Darzi, S.A.; Singh, T.U.; Parida, S.; Dash, J.R.; Singh, V.; et al. Atorvastatin along with imipenem attenuates acute lung injury in sepsis through decrease in inflammatory mediators and bacterial load. *Eur. J. Pharmacol.* **2015**, *765*, 447–456. [CrossRef] [PubMed]

207. Siempos, I.I.; Maniatis, N.A.; Kopterides, P.; Magkou, C.; Glynos, C.; Roussos, C.; Armaganidis, A. Pretreatment with atorvastatin attenuates lung injury caused by high-stretch mechanical ventilation in an isolated rabbit lung model. *Crit. Care Med.* **2010**, *38*, 1321–1328.

208. Singla, S.; Zhou, T.; Javaid, K.; Abbasi, T.; Casanova, N.; Zhang, W.; Ma, S.F.; Wade, M.S.; Noth, I.; Sweiss, N.J.; et al. Expression profiling elucidates a molecular gene signature for pulmonary hypertension in sarcoidosis. *Pulm. Circ.* **2016**, *6*, 465–471. [CrossRef] [PubMed]

209. Akamatsu, H.; Asada, M.; Komura, J.; Asada, Y.; Niwa, Y. Effect of doxycycline on the generation of reactive oxygen species: A possible mechanism of action of acne therapy with doxycycline. *Acta Dermato-Venereol.* **1992**, *72*, 178–179.

210. Dalm, D.; Palm, G.J.; Aleksandrov, A.; Simonson, T.; Hinrichs, W. Nonantibiotic properties of tetracyclines: Structural basis for inhibition of secretory phospholipase a2. *J. Mol. Biol.* **2010**, *398*, 83–96. [CrossRef] [PubMed]

211. Fainaru, O.; Adini, I.; Benny, O.; Bazinet, L.; Pravda, E.; D'Amato, R.; Folkman, J. Doxycycline induces membrane expression of ve-cadherin on endothelial cells and prevents vascular hyperpermeability. *FASEB J.* **2008**, *22*, 3728–3735. [CrossRef] [PubMed]

212. Golub, L.M.; Lee, H.M.; Ryan, M.E.; Giannobile, W.V.; Payne, J.; Sorsa, T. Tetracyclines inhibit connective tissue breakdown by multiple non-antimicrobial mechanisms. *Adv. Dent. Res.* **1998**, *12*, 12–26. [CrossRef] [PubMed]

213. Krakauer, T.; Buckley, M. Doxycycline is anti-inflammatory and inhibits staphylococcal exotoxin-induced cytokines and chemokines. *Antimicrob. Agents Chemother.* **2003**, *47*, 3630–3633. [CrossRef] [PubMed]

214. Dalhoff, A. Immunomodulatory activities of fluoroquinolones. *Infection* **2005**, *33* (Suppl. 2), 55–70. [CrossRef] [PubMed]

215. Bosma, K.J.; Lewis, J.F. Emerging therapies for treatment of acute lung injury and acute respiratory distress syndrome. *Expert Opin. Emerg. Drugs* **2007**, *12*, 461–477. [CrossRef] [PubMed]

216. Jain, R.; DalNogare, A. Pharmacological therapy for acute respiratory distress syndrome. *Mayo Clin. Proc.* **2006**, *81*, 205–212. [CrossRef] [PubMed]

217. Liu, K.D.; Matthay, M.A. Advances in critical care for the nephrologist: Acute lung injury/ards. *Clin. J. Am. Soc. Nephrol.* **2008**, *3*, 578–586. [CrossRef] [PubMed]

218. Raghavendran, K.; Pryhuber, G.S.; Chess, P.R.; Davidson, B.A.; Knight, P.R.; Notter, R.H. Pharmacotherapy of acute lung injury and acute respiratory distress syndrome. *Curr. Med. Chem.* **2008**, *15*, 1911–1924. [CrossRef] [PubMed]

219. D'Elia, R.V.; Harrison, K.; Oyston, P.C.; Lukaszewski, R.A.; Clark, G.C. Targeting the "cytokine storm" for therapeutic benefit. *Clin. Vaccine Immunol. CVI* **2013**, *20*, 319–327. [CrossRef] [PubMed]

220. Mabley, J.G.; Pacher, P.; Szabo, C. Activation of the cholinergic antiinflammatory pathway reduces ricin-induced mortality and organ failure in mice. *Mol. Med.* **2009**, *15*, 166–172. [CrossRef] [PubMed]

221. Cara, D.C.; Kaur, J.; Forster, M.; McCafferty, D.M.; Kubes, P. Role of p38 mitogen-activated protein kinase in chemokine-induced emigration and chemotaxis in vivo. *J. Immunol.* **2001**, *167*, 6552–6558. [CrossRef] [PubMed]

222. McDonald, B.; Pittman, K.; Menezes, G.B.; Hirota, S.A.; Slaba, I.; Waterhouse, C.C.; Beck, P.L.; Muruve, D.A.; Kubes, P. Intravascular danger signals guide neutrophils to sites of sterile inflammation. *Science* **2010**, *330*, 362–366. [CrossRef] [PubMed]

223. Druey, K.M.; Greipp, P.R. Narrative review: The systemic capillary leak syndrome. *Ann. Intern. Med.* **2010**, *153*, 90–98. [CrossRef] [PubMed]

224. Mathison, R.D.; Christie, E.; Davison, J.S. The tripeptide feg inhibits leukocyte adhesion. *J. Inflamm. (Lond.)* **2008**, *5*, 6. [CrossRef] [PubMed]

225. Dery, R.E.; Ulanova, M.; Puttagunta, L.; Stenton, G.R.; James, D.; Merani, S.; Mathison, R.; Davison, J.; Befus, A.D. Frontline: Inhibition of allergen-induced pulmonary inflammation by the tripeptide feg: A mimetic of a neuro-endocrine pathway. *Eur. J. Immunol.* **2004**, *34*, 3315–3325. [CrossRef] [PubMed]

226. Mathison, R.D.; Davison, J.S.; Befus, A.D.; Gingerich, D.A. Salivary gland derived peptides as a new class of anti-inflammatory agents: Review of preclinical pharmacology of c-terminal peptides of smr1 protein. *J. Inflamm. (Lond.)* **2010**, *7*, 49. [CrossRef] [PubMed]

227. Elder, A.S.; Bersten, A.D.; Saccone, G.T.; Dixon, D.L. Prevention and amelioration of rodent endotoxin-induced lung injury with administration of a novel therapeutic tripeptide feg. *Pulm. Pharmacol. Ther.* **2013**, *26*, 167–171. [CrossRef] [PubMed]

228. Panos, R.J.; Bak, P.M.; Simonet, W.S.; Rubin, J.S.; Smith, L.J. Intratracheal instillation of keratinocyte growth factor decreases hyperoxia-induced mortality in rats. *J. Clin. Investig.* **1995**, *96*, 2026–2033. [CrossRef] [PubMed]

229. Yano, T.; Deterding, R.R.; Simonet, W.S.; Shannon, J.M.; Mason, R.J. Keratinocyte growth factor reduces lung damage due to acid instillation in rats. *Am. J. Respir. Cell Mol. Biol.* **1996**, *15*, 433–442. [CrossRef] [PubMed]

230. Johnson, C.L.; Soeder, Y.; Dahlke, M.H. Concise review: Mesenchymal stromal cell-based approaches for the treatment of acute respiratory distress and sepsis syndromes. *Stem Cells Transl. Med.* **2017**, *6*, 1141–1151. [CrossRef] [PubMed]

toxins

MDPI

Article

Total Body Irradiation Mitigates Inflammation and Extends the Therapeutic Time Window for Anti-Ricin Antibody Treatment against Pulmonary Ricinosis in Mice

Yoav Gal [†], Anita Sapoznikov [†], Reut Falach, Sharon Ehrlich, Moshe Aftalion, Chanoch Kronman * and Tamar Sabo

Department of Biochemistry and Molecular Genetics, Israel Institute for Biological Research, Ness-Ziona 76100, Israel; yoavg@iibr.gov.il (Y.G.); anitas@iibr.gov.il (A.S.); reutf@iibr.gov.il (R.F.); sharone@iibr.gov.il (S.E.); moshea@iibr.gov.il (M.A.); tamars@iibr.gov.il (T.S.)
* Correspondence: chanochk@iibr.gov.il; Tel.: +972-8-938-1522
† These authors contributed equally to this work.

Academic Editors: Daniel Gillet and Julien Barbier
Received: 16 August 2017; Accepted: 9 September 2017; Published: 11 September 2017

Abstract: Ricin, a highly toxic plant-derived toxin, is considered a potential weapon in biowarfare and bioterrorism due to its pronounced toxicity, high availability, and ease of preparation. Pulmonary exposure to ricin results in the generation of an acute edematous inflammation followed by respiratory insufficiency and death. Massive neutrophil recruitment to the lungs may contribute significantly to ricin-mediated morbidity. In this study, total body irradiation (TBI) served as a non-pharmacological tool to decrease the potential neutrophil-induced lung injury. TBI significantly postponed the time to death of intranasally ricin-intoxicated mice, given that leukopenia remained stable following intoxication. This increase in time to death coincided with a significant reduction in pro-inflammatory marker levels, and led to marked extension of the therapeutic time window for anti-ricin antibody treatment.

Keywords: ricin; total body irradiation; leukopenia; neutropenia; inflammation

---

1. Introduction

Ricin, a type II ribosome inactivating protein (RIP), is a toxin derived from the seeds of *Ricinus communis* (the castor oil plant). The holotoxin consists of two polypeptide chains, ricin toxin B (RTB) and ricin toxin A (RTA), linked by a single disulfide bond. RTB binds to galactose residues on the surface of cells and mediates the toxin's cellular internalization, whereas RTA possesses the catalytic activity of ricin [1].

Due to the high availability of the toxin and the relative ease of its production, ricin is considered a biological threat agent [2]. The toxicity of ricin depends on the route of exposure, the inhalational route being considered most fatal [3]. Pulmonary ricinosis comprises two pathological processes, ribosomal depurination and pulmonary inflammation, occurring at the molecular and cellular/tissue levels, respectively.

The RTA-dependent depurination is due to RNA N-glycosidase activity which cleaves a specific adenine residue located within the 28S rRNA of the mammalian 60S ribosome subunit [4–7]. This site-specific depurination event prevents binding of elongation factor-2 to the ribosome, thereby causing translational arrest and cell death [1,8]. In our laboratory, we have recently developed a method for quantitation of lung tissue depurination in intranasally ricin-intoxicated mice, and assessed the toxin's effect on different pulmonary cell populations. Using this approach, we have demonstrated that intranasal ricin intoxication leads to massive depurination-derived damage of the pulmonary epithelium cells [9], which play a key role in maintaining the integrity and function of the respiratory system [10].

In addition to depurination-induced cell death and tissue damage, pulmonary ricin exposure results in a severe localized inflammation, associated with an intrapulmonary cytokine storm, massive neutrophil recruitment, and pulmonary edema, subsequently resulting in respiratory failure and death [3,11,12]. The clinical manifestation of ricin-induced respiratory damage in swine complies with the accepted diagnostic criteria for acute respiratory distress syndrome (ARDS) [13].

Deciphering the relative contribution of each of these processes, cell damage stemming directly from depurination, versus acute severe inflammation, to the pathology and prognosis of pulmonary ricin intoxication, is of considerable importance and is relevant for anti-ricin treatment development. To this end, we focused in the present study on the relative contribution of the inflammatory process to pulmonary ricinosis.

We have previously demonstrated that co-administration of the anti-inflammatory drugs doxycycline or ciprofloxacin with anti-ricin antibodies to mice intranasally exposed to lethal doses of ricin, significantly attenuated lung injury and improved treatment outcome in comparison to mice treated with anti-ricin antibodies alone. This improved protection was attributed to a significant quantitative reduction of various inflammatory related markers—cytokine response, neutrophil infiltration, and vascular hyperpermeability [11,14]. The anti-inflammatory effects of doxycycline or ciprofloxacin were significant, yet limited, and their application without co-administration of anti-ricin antibody did not confer protection. Furthermore, unlike doxycycline and ciprofloxacin, dexamethasone—a classical, potent first-line anti-inflammatory drug—was incapable of conferring protection when co-administered with anti-ricin antibody at 24 h post exposure [11]. This atypical anti-inflammatory response, as well as the inability of a drug-alone treatment to confer protection, prompted us to search for a non-pharmacologic anti-inflammatory tool in order to systematically analyze the contribution of the inflammatory process to the morbidity associated with pulmonary ricinosis.

Charting the progression of pulmonary ricin-induced toxicity in neutrophil-depleted mice may serve to define the role of inflammation in pulmonary ricinosis. As previously demonstrated, extensive time-evolved neutrophil infiltration takes place following pulmonary ricin exposure [15–17]. Neutrophils have a crucial role in the development and progression of both sterile [18–20] and infectious [21–23] lung injuries. These polymorphonuclear cells actively contribute to oxidative and proteolytic damages via secretion of reactive oxygen species, metalloproteinases, as well as other damage mediating factors [24,25]. Neutrophil depletion was shown to mitigate several forms of acute lung injury [18,21], and vice versa—effective manipulations/treatments that reduced lung injury parameters were shown to correlate with reduced pulmonary neutrophil counts [26]. The meager response to dexamethasone in pulmonary ricinosis previously mentioned, may also be relevant to the contribution of neutrophils to ricin-induced morbidity and mortality, since inflammation-activated neutrophils synthesize functionally inactive beta isoforms of the glucocorticoid receptor, rendering them less corticosteroid-sensitive [27]. Moreover, neutrophils are relatively insensitive to corticosteroid-mediated apoptosis [28,29].

In the present study, total body irradiation (TBI) served as a tool to prevent neutrophil recruitment to the lungs. TBI is a highly efficient immunosuppressive tool, extensively used in the research of bone marrow- [30,31] and hematopoietic stem cell- [32] transplantations, as well as in myelosuppression research [33] in mice. Generally, TBI induces not only neutropenia but also leukopenia, namely total white cell count depletion, and in some cases also erythrocyte/platelet depletion [33]. However, in our system, in which ricin in itself reduces T and B lymphocytes and NK cells while in parallel brings about a significant increase in neutrophil counts in lungs—TBI, which attenuates neutrophil migration to the lungs—could serve as a useful tool for elucidating their contribution to the pathology associated with ricin intoxication.

Here we show that TBI-induced leukopenia/neutropenia considerably increased the mean time to death and alleviated the inflammatory response of mice following intranasal ricin intoxication, thereby expanding the therapeutic time window for anti-ricin antibody-based treatment in a significant manner.

2. Results

2.1. Effects of TBI on Peripheral Blood Count and Pulmonary Hematopoietic Cell Count Following Intranasal Ricin Intoxication

A single sublethal irradiation dose of 6.5 Gy was previously reported to induce a stable severe leukopenia in naive CD-1 mice for at least 14 days, the nadir (minimal white blood cell count) observed at 3–9 days following irradiation, after which, an increase in white blood count was demonstrated between 9–14 days post-TBI, while full recovery was reached at 3–4 weeks following TBI [34]. Accordingly, in order to attain a stable neutropenia following ricin exposure, mice were subjected to TBI at a single dose of 6.5 Gy and were intranasally challenged three days later with a lethal dose of ricin. Peripheral blood leukopenia (Figure 1A), and specifically neutropenia (Figure 1B), were found to be stable in all mice subjected to TBI, whether exposed to ricin or not, until day 7 post ricin-exposure (at this time point, most of the ricin-intoxicated mice have died) in contrast to the prominent leukocytosis and neutrophilia observed in the non-irradiated ricin-intoxicated group. Quantitative analysis of pulmonary cells in non-irradiated mice revealed a time-dependent increase in hematopoietic cell counts from 10 to 70×10^6 cells/lungs during the 72 h following intranasal ricin intoxication, while in irradiated ricin-intoxicated mice, the pulmonary hematopoietic count did not alter significantly during this period of time (5–10×10^6 cells/lungs at all examined time points) and values were lower than those measured in naive non-irradiated mice (Figure 1C).

FACS analysis of pulmonary hematopoietic cells demonstrated that following ricin-intoxication of non-irradiated mice, lymphocyte counts (T and B) were slightly reduced (Figure 1D,E), macrophage counts were reduced to a greater extent (Figure 1F), whereas neutrophil counts were extensively amplified (Figure 1G), all in line with previous observations at our laboratory [35]. In contrast, in irradiated mice, where the lymphocyte counts were dramatically reduced (Figure 1D,E), lymphocytopenia remained stable at all time points examined following ricin-intoxication and most importantly, neutrophil recruitment to the lungs was virtually abolished. Thus, at 72 h post-exposure, the pulmonary neutrophil count was 5×10^6 cells/lungs in irradiated mice, as opposed to 5×10^7 neutrophils/lungs in non-irradiated mice (Figure 1G). These results clearly indicate stable neutropenia following ricin intoxication of irradiated mice.

2.2. Lack of Additive Toxicity Following TBI in Ricin Intoxicated Mice

Since TBI per se may induce pulmonary damage, we next ruled out the possibility that excessive additive toxicity occurs following ricin intoxication of irradiated mice. Naive or irradiated mice were intranasally intoxicated with ricin (three days following TBI in the relevant group), body weights were monitored, and in addition, lungs were harvested at 0–72 h post exposure for cell quantification. Following intranasal ricin intoxication, 25–30% reductions in body weights were observed in both irradiated or naive mice subjected to ricin intoxication, without any noticeable difference between the two groups (Figure 2A). Likewise, parenchymal cell counts were similar in both irradiated and non-irradiated mice (Figure 2B), and a significant decrease in both endothelial (Figure 2C) and epithelial (Figure 2D) cells was displayed at 48 and 72 h post exposure, however, without significant additive cytotoxicity in the irradiated mice group. The epithelial cell count in the mice group subjected to TBI was somewhat lower than that measured for the non-irradiated mice before intoxication and at 24 h following intoxication, apparently due to the increased sensitivity of these cells to irradiation [36–41]. Overall, these results indicate for the lack of overt additive toxicity due to both irradiation and ricin intoxication, in comparison to ricin intoxication alone, enabling evaluation of the effect of neutrophil depletion in pulmonary ricinosis.

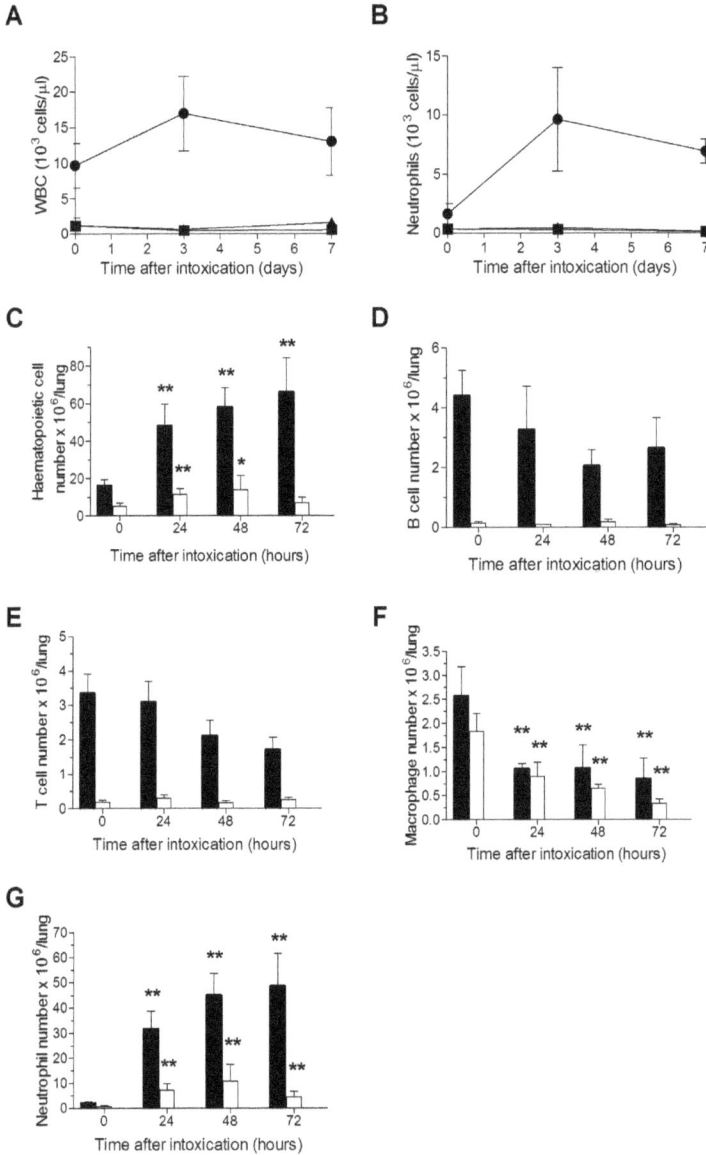

Figure 1. TBI-induced leukopenia in ricin intoxicated mice: Blood counts in mice following ricin intoxication. Naive or irradiated mice were intranasally intoxicated with ricin (circles and triangles, respectively), blood samples were withdrawn at the indicated time points and WBCs (**A**) and neutrophils (**B**) were quantified. Irradiated non-intoxicated mice served as control (squares). Irradiated (white bars) and non-irradiated (black bars) mice were intoxicated with ricin, lungs were harvested at different time-points following intoxication, and cells were quantified by flow cytometric analysis. Hematopoietic cells (**C**); B lymphocytes (**D**); T lymphocytes (**E**); macrophages (**F**); neutrophils (**G**). Number of animals per experimental group: A. $n = 7$–8 mice per group for all ricin-intoxicated mice; $n = 2$ mice per group for irradiated mice that were not intoxicated. B. $n = 3$ (irradiated) or 10 (non-irradiated) mice per group.* $p < 0.05$ in comparison to non-intoxicated mice; ** $p < 0.01$ in comparison to non-intoxicated mice.

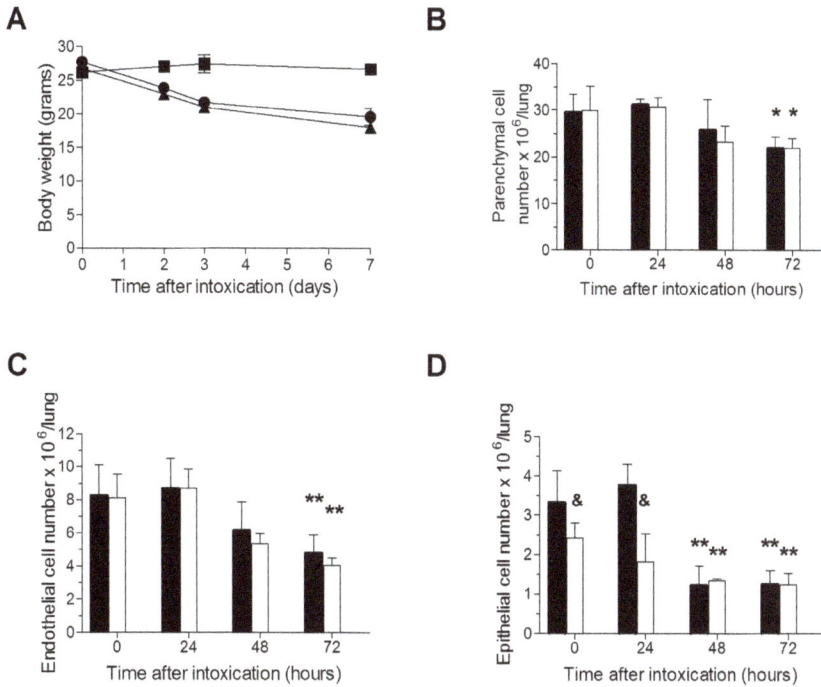

Figure 2. Lack of additive toxicity following TBI in ricin intoxicated mice: (**A**) Non-irradiated or irradiated mice (circles and triangles, respectively) were intranasally intoxicated with ricin, and body weights were determined at the indicated time points. Irradiated-non-intoxicated-mice served as control (squares). Irradiated (white bars) or non-irradiated (black bars) mice were intoxicated with ricin, lungs were harvested at the indicated time points and single cell suspensions were subjected to flow cytometric analysis for cell quantification; parenchymal cells (**B**); endothelial (**C**); and epithelial (**D**) cells. Number of animals per experimental group: A. $n = 2$ mice per group. B. $n = 3$ (irradiated) or 10 (non-irradiated) mice per group. * $p < 0.05$ in comparison to non-intoxicated mice; ** $p < 0.01$ in comparison to non-intoxicated mice; & $p < 0.05$ in comparison to non-irradiated mice at the same time point.

2.3. TBI-Induced Extension of the Mean Time to Death in Ricin Intoxicated Mice

To determine the effect of TBI on the time-course of pulmonary ricin-intoxication, naive and TBI-mice were intranasally intoxicated with ricin three days after irradiation, and the mean time to death (MTTD) of the two mice groups was monitored over a 14 day period. As seen, a significant increase in MTTD, from 6.3 days in non-irradiated mice to 8.7 days in the TBI group, was achieved (Figure 3).

As mentioned above [34], leukopenia is firmly stable in CD-1 mice during the nine days following irradiation, a gradual and moderate recovery of neutrophils is observed during the next four days, whereas between 14–28 days post-TBI, neutrophil counts increase gradually until full regeneration is achieved. The significant increase in MTTD of ricin-intoxicated mice described above, occurred when irradiation preceded intoxication by three days. To determine whether recovery of neutrophil production in irradiated mice reverses the apparent effect of TBI on MTTD, we extended our study to include groups of mice exposed to a lethal dose of ricin at later time points after irradiation, 9 days or 18 days post-TBI. Although the mice of both irradiated groups were similarly neutropenic at the time of intoxication, subsequent exposure to ricin caused a sharp increase in peripheral blood neutrophil counts, from 0.35 to 1.36 × 10³ cells/μL, only in the mice intoxicated at the latest time point, 18 days following TBI (Table 1). Furthermore, the MTTD value calculated for the mice exposed to ricin at 18 days post-TBI, 5.6 days, was markedly lower than the MTTD exhibited by mice exposed to ricin at 3 or 9 days post-TBI (8.1 and 8.4 days, respectively) and was similar to the MTTD value determined for non-irradiated ricin-intoxicated mice (6.3 days). These two parallel observations, the ability of bone marrow to produce neutrophils in mice exposed to ricin at day 18 post-TBI and the markedly lower MTTD exhibited by these mice following intoxication, suggests a causal linkage between the renewal of peripheral neutrophilia and the cancellation of TBI-induced MTTD prolongation. It should be mentioned that counts of leukocytes other than neutrophils, following intoxication at day 9 and 18 post irradiation were similar (data not shown), supporting our hypothesis that the noted changes in MTTD values following TBI are governed mainly by the presence or absence of neutrophils.

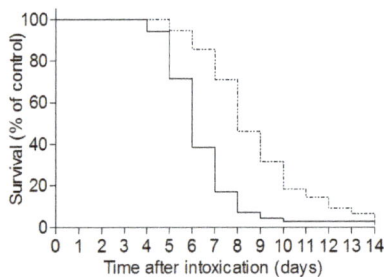

Figure 3. TBI-induced leukopenia extends the mean time to death in ricin intoxicated mice: irradiated (dashed line, *n* = 76) or naive mice (solid line, *n* = 70) were intranasally intoxicated with ricin (7 μg/kg). Mice survival was monitored until day 14 following intoxication.

Table 1. TBI-induced stable leukopenia extends the mean time to death in ricin intoxicated mice. Mice were intoxicated at the indicated time-points following irradiation (Intoxication Time). Blood neutrophil counts were determined prior to intoxication and three days following intoxication. MTTD was determined. * $p < 0.05$ in comparison to neutrophil counts at intoxication day. ** $p < 0.01$ in comparison to neutrophil counts at intoxication day.

Intoxication Time	Blood Neutrophil Count (10³ Cells/μL)		MTTD (Days)
	At Exposure Day	**3 Days Post Exposure**	
Non irradiated mice	1.53 ± 0.60 (*n* = 12)	13.43 ± 1.56 ** (*n* = 3)	6.3 (*n* = 18)
3 days post irradiation	0.18 ± 0.09 (*n* = 9)	0.40 ± 0.45 (*n* = 4)	8.1 (*n* = 11)
9 days post irradiation	0.29 ± 0.26 (*n* = 3)	0.50 ± 0.23 (*n* = 2)	8.4 (*n* = 14)
18 days post irradiation	0.35 ± 0.12 (*n* = 3)	11.36 ± 7.25 * (*n* = 4)	5.6 (*n* = 9)

2.4. Anti-Inflammatory Effects Induced by TBI Following Intranasal Ricin Intoxication

To further characterize the effects of TBI on pulmonary ricinosis, we compared the occurrence and progression of various ricin-intoxication-associated inflammatory-related processes in irradiated and non-irradiated mice. To this end, bronchoalveolar lavage fluids (BALFs) were collected before and at 72 h post-intoxication, and pro-inflammatory cytokines, as well as edema and tissue degradation markers were measured. The pro-inflammatory cytokines examined were TNFα, IL-1β, and IL-6, which were previously determined to be elevated in mice lungs following ricin exposure [11,16]. The edema markers measured comprise total protein, as well as cholinesterase (ChE), the latter being a serum-resident enzyme which is normally confined to the bloodstream and appears at elevated levels in the lungs only in the case where the pulmonary epithelial–endothelial barrier was disrupted [11]. The tissue degradation markers measured included secretory phospholipase A2 (sPLA₂), and xanthine oxidase (XO), representing lipolytic and oxidant activities, respectively. TNFα levels were found to be similarly elevated at 72 h post exposure in both irradiated and non-irradiated mice. In contrast, IL-1β levels in the non-irradiated and irradiated groups were 100 and 5 pg/mL, respectively, and IL-6 levels were 1700 and 360 pg/mL (~80% reduction), respectively, indicative of a reduced inflammatory response [11]. Likewise, significant attenuation of damage markers was noted in the irradiated mice. In the case of edema markers, approximately 50% reductions were determined in the irradiated mice (130 versus 270 mU/mL ChE; and 3 versus 5 mg/mL protein, in the BALFs of irradiated and not irradiated, ricin-intoxicated mice, respectively). The oxidative stress marker XO was also found to be lower, by nearly half, in the irradiated mice (2.5 and 4 mU/mL in TBI and non-irradiated groups, respectively). Thus, while a three-fold increase in sPLA$_2$ was measured at 72 h after exposure to ricin in the BALFs of non-irradiated mice, the levels of this marker were very low already at the time of intoxication in the BALFs of the irradiated mice, and remained so until 72 h post exposure (Table 2). Taken together, these findings suggest that TBI-induced neutrophil depletion leads to a wide-range of anti-inflammatory effects following ricin intoxication.

Table 2. Pro-inflammatory markers in the BALFs of irradiated and non-irradiated mice following ricin intoxication. Irradiated (TBI) or non-irradiated (none) mice were intranasally intoxicated with ricin (7 µg/kg) and BALF samples collected before (0) or 72 h (72) after intoxication were monitored for TNFα, IL-1β, IL-6, cholinesterase (ChE), protein, xanthine oxidase (XO), and secretory phospholipase A2 (sPLA$_2$). n = 3–5. * $p < 0.05$ between tested group and control (t = 0). ** $p < 0.01$ between tested group and control. # $p < 0.05$ in comparison to parallel tested group of non-irradiated mice at the same time point post exposure. ## $p < 0.01$ in comparison to parallel tested group of non-irradiated mice at the same time point post exposure.

Marker	Pre-Intoxication Treatment	Time after Exposure (h)	
		0	72
TNFα (pg/mL)	none	37 ± 16	80 ± 27 *
	TBI	35 ± 5	106 ± 32 *
IL-1β (pg/mL)	none	0 ± 0	108 ± 76
	TBI	3 ± 4	4 ± 6 #
IL-6 (pg/mL)	none	3 ± 2	1695 ± 711 **
	TBI	0 ± 0	363 ± 99 *##
ChE (mU/mL)	none	3 ± 1	273 ± 38 **
	TBI	5 ± 1	130 ± 38 **##
Protein (mg/mL)	none	0.4 ± 0.1	5.2 ± 1.1 **
	TBI	0.7 ± 0.1	2.8 ± 0.5 **##
XO (mU/mL)	none	0.3 ± 0.3	3.7 ± 0.6 **
	TBI	0.2 ± 0.1	2.4 ± 0.6 **##
sPLA$_2$ (U/mL)	none	7.8 ± 5.8	24.5 ± 8.7 *
	TBI	1.8 ± 1.8	1.6 ± 3.2 ##

To further probe the immunomodulatory effects of TBI, we next examined whether the irradiation allows extension of the therapeutic time window for anti-ricin antibody treatment of mice following ricin intoxication. To this end, non-irradiated and irradiated mice intranasally intoxicated with ricin (three days following TBI in the latter group), were treated with anti-ricin antibody at 48 or 72 h following intoxication, and survival rates were monitored over a period of 14 days. When antibody treatment was applied at 48 h post-exposure, survival rates of the irradiated mice were approximately 40%, while only negligible survival levels (4%) were documented for the non-irradiated mice treated with anti-ricin antibodies at this time point. Significant surviving rates (36%) were observed for irradiated mice treated at 72 h post exposure as well. At this late time point, antibody treatment offered no protection to the non-irradiated ricin-intoxicated mice. Interestingly, the irradiated mice exhibited similarly prolonged MTTD values in comparison to non-irradiated mice (approximately two days longer), whether or not they were subjected to antibody treatment (Table 3).

Table 3. Survival rates and mean time to death values of irradiated and non-irradiated ricin-intoxicated mice subjected to post-exposure anti-ricin antibody treatment. Irradiated (TBI) or non-irradiated (none) mice were intranasally intoxicated with ricin (7 μg/kg) and then treated or not (—) with anti-ricin antibody at 48 or 72 h post exposure.

Anti-Ricin Ab Treatment Following i.n. Exposure (2LD$_{50}$)					
Time of Ab Treatment (h)	—		48		72	
Pretreatment of Mice	none	TBI	none	TBI	none	TBI
% survival	1	1	4	42	0	36
(survivors/total)	(1/70)	(1/76)	(1/23)	(10/24)	(0/7)	(4/11)
MTTD (days)	6.4	9.4	6.0	8.7	3.6	8.4

As mentioned, ricin-related pathology could be attributed to both direct depurination-mediated cell damage and death, and on the other hand, to the indirect damage stemming from the raging inflammatory process, which in turn leads to respiratory insufficiency. The TBI-related dampening of ricin toxicity detailed above seems to be associated to with the mitigation of the inflammatory process. To evaluate whether irradiation of mice affects the catalytic performance of the toxin, irradiated and non-irradiated mice were intoxicated with ricin, lungs were harvested, and depurination ratios in the lung cells were calculated (Figure 4). Nearly equal depurination levels were observed for irradiated and non-irradiated mice at all time points examined, clearly demonstrating that ribosome depurination was not affected by the irradiation. Thus, one may conclude that the TBI-related beneficial effects observed in context of pulmonary ricinosis were entirely of an anti-inflammatory nature.

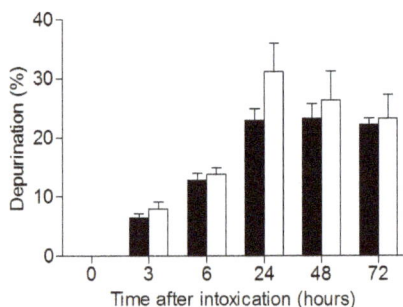

Figure 4. Catalytic damage of 28S rRNA in lungs of irradiated and non-irradiated mice intoxicated with ricin. Irradiated (white bars) and non-irradiated (black bars) mice were intranasally intoxicated with ricin (7 μg/kg), lungs were harvested at different time points following intoxication and catalytic damage (depurination) was quantified. $n = 4$.

3. Discussion

Pulmonary ricin intoxication is characterized by a confined lung pathology associated with cytokine storm, massive neutrophil infiltration, and ultimately, severe edema formation, leading to respiratory failure and death. Neutrophils are considered a major hallmark of ricin- [11,14,16], as well as non-ricin- [19,20,22,23] mediated lung injuries, where aggressive or prolonged neutrophil responses result in deleterious inflammatory conditions and tissue destruction, while on the other hand, decreased pulmonary neutrophil infiltration is associated with attenuation of injury severity [18,21,26].

In this work, TBI was employed to achieve a stable, long lasting neutropenia, aiming at selectively attenuating the inflammatory-related processes following ricin intoxication, without affecting the ribosomal damage induced by the catalytic activity of the toxin (28S rRNA depurination). Since pulmonary exposure to ricin induces neutrophil activation and recruitment, we first verified that irradiation intensity of 6.5 Gy, previously reported to induce a stable neutropenia in CD-1 mice [34], promotes stable neutropenia even when mice are subjected to intranasal ricin intoxication. Indeed, leukopenia/neutropenia remained stable in irradiated ricin-intoxicated mice at least for a week in the peripheral blood. The TBI-induced leukopenia was characterized by substantial reductions in pulmonary T- and B-lymphocyte counts (Figure 1D,E), and most notably in neutrophil counts. Although both pulmonary lymphocytopenia and neutropenia were observed at all time points tested following intoxication (Figure 1G), the TBI-induced effect is mainly related to the neutrophil depletion, since lymphocyte count decrease occurs following ricin intoxication regardless of irradiation, whereas neutrophil counts increase extensively following extravasation into the lungs, in non-irradiated mice intranasally exposed to ricin. It should be noted in this context, that while most hematopoietic-derived inflammatory cells are highly susceptible to ricin-induced protein synthesis arrest and cell death, ricin binds poorly if at all to neutrophils and hence, this cell type is not directly affected by exposure to the toxin [35].

Since irradiation in itself may injure various organs, including the lungs [36], we determined that under the experimental conditions of the present study, TBI did not inflict any notable damage to the ricin-intoxicated mice. Indeed, as reflected by the similar patterns of weight loss (Figure 2A) and parenchymal cell injury (Figure 2B) in both irradiated and non-irradiated mice, the stable leukopenia obtained by irradiation seems to be without overt toxicity. The single TBI-related deleterious effect noticed was a minor reduction in epithelial cell counts, in comparison to the non-irradiated mice (Figure 2C), whereas no observed TBI-mediated endothelial cell injury was noticed (Figure 2D).

While TBI did not improve survival rates following ricin intoxication, a significant increase in MTTD, from 6.3 to 8.7 days, was measured in ricin intoxicated mice subjected to TBI (Figure 3). This prolongation in time to death required a stable TBI-induced neutropenia, as MTTD prolongation was reversed when intoxication conditions allowed for peripheral blood neutrophilia formation in ricin intoxicated mice. Thus, prolonged MTTDs were measured in irradiated mice that were intoxicated at three and nine days following TBI, in which case peripheral neutropenia remained stable at three days following intoxication. In contrast, when mice were intoxicated at 18 days following TBI, a marked neutrophilia was exhibited at three days following intoxication, and indeed these mice displayed a considerably lower MTTD value, similar to that demonstrated by non-irradiated ricin-intoxicated mice (Table 1). Hence, it seems that the regeneration of neutrophil production capability by the bone marrow is critical for the development of full-blown pulmonary ricinosis.

In addition to neutrophil depletion, our findings clearly illustrate that TBI induces pronounced anti-inflammatory effects in ricin intoxicated mice (Table 2), as reflected by reduced cytokine response and damage markers in the BALFs collected at 72 h post exposure. IL-6, a prognostic biomarker of acute lung injury severity in pulmonary ricinosis [11] as well as in pulmonary abrinosis [42], was dramatically reduced by ~80%, in irradiated mice intranasally exposed to ricin, compared to the non-irradiated corresponding group. In addition, IL-1β levels of irradiated mice returned to the basal levels (namely the levels measured in naive mice) at 72 h after ricin intoxication, while significantly higher levels were measured in the ricin intoxicated, non-irradiated mice at this time point. It was previously demonstrated that IL-1β production is critical for the lung injury progression in pulmonary ricinosis, as markedly attenuated lung injury severity was observed in IL-1β depleted mice following intratracheal ricin exposure [16]. TNFα plays a critical role in lung pathologies, including ARDS [43]. However, the levels detected in the irradiated mice were similar to those measured in the non-irradiated mice following intoxication. TNFα is an early cytokine secreted primarily by activated monocytes or macrophages, which were shown to be strongly and rapidly affected by direct ricin activity [9,35] and therefore secretion of TNFα, which is not related to neutrophil activity, is not affected by irradiation.

Neutrophils play a crucial role in the development of pulmonary vascular hyperpermeability [22,44]. Accordingly, edema markers were assessed in irradiated mice following intranasal ricin exposure. In comparison to non-irradiated ricin-intoxicated mice, ricin-intoxicated TBI-mice exhibited markedly reduced ChE and total protein levels (~40% reduction) in the BALF. Likewise, XO levels in BALFs of ricin-intoxicated mice were found significantly lower following irradiation. XO is an important source of reactive oxygen species (ROS) in lung pathologies [45,46]. In particular, XO was shown to actively contribute to pulmonary edema formation following ischemia/reperfusion (I/R), via direct ROS-induced pulmonary endothelium injury [47], as well as following LPS administration [48]. Levels of the lipolytic enzyme sPLA$_2$, a potent mediator of inflammation responsible for hydrolysis and degradation of surfactant phospholipids [49,50], rose significantly after ricin intoxication only in non-irradiated mice. In contrast, its levels in the irradiated mice, which were lower than the levels measured in naive mice, did not alter following ricin intoxication at all time points examined. The markedly low levels of sPLA$_2$ observed in irradiated mice are presumed to be a direct outcome of the ensuing leukopenia formation, since this enzyme is secreted mainly by the white blood cells macrophages, lymphocytes, NK cells, and neutrophils [50–54].

According to the literature, lung neutrophil infiltration is induced by IL-1β [16], IL-6 [55–57], XO [58–60], and sPLA$_2$ [61,62]. The finding that these markers are significantly reduced in neutrophil-depleted mice intoxicated with ricin, may suggest that neutrophils operate upstream to these markers in a positive feedback loop, enhancing damage propagation in lung injuries, rather than being downstream responders to pro-inflammatory signals.

Previously, we have shown that immunomodulators attenuate the inflammatory response and improve survival rates when co-administered to ricin-intoxicated mice with anti-ricin antibody. The results obtained in this work provide supplementary data regarding the clinical importance of inflammatory mitigation in the course of pulmonary ricinosis, since TBI expands the therapeutic window in irradiated animals treated with anti-ricin antibody. While it was impossible to rescue non-irradiated mice by anti-ricin antibody treatment at 48 h post exposure, the survival levels of TBI-subjected mice were as high as ~40% at this point of care. Moreover, treating irradiated mice as late as 72 h following intoxication resulted in a similar survival rate, attesting to the significant anti-inflammatory effect of TBI (Table 3).

We are fully aware that in addition to neutropenia, TBI induced stable lymphocytopenia and in our system the role of these cells could not be excluded. However, the prolongation of MTTD was shown to be dependent on neutrophil production, and not T- and B-cell production. We would like to point out in this context that since this study was performed with outbred mice, employing an adoptive transfer approach to prove unequivocally the role of the neutrophils in promotion of toxicity is not possible. Understandably, TBI is not an applicable clinical countermeasure against ricin intoxication, however, the results obtained in this work strongly strengthen our findings from the past, indicating that attenuation of inflammation is critical for optimal treatment of pulmonary ricinosis, especially if antibody-based treatment begins at late time points following intoxication [11,14]. Furthermore, the present study strongly suggests that clinical approaches aimed to attenuate neutrophil-mediated damage could be implemented in the future in the development of an effective anti-ricin medical countermeasure. Some drug candidates aiming at suppressing neutrophil activity are clinically available, and others are under pre-clinical evaluation [63]. Moreover, since TBI induces an immunosuppressive effect, immunosuppressant agents could also be tested as a treatment for pulmonary ricinosis, pending adverse reactions and toxicity. The present study shows that improved survival rates of ricin-intoxicated mice could be reached only following irradiation in conjunction with anti-ricin antibody treatment. The fact that survival rates were not increased by attenuation of inflammation via irradiation in itself seems to stem on one hand from the short term effect of irradiation and on the other hand, from the direct non-inflammogenic effects of ricin, namely depurination induced tissue destruction, which is not affected by TBI.

4. Materials and Methods

4.1. Ricin Preparation

Crude ricin was prepared from seeds of endemic *Ricinus communis*, essentially as described before [64]. Briefly, seeds were homogenized in a blender (Waring, Torrington, CT, USA) in 5% acetic acid (Merck, Darmstadt, Germany)/phosphate buffer (Na_2HPO_4, pH 7.4, Sigma-Aldrich, Rehovot, Israel). The homogenate was centrifuged and the clarified supernatant containing the toxin was subjected to ammonium sulfate (Merck, Darmstadt, Germany) precipitation (60% saturation). The precipitate was dissolved in PBS (Bioligical Industries, Beit Haemek, Israel) and dialyzed extensively against the same buffer. The toxin preparation appeared on a Coomassie Blue (Bio-Rad, Rishon Le Zion, Israel) stained non-reducing 10% polyacrylamide gel (ThermoFisher Scientific, Carlsbad, CA, USA) as two major bands of molecular weight of approximately 65 kDa (=ricin toxin, ~80%) and 120 kDa (=*ricinus communis* agglutinin, RCA, ~20%). Protein concentration was determined as 2.86 mg/mL by 280 nm absorption (NanoDrop 2000, Thermo Fisher Scientific, Waltham, MA, USA).

4.2. Anti-Ricin Antibodies

Rabbit polyclonal anti-ricin antibodies were prepared as described before [11].

4.3. Animal Studies

Animal experiments were performed in accordance with the Israeli law and were approved by the Ethics Committee for animal experiments at the Israel Institute for Biological Research (project identification code M-01-2012, date of approval 2 January 2012). Treatment of animals was in accordance with regulations outlined in the USDA Animal Welfare Act and the conditions specified in the National Institute of Health Guide for Care and Use of Laboratory Animals.

All animals in this study were female CD-1 mice (Charles River Laboratories Ltd., Margate, UK) weighing 27–32 grams. Prior to exposure, animals were habituated to the experimental animal unit for five days. All mice were housed in filter-top cages in an environmentally controlled room and maintained at 21 $\pm$ 2 °C and 55 $\pm$ 10% humidity. Lighting was set to mimic a 12/12 h dawn to dusk cycle. Animals had access to food and water ad libitum.

For intoxication, mice were anesthetized by an intraperitoneal injection of ketamine (1.9 mg/mouse, Vetoquinol, Lure, France) and xylazine (0.19 mg/mouse, Eurovet Animal Health, AD Bladel, The Netherlands). Crude ricin (50 µL; 7 µg/kg diluted in PBS) was applied intranasally (2 × 25 µL) and mortality was monitored over 14 days. Preceding these studies, we determined that 3.5 µg crude ricin/kg body weight is approximately equivalent to one mouse (intranasal) LD_{50} (95% confidence intervals of 2.3 to 4.5 µg/kg body weight).

A volume of 100 µL of anti-ricin antibody preparation was delivered intravenously at the indicated times following intoxication.

4.4. Total Body Irradiation (TBI) Protocol

Mice were subjected to a single dose of 6.5 Gy whole body irradiation. For this purpose, the X-ray biological irradiator XRAD 320 (Precision X-ray, North Branford, CT, USA) at the Weizmann Institute of Science was used. The irradiation time was approximately 520 s.

4.5. Peripheral Blood Counts

White blood cells- and neutrophil-counts were determined in peripheral blood. Samples at a volume of 50 µL were collected from the tail vein of mice into EDTA containing tubes (BD, Franklin Lakes, NJ, USA) and were analyzed using Veterinary Multi-species Hematology System Hemavet 850 (Drew Scientific, Miami Lakes, FL, USA).

4.6. Flow Cytometry of the Lungs

Lungs were harvested, cut into small pieces, and digested for 2 h at 37 °C with 4 mg/mL collagenase D (Roche, Mannheim, Germany) in PBS Ca^{+2} Mg^{+2} (Biological Industries, Beit Haemek, Israel). The tissue was then meshed through a 40 μm cell strainer and red blood cells were lysed. Cells were co-stained for surface markers in a flow cytometry buffer (PBS with 2% FCS, 0.1% sodium azide, and 5 mM EDTA) as previously described [35] and analyzed using FACSCalibur (BD Biosciences, San Jose, CA, USA) and FlowJo software (version 7.1.2, Tree Star, Ashland, OR, USA, 2007).

4.7. Bronchoalveolar Lavage Fluid (BALF) Preparation and Analysis

BALFs, collected by instillation of 1 mL PBS at room temperature, were centrifuged at 3000 rpm at 4 °C for 10 min. Supernatants were collected and stored at −20 °C until further use.

BALF levels of TNFα, IL-1β, and IL-6 were determined by ELISA (R&D systems, Minneapolis, MN, USA).

Cholinesterase (ChE) enzymatic activity was measured according to Ellman [65]. Assays were performed in the presence of 0.5 mM acetylthiocholine (Sigma-Aldrich, Rehovot, Israel), 50 mM sodium phosphate buffer pH 8.0 (Sigma-Aldrich, Rehovot, Israel), 0.1 mg/mL BSA (Sigma-Aldrich, Rehovot, Israel), and 0.3 mM 5,5′-dithiobis-(2-nitrobenzoic acid, Sigma-Aldrich, Rehovot, Israel). The assay was carried out at 27 °C and monitored by a Thermomax microplate reader (Molecular Devices, Ramsey, MN, USA). Protein levels in BALF were determined by 280 nm absorption (NanoDrop 2000, ThermoFisher Scientific, Waltham, MA, USA). Xanthine oxidase (XO, Molecular Probes, Eugene, OR, USA) in BALF was determined by an activity assay kit.

Levels of secretory phospholipase A2 (sPLA$_2$) were determined by an activity assay kit (Assay Designs, Ann Arbor, MI, USA).

4.8. Depurination Assay

Depurination was quantified as previously described [9]. Briefly, reverse transcriptase (RT) reaction was conducted with two oligonucleotids primers: the first, R-1-GGTAGACACCCTAATACT marked with FAM, and the second, R-HEX-CTTTGATTGGTCCTAAGGGAGTCATT marked with HEX. The primers and RNA were incubated with a RT mix containing M-MLV RT, DTT, dNTPs, and RNasin (Promega, Madison, WI, USA), for 20 min incubation at 37 °C and then another 20 min at 48 °C. The cDNA that was produced in the reaction was later separated by electrophoresis in the GeneScan (GeneScan, ABI PRISM 310 Genetic Analyzer, Applied Biosystems, Thermo Fisher Scientific, Waltham, MA, USA). ROX (MapMarker 400, BioVentures, Wellesley, MA, USA) served as a size marker.

4.9. Statistical Analysis

Individual groups were compared using unpaired *t*-test analysis. To estimate *p* values, all statistical analyses were interpreted in a two-tailed manner. Values of $p < 0.05$ were considered to be statistically significant. Kaplan–Meier analysis was performed for survival curves. All data is presented as means ± SEM.

Author Contributions: Conceived and designed the study: Y.G., A.S., C.K., and T.S.; Acquired data: Y.G., A.S., R.F., S.E., M.A., and T.S.; Analyzed and interpreted data: Y.G., A.S., R.F., C.K., and T.S.; Wrote manuscript: Y.G., A.S., C.K., and T.S.; All authors read and approved the manuscript prior to submission.

Conflicts of Interest: The authors declare no conflict of interest.

References

1. Olsnes, S.; Kozlov, J.V. Ricin. *Toxicon* **2001**, *39*, 1723–1728. [CrossRef]
2. Greenfield, R.A.; Brown, B.R.; Hutchins, J.B.; Iandolo, J.J.; Jackson, R.; Slater, L.N.; Bronze, M.S. Microbiological, biological, and chemical weapons of warfare and terrorism. *Am. J. Med. Sci.* **2002**, *323*, 326–340. [CrossRef] [PubMed]

3. Audi, J.; Belson, M.; Patel, M.; Schier, J.; Osterloh, J. Ricin poisoning: A comprehensive review. *JAMA* **2005**, *294*, 2342–2351. [CrossRef] [PubMed]

4. Endo, Y.; Tsurugi, K. Rna n-glycosidase activity of ricin a-chain. Mechanism of action of the toxic lectin ricin on eukaryotic ribosomes. *J. Biol. Chem.* **1987**, *262*, 8128–8130. [PubMed]

5. Hartley, M.R.; Lord, J.M. Cytotoxic ribosome-inactivating lectins from plants. *Biochim. Biophys. Acta* **2004**, *1701*, 1–14. [CrossRef] [PubMed]

6. Olmo, N.; Turnay, J.; Gonzalez de Buitrago, G.; Lopez de Silanes, I.; Gavilanes, J.G.; Lizarbe, M.A. Cytotoxic mechanism of the ribotoxin alpha-sarcin. Induction of cell death via apoptosis. *Eur. J. Biochem.* **2001**, *268*, 2113–2123. [CrossRef] [PubMed]

7. Olsnes, S.; Fernandez-Puentes, C.; Carrasco, L.; Vazquez, D. Ribosome inactivation by the toxic lectins abrin and ricin. Kinetics of the enzymic activity of the toxin a-chains. *Eur. J. Biochem.* **1975**, *60*, 281–288. [CrossRef] [PubMed]

8. Olivieri, F.; Prasad, V.; Valbonesi, P.; Srivastava, S.; Ghosal-Chowdhury, P.; Barbieri, L.; Bolognesi, A.; Stirpe, F. A systemic antiviral resistance-inducing protein isolated from clerodendrum inerme gaertn. Is a polynucleotide:Adenosine glycosidase (ribosome-inactivating protein). *FEBS Lett.* **1996**, *396*, 132–134. [CrossRef]

9. Falach, R.; Sapoznikov, A.; Gal, Y.; Israeli, O.; Leitner, M.; Seliger, N.; Ehrlich, S.; Kronman, C.; Sabo, T. Quantitative profiling of the in vivo enzymatic activity of ricin reveals disparate depurination of different pulmonary cell types. *Toxicol. Lett.* **2016**, *258*, 11–19. [CrossRef] [PubMed]

10. Knight, D.A.; Holgate, S.T. The airway epithelium: Structural and functional properties in health and disease. *Respirology* **2003**, *8*, 432–446. [CrossRef] [PubMed]

11. Gal, Y.; Mazor, O.; Alcalay, R.; Seliger, N.; Aftalion, M.; Sapoznikov, A.; Falach, R.; Kronman, C.; Sabo, T. Antibody/doxycycline combined therapy for pulmonary ricinosis: Attenuation of inflammation improved survival of ricin-intoxicated mice. *Toxicol. Rep.* **2014**, *1*, 496–504. [CrossRef]

12. Wilhelmsen, C.L.; Pitt, M.L. Lesions of acute inhaled lethal ricin intoxication in rhesus monkeys. *Vet. Pathol.* **1996**, *33*, 296–302. [CrossRef] [PubMed]

13. Katalan, S.; Falach, R.; Rosner, A.; Goldvaser, M.; Brosh-Nissimov, T.; Dvir, A.; Mizrachi, A.; Goren, O.; Cohen, B.; Gal, Y.; et al. A novel swine model of ricin-induced acute respiratory distress syndrome. *Dis. Model. Mech.* **2017**, *10*, 173–183. [CrossRef] [PubMed]

14. Gal, Y.; Sapoznikov, A.; Falach, R.; Ehrlich, S.; Aftalion, M.; Sabo, T.; Kronman, C. Potent antiedematous and protective effects of ciprofloxacin in pulmonary ricinosis. *Antimicrob. Agents Chemother.* **2016**, *60*, 7153–7158. [PubMed]

15. DaSilva, L.; Cote, D.; Roy, C.; Martinez, M.; Duniho, S.; Pitt, M.L.; Downey, T.; Dertzbaugh, M. Pulmonary gene expression profiling of inhaled ricin. *Toxicon* **2003**, *41*, 813–822. [CrossRef]

16. Lindauer, M.L.; Wong, J.; Iwakura, Y.; Magun, B.E. Pulmonary inflammation triggered by ricin toxin requires macrophages and il-1 signaling. *J. Immunol.* **2009**, *183*, 1419–1426. [CrossRef] [PubMed]

17. Poli, M.A.; Rivera, V.R.; Pitt, M.L.; Vogel, P. Aerosolized specific antibody protects mice from lung injury associated with aerosolized ricin exposure. *Toxicon* **1996**, *34*, 1037–1044. [CrossRef]

18. Abraham, E.; Carmody, A.; Shenkar, R.; Arcaroli, J. Neutrophils as early immunologic effectors in hemorrhage- or endotoxemia-induced acute lung injury. *Am. J. Physiol. Lung Cell. Mol. Physiol.* **2000**, *279*, L1137–L1145. [PubMed]

19. Till, G.O.; Johnson, K.J.; Kunkel, R.; Ward, P.A. Intravascular activation of complement and acute lung injury. Dependency on neutrophils and toxic oxygen metabolites. *J. Clin. Investig.* **1982**, *69*, 1126–1135. [CrossRef] [PubMed]

20. Yan, B.; Chen, F.; Xu, L.; Xing, J.; Wang, X. HMGB1-TLR4-IL23-IL17A axis promotes paraquat-induced acute lung injury by mediating neutrophil infiltration in mice. *Sci. Rep.* **2017**, *7*, 597. [CrossRef] [PubMed]

21. Gao, X.P.; Liu, Q.; Broman, M.; Predescu, D.; Frey, R.S.; Malik, A.B. Inactivation of cd11b in a mouse transgenic model protects against sepsis-induced lung pmn infiltration and vascular injury. *Physiol. Genom.* **2005**, *21*, 230–242. [CrossRef] [PubMed]

22. Koma, T.; Yoshimatsu, K.; Nagata, N.; Sato, Y.; Shimizu, K.; Yasuda, S.P.; Amada, T.; Nishio, S.; Hasegawa, H.; Arikawa, J. Neutrophil depletion suppresses pulmonary vascular hyperpermeability and occurrence of pulmonary edema caused by hantavirus infection in c.B-17 scid mice. *J. Virol.* **2014**, *88*, 7178–7188. [CrossRef] [PubMed]

23. Sercundes, M.K.; Ortolan, L.S.; Debone, D.; Soeiro-Pereira, P.V.; Gomes, E.; Aitken, E.H.; Neto, A.C.; Russo, M.; MR, D.I.L.; Alvarez, J.M.; et al. Targeting neutrophils to prevent malaria-associated acute lung injury/acute respiratory distress syndrome in mice. *PLoS Pathog.* **2016**, *12*, e1006054. [CrossRef] [PubMed]

24. Segel, G.B.; Halterman, M.W.; Lichtman, M.A. The paradox of the neutrophil's role in tissue injury. *J. Leukoc. Biol.* **2011**, *89*, 359–372. [CrossRef] [PubMed]

25. Wright, H.L.; Moots, R.J.; Bucknall, R.C.; Edwards, S.W. Neutrophil function in inflammation and inflammatory diseases. *Rheumatology* **2010**, *49*, 1618–1631. [CrossRef] [PubMed]

26. Zhao, Y.; Sharma, A.K.; LaPar, D.J.; Kron, I.L.; Ailawadi, G.; Liu, Y.; Jones, D.R.; Laubach, V.E.; Lau, C.L. Depletion of tissue plasminogen activator attenuates lung ischemia-reperfusion injury via inhibition of neutrophil extravasation. *Am. J. Physiol. Lung Cell. Mol. Physiol.* **2011**, *300*, L718–L729. [CrossRef] [PubMed]

27. Strickland, I.; Kisich, K.; Hauk, P.J.; Vottero, A.; Chrousos, G.P.; Klemm, D.J.; Leung, D.Y. High constitutive glucocorticoid receptor beta in human neutrophils enables them to reduce their spontaneous rate of cell death in response to corticosteroids. *J. Exp. Med.* **2001**, *193*, 585–593. [CrossRef] [PubMed]

28. Cox, G. Glucocorticoid treatment inhibits apoptosis in human neutrophils. Separation of survival and activation outcomes. *J. Immunol.* **1995**, *154*, 4719–4725. [PubMed]

29. Daffern, P.J.; Jagels, M.A.; Hugli, T.E. Multiple epithelial cell-derived factors enhance neutrophil survival. Regulation by glucocorticoids and tumor necrosis factor-alpha. *Am. J. Respir. Cell Mol. Biol.* **1999**, *21*, 259–267. [CrossRef] [PubMed]

30. Cui, Y.Z.; Hisha, H.; Yang, G.X.; Fan, T.X.; Jin, T.; Li, Q.; Lian, Z.; Ikehara, S. Optimal protocol for total body irradiation for allogeneic bone marrow transplantation in mice. *Bone Marrow Transplant.* **2002**, *30*, 843–849. [CrossRef] [PubMed]

31. Schwartz, E.; Lapidot, T.; Gozes, D.; Singer, T.S.; Reisner, Y. Abrogation of bone marrow allograft resistance in mice by increased total body irradiation correlates with eradication of host clonable t cells and alloreactive cytotoxic precursors. *J. Immunol.* **1987**, *138*, 460–465. [PubMed]

32. Shen, H.; Yu, H.; Liang, P.H.; Cheng, H.; XuFeng, R.; Yuan, Y.; Zhang, P.; Smith, C.A.; Cheng, T. An acute negative bystander effect of gamma-irradiated recipients on transplanted hematopoietic stem cells. *Blood* **2012**, *119*, 3629–3637. [CrossRef] [PubMed]

33. Neelis, K.J.; Visser, T.P.; Dimjati, W.; Thomas, G.R.; Fielder, P.J.; Bloedow, D.; Eaton, D.L.; Wagemaker, G. A single dose of thrombopoietin shortly after myelosuppressive total body irradiation prevents pancytopenia in mice by promoting short-term multilineage spleen-repopulating cells at the transient expense of bone marrow-repopulating cells. *Blood* **1998**, *92*, 1586–1597. [PubMed]

34. Heissig, B.; Rafii, S.; Akiyama, H.; Ohki, Y.; Sato, Y.; Rafael, T.; Zhu, Z.; Hicklin, D.J.; Okumura, K.; Ogawa, H.; et al. Low-dose irradiation promotes tissue revascularization through vegf release from mast cells and mmp-9-mediated progenitor cell mobilization. *J. Exp. Med.* **2005**, *202*, 739–750. [CrossRef] [PubMed]

35. Sapoznikov, A.; Falach, R.; Mazor, O.; Alcalay, R.; Gal, Y.; Seliger, N.; Sabo, T.; Kronman, C. Diverse profiles of ricin-cell interactions in the lung following intranasal exposure to ricin. *Toxins* **2015**, *7*, 4817–4831. [CrossRef] [PubMed]

36. Ghafoori, P.; Marks, L.B.; Vujaskovic, Z.; Kelsey, C.R. Radiation-induced lung injury. Assessment, management, and prevention. *Oncology* **2008**, *22*, 37–47; discussion 52–53. [PubMed]

37. Marks, L.B.; Yu, X.; Vujaskovic, Z.; Small, W., Jr.; Folz, R.; Anscher, M.S. Radiation-induced lung injury. *Semin. Radiat. Oncol.* **2003**, *13*, 333–345. [CrossRef]

38. Medhora, M.; Gao, F.; Jacobs, E.R.; Moulder, J.E. Radiation damage to the lung: Mitigation by angiotensin-converting enzyme (ace) inhibitors. *Respirology* **2012**, *17*, 66–71. [CrossRef] [PubMed]

39. Mehta, V. Radiation pneumonitis and pulmonary fibrosis in non-small-cell lung cancer: Pulmonary function, prediction, and prevention. *Int. J. Radiat. Oncol. Biol. Phys.* **2005**, *63*, 5–24. [CrossRef] [PubMed]

40. Pan, J.; Su, Y.; Hou, X.; He, H.; Liu, S.; Wu, J.; Rao, P. Protective effect of recombinant protein sod-tat on radiation-induced lung injury in mice. *Life Sci.* **2012**, *91*, 89–93. [CrossRef] [PubMed]

41. Tsoutsou, P.G.; Koukourakis, M.I. Radiation pneumonitis and fibrosis: Mechanisms underlying its pathogenesis and implications for future research. *Int. J. Radiat. Oncol. Biol. Phys.* **2006**, *66*, 1281–1293. [CrossRef] [PubMed]

42. Sabo, T.; Gal, Y.; Elhanany, E.; Sapoznikov, A.; Falach, R.; Mazor, O.; Kronman, C. Antibody treatment against pulmonary exposure to abrin confers significantly higher levels of protection than treatment against ricin intoxication. *Toxicol. Lett.* **2015**, *237*, 72–78. [CrossRef] [PubMed]

43. Mukhopadhyay, S.; Hoidal, J.R.; Mukherjee, T.K. Role of tnfalpha in pulmonary pathophysiology. *Respir. Res.* **2006**, *7*, 125. [CrossRef] [PubMed]

44. Finsterbusch, M.; Voisin, M.B.; Beyrau, M.; Williams, T.J.; Nourshargh, S. Neutrophils recruited by chemoattractants in vivo induce microvascular plasma protein leakage through secretion of tnf. *J. Exp. Med.* **2014**, *211*, 1307–1314. [CrossRef] [PubMed]

45. Komaki, Y.; Sugiura, H.; Koarai, A.; Tomaki, M.; Ogawa, H.; Akita, T.; Hattori, T.; Ichinose, M. Cytokine-mediated xanthine oxidase upregulation in chronic obstructive pulmonary disease's airways. *Pulm. Pharmacol. Ther.* **2005**, *18*, 297–302. [CrossRef] [PubMed]

46. Wright, R.M.; Ginger, L.A.; Kosila, N.; Elkins, N.D.; Essary, B.; McManaman, J.L.; Repine, J.E. Mononuclear phagocyte xanthine oxidoreductase contributes to cytokine-induced acute lung injury. *Am. J. Respir. Cell Mol. Biol.* **2004**, *30*, 479–490. [CrossRef] [PubMed]

47. Grosso, M.A.; Brown, J.M.; Viders, D.E.; Mulvin, D.W.; Banerjee, A.; Velasco, S.E.; Repine, J.E.; Harken, A.H. Xanthine oxidase-derived oxygen radicals induce pulmonary edema via direct endothelial cell injury. *J. Surg. Res.* **1989**, *46*, 355–360. [CrossRef]

48. Faggioni, R.; Gatti, S.; Demitri, M.T.; Delgado, R.; Echtenacher, B.; Gnocchi, P.; Heremans, H.; Ghezzi, P. Role of xanthine oxidase and reactive oxygen intermediates in lps- and tnf-induced pulmonary edema. *J. Lab. Clin. Med.* **1994**, *123*, 394–399. [PubMed]

49. Arbibe, L.; Koumanov, K.; Vial, D.; Rougeot, C.; Faure, G.; Havet, N.; Longacre, S.; Vargaftig, B.B.; Bereziat, G.; Voelker, D.R.; et al. Generation of lyso-phospholipids from surfactant in acute lung injury is mediated by type-ii phospholipase a2 and inhibited by a direct surfactant protein a-phospholipase a2 protein interaction. *J. Clin. Investig.* **1998**, *102*, 1152–1160. [CrossRef] [PubMed]

50. Touqui, L.; Arbibe, L. A role for phospholipase a2 in ards pathogenesis. *Mol. Med. Today* **1999**, *5*, 244–249. [CrossRef]

51. Costa-Junior, H.M.; Hamaty, F.C.; da Silva Farias, R.; Einicker-Lamas, M.; da Silva, M.H.; Persechini, P.M. Apoptosis-inducing factor of a cytotoxic t cell line: Involvement of a secretory phospholipase a2. *Cell Tissue Res.* **2006**, *324*, 255–266. [CrossRef] [PubMed]

52. Marshall, L.A.; Bolognese, B.; Roshak, A. Respective roles of the 14 kda and 85 kda phospholipase a2 enzymes in human monocyte eicosanoid formation. *Adv. Exp. Med. Biol.* **1999**, *469*, 215–219. [PubMed]

53. Rosenthal, M.D.; Gordon, M.N.; Buescher, E.S.; Slusser, J.H.; Harris, L.K.; Franson, R.C. Human neutrophils store type ii 14-kda phospholipase a2 in granules and secrete active enzyme in response to soluble stimuli. *Biochem. Biophys. Res. Commun.* **1995**, *208*, 650–656. [CrossRef] [PubMed]

54. Von Allmen, C.E.; Schmitz, N.; Bauer, M.; Hinton, H.J.; Kurrer, M.O.; Buser, R.B.; Gwerder, M.; Muntwiler, S.; Sparwasser, T.; Beerli, R.R.; et al. Secretory phospholipase a2-iid is an effector molecule of cd4+cd25+ regulatory t cells. *Proc. Natl. Acad. Sci. USA* **2009**, *106*, 11673–11678. [CrossRef] [PubMed]

55. Jones, M.R.; Quinton, L.J.; Simms, B.T.; Lupa, M.M.; Kogan, M.S.; Mizgerd, J.P. Roles of interleukin-6 in activation of stat proteins and recruitment of neutrophils during escherichia coli pneumonia. *J. Infect. Dis.* **2006**, *193*, 360–369. [CrossRef] [PubMed]

56. Leemans, J.C.; Vervoordeldonk, M.J.; Florquin, S.; van Kessel, K.P.; van der Poll, T. Differential role of interleukin-6 in lung inflammation induced by lipoteichoic acid and peptidoglycan from staphylococcus aureus. *Am. J. Respir. Crit. Care Med.* **2002**, *165*, 1445–1450. [CrossRef] [PubMed]

57. Rijneveld, A.W.; van den Dobbelsteen, G.P.; Florquin, S.; Standiford, T.J.; Speelman, P.; van Alphen, L.; van der Poll, T. Roles of interleukin-6 and macrophage inflammatory protein-2 in pneumolysin-induced lung inflammation in mice. *J. Infect. Dis.* **2002**, *185*, 123–126. [CrossRef] [PubMed]

58. Granell, S.; Gironella, M.; Bulbena, O.; Panes, J.; Mauri, M.; Sabater, L.; Aparisi, L.; Gelpi, E.; Closa, D. Heparin mobilizes xanthine oxidase and induces lung inflammation in acute pancreatitis. *Crit. Care Med.* **2003**, *31*, 525–530. [CrossRef] [PubMed]

59. Nielsen, V.G.; Tan, S.; Weinbroum, A.; McCammon, A.T.; Samuelson, P.N.; Gelman, S.; Parks, D.A. Lung injury after hepatoenteric ischemia-reperfusion: Role of xanthine oxidase. *Am. J. Respir. Crit. Care Med.* **1996**, *154*, 1364–1369. [CrossRef] [PubMed]

60. Terada, L.S.; Dormish, J.J.; Shanley, P.F.; Leff, J.A.; Anderson, B.O.; Repine, J.E. Circulating xanthine oxidase mediates lung neutrophil sequestration after intestinal ischemia-reperfusion. *Am. J. Physiol.* **1992**, *263*, L394–L401. [PubMed]

61. Lee, Y.M.; Hybertson, B.M.; Terada, L.S.; Repine, A.J.; Cho, H.G.; Repine, J.E. Mepacrine decreases lung leak in rats given interleukin-1 intratracheally. *Am. J. Respir. Crit. Care Med.* **1997**, *155*, 1624–1628. [CrossRef] [PubMed]
62. Munoz, N.M.; Meliton, A.Y.; Meliton, L.N.; Dudek, S.M.; Leff, A.R. Secretory group v phospholipase a2 regulates acute lung injury and neutrophilic inflammation caused by lps in mice. *Am. J. Physiol. Lung Cell. Mol. Physiol.* **2009**, *296*, L879–L887. [CrossRef] [PubMed]
63. Burgos, R.A.; Hidalgo, M.A.; Figueroa, C.D.; Conejeros, I.; Hancke, J.L. New potential targets to modulate neutrophil function in inflammation. *Mini Rev. Med. Chem.* **2009**, *9*, 153–168. [CrossRef] [PubMed]
64. Lin, J.Y.; Liu, S.Y. Studies on the antitumor lectins isolated from the seeds of ricinus communis (castor bean). *Toxicon* **1986**, *24*, 757–765. [CrossRef]
65. Ellman, G.L.; Courtney, K.D.; Andres, V., Jr.; Feather-Stone, R.M. A new and rapid colorimetric determination of acetylcholinesterase activity. *Biochem. Pharmacol.* **1961**, *7*, 88–95. [CrossRef]

toxins

MDPI

Article

Production, Characterisation and Testing of an Ovine Antitoxin against Ricin; Efficacy, Potency and Mechanisms of Action

Sarah J. C. Whitfield, Gareth D. Griffiths, Dominic C. Jenner, Robert J. Gwyther, Fiona M. Stahl, Lucy J. Cork, Jane L. Holley, A. Christopher Green * and Graeme C. Clark *

Chemical, Biological and Radiological Division, Dstl, Porton Down, Salisbury SP4 0JQ, UK;
SJWHITFIELD@dstl.gov.uk (S.J.C.W.); dcjenner@dstl.gov.uk (D.C.J.); RJGWYTHER@dstl.gov.uk (R.J.G.);
fmstahl@dstl.gov.uk (F.M.S.); ljcork@dstl.gov.uk (L.J.C.); jlholley@dstl.gov.uk (J.L.H.)
* Correspondence: acgreen@dstl.gov.uk (A.C.G.); gcclark@dstl.gov.uk (G.C.C.)

Academic Editors: Julien Barbier and Daniel Gillet
Received: 4 September 2017; Accepted: 13 October 2017; Published: 18 October 2017

Abstract: Ricin is a type II ribosome-inactivating toxin that catalytically inactivates ribosomes ultimately leading to cell death. The toxicity of ricin along with the prevalence of castor beans (its natural source) has led to its increased notoriety and incidences of nefarious use. Despite these concerns, there are no licensed therapies available for treating ricin intoxication. Here, we describe the development of a F(ab')$_2$ polyclonal ovine antitoxin against ricin and demonstrate the efficacy of a single, post-exposure, administration in an in vivo murine model of intoxication against aerosolised ricin. We found that a single dose of antitoxin afforded a wide window of opportunity for effective treatment with 100% protection observed in mice challenged with aerosolised ricin when given 24 h after exposure to the toxin and 75% protection when given at 30 h. Treated mice had reduced weight loss and clinical signs of intoxication compared to the untreated control group. Finally, using imaging flow cytometry, it was found that both cellular uptake and intracellular trafficking of ricin toxin to the Golgi apparatus was reduced in the presence of the antitoxin suggesting both actions can contribute to the therapeutic mechanism of a polyclonal antitoxin. Collectively, the research highlights the significant potential of the ovine F(ab')$_2$ antitoxin as a treatment for ricin intoxication.

Keywords: ricin; antibody; antitoxin; efficacy; intracellular trafficking

1. Introduction

Ricin is a toxic protein obtained from the seeds of the castor oil plant (*Ricinus communis*) which is grown commercially in many parts of the developing world for castor oil production. Although approximately 1000-fold less toxic than the botulinum toxins, ricin is considered a potential biological warfare agent because of the ease and rapidity with which large quantities can be produced and the wide availability of castor beans [1,2]. Historically, it has been employed in many criminal activities and recently it has been considered a weapon of choice for extremist and terrorist groups [3]. A significant number of fatal and sub-lethal cases of human intoxication have been reported following accidental or deliberate ingestion or from parenteral exposure to ricin [4]. The development of prophylactic and post-exposure therapies for ricin intoxication has been on-going for many years [5,6] and a continued requirement for the development of medical countermeasures for ricin has been identified [7]. At present, however, there is no specific prophylactic or post-exposure therapy available for the clinical management of individuals exposed to ricin.

Ricin is a 66 kilodalton (KDa) glycoprotein cytotoxin consisting of two polypeptide chains, termed the A-chain and the B-chain, which are linked by an easily reduced disulphide bond [8–10]. The toxin binds via its B chain to terminal galactose residues found on the surface of many cells and also via both the A and B chain to mannose receptors found on specific cell populations [11–13]. Following intra-nasal administration of ricin, the toxin binds initially to lung epithelia and subsequently to immune cells (e.g., macrophages and dendritic cells) [14]. This binding event triggers the endocytic uptake of the toxin into the cell. Once internalised the ricin is transported via the trans-Golgi network into the endoplasmic reticulum lumen where the enzymatically active A chain is translocated into the cytosol [15]. The toxin then enzymatically inactivates the ribosome though depurination of a single adenine residue in the 28 S ribosomal subunit. This event inactivates the 60 S ribosomal subunit, disrupts protein synthesis and results in cell death [1,6]. A single ricin molecule has been hypothesised to have the potential to inactivate multiple ribosomes and consequently killing a cell [16–18]. Ricin also causes wider pathological issues such as vascular leakage, pulmonary oedema (in part though uncontrolled cellular recruitment), cytokine storm and, ultimately, the death of the exposed individual though an acute respiratory distress syndrome (ARDS)-like condition [19–21]. Mitigating these wider impacts of acute lung injury caused by ricin intoxication is important; has led to testing of novel immunotherapeutic approaches; and has been linked to the efficacy of candidate therapies [22,23].

The various animal models available to study ricin intoxication were reviewed by Roy et al. [24]. The pathological effects and subsequent clinical signs of ricin intoxication depend on the route of exposure, as this dictates the subsequent tissue distribution of the toxin [25]. Following intravenous or intramuscular administration, lesions eventually develop in the spleen, liver and kidneys [26,27] whilst the lung remains unaffected [25]. After oral ingestion, the gastrointestinal tract is severely affected [28,29]. Inhalational exposure produces effects that are mainly confined to the respiratory tract [30,31]. Rhesus macaques exposed to a sub-lethal challenge with ricin exhibited dyspnea, tachypnea, laboured breathing and anorexia. These signs decreased substantially within 48–60 h and pathologically resulted in a sub-acute/chronic reparative process in the lung [31]. Retention of radio-labelled toxin within the lung has also been demonstrated using murine models of exposure perhaps explaining the extensive localised tissue damage caused by ricin [27]. However, despite often irreversible damage having been caused to cells there are often no visible signs of intoxication in animals exposed to low doses of ricin. The presence of a lag phase in the appearance of signs of intoxication is related to the dose of ricin and by the route by which the toxin was administered [25,28,29]. The delayed appearance of clinical signs and a narrow window for the post-exposure administration of antitoxin therapies, which are key aspects for an effective medical response to a ricin incident, means the diagnosis and subsequent treatment of ricin intoxication are technically challenging.

Neutralising antibodies, raised in animals, are often used for treatment of intoxication and are effective if passive immunisation occurs prophylactically or shortly after exposure to the toxin [1,32,33]. Prophylactic administration of IgG or antibody-fragments (e.g., F(ab')$_2$ and Fab'), which often have an improved bioavailability compared to the whole IgG molecule, can provide protection against subsequent toxin exposure [32–35]. In addition, antibody-fragments typically have a reduced incidence of adverse anaphylactoid or anaphylactic reactions on single or repeat administration of antitoxin than do IgG-based antitoxins [33,36,37].

The window of opportunity (WOO) for administration of therapeutic antibodies against ricin has been shown to be between 0 and ~8 h for full protection [38–40] and longer, 18–72 h for partial protection [41,42]. Here, we have evaluated a purified, despeciated, (Fab')$_2$ polyclonal ovine antibody made to Good Manufacturing Practice (GMP) which has also demonstrated a long window of opportunity for an antibody-based therapeutic; offering full protection up to 24 h and partial protection when administered up to 30 h after an inhaled aerosol ricin challenge. This route of intoxication is considered to be a key route of exposure for which treatment is required in a biodefense context [1,3,5]. In the process of conducting these studies, we have also established in vitro and in vivo neutralization assays to assess the potency of the antitoxin for the determination of the consistency of the manufacturing process and for batch release and stability studies. In parallel, we used cutting edge imaging flow cytometry methods to provide new insights in to the effects of this antitoxin on the uptake and intracellular transport of ricin that may contribute to its protective actions.

2. Results

2.1. Determination of the Ricin Neutralising Activity of the Antitoxin In Vitro

Prevention of the in vitro cytotoxic effects of ricin was used as a measure of the neutralizing activity of the antitoxin. Ricin at increasing concentrations and in the absence of the antitoxin produced a concentration-dependent cell death over 48 h in Vero cell cultures as measured by the WST-1 cell viability assay (Figure 1A). The LC_{50} (the concentration of ricin which kills 50% of the cell population) was determined to be 0.051 ng·mL^{-1}, with a 95% confidence interval (CI) of 0.030 to 0.089 ng·mL^{-1} (Figure 1A). Although the cell viability data were routinely normalized to control (100%) values, inconsistencies across the culture plates meant that on occasions more accurate and reliable estimates of the IC_{50} of ricin and EC_{50} of antitoxin were obtained using the top plateau of the viability data to define the maximum viability in an experiment (Figure 1A). It was also noted that in these assays ricin did not appear to result in complete loss of cell viability with a small ~10–20% WST 1 signal persisting at high ricin concentrations (Figure 1A,B) either because of a minor proportion of resistant cells or residual active dehydrogenase retained in cellular fragments. Premixing the ricin with antitoxin prior to addition of the mixture to the cells resulted in a very steep concentration-dependent protection of the cells from the cytotoxic effects of ricin (Figure 1B). The EC_{50} in this case was 377 ng·mL^{-1} (95% CI of 337 to 422 ng·mL^{-1}). Assuming that at the EC_{50} concentration the amount of free ricin must be equal to that causing 50% cell death, i.e., equivalent to the ricin LC_{50}, a neutralising ratio (µg of antitoxin required to neutralize 1 µg of ricin) can be calculated using Equation (1). On this occasion, the neutralising ratio was 106:1. Although characterisation using an ELISA approach suggests approximately 40% of the antitoxin is anti-ricin F(ab')$_2$, the neutralising ratios presented have not been corrected and are based upon the total protein content of the antitoxin material.

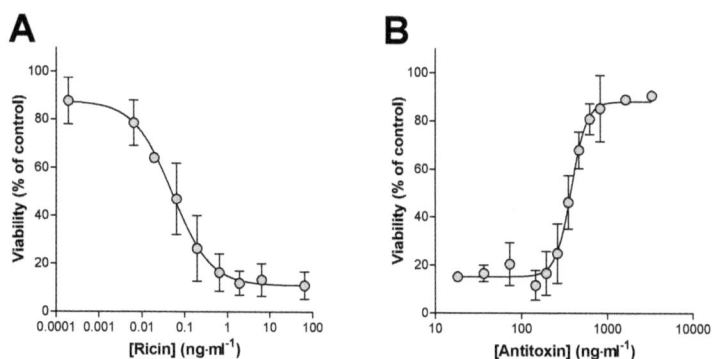

Figure 1. In vitro cytotoxicity of ricin and its neutralization by antitoxin. (**A**) Ricin caused a concentration dependent cytotoxicity in Vero cell cultures. (**B**) Pre-incubation of ricin (3.24 ng·mL^{-1}) with the indicated concentration of antitoxin resulted in neutralization of the cytotoxic effects of ricin. LC_{50} and EC_{50} values were determined from the 4-parameter logistical curve fits shown and used to calculate the neutralizing ratio as described in the main text. Data are shown as mean ± SD from three independent experiments in each case.

This cytotoxicity assay was characterized and validated sufficiently to support the production of antitoxin to Good Manufacturing Practice (GMP) standards and it enabled the evaluation of the ricin antitoxin potency and its stability during production and storage. The stability of the antitoxin, when stored at 2–8 °C, was tested using this assay at regular intervals from 0 to 72 months. Overall, there was no decrease in potency over this time period (data not shown), and the mean neutralising ratio, calculated from all 24 determinations, was 108:1.

2.2. Determination of the Ricin Neutralising Activity of the Antitoxin In Vivo

The ricin neutralising activity of the antitoxin was also determined in vivo by measuring the protection afforded in mice following intraperitoneal (i.p.) administration of ricin alone and of ricin premixed with increasing amounts of antitoxin. Increasing doses of ricin resulted in earlier deaths (Figure 2A) and a progressive increase in the probability of death (Figure 2A,B). Analysis of the dose–lethality data using a probit response model gave an LD_{50} for i.p. ricin of 0.377 µg·mouse^{-1} (95% CI 0.358–0.397 µg·mouse^{-1}). The LD_{99} dose of ricin was estimated to be 0.61 µg·mouse^{-1} (95% CI 0.549–0.705 µg·mouse^{-1}). Doses of ricin in excess of this would be expected to kill all animals challenged and therefore the administration of a challenge dose of 2 µg (approximately $6LD_{50}$) was selected for use during the neutralisation study. The majority of mice challenged with this dose of ricin died within two days with weight loss and signs of intoxication being apparent after only 24 h (Figure 3C,D).

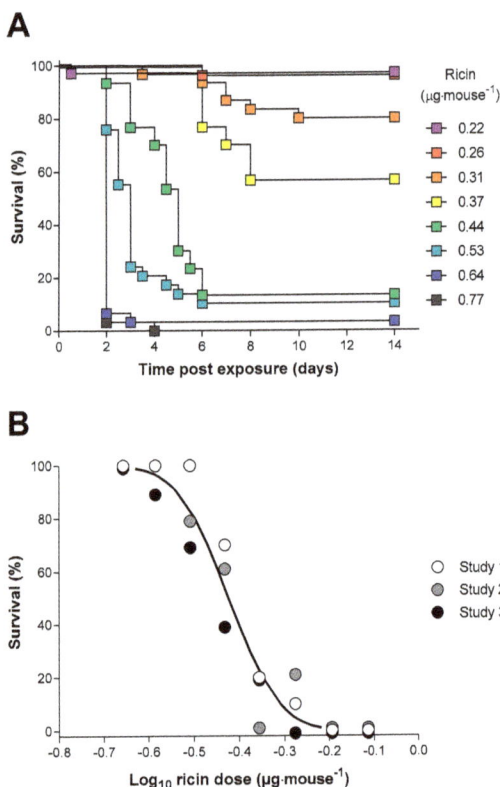

Figure 2. Ricin toxicity via the intraperitoneal (i.p.) route in the Balb/C mouse. (**A**) Kaplan–Meier plot of mouse survival following administration of the indicated concentrations of ricin. Data are shown for the combined studies, *n* = 30 for each dose. (**B**) Probit analysis of percentage survival of mice against the dose of ricin received: the curve was reconstructed from the output of the Minitab probit analysis, which was fitted to the combined dataset. Data points are shown as survival at 14 days from three independent experiments each with 10 mice per dose.

Increasing amounts of antitoxin, premixed with the ricin prior to i.p. administration, resulted in a delayed time to death and a protection from lethality which was dependent on the dose of antitoxin administered with the ricin challenge (Figure 3A,B). There was also a dose dependent decrease in the rate, and absolute amount, of weight loss (Figure 3C), and in a reduced score for clinical signs of intoxication (Figure 3D). For all but the highest doses of antitoxin, surviving animals exhibited signs of intoxication which persisted for 14 days after challenge at which point the experiment was terminated. In these studies, a small number of animals survived at low antitoxin doses probably because of misplaced i.p. injections [43]. Data from these animals was however included in the probit analysis which determined an ED_{50} for the antitoxin of 80 μg·mouse^{-1} (95% CI 79–81 μg·mouse^{-1}) and probit slope of 33 demonstrating a very steep dose–response relationship. The neutralising ratio (NR) can again be calculated based on the principles previously described (Section 2.1) and using Equation (2). Using this in vivo method, the neutralising ratio was determined to be 49:1.

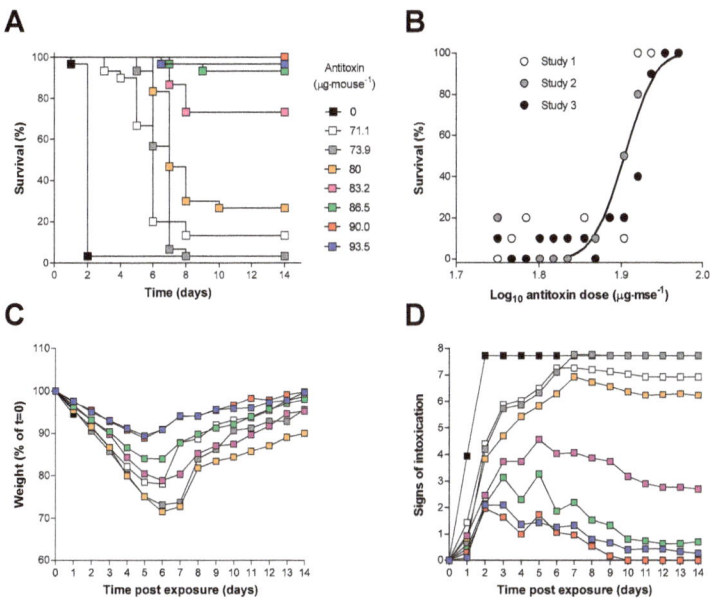

Figure 3. **An in vivo neutralising assay in Balb/C mice.** (**A**) Kaplan–Meier plot of survival in mice administered ricin (2 μg) pre-incubated with the indicated dose of antitoxin (the key shown in (**A**) also applies to (**C,D**)). Data are combined for three independent experiments (n = 30). (**B**) Probit analysis of survival at 14 days: the curve was reconstructed from the output of the Minitab probit analysis which was fitted to the combined dataset. Data points are shown for survival at 14 days from three independent experiments. (**C,D**) Body weight and signs of intoxication following administration of ricin and the indicated amount of antitoxin are shown. Data are combined from three independent experiments. Error bars have been omitted for clarity; the SDs are generally <10% and <2 units, respectively, for n = 30 mice in each dose group at the start of the study. Body weights are reported for all surviving animals at a particular time point. Signs of intoxication were scored according to Table S1.

2.3. Inhalation Toxicity of Ricin in Balb/C Mice

The efficacy of the antitoxin against an inhalation challenge to aerosolised ricin was determined in mice. Exposure of mice, head only, to increasing doses of ricin resulted in shorter times to death and increased probability of death (Figure 4A,B). Analysis of this dose–lethality data using a logistic model gave an estimate for the LCt_{50} of 7.2 mg·min·m^{-3} (95% CI 6.2–8.3 mg·min·m^{-3}).

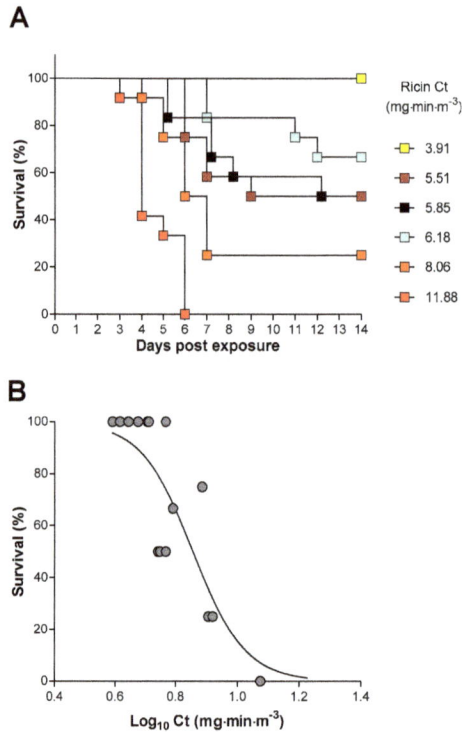

Figure 4. Characterisation of an inhalation model of ricin intoxication in Balb/C mice. (A) Kaplan–Meier plot of survival in groups of 12 mice following the head only exposure to the indicated Ct of ricin aerosol. (B) Dose–response analysis of survival at 14 days following inhalation of the indicated Ct of ricin aerosol. Each data point is from a group of 12 mice.

A similar analysis using data corrected for total ventilatory volume during exposure, obtained from plethysmography, gave an LD_{50} for the total inhaled dose of 10.4 $\mu g \cdot kg^{-1}$ (95% CI 9.3–11.6 $\mu g \cdot kg^{-1}$) (data not shown).

The estimated LCt_{99}, 17.7 $mg \cdot min \cdot m^{-3}$ (95% CI 9.0–24.8 $mg \cdot min \cdot m^{-3}$) was less than the challenge dose selected for the efficacy studies, $3LCt_{50}$, 21.5 $mg \cdot min \cdot m^{-3}$, and thus all animals exposed to this challenge dose were expected to die as a result of the ricin exposure. Indeed, all animals exposed to $3LCt_{50}$ of ricin succumbed to its lethal effects within 3–5 days of exposure (Figures 5A and 6A) and during this time they lost weight and exhibited typical signs of intoxication which were apparent in some mice by 24 h after exposure (Figure 5B,C).

2.4. Antitoxin Protection against an Inhalational Ricin Challenge

The efficacy of the antitoxin against an inhalation challenge to $3LCt_{50}$ of aerosolised ricin was assessed in mice. Antitoxin was administered via the intravenous route (i.v.) at 16, 24 or 30 h after the ricin challenge to determine the window of opportunity for effective antitoxin therapy. Survival was seen in all mice given a single 2.5 mg dose of antitoxin at the 16 and 24 h time points after ricin exposure. This antitoxin dose was selected because it was found to be an effective dose in preliminary and limited dose–response experiments (data not shown). A "breakthrough" in protection was observed when the ricin antitoxin was given 30 h after the ricin aerosol challenge, where 75% protection was achieved (Figure 5A). Greater weight loss was seen in mice where ricin antitoxin was given at later time points (Figure 5B) suggesting that the quality of protection in surviving animals also decreased as antitoxin administration was delayed. Antitoxin administration had little effect on the observed signs of intoxication up to 48 h after ricin challenge however, whereas the control mice continued to develop more severe signs of intoxication, those treated with antitoxin gradually recovered over the next 12 days (Figure 5C). Control mice exposed to aerosolized vehicle only and given ricin antitoxin via the i.v. route after 16 h showed no weight loss or observed signs of intoxication (Figure 5B,C).

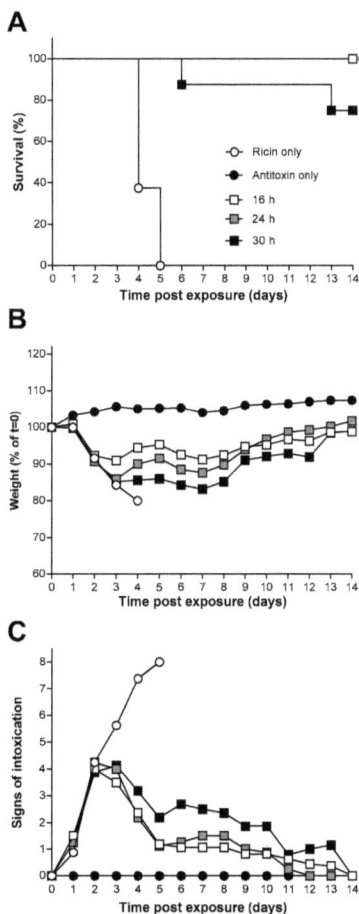

Figure 5. The window of opportunity for effective ricin antitoxin therapy following inhalational exposure of Balb/C mice to ricin. (**A**) Kaplan–Meier plot of survival in treated with antitoxin (2.5 mg·mouse^{-1}) at the indicated times after challenge with 3LCt$_{50}$ of ricin (the key shown in (**A**) also applies to (**B,C**). Data are from groups of eight mice in each case. (**B,C**) Body weight and signs of intoxication following administration of ricin and the indicated amount of antitoxin are shown. Data are combined from three independent experiments. Error bars have been omitted for clarity; the SDs are generally <10% and <2 units, respectively, for n = 8 mice in each group. Signs of intoxication were scored according to Table S1.

For assessment of the dose-dependence of the antitoxin-mediated protection, antitoxin or vehicle was administered to mice that had been exposed to ricin by the inhalational route 20 h earlier. This time point was chosen, as it was logistically more convenient for the experimental design than a 24 h administration time. Control animals challenged with ricin and administered antitoxin vehicle died or reached the humane endpoint within five days of exposure as expected (Figure 6A). Antitoxin administration produced a dose-dependent increase in the time to death (Figure 6A) and a dose dependent increase in survival at 14 days after challenge (Figure 6A,B). Antitoxin-mediated protection had an ED_{50} of 162 µg·mouse^{-1} (95% CI 133 to 196 µg·mouse^{-1}) by probit analysis.

Consistent with the affects observed within the intraperitoneal model, it was found that there was a dose-dependent decrease in both visible signs and weight loss associated increasing antitoxin dose administered (Figure 6C,D). Early signs of intoxication could be observed within a small number of animals after 24 h following challenge. However, between 24 h and 48 h, there is a rapid increase in the signs observed within the experimental animals representing a localised toxin-associated tissue damage triggering wider systemic issues (Figure 6D).

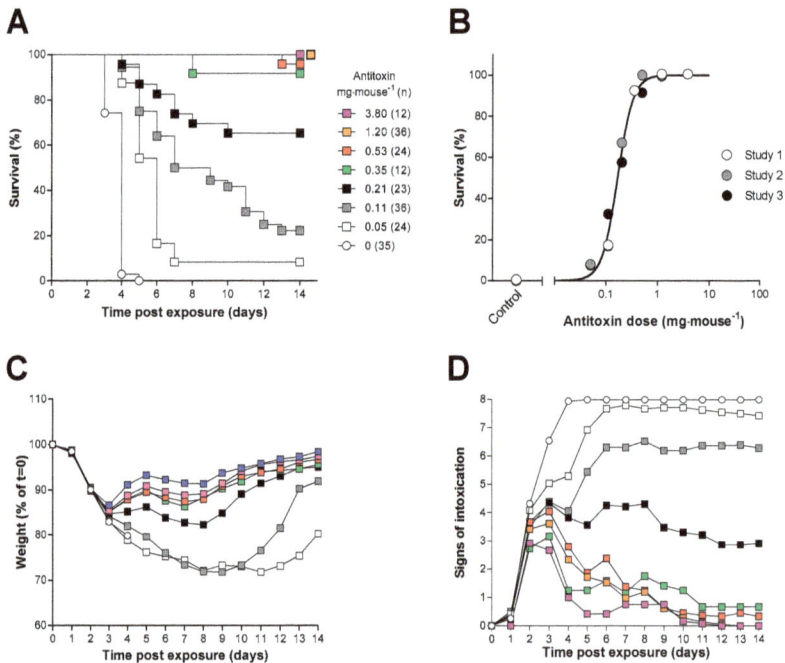

Figure 6. **Characterisation of the dose–response of ovine F(ab')$_2$ in the mouse ricin inhalation model.** (**A**) Kaplan–Meier plot of survival in mice administered ricin (3LCt$_{50}$, 21.54 mg·min·m^{-3}) and treated 20 h later with the indicated dose of antitoxin i.v. (the key in (**A**) also applies to (**C**,**D**)). Data have been combined from three independent experiments each of which used a group size of 12 mice. (**B**) Probit analysis of survival at 14 days, the curve was reconstructed from the output of the Minitab probit analysis which was fitted to the combined dataset. Data points are shown for survival at 14 days from three independent experiments. (**C**) Body weight; and (**D**) signs of intoxication following administration of ricin and the indicated amount of antitoxin are shown. Data are combined from three independent experiments. Error bars have been omitted for clarity; the SDs are generally <10% and <2 units, respectively, for n = 12–36 mice. Signs of intoxication were scored according to Table S1.

2.5. Antitoxin Can Affect the Uptake and Intracellular Trafficking of Ricin in A549 Cells

Cellular uptake and intracellular trafficking of the toxin in the presence or absence of antitoxin was examined to determine the mechanism of action of the antitoxin. A549 human lung epithelial cells were used and labelled ricin was tracked using Imaging Flow Cytometry (IFC), allowing transport of the toxin to both the cytoplasm and the Golgi apparatus to be observed in the presence and absence of antitoxin. For this, ricin was either added alone or was pre-incubated with antitoxin (1 h at 37 °C) at two different ratios (1:20 and 1:50) prior to exposure of the cells for 15, 30 or 60 min (Figure 7A,B). Ricin was rapidly taken up by the cells appearing within the cytoplasmic mask of approximately 90% of the cells by 60 min (Figure 7A). Ricin antitoxin caused a concentration dependent reduction in the percentage of cells taking up ricin with the 1:50 ratio of ricin to antitoxin showing a 50% reduction in ricin positive cells at 60 min (Figure 7A).

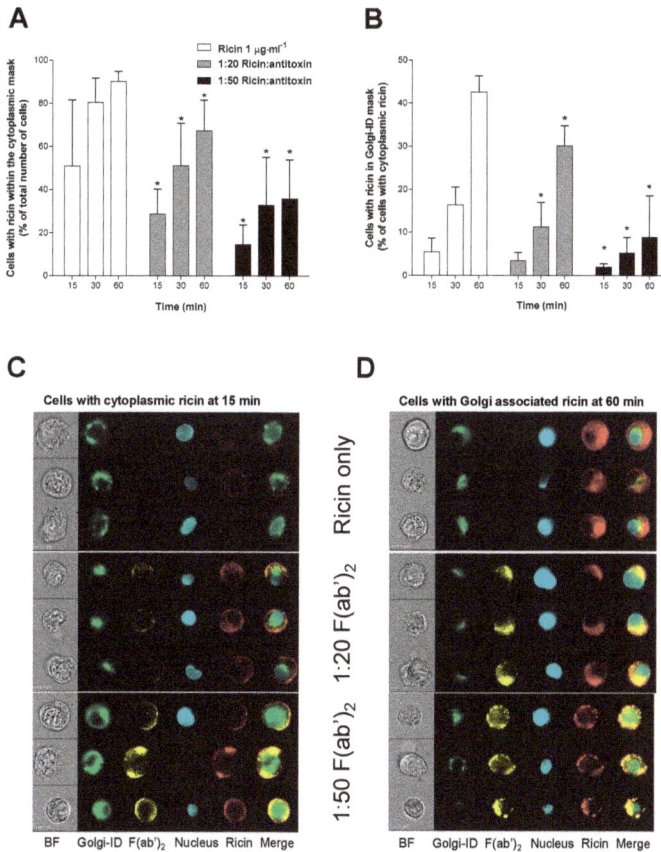

Figure 7. Effect of antitoxin on uptake and trafficking of ricin in human cultured A549 cells. (**A**) Antitoxin reduces cellular uptake and appearance of ricin within the cellular cytoplasmic mask in a concentration dependent manner. (**B**) In those cells that are positive for cytoplasmic ricin, antitoxin reduces the percentage of cells that have Golgi-associated ricin compared to control. Data in (**A,B**) are shown as mean $\pm$ SD from three independent experiments. * $p < 0.05$ (two-way ANOVA with multiple comparisons). (**C,D**) Example images of F(ab')$_2$ and ricin interactions at 15 and 60 min as shown as part of the ImageStreamX analysis (BF = Brightfield).

In the subset of cells that were ricin positive, the toxin became associated with the Golgi apparatus slowly over the time course of the study with only 40% of the cells with cytoplasmic ricin having the toxin also co-localising to the Golgi apparatus by 60 min (Figure 7B). Administration of the ricin-antitoxin mixture markedly reduced the co-localisation of ricin and the Golgi apparatus in the cytoplasmic ricin positive cells, with statistically significant reductions at 30 and 60 min for the 1:20 ratio and at all time points for the 1:50 ratio (Figure 7B, $p < 0.05$ 2-way ANOVA) suggesting that the antitoxin affects intracellular trafficking of ricin to the Golgi apparatus. It can be seen from the flow cytometry images that the A549 cells accumulate both the ricin and antitoxin over the time course of the experiment and that these remain co-localised within the cell (Figure 7C,D).

3. Discussion

We have produced an antitoxin for treatment of ricin intoxication which is based on a ovine polyclonal anti-ricin IgG, despeciated to its F(ab')$_2$ fragments. The antitoxin was characterised in vitro and in vivo to determine its ability to neutralise ricin toxin and was shown to be effective in protecting mice against an inhalational challenge with aerosolised ricin. Importantly, the antitoxin demonstrated effective protection against the lethal effects of ricin when it was administered up to 30 h after the ricin challenge. Full protection was seen at 24 h and partial, 75%, protection was seen at 30 h. Efficacy at such delayed time points is likely to be necessary for effective treatment of ricin intoxication given that there is a delay in the appearance of signs of intoxication making confirmation of exposure, diagnosis of intoxication and the subsequent medical response, technically and logistically challenging. This is especially true considering that the potential use of ricin in military or civilian scenarios is unlikely to be predicted [1–3]. These data are therefore encouraging in that such long WOO may enable an effective medical response to a ricin incident, provided that this WOO is reproduced in other animal species and is predicted in man.

Whilst our data provide robust evidence that there is a post-exposure WOO for treatment of ricin intoxication, other authors have also described effective post-exposure treatment of ricin intoxication with anti-ricin antibodies. Many of these reports have highlighted the limited WOO for protection against the toxin, typically found to be up to ~8 h after ricin exposure [22,38–40]. There have however been reports of WOO similar to that described here however our research if the first to demonstrate full protection at 24 h. For example, Pratt et al. [41] used an oropharyngeal aspiration model to challenge mice with ricin. They were able to protect all of the ricin-exposed mice with polyclonal anti-ricin antibodies raised against deglycosylated ricin A chain when the antibodies were administered 18 h after challenge. When administered at 24 h after challenge this antibody preparation protected only 30% of the mice. RAC 18 is a monoclonal antibody also raised against ricin A chain. This antibody protected ~60% of ricin-challenged mice when administered at 18 h and ~50% when given after 24 h from intoxication [41]. Chimeric high affinity IgG antibodies originally isolated from rhesus macaques immunised with ricin holotoxin or subunit vaccine [44] have also been shown to provide an extended window of opportunity in an intranasal mouse model of ricin intoxication [42]. When administered 24 h after a 2LD$_{50}$ ricin challenge, administration of the individual antibodies gave survival rates of between 62–89%. Furthermore, a cocktail of three antibodies, one directed at ricin A chain and two directed at the B chain resulted 80% survival at 24 h and statistically significant protections, 73% and 36%, when administered at 48 h or 72 h respectively [42]. It should be noted that direct comparison of the window of opportunity for treatment with antibody-based therapeutics is complicated by the different challenge and treatment doses and routes of exposure used. The lower challenge dose of ricin (i.e., 2LD$_{50}$ via the intranasal route) used in an number of studies may represent a less stringent challenge model for assessing antitoxin treatment and may explain the longer times to death of ~4–9 days [42,44] when compared to the 3LCt$_{50}$ inhalation challenge used in our experiments which results in death or humane end-point at 3–5 days post exposure. For our polyclonal antitoxin doses of 2.5 mg·mouse^{-1} (approximately 125 mg·kg^{-1}) were used to provide protection at 24 and 30 h and doses in excess of 0.8 mg·mouse^{-1} (approximately 40 mg·kg^{-1}) are predicted to provide full protection when administered at 20 h after intoxication (Figure 6). These doses are in excess of those typically required of monoclonal neutralising antibodies which, in the examples above, range from ~3 to ~17 mg·kg^{-1}. There is however little information available to inform how the dose of antitoxin required for protection is related to the WOO making direct comparisons difficult. Irrespective of the differences in ricin challenge, our results and those detailed above demonstrate that following a pulmonary challenge with ricin, either directly with aerosolised ricin in the case of the data reported here or via intranasal instillation, there is a relatively long window of opportunity for effective treatment in mice that extends for 24 h or longer. Given that the antitoxin must physically interact with the ricin to bring about neutralisation of toxicity, these observations imply that toxicologically relevant amounts of ricin are accessible to antitoxin at these time points after exposure. This is perhaps unexpected as ricin is known to bind and be taken up rapidly by cells in culture [45]. This suggests that either the kinetics of ricin cellular uptake and trafficking, observed in vitro, are not reflected in vivo or that that the assumption of a purely extracellular mode of neutralisation is an incomplete explanation of the therapeutic mechanism of these antitoxins.

To examine these possibilities, in vitro studies to track labelled ricin and antitoxin were conducted in A549 pulmonary epithelial cells. As expected, ricin alone rapidly associated with and became internalised in the cells. The polyclonal antibody, when pre-incubated with ricin, reduced toxin uptake, demonstrating that interference with this process could contribute to protection. Ricin cellular binding and uptake is mediated by the ricin B chain, and this has typically been targeted by neutralising antibody strategies. B chain binders have previously been shown to be responsible for ricin vaccine efficacy [46] and to neutralise ricin in vitro and provide protection in vivo [47,48] with some B chain binders being superior to A chain binders in these assays [49]. Nevertheless, the A chain binders have been shown to be effective treatments for ricin intoxication [41,42,44] and thus can clearly contribute to protection.

The antitoxin used in our studies was raised in sheep against a formaldehyde toxoid of ricin holotoxin and will contain a polyclonal mix of both A and B chain binders. Interestingly it has been observed that both A and B chain directed anti-ricin antibodies have the ability to neutralise ricin activity intracellularly and can disrupt the intracellular trafficking of ricin [50–53]. Our data, from imaging flow cytometry studies, suggest that such effects can also be mediated by polyclonal antibody-fragment antitoxins and that such block or miss-direct of normal intracellular trafficking of ricin towards proteolytic or other non-toxic routes could contribute to the long therapeutic window for this antitoxin.

Anaphylactoid reactions, anaphylaxis and serum sickness are recognised complications that can result from passive immunisation with foreign antibody products [33,37], particularly if a large dose is needed to provide protection, as is likely to be the case with our antitoxin. Thus, a further difference between our antitoxin and monoclonal chimeric/humanised antibody approaches used by others [42,44,48] is that despeciation to create F(ab')$_2$ was considered necessary to reduce the likelihood of adverse immune reactions in any recipient [35,54]. In addition, an ovine source of hyperimmune plasma was selected as ovine antibodies are considered to be less immunogenic than those of equine or caprine origin [55,56]. Importantly, the protective efficacy of the F(ab')$_2$ antitoxin was found to be comparable with the parent IgG molecule [57] demonstrating that the Fc-mediated properties of the whole IgG molecule did not appear to be important in the overall protective mechanism. In the studies reported here, no immunotoxicological indices were monitored in the mice administered the 2.5 mg dose of antitoxin alone, however, these animals did show a normal body weight profile and no observed signs of intoxication (many of which are general indicators of health and wellbeing) over a subsequent 14 day observation period (Figure 5). It should be noted that our polyclonal antitoxin is currently in advanced development as a treatment for inhalational ricin intoxication. Whilst it has been manufactured to GMP and has shown efficacy, assessment of the safety, in animals, of the predicted protective dose is planned but has yet been undertaken and this will represent a regulatory challenge for the clinical development of the antitoxin. However, as there are no available treatments for ricin intoxication, the polyclonal antitoxin approach could still represent an effective and pragmatic approach for treatment of ricin inhalation, although consideration would need to be given to the management of side effects should they occur.

One of the key aims of this research was to reduce the numbers of animals required during the development of the antitoxin which requires potency determinations assessment of their stability at multiple time points and storage temperatures. For antitoxins, this has traditionally involved in vivo methodologies. However, Sesardic et al. [58] examined the antibodies raised against diphtheria toxoid in an in vitro Vero cell assay and compared potency within the guinea pigs concluding that there was a strong correlation between the potencies in both models. These models are now the basis of an international standard for characterising batches of human diphtheria antitoxin [59].

Here, work was undertaken to investigate the utility of an in vitro model using Vero cells for assessing potency and stability of the ricin antitoxin. The in vitro approach was used routinely and reproducibly in GMP processes to produce the antitoxin. Importantly, it demonstrated that the $F(ab')_2$ antitoxin product maintained its neutralising activity for at least 72 months after manufacture, a finding which was consistent with the physico-chemical properties of the antitoxin which were also unchanged over this period (data not shown). Comparison of the in vitro and in vivo neutralising assay has been conducted for the $F(ab')_2$ antitoxin with the ratio (antitoxin:ricin) being ~2 fold lower in vivo than in vitro. That the values are not identical may be due to the different factors involved in ricin toxicity in vitro and in vivo. For example, in vivo, clearance and binding to non-critical sites (sites that do not result in cytotoxicity or that do not link directly to lethality) may also contribute to protection and therefore less antitoxin might be needed to neutralise the remaining toxin. Whereas in vitro there is no clearance and all cytotoxicity counts towards the final assay result. Further studies with other antitoxins will be required to determine if this observation is a consistent feature of the in vitro and in vivo neutralising assay methods.

Antitoxins and other approaches to treatment of ricin intoxication such as small molecule retrograde transport inhibitors [60–62] or combinations of therapies such as antibiotics and antitoxins [63] could increase the WOO for effective treatment. Efficacy data in the mouse have been demonstrated for these approaches and, additionally, the research has also provided important mechanistic insights with regard to the properties of an effective therapy. However, as will be necessary for the antibody-based antitoxin developed here, further safety and efficacy studies will be required in other (higher) animal species for these approaches to gain approval for use in humans. Our current data add significantly to the current research field and provide evidence that a practicable mitigation for exposure to this potentially lethal, potent and readily-available ricin toxin should be achieved in the near future.

4. Conclusions

Passive immunity is a highly effective strategy for treating intoxications caused by ribosome inactivating proteins such as the ricin toxin. The development of antibody-based antitoxins that are effective when used many hours after exposure to the toxin will represent a step forward in the treatment of ricin-associated intoxication. We have produced an ovine polyclonal $F(ab')_2$ that can fully protect mice from the lethal effects of a $3LCt_{50}$ inhaled aerosol challenge with ricin at prolonged times that are comparable to those achieved with other polyclonal and/or monoclonal antitoxins. Full protection was observed at 24 h following challenge and partial protection at 30 h demonstrating that this polyclonal antitoxin has a wide window of opportunity commensurate with that required to represent a practicable approach to mitigate the effects of ricin intoxication.

5. Materials and Methods

5.1. Preparation of Ricin and Ricin Toxoid

Ricin toxin was prepared from the seeds of *Ricinus communis subspecies zanzibariensis* as previously described [64]. Ricin toxoid was produced by adding formaldehyde (Sigma Aldrich, Poole, UK) to a final concentration of 2.5% (w/v), (pH 7.10–7.30) and incubating at 37 °C for 14 days. Excess formaldehyde was quenched by the addition of lysine hydrochloride (Sigma Aldrich, Poole, UK) to a final concentration of 0.1 M. The quenched material was then desalted into phosphate buffered saline (PBS) (Gibco, ThermoFisher Scientific, Loughborough, UK) using a 5.0 × 50 cm G25M Sephadex column. The protein concentration determined by BCA protein assay (ThermoFisher Scientific, Loughborough, UK), and formaldehyde added back to final concentration of 0.025% (w/v).

5.2. Preparation of Ovine Ricin Antitoxin and Despeciated Antitoxin Fragment F(ab')₂

Briefly, a polyclonal antibody (IgG) was produced from hyper-immune plasma that was raised in sheep (Selborne Biological Services, Tasmania, Australia) following immunisation with ricin toxoid at 6 weekly intervals. Sheep received a priming dose of ricin toxoid in Freund's Complete Adjuvant (Sigma Aldrich, Poole, UK) followed by booster doses of ricin toxoid in Freund's Incomplete Adjuvant at 6 weekly intervals (100 μg ricin toxoid per dose).

The production of F(ab')₂ was undertaken by International Therapeutic Proteins (ITP Ltd, Tasmania, Australia). In brief, IgG was purified from selected production bleeds that were taken between Week 14 and Week 38 of the immunisation schedule. The IgG was digested with pepsin (Sigma Aldrich, Poole, UK) to produce F(ab')₂ which was lyophilized to form a stable pellet and was stored at 2–8 °C. Prior to use the antitoxin was reconstituted with distilled water (Gibco, ThermoFisher Scientific, Loughborough, UK) and aliquots stored at −20 °C.

5.3. In Vitro Cytotoxicity Assay

Vero cells (ECACC 84113001) were obtained from the European Collection of Animal Cell Cultures (ECACC) (Public Health England, Salisbury, UK). Cells were maintained in culture medium consisting of DMEM (Sigma Aldrich, Poole, UK) with 10% (v/v) foetal calf serum (Sigma Aldrich, Poole, UK), 1% penicillin/streptomycin solution (containing 100 units·mL^{-1} penicillin and 0.01 mg·mL^{-1} streptomycin), and 1% (w/v) L-glutamine (Sigma Aldrich, Poole, UK) 2 mM. Vero cells were grown in 150 cm² flasks in a humidified atmosphere of 5% CO_2 in air at 37 °C and removed from the flask surface using incubation with trypsin (0.05% w/v) (Sigma Aldrich, Poole, UK) containing EDTA (0.03% w/v) (Sigma Aldrich, Poole, UK) on achieving 70–90% confluency for seeding into 96 well test plates (cell density of 5 × 10³ cells per well). The cells were allowed to adhere to the culture plates for 24 h before use.

For toxicity assessment, ricin toxin was diluted to 100 ng·mL^{-1} in culture medium and filtered using a 0.2 μm sterile filter before further dilution in culture medium and addition to the assay plate in triplicate. The plates were then incubated for 48 h prior to the addition of 10 μL of Roche Cell Proliferation Reagent WST-1 (Sigma Aldrich, Poole,UK). After 3 h the absorbance was read on a Thermo Multiskan plate reader (Thermo Fisher Scientific, Loughborough, UK) at 450 nm to assess cell viability.

5.4. In Vitro Neutralising Activity of Anti-Ricin Antitoxin

For in vitro neutralisation assays using the F(ab')₂ antitoxin, Vero cells were seeded into 96-well plates (5 × 10³ per well) and incubated overnight 37 °C ± 5% CO_2. The F(ab')₂ antitoxin (73.4 ng·mL^{-1} to 3300 ng·mL^{-1}) was mixed with a fixed concentration of ricin (3.24 ng·mL^{-1}) for 1 h prior to the addition of the mixture to cells in the plate. A ricin cytotoxicity curve (3.56 ng·mL^{-1} to 0.001 ng·mL^{-1} ricin) was included on the plate in addition to media only (negative) controls. After 48 h incubation, 10 μL of WST-1 reagent (Sigma Aldrich, Poole, UK) was added and plates were incubated for a further 4 h at 37 °C at 5% CO_2. The absorbance was read at 450 nm.

Cytotoxicity data were normalised to the negative control values and analysed with a 4-parameter logistic fit (GraphPad Prism for Windows version 6.02). The neutralising ratio was determined from the calculated ricin LC$_{50}$, the antitoxin EC$_{50}$ and the concentration of ricin used in the assay (3.24 ng·mL^{-1}) according to Equation (1).

$$\text{Neutralising ratio} = \frac{\text{Antitoxin EC}_{50} \left(\text{ng}\cdot\text{mL}^{-1}\right)}{3.24 - \text{Ricin LC}_{50} \left(\text{ng}\cdot\text{mL}^{-1}\right)}, \tag{1}$$

5.5. Animal Husbandry

All investigations involving animals conformed to the UK Animal (Scientific Procedures) Act 1986 and also the UK Codes of Practise for the Housing and Care of Animals Used in Scientific Procedures 1989 in an established process required to gain the UK Home Office Project Licence that covered this research (granted 30 October 2012). Age-matched female Balb/C mice (6–7 weeks old, Charles River Laboratories Ltd, Margate, Kent, UK) were used in all in vivo studies. On receipt, animals were acclimatised to the facility prior to being used on study. Mice were housed in rooms maintained at 21 °C ± 2 °C on a 12/12 h dawn to dusk cycle. Humidity was maintained at 55 ± 10% with airflow of 15–18 changes·h^{-1} and given water and food and water ad libitum. Mice were fed a standard pelleted Teklad TRM 19% protein irradiated diet (Harlan Teklad, Bicester, UK).

5.6. In Vivo Neutralisation Assay in Mice

Mice (10 per group) were randomly assigned to control or treatment groups and weighed on Day 0. The target weight range for animals on all studies was 17–21 g and no correction for body weight was made when dosing with ricin or antitoxin. Doses are therefore provided on a per mouse basis. On day zero, mice received one of a range of doses of ricin toxin between 0.22 and 0.77 µg per mouse (100 µL) via the i.p. route. The data were analysed with a probit model using Minitab (v17) to determine the LD$_{50}$. For in vivo neutralisation studies, a dose range of 56.2 to 93.6 µg antitoxin·mouse^{-1} F(ab')$_2$ was pre-incubated with 2 µg ricin·mouse^{-1} for 1 h and administered to Balb/C mice (10 per group) via the intraperitoneal route. Mice were observed at least twice daily for 14 days and the number of live and dead/moribund mice recorded at each observation. Moribund mice were culled upon reaching the humane end point (Table S1). Weight and visible signs of intoxication were also observed and recorded for 14 days. Dose–response data were analysed with a probit model using Minitab (v17) and the neutralizing ratio calculated from the antitoxin ED$_{50}$, the ricin LD$_{50}$ value and the challenge dose of ricin (2 µg·mouse^{-1}) used according to Equation (2).

$$\text{Neutralising ratio} = \frac{\text{Antitoxin ED}_{50} \left(\text{ng}\cdot\text{mL}^{-1}\right)}{2 - \text{Ricin LD}_{50} \left(\text{ng}\cdot\text{mL}^{-1}\right)}, \tag{2}$$

5.7. Determination of Inhalational Toxicity of Ricin in Mice

Balb/C mice were exposed to aerosols of ricin (head only exposure) generated using a system previously described [5,30]. Each inhalation run had the capacity for the exposure of 12 mice. Animals were enclosed in whole body plethysmography tubes attached to the exposure line allowing individual respiratory parameters to be recorded. Mice (*n* = 12 per group) were exposed to targeted inhaled concentrations of ricin and survival and visible signs of intoxication (Table S1) were monitored for 14 days.

5.8. Efficacy of Ricin Antitoxins against Lethal Inhalation Ricin Challenge

Groups of mice (*n* = 8–12) received approximately a 3LCt$_{50}$ aerosol challenge of ricin and 20 h following challenge, the indicated dose of ricin antitoxin was given by an intravenous (i.v.) injection. Mice were observed at least twice daily for 14 days and the number of live and dead/moribund mice recorded at each observation. Moribund mice were those that had reached the humane end point (severe piloerection, severe abdominal pinching and unable to move) and these mice were humanely killed. Weights and visible signs of intoxication were recorded daily for 14 days. Three independent experiments were conducted (with up to 12 mice for each antitoxin dose group in each experiment). Control ricin-exposed mice received a vehicle (110 mM sodium chloride, 25 mM glycine and 1% v/v sucrose, pH 6.5–7.2) (all Sigma Aldrich, UK) injection. Mortality data were used to produce a dose–response relationship, which was analysed with a logistic model (GraphPad Prism for Windows version 6.02).

5.9. Intracellular Trafficking of Ricin and Fluorochome Labelled F(ab')₂

Ricin was labelled and experiments were conducted essentially as previously described (Jenner et al., 2017; manuscript submitted). F(ab')$_2$ was labelled using a microscale AF555 labelling kit (Thermo Life Sciences, Loughborough, UK). Briefly, F(ab')$_2$ was mixed with AF555 succinimidyl ester in a molar ratio of 13:1 in PBS buffer pH 8. The mix a incubated for 15 min at room temperature and the labelled F(ab')$_2$ purified from unbound dye using a NAP-5 column (GE Healthcare, Little Chalfont, UK). A549 cells were grown in Dulbecco's Modified Eagles Medium (DMEM; Gibco, ThermoFisher, Loughborough, UK) supplemented with 10% fetal bovine serum (Gibco,ThermoFisher, Loughborough, UK) and 2 mM glutamine (Gibco, ThermoFisher, Loughborough, UK) at 37 °C in a 5% CO_2 incubator with ~95% humidity. Cells were harvested with trypsin, enumerated and plated into a 24-well plate (Corning Costar) at a density of 5×10^5 cells·well^{-1}. Cells were incubated at 37 °C in a 5% CO_2 incubator with ~95% humidity for ~18 h to allow cell adherence to the plate. Media was removed from the wells and replaced with 200 µL of either a 1 µg·mL^{-1} ricin-AF647 solution or mixtures of the ricin with AF555-F(ab')$_2$, (equating to neutralising ratios of 1:20 (1 µg ricin, 20 µg F(ab')$_2$) or 1:50 (1 µg ricin, 50 µg F(ab')$_2$) for 1 h at 37 °C before being added to plated cells. The 1 µg·mL^{-1} concentration of ricin was used to enable visualisation of the fluorescence signal at the cellular level. This concentration of ricin is toxic to the cells (Figure 1), but is slightly less than concentrations used by others for confocal imaging of labelled ricin [51,52].

Exposed cells were then incubated for a range of times (15, 30 and 60 min) after which they were washed once with PBS before 200 µL detachin (AMS Biotechnology Limited, Abingdon, UK) was added. Cells were incubated at 37 °C for 3 min to allow detachment from the plate surface. Cells were then harvested and centrifuged at $300 \times g$ for 5 min and resuspended in 60 µL 4% paraformaldehyde. Cells were then counter stained with 1 µg Hoechst 33342 (ThermoFisher, Loughborough, UK) before data capture using Imaging Flow Cytometry.

5.10. Staining of Golgi Using Golgi-ID

Staining using Golgi-ID (Enzo Life Sciences, Exeter, UK) was achieved according to the manufacturer's instructions. Briefly, A549 cells were plated as previously described. Cells were incubated overnight and washed with 100 µL of assay solution before the addition of a 200 µL of Golgi-ID stain (1 in 100 dilution of provided stock solution) per well. Cells were then incubated at 4 °C for 30 min before removal of the Golgi-ID stain and washing of the cells with 100 µL assay buffer. A final 1 mL of DMEM was added to the cells and they were incubated for 30 min at 37 °C in a 5% CO_2 incubator with ~95% humidity before use in the ricin assays.

5.11. Imaging Flow Cytometry Data Collection and Analysis

IFC data were acquired using an ImageStream X MkII (ISX, Amnis, Seattle, WA, USA) equipped with dual cameras and 405, 488, 561 and 633 nm excitation lasers. All samples were acquired at ×60 magnification with a 0.9 NA objective, giving a 2.5 µm optical slice image and allowing colocalisation studies to be undertaken with this approach [65,66] A minimum of 7000 in-focus single cell events were collected for each sample. Only data from relevant channels were collected including Channel 01 (Ch01, brightfield camera 1), Channel 02 (Ch02, Golgi-ID, 488 nm laser power: 100 mW), Ch03 (AF555 F(ab')₂ 561 nm laser 200 mW), Channel 06 (Ch06, side scatter 785 nm laser power: 10 mW), Channel 07 (Ch07, Hoescht fluorescence 405 nm 120 mA laser power), Channel 09 (Ch09, brightfield camera 2) and Ch11 (AF647-ricin fluorescence 642 nm laser power: 150 mW). Data from samples with only single stains were also captured to calculate the compensation matrix required to account for spectral overlap between the chosen fluorophores. All images shown are pseudo-coloured according to the following: nucleus = blue, Golgi = green, ricin = red, F(ab')$_2$ = yellow.

5.12. IFC Data Analysis

Analysis of IFC data was achieved using IDEAS® software (version 6.1). IDEAS® utilizes two main principles to make calculations from the images acquired these are masks and features. Masks are used to define regions of interest within the cell or fluorescence image. Those masks are used to calculate quantitative measurements from or within a masked region.

A mask was used to identify the cytoplasmic region of cells. The cytoplasmic mask was created by making an erode (M01, 8) mask (the default mask M01, eroded by 8 pixels around) and morphology (M07) mask of the nuclear channel. These two masks were then combined using Boolean logic (AND NOT) to make a mask that is specific for the cytoplasm. The full mask nomenclature for the cytoplasmic mask is: Erode (M01, 8) AND NOT Morphology (M07). When using Golgi-ID the default mask for the Golgi apparatus is not very specific, a more stringent mask was required to accurately reflect the Golgi location. To achieve this, a threshold mask is used; this masks a percentage of the brightest pixels within a given starting mask. Here, we have masked the top 70% of pixels within the default M02 mask. The nomenclature for this mask is Threshold (M02, Ch02, 70). All data shown is a percentage population using the in-focus single cells as the parent population. All modelling and statistical analysis was performed in GraphPad Prism version 6.02.

Supplementary Materials: The following are available online at www.mdpi.com/2072-6651/9/10/329/s1, Table S1: Scoring of observable signs of ricin intoxication in the mouse.

Acknowledgments: This research was funded by the UK Ministry of Defence. We wish to thank the animal services and veterinary staff at Dstl-Porton Down for enabling these studies to be conducted.

Author Contributions: G.D.G., S.J.C.W., D.C.J., G.C.C. and A.C.G. wrote the manuscript. R.J.G. provided the statistical analysis during the research. A.C.G. and J.L.H. reviewed the manuscript. All authors contributed to the research and/or data analysis.

Conflicts of Interest: The authors declare no conflict of interest.

References

1. Franz, D.R.; Jaax, N.K. Ricin toxin. In *Medical Aspects of Chemical and Biological Warfare*; Borden Institute, Walter Reed Army Medical Center: Washington, DC, USA, 1997; Chapter 32; pp. 631–642.
2. Shoham, D. Iraqs biological warfare agents: A comprehensive analysis. *Crit. Rev. Microbiol.* **2000**, *26*, 179–204. [CrossRef] [PubMed]
3. Roxas-Duncan, V.I.; Smith, L.A. Ricin: Perspective in bioterrorism. In *Bioterrorism*; Morse, S., Ed.; InTech: Rijeka, Croatia, 2012; Volume 7, pp. 133–158.
4. Worbs, S.; Köhler, K.; Pauly, D.; Avondet, M.-A.; Schaer, M.; Dorner, M.B.; Dorner, B.G. *Ricinus communis* intoxications in human and veterinary medicine-a summary of real cases. *Toxins* **2011**, *3*, 1332–1372. [CrossRef] [PubMed]
5. Griffiths, G.D.; Phillips, G.J.; Holley, J.L. Inhalation toxicology of ricin preparations: Animal models, prophylactic and therapeutic approaches to protection. *Inhal. Toxicol.* **2007**, *19*, 873–887. [CrossRef] [PubMed]
6. Lord, J.M.; Griffiths, G.D. Ricin: Chemistry, sources, exposures, toxicology and medical aspects. In *General, Applied and Systems Toxicology*; Wiley Ltd.: Chichester, UK, 2009. [CrossRef]
7. Reisler, R.B.; Smith, L.A. The need for continued development of ricin countermeasures. *Adv. Prev. Med.* **2012**, *1*, 1–4. [CrossRef] [PubMed]
8. Barbieri, L.; Baltelli, M.; Stirpe, F. Ribosomes-inactivating proteins from plants. *Biochem. Biophys. Acta* **1993**, *1154*, 237–282. [CrossRef]
9. Lord, J.M.; Roberts, L.M.; Robertus, J.D. Ricin: Structure, mode of action and some current application. *FASEB J.* **1994**, *8*, 201–208. [CrossRef] [PubMed]

10. Rutenber, E.; Robertus, J.D. Structure of ricin B-chain at 2.5 A resolution. *Proteins Struct. Funct. Genet.* **1991**, *10*, 260–269. [CrossRef] [PubMed]

11. Khan, T.; Waring, P. Macrophage adherence prevents apoptosis induced by ricin. *Eur. J. Cell Biol.* **1993**, *62*, 406–414. [PubMed]

12. Simmons, B.M.; Stahl, P.D.; Russell, J.H. Mannose receptor mediated uptake of ricin toxin and ricin A-chain by macrophages. Multiple intracellular airways for A chain translocation. *J. Biol. Chem.* **1986**, *261*, 7912–7920. [PubMed]

13. Magnusson, S.; Berg, T. Endocytosis of ricin by rat liver cells in vivo and in vitro is mainly mediated by mannose receptors on sinusoidal endothelial cells. *Biochem. J.* **1994**, *291*, 749–755. [CrossRef]

14. Sapoznikov, A.; Falach, R.; Mazor, O.; Alcalay, R.; Gal, Y.; Seliger, N.; Sabo, T.; Kronman, C. Diverse profiles of ricin-cell interactions in the lung following intranasal exposure to ricin. *Toxins* **2015**, *7*, 4817–4831. [CrossRef] [PubMed]

15. Lord, J.M.; Smith, D.C.; Roberts, L.M. Toxin entry: How bacterial proteins get into mammalian cells. *Cell Microbiol.* **1991**, *1*, 85–91. [CrossRef]

16. Robertus, J. The structure and action of ricin: A cytotoxic N-glycosidase. *Cell Biol.* **1991**, *2*, 23–30.

17. Eiklid, K.; Olsnes, S.; Pihl, A. Entry of lethal doses of abrin, ricin and modeccin into the Cytosol of HeLa cells. *Exp. Cell Res.* **1980**, *126*, 321–326. [CrossRef]

18. Olsnes, S.; Kozlov, J.V. Ricin. *Toxicon* **2001**, *39*, 1723–1728. [CrossRef]

19. Soler-Rodriguez, A.-M.; Ghetie, M.-A.; Oppenheimer-Marks, N.; Uh, J.W.; Vitetta, E.S. Ricin A-chain and ricin A-chain immunotoxins rapidly damage human endothelial cells, Implications for vascular leak syndrome. *Exp. Cell Res.* **1993**, *206*, 227–234. [CrossRef] [PubMed]

20. Lindauer, M.L.; Wong, J.; Iwakura, Y.; Magun, B.E. Pulmonary inflammation triggered by ricin toxin requires macrophages and IL-1 signalling. *J. Immunol.* **2009**, *183*, 1419–1426. [CrossRef] [PubMed]

21. Mabley, J.G.; Pacher, P.; Szabo, C. Activation of cholinergic anti-inflammatory pathway reduces ricin-induced mortality and organ failure in mice. *Mol. Med.* **2009**, *15*, 166–172. [CrossRef] [PubMed]

22. Gal, Y.; Mazor, O.; Alcalay, R.; Seliger, N.; Aftalion, M.; Sapoznikov, A.; Falach, R.; Kronman, C.; Sabo, T. Antibody/doxycycline combined therapy for pulmonary ricinosis: Attenuation of inflammation improves survival of ricin-intoxicated mice. *Toxicol. Rep.* **2014**, *1*, 496–504. [CrossRef] [PubMed]

23. Sabo, T.; Gal, Y.; Elhanany, E.; Sapoznikov, A.; Falach, R.; Mazor, O.; Kronman, C. Antibody treatment against pulmonary exposure to abrin confers significantly higher levels of protection than treatment against ricin intoxication. *Toxicol. Lett.* **2015**, *237*, 72–78. [CrossRef] [PubMed]

24. Roy, C.J.; Song, K.; Sivasubramani, S.K.; Gardner, D.J.; Pincus, S.H. Animal models of ricin toxicosis. *Curr. Top. Microbiol. Immunol.* **2012**, *357*, 243–257. [PubMed]

25. Godal, A.; Fodstad, O.; Ingebrigtsen, K.; Pihl, A. Pharmacological studies of ricin in mice and humans. *Cancer Chemother. Pharmacol.* **1984**, *13*, 157–163. [CrossRef] [PubMed]

26. Fodstad, O.; Olsnes, S.; Pihl, A. Toxicity, distribution and elimination of the cancerostatic lectins abrin and ricin after parenteral injections into mice. *Br. J. Cancer* **1976**, *32*, 418–425. [CrossRef]

27. Griffiths, G.D.; Leek, M.D.; Gee, D.J. The toxic plant proteins ricin and abrin induce apoptotic changes in mammalian lymphoid tissues and intestines. *J. Pathol.* **1987**, *151*, 221–229. [CrossRef] [PubMed]

28. Balint, G.A. Ricin: The toxic protein of castor oil seeds. *Toxicology* **1974**, *2*, 77–102. [CrossRef]

29. Muldoon, D.F.; Stoh, S.J. Modulation of ricin toxicity in mice by biologically active substances. *J. Appl. Toxicol.* **1994**, *14*, 81–86. [CrossRef] [PubMed]

30. Griffiths, G.D.; Rice, P.; Allenby, A.C.; Bailey, S.C.; Upshall, D.G. Inhalation toxicology and histopathology of ricin and abrin toxins. *Inhal. Toxicol.* **1995**, *7*, 269–288. [CrossRef]

31. Bhaskaran, M.; Didier, P.J.; Sivasubramani, S.K.; Doyle, L.A.; Holley, J.; Roy, C.J. Pathology of lethal and sublethal doses of aerosolized ricin in Rhesus Macaques. *Toxicol. Pathol.* **2014**, *42*, 573–581. [CrossRef] [PubMed]

32. Mayers, C.N.; Holley, J.L.; Brooks, T. Antitoxin therapy for botulinum intoxication. *Rev. Med. Microbiol.* **2001**, *12*, 1–9. [CrossRef]

33. Casadevall, A. Passive antibody administration as a specific defence against biological weapons. *Emerg. Infect. Dis.* **2002**, *8*, 833–841. [CrossRef] [PubMed]

34. Chippaux, J.P.; Goyffon, M. Venoms, anti-venoms and immunotherapy. *Toxicon* **1998**, *36*, 823–846. [CrossRef]

35. Sedlacek, H.H.; Gronski, P.; Hofstaetter, E.J.; Kanzy, E.J.; Schlorlemmer, H.U.; Seiler, F.R. The biological properties of Immunoglobulin G and its split products (Fab')2 and Fab. *Klin. Wochenschr.* **1983**, *61*, 723–736. [CrossRef] [PubMed]

36. Nydegger, U.E.; Sturzenegger, M. Adverse effects of intravenous immunoglobulinum therapy. *Drug Saf.* **1999**, *21*, 171–185. [CrossRef] [PubMed]

37. Morais, V.M.; Massaldi, H. Snake antivenoms: Adverse reactions and production technology. *Venom. Anim. Toxins Incl. Trop. Dis.* **2009**, *15*, 2–18. [CrossRef]

38. O'Hara, J.M.; Whaley, K.; Pauly, M.; Zeitlin, L.C.; Mantis, N. Plant-based expression of a partially humanized neutralizing monoclonal IgG directed against an immunodominant epitope on the ricin toxin A subunit. *Vaccine* **2012**, *30*, 1239–1243. [CrossRef] [PubMed]

39. Sully, E.K.; Whaley, K.J.; Bohorova, N.; Goodman, C.; Kim, D.H.; Pauly, M.H.; Velasco, J.; Hiatt, E.; Morton, J.; Swope, K.; et al. Chimeric plantibody passively protects mice against aerosolized ricin challenge. *Clin. Vaccine Immunol.* **2014**, *21*, 777–782. [CrossRef] [PubMed]

40. Van Slyke, G.; Sully, E.K.; Bohorova, N.; Bohorov, O.; Kim, D.; Pauly, M.; Whaley, K.; Zeitlin, L.; Mantis, N.J. A humanized monoclonal antibody that passively protects mice 3 against systemic and intranasal ricin toxin challenge. *Clin. Vaccine Immunol.* **2016**. [CrossRef] [PubMed]

41. Pratt, T.S.; Pincus, S.H.; Hale, M.L.; Moreira, A.L.; Roy, C.J.; Tchou-Wong, K.-M. Oropharyngeal aspiration of ricin as a lung challenge model for evaluation of the therapeutic index of antibodies against ricin A chain for post-exposure treatment. *Exp. Lung Res.* **2007**, *33*, 459–481. [CrossRef] [PubMed]

42. Noy-Porat, T.; Alcalay, R.; Epstein, E.; Sabo, T.; Kronman, C. Extended therapeutic window for post exposure treatment of ricin intoxication conferred by the use of high-affinity antibodies. *Toxicon* **2017**, *127*, 100–105. [CrossRef] [PubMed]

43. Miner, N.A.; Koehler, J. Intraperitoneal injection of mice. *Appl. Microbiol.* **1968**, *16*, 1418–1419.

44. Noy-Porat, T.; Rosenfeld, R.; Ariel, N.; Epstein, E.; Alcalay, R.; Zvi, A.; Kronman, C.; Ordentlich, A.; Mazor, O. Isolation of anti-ricin protective antibodies exhibiting high affinity from immunized non-human primates. *Toxins* **2016**, *8*, 64. [CrossRef] [PubMed]

45. Griffiths, G.D.; Lindsay, C.D.; Upshall, D.G. Examination of the toxicity of several protein toxins of plant origin using bovine pulmonary endothelial cells. *Toxicology* **1994**, *90*, 11–27. [CrossRef]

46. Yermakova, A.; Mantis, N.J. Protective immunity to ricin toxin conferred by antibodies against the toxin's binding subunit (RTB). *Vaccine* **2011**, *29*, 7925–7935. [CrossRef] [PubMed]

47. Wei, G.; Hu, W.-G.; Yin, J.; Chau, D.; Hu, C.C.; Lillico, D.; Yu, J.; Negrych, L.M.; Cherwonogrodzky, J.W. Conformation-dependent high-affinity potent ricin-neutralizing monoclonal antibodies. *Biomed. Res. Int.* **2013**. [CrossRef]

48. Hu, W.G.; Yin, J.; Chau, D.; Negreych, L.M.; Cherwonogrodzky, J.W. Humanization and characterization of an anti-ricin neutralization monoclonal antibody. *PLoS ONE* **2012**, *7*, e45595. [CrossRef] [PubMed]

49. Prigent, J.; Panigai, L.; Lamourette, P.; Sauvaire, D.; Devilliers, K.; Plaisance, M.; Volland, H.; Creminon, C.; Simon, S. Neutralising antibodies against ricin toxin. *PLoS ONE* **2009**, *6*, e20166. [CrossRef] [PubMed]

50. Herrera, C.; Klokk, T.; Cole, R.; Sandvig, K.; Mantis, N.J. A bispecific antibody promotes aggregation of ricin toxin on cell surfaces and alters dynamics of toxin internalization and trafficking. *PLoS ONE* **2016**, *11*, e0156893. [CrossRef] [PubMed]

51. Song, K.; Mize, R.R.; Marrero, L.; Corti, M.; Kirk, J.M.; Pincus, S.H. Antibody to ricin A chain hinders intracellular routing of toxin and protects cells even after toxin has been internalized. *PLoS ONE* **2013**, *8*, e62417. [CrossRef] [PubMed]

52. Yermakova, A.; Klokk, T.I.; Cole, R.; Sandvig, K.; Mantis, N.J. Antibody-mediated inhibition of ricin toxin retrograde transport. *mBio* **2014**, *5*, e00995. [CrossRef] [PubMed]

53. Yermakova, A.; Klokk, T.I.; O'Hara, J.M.; Cole, R.; Sandvig, K.; Mantis, N.J. Neutralizing monoclonal antibodies against disparate epitopes on ricin toxin's enzymatic subunit interfere with intracellular toxin transport. *Sci. Rep.* **2016**, *6*, 22721. [CrossRef] [PubMed]

54. Gawarammana, I.; Keyler, D. Dealing with adverse reactions to snake venom. *Ceylon Med. J.* **2011**, *56*, 87–90. [CrossRef] [PubMed]

55. Theakston, R.D.G.; Smith, D.C. Therapeutic antibodies to snake venoms. In *Therapeutic Antibodies*; Landon, J., Chard, T., Eds.; Springer: London, UK, 1995; Chapter 6, pp. 109–133.

56. Mayers, C.N.; Veall, S.; Bedford, R.J.; Holley, J.L. Anti-immunoglobulin responses to IgG, F(ab')2 and Fab botulinum antitoxins in mice. *Immunopharmacol. Immunotoxicol.* **2003**, *25*, 397–408. [CrossRef] [PubMed]

57. Holley, J.L.; Poole, S.J.C.; Cooper, I.A.M.; Griffiths, G.D.; Simpson, A.J. The production and evaluation of ricin antitoxin. In *Defence against the Effects of Chemical Hazards: Toxicology, Diagnosis and Medical Countermeasures*; NATO Meeting Proceedings RTO-MP-HFM-149; NATO: Neuilly-sur-Seine, France, 2007; pp. 12-1–12-8.

58. Sesardic, D.; Winsnes, R.; Rigsby, P.; Beh-Gross, M.-E. Collaborative study for the calibration of serological methods for potency testing of diphtheria toxoid vaccines. Extended study: Correlation of serology with in vivo toxin neutralisation. *Pharmeuropa Bio* **2003**, *2*, 69–75.

59. *First WHO International Standard for Human Diphtheria Antitoxin*; WHO Expert Committee on Biological Standardization: 63rd Report; World Health Organization: Geneva, Switzerland, 2013; Section 4.1.3, p. 29.

60. Stechmann, B.; Bai, S.K.; Gobbo, E.; Lopez, R.; Merer, G.; Pinchard, S.; Panigai, L.; Tenza, D.; Raposo, G.; Beaumelle, B.; et al. Inhibition of retrograde transport protects mice from lethal ricin challenge. *Cell* **2010**, *141*, 231–242. [CrossRef] [PubMed]

61. Barbier, J.; Bouclier, C.; Johannes, L.; Gillet, D. Inhibitors of the cellular trafficking of ricin. *Toxins* **2012**, *4*, 15–27. [CrossRef] [PubMed]

62. Gupta, N.; Pons, V.; Noel, R.; Buisson, D.A.; Uisson, D.A.; Michau, A.; Johannes, L.; Gillet, D.; Barbier, J.; Cintrat, J.-C. (S)-N-Methyldihydroquinazolinones are the Active Enantiomers of Retro-2 Derived Compounds against toxins. *ACS Med. Chem. Lett.* **2014**, *5*, 94–97. [CrossRef] [PubMed]

63. Gal, Y.; Sapoznikov, A.; Falach, R.; Ehrlich, S.; Aftalion, M.; Sabo, T.; Kronman, C. Potent antiedematous and protective effects of ciprofloxacin in pulmonary ricinosis. *Antimicrob. Agents Chemother.* **2016**, *60*, 7153–7158. [PubMed]

64. Thullier, P.; Griffiths, G.D. Broad recognition of ricin toxins prepared from a range of *Ricinus* cultivars using immunochomatographic tests. *Clin. Toxicol.* **2009**, *47*, 643–650. [CrossRef] [PubMed]

65. Pugsley, H.R. Quantifying autophagy: Measuring LC3 puncta and autolysosome formation in cells using multispectral imaging flow cytometry. *Methods* **2017**, *112*, 147–156. [CrossRef] [PubMed]

66. Rajan, R.; Karbowniczek, M.; Pugsley, H.R.; Sabnani, M.K.; Astrinidis, A.; La-Beck, N.M. Quantifying autophagosomes and autolysosomes in cells using imaging flow cytometry. *Cytometry A* **2015**, *87*, 451–458. [CrossRef] [PubMed]

Article

A Supercluster of Neutralizing Epitopes at the Interface of Ricin's Enzymatic (RTA) and Binding (RTB) Subunits

Amanda Y. Poon [1], David J. Vance [2], Yinghui Rong [2], Dylan Ehrbar [2] and Nicholas J. Mantis [1,2,*]

[1] Department of Biomedical Sciences, University at Albany School of Public Health, Albany, NY 12201, USA; aypoon@albany.edu

[2] Division of Infectious Disease, Wadsworth Center, New York State Department of Health, Albany, NY 12208, USA; david.vance@health.ny.gov (D.J.V.); yinghui.rong@health.ny.gov (Y.R.); dylan.ehrbar@health.ny.gov (D.E.)

* Correspondence: nicholas.mantis@health.ny.gov; Tel.: +1-518-473-7487

Academic Editors: Julien Barbier and Daniel Gillet
Received: 5 September 2017; Accepted: 18 November 2017; Published: 23 November 2017

Abstract: As part of an effort to engineer ricin antitoxins and immunotherapies, we previously produced and characterized a collection of phage-displayed, heavy chain-only antibodies (V$_H$Hs) from alpacas that had been immunized with ricin antigens. In our initial screens, we identified nine V$_H$Hs directed against ricin toxin's binding subunit (RTB), but only one, JIZ-B7, had toxin-neutralizing activity. Linking JIZ-B7 to different V$_H$Hs against ricin's enzymatic subunit (RTA) resulted in several bispecific antibodies with potent toxin-neutralizing activity in vitro and in vivo. JIZ-B7 may therefore be an integral component of a future V$_H$H-based neutralizing agent (VNA) for ricin toxin. In this study, we now localize, using competitive ELISA, JIZ-B7's epitope to a region of RTB's domain 2 sandwiched between the high-affinity galactose/N-acetylgalactosamine (Gal/GalNAc)-binding site and the boundary of a neutralizing hotspot on RTA known as cluster II. Analysis of additional RTB ($n = 8$)- and holotoxin ($n = 4$)-specific V$_H$Hs from a recent series of screens identified a "supercluster" of neutralizing epitopes at the RTA-RTB interface. Among the V$_H$Hs tested, toxin-neutralizing activity was most closely associated with epitope proximity to RTA, and not interference with RTB's ability to engage Gal/GalNAc receptors. We conclude that JIZ-B7 is representative of a larger group of potent toxin-neutralizing antibodies, possibly including many described in the literature dating back several decades, that recognize tertiary and possibly quaternary epitopes located at the RTA-RTB interface and that target a region of vulnerability on ricin toxin.

Keywords: ricin; antibody; neutralizing; epitope

1. Introduction

Ricin toxin, a product of the castor bean plant (*Ricinus communis*), is the archetype Type II ribosome-inactivating protein (RIP) [1]. Ricin is initially synthesized as preprotein, but accumulates in storage vesicles as a mature, 65 kDa glycosylated protein in which the two subunits, RTA and RTB, are joined by a single disulfide bond [2–4]. Ricin's enzymatic subunit (RTA) is an RNA N-glycosidase (EC 3.2.2.22) that catalyzes the hydrolysis of a conserved adenine residue within the sarcin/ricin loop (SRL) of 28S rRNA [5–7]. It is a globular protein with 10 β-strands (a–j) and seven α-helices (A–G) [8,9]. RTA's active site forms a shallow pocket that includes Tyr80, Tyr123, Glu177, Arg180, and Trp211 (Figure 1). RTA's C-terminus (residues 211–267) forms a protruding element that interacts with RTB [8]. Residue Cys259 of RTA forms a disulfide bond with Cys4 of RTB.

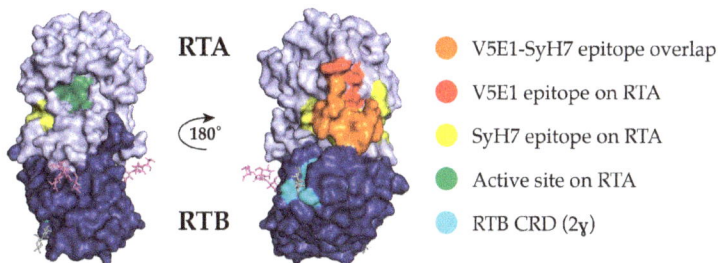

Figure 1. V5E1 and SyH7 epitope localized on surface depiction of ricin toxin. Ricin holotoxin (PDB ID 2AAI) displayed in PyMol showing ricin's enzymatic subunit (RTA; light blue) and ricin toxin's binding subunit (RTB; dark blue). The overlap between SyH7 and V5E1 epitopes is shown in orange; SyH7's additional epitope coverage is in yellow, and V5E1's additional epitope coverage is in red. Also highlighted are RTA's active site (green), and RTB's 2γ Gal/GalNAc binding pocket (sky blue) with lactose molecule (gray) (right image).

RTB (262 residues) is a galactose (Gal)- and N-acetylgalactosamine (GalNAc)-specific lectin responsible for attachment to glycolipid and glycoproteins on the surfaces of mammalian cells, including lung epithelial cells [10,11]. Structurally, RTB consists of two globular domains with identical folding topologies [8]. Each of the two domains (1 and 2) is comprised of three homologous sub-domains (α, β, γ) that arose by gene duplication from a primordial carbohydrate recognition domain (CRD) [8,10,12]. Only the external sub-domains, 1α and 2γ, retain carbohydrate recognition activity: sub-domain 1α (residues 17–59) is specific for Gal, whereas 2γ (residues 228–262) recognizes Gal and GalNac [9,13,14]. Although RTB's overall affinity for monosaccharides is quite low (K_D in the range 10^{-3} to 10^{-4} M), its affinity for complex sugars on the surface of cells is 3–4 magnitudes greater [15].

As part of an effort to engineer ricin antitoxins and immunotherapies, we have produced and characterized libraries of phage-displayed, heavy chain-only antibodies (V_HHs) from alpacas hyperimmunized with non-toxic mixtures of recombinant RTA, RTB, and ricin toxoid [16]. We recently described the full-length DNA sequences, binding affinities, epitope specificities, and neutralizing activities of 68 unique V_HHs: 31 against RTA, 33 against RTB, and four against ricin holotoxin [17]. Epitope positioning of these V_HHs was achieved by cross-competition ELISAs with a panel of toxin-neutralizing monoclonal antibodies (mAbs) against known (or postulated) epitopes on RTA [18,19] and RTB [20]. The 68 V_HHs grouped into >20 different competition bins, revealing the diversity of antibody binding sites on ricin toxin.

The RTA-specific V_HHs with the strongest toxin-neutralizing activities were confined to bins that overlapped with two previously known neutralizing hotspots, the so-called clusters I and II. Cluster I, defined by mAb PB10, is focused around RTA's α-helix B (residues 98–106), a secondary element conserved among RIPs [19,21–23]. Cluster II, defined by mAb SyH7, centers around RTA's α-helix F and the F-G loop situated on RTA's "backside", relative to the active site [19,22]. In the context of ricin holotoxin, cluster II epitopes on RTA are in close proximity to RTB domain 2. In fact, the recently described crystal structure of V_HH V5E1 bound to RTA (PDB ID 5J57) indicates that the V5E1's complementarity determining regions (CDR) 1 and 2 make contact with RTA, while CDR3 could likely interact with RTB in the context of ricin holotoxin (Figure 2) [24]. Thus, V5E1 recognizes a quaternary epitope that encompasses parts of RTA cluster II and RTB domain 2.

JIZ-B7 (also known as RTB-B7) is an RTB-specific V_HH identified in our first series of pannings of the so-called HobJo alpaca library [16]. Among the original nine RTB-specific V_HHs that were identified, JIZ-B7 was the only one that had in vitro toxin-neutralizing activity. For that reason, it was chosen as the de facto partner to pair with different RTA-specific V_HHs in the design of bispecific antitoxins [16,25,26]. The resulting V_HH heterodimers proved to be highly effective at neutralizing ricin toxin in vitro and in vivo, possibly because of their propensity to induce toxin aggregation in solution and on cell surfaces [27].

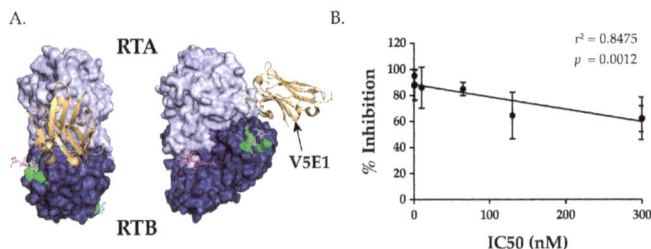

Figure 2. Relationship between V5E1 competition and toxin-neutralizing activity. (**A**) The crystal structure of V5E1 (gold) in complex with RTA (light blue) and superimposed on ricin holotoxin (RTB, dark blue), from [24]. RTB's 2γ Gal/GalNAc binding pocket is colored green; (**B**) Linear regression analysis of IC_{50}s of Tier I and Tier II V_HHs versus competition (% inhibition) with V5E1 ($n = 8$; $r^2 = 0.8475$; $p = 0.0012$). Tier III V_HHs were not included in this analysis since they did not produce quantifiable IC_{50}s. Each point on the graph represents a V_HH. IC_{50} values are the mean of at least three technical replicates, while % inhibition values are the mean and SD of three technical replicates. Refer to the Material and Methods for additional details.

However, pinpointing JIZ-B7's actual binding site on ricin toxin has proven difficult: JIZ-B7 did not react with an RTB peptide array, nor was it competitively inhibited from binding to ricin by any of the RTB-specific mAbs that are in our collection [16,25]. It was therefore fortuitous to discover that JIZ-B7 was competitively inhibited from binding to ricin by SyH7 [17]. As noted above, SyH7 recognizes an epitope within cluster II on the backside of RTA, in close proximity with the interface of RTB domain 2 (Figure 1). We therefore hypothesize that JIZ-B7's epitope on RTB is located near the cluster II footprint on RTA, at the border of RTA and RTB.

In the current study, we have now tentatively localized, using competition ELISA, JIZ-B7's epitope to RTB subdomain 2γ, which is located in close proximity to SyH7's epitope on RTA. Moreover, we have positioned the epitopes that are recognized by an additional panel of toxin-neutralizing and non-neutralizing RTB- ($n = 8$) and holotoxin-specific ($n = 4$) V_HHs. Overall, the results are indicative of there being a "supercluster" of epitopes at the RTA-RTB interface with a neutralizing hotspot at or near its core. Among the V_HHs tested, toxin-neutralizing activity was most closely associated with epitope proximity to RTA, not interference with RTB's ability to engage Gal/GalNAc receptors. These and other results suggest that antibody engagement with epitopes within this supercluster neutralize ricin by perturbing toxin uptake and/or intracellular trafficking, not blocking the attachment to cell surfaces.

2. Results

2.1. Characterization of RTB- and Holotoxin-Specific V_HHs that Compete with SyH7

As noted above, we recently discovered that JIZ-B7 was unable to bind ricin holotoxin when captured in a sandwich ELISA by SyH7 [17]. SyH7 recognizes an epitope on the backside of RTA near RTB's domain 2 (Figure 1; Table S1), suggesting that JIZ-B7's epitope may be in the vicinity of the RTA-RTB interface. JIZ-B7 is not unique, as we recently identified an additional 12 RTB- and holotoxin-specific V_HHs whose binding to ricin was also impacted negatively in a SyH7 sandwich ELISA (Table S1) [17]. By direct ELISA, nine V_HHs recognize RTB (JIZ-B7, as well as V5E4, V2C11, V5G1, V2D4, V4A1, V5H2, V6B9, V8D12) and four recognize ricin holotoxin, but not the individual RTA or RTB subunits (V5D1, V1B4, V5G6, V5G12). The SyH7 competition results by sandwich ELISA were confirmed using a slightly different assay, known as EPICC (see Materials and Methods). In the EPICC assay, biotinylated-ricin was incubated in solution with competitor V_HHs (e.g., JIZ-B7), and then applied to SyH7-coated microtiter plates and detected with streptavidin-HRP. This competition strategy (unlike the sandwich ELISA) ensured that the query V_HHs had access to their epitopes prior to competition with SyH7. Overall, there was good agreement between the two assays (Table S1), although the absolute degree of competition varied considerably, probably because of the large differences in the relative affinities. Two RTB-specific control V_HHs (V5D5, V5H6) whose ability to bind ricin was not affected by SyH7 were included in the panel. Thus, in total, we have a collection of 13 V_HHs that tentatively recognize epitopes at the RTA-RTB border, which we will refer to as Supercluster II (SCII).

As shown in Table 1, we next rank ordered the 13 SCII V_HHs, plus two control V_HHs, by their in vitro toxin-neutralizing activities (TNA). V_HHs were categorized as Tier I if their IC_{50}s were <10 nM, Tier II if their IC_{50}s were between 10 nM and 300 nM, and Tier III if their IC_{50}s were >300 nM. V_HH binding affinities for ricin holotoxin are also provided in Table 1. Tier I contains JIZ-B7 plus three additional V_HHs (V5E4, V2C11, V5G1). These four V_HHs are specific for RTB but are not clonally related, as evidenced by different CDR3 lengths and primary amino acid sequences (Figure S1A). Tier II also contains four V_HHs, with IC_{50}s ranging from 65–300 nM and equilibrium dissociation constants (K_D) between 0.19–3.2 nM. One of the Tier II V_HHs is specific for ricin holotoxin, and the other three for RTB. Finally, in Tier III, there are five V_HHs in which three are specific for ricin holotoxin and two for RTB. The Tier III V_HHs are, by definition, lacking in TNA (i.e., IC_{50} > 300 nM). The two control V_HHs (V5D5, V5H2) are specific for RTB, but have relatively poor binding affinities for ricin toxin and no detectable TNA. It is of interest to note that the CDR3s of V2C11 (Tier I), V2D4 (Tier II), and V5H2 (Tier III) are identical in length and have considerable primary amino acid sequence identity, indicating that they likely derived from the same B cell lineage (Figure S1B). In addition, in a recent report, we noted that V_HHs with the highest binding affinity (K_D) generally had the strongest TNA, and that TNA was associated with a slower off rate (k_d) rather than faster on rate (k_a) [17]. Others have noted a similar relationship [28].

Table 1. Classification and characterization of SCII V_HHs.

Tier [a]	V_HH	Target	K_D [b] [nM]	TNA [c] (IC_{50})	V5E1 [d–f]	Bin [g]
	V5E4	RTB	0.006	0.3	87 ± 12	3
I	**JIZ-B7**	RTB	0.0306	0.6	95 ± 2	4
	V2C11	RTB	0.0154	0.8	88 ± 12	4
	V5G1	RTB	0.106	5	86 ± 16	5
	V5D1	Holotoxin	0.216	65	85 ± 5	1
II	V2D4	RTB	3.2	130	64 ± 17 *	6
	V4A1	RTB	0.666	300	62 ± 16 *	6
	V8D12	RTB	0.194	300	62 ± 10 *	n.a.
	V1B4	Holotoxin	4.09	-	85 ± 7	2
	V5G6	Holotoxin	1.24	-	37 ± 6 **	7
III	V5G12	Holotoxin	0.993	-	42 ± 12 **	2
	V5H2	RTB	2.17	-	38 ± 3 **	7
	V6B9	RTB	1.61	-	40 ± 14 **	n.a.
Control	V5D5	RTB	5.54	-	7 ± 12 **	10
	V5H6	RTB	18.3	-	9 ±16 **	8

[a] V_HHs are grouped based on TNA (IC_{50}): Tier I: IC_{50} < 10 nM; Tier II: IC_{50} 10–300 nM; Tier III: IC_{50} > 300 nm; [b] Binding affinities (nM) were determined by SPR and reported in a separate study [17]; [c] TNA was determined in a Vero cell cytotoxicity assay and repeated at least three times; [d,e] V5E1 competition was determined using biotinylated ricin competition assays, as described in the Materials and Methods. V5E1 competition values represented in % inhibition; [f] One-way ANOVA was performed on V5E1 competition (% inhibition) and was followed up with a Dunnet Test, producing p-values, comparing V5E1 vs. itself (control). * p-value < 0.05; ** p-value < 0.0001. JIZ-B7 is bolded for emphasis. The V2C11 family is underlined; [g] Bins are competition profiles based on mAb competitions as adapted from [17].

2.2. Epitope Positioning by V5E1 Competition

As a strategy to further refine the epitopes in Supercluster II, we performed competition assays with V5E1, an RTA-specific $V_H H$ that binds ricin with high affinity (0.02 nM) and has potent toxin-neutralizing activity (0.5 nM IC_{50}) [24]. The X-ray crystal structure of V5E1 bound to RTA (PDB ID 5J57) indicates that V5E1's complementarity determining regions (CDR) 1 and 2 make contact with RTA, while CDR3 likely interacts with RTB's subdomain 2γ (Figure 2A). Microtiter plates were coated with V5E1 and then assessed for the ability to capture biotinylated ricin that was pre-incubated with competitor $V_H H$s in solution. As shown in Table 1, all 13 $V_H H$s in Tiers I-III were competed to varying degrees with V5E1, with the most notable competitor being JIZ-B7 (95%). The remaining three Tier I $V_H H$s were also potent competitors of V5E1 (86–88%), as was a single Tier II antibody, V5D1 (85%). The remaining three Tier II $V_H H$s were good competitors (62–64%), while the Tier III $V_H H$s were weak to moderate inhibitors (37–42%), with the notable exception of V1B4 (85%) (see below). Control antibodies (V5D5 and V5H6) did not compete with V5E1 (<10%).

Within the Tier I and II $V_H H$s, there was a statistically significant linear relationship between toxin-neutralizing activity (IC_{50}) and competition with V5E1, as demonstrated in Figure 2B. A Pearson's correlation coefficient of -0.9206 ($p = 0.0012$) was calculated between IC_{50} values and competition with V5E1, as well. Thus, we propose that there is a neutralizing focal point at the RTA-RTB interface, which is accessible by making primary contact with RTA (e.g., V5E1) or RTB (e.g., JIZ-B7). Holotoxin-specific antibodies like V5D1 may straddle that divide. V1B4, however, is somewhat of an anomaly in that it has a relatively weak affinity for ricin holotoxin (4 nM) and lacks TNA, even though it is as effective (or nearly as effective) as the Tier I $V_H H$s (e.g., V2C11, V5G1) at inhibiting V5E1 from capturing ricin. We speculate that V1B4 may recognize a "cold spot" within Supercluster II, although additional competition and differential binding assays are needed to localize V1B4's epitope with real confidence.

2.3. Cross-Competition ELISAs Indicate That Tier I $V_H H$s Recognize Overlapping Epitopes on RTB

Based on their capacities to compete with V5E1, we hypothesized that the Tier I $V_H H$s (V5E4, JIZ-B7, V2C11, V5G1) likely recognize overlapping epitopes on RTB. To test this hypothesis, the $V_H H$s were subject to cross-competition binding assays (Table 2). As expected, each $V_H H$ was able to compete with itself, as evidenced by >94% reduction in the binding of the analyte. Pairwise competitions with the Tier I $V_H H$s indicated that V5E4, JIZ-B7, V2C11, and V5G1 do indeed compete with each other. However, there was non-reciprocal competition in the case of JIZ-B7. As the capture antibody, JIZ-B7 partially affected Tier I $V_H H$s from binding ricin (43–48%). However, as a competitor antibody (in solution), JIZ-B7 was able to compete with the three other Tier I $V_H H$s (94–96%). This disparity is not easily explained by binding affinity, since the Tier I antibodies have equilibrium dissociation constant (K_D) constants ranging from 0.006 to 0.106 nM. In the case of several HIV-1 specific mAbs like PGT135, non-reciprocal competition ELISAs were attributed to the induction of conformational/allosteric changes in the capture antigen [29]. Preliminary experimental results examining the kinetics of ricin-receptor interactions in the presence of JIZ-B7 are not incompatible with the possibility that JIZ-B7 exerts an allosteric effect on ricin, possibly capturing the toxin in a slightly altered conformation that affects the structure of nearby epitopes (A. Poon and N. Mantis, *unpublished results*).

Table 2. Epitope localization of SCII V$_H$Hs by competition ELISA.

Tier	Analyte [a]	Capture V$_H$Hs			
		V5E1	JIZ-B7	V2C11	V5G1
I	V5E4	96 ± 2 [b]	48 ± 11	96 ± 1	95 ± 2
	JIZ-B7	96 ± 1	96 ± 1	96 ± 2	95 ± 2
	V2C11	96 ± 2	42 ± 9	95 ± 3	95 ± 2
	V5G1	94 ± 3	43 ± 4	94 ± 3	94 ± 3
II	V5D1 *	80 ± 10	82 ± 8	79 ± 12	82 ± 12
	V2D4	83 ± 1	35 ± 11	82 ± 4	89 ± 2
	V4A1	83 ± 3	29 ± 8	88 ± 1	88 ± 2
	V8D12	84 ± 16	31 ± 10	89 ± 8	88 ± 10
III	V1B4 *	43 ± 12	45 ± 4	40 ± 11	70 ± 6
	V5G6 *	22 ± 2	10 ± 2	29 ± 7	44 ± 11
	V5G12 *	9 ± 3	13 ± 1	6 ± 5	29 ± 14
	V5H2	47 ± 5	16 ± 6	53 ± 3	67 ± 7
	V6B9	68 ± 4	27 ± 7	68 ± 13	79 ± 4
Controls	V5D5	2 ± 2	2 ± 1	3 ± 2	19 ± 14
	V5H2	2 ± 3	1 ± 2	4 ± 5	10

[a] Indicated V$_H$Hs (column) were mixed in solution with biotinylated ricin (80 ng/mL) and applied to microtiter plates coated with indicated capture V$_H$Hs (top row). [b] Values represent % inhibition of ricin binding ± standard deviation from at least 3 replicates. The asterisks (*) denote V$_H$Hs whose epitopes are holotoxin-specific. JIZ-B7 is bolded for emphasis. Underlines indicate members of the V2C11 family of V$_H$Hs.

The Tier II V$_H$Hs were also effective competitors of the Tier I V$_H$Hs (range 79–89%), suggesting (in general) that the Tier I and Tier II epitopes are in close proximity to each other. The JIZ-B7 profile was interesting as it was a good competitor of V5D1 and a weak competitor against the other three Tier II V$_H$Hs (V2D4, V4A1, and V8D12), suggesting that V5D1 and JIZ-B7 have considerable epitope overlap. V5D1 was also inhibited by V5E1 (Table 1), which further aids us in locating V5D1's epitope relative to the other Supercluster II antibodies. Finally, the interaction between the Tier I and Tier III V$_H$Hs revealed a wide range of competition profiles, ranging from moderate (68%) to effectively null (6%). We will interpret these profiles later when proposing the relative localization of the Tier III antibody epitopes.

2.4. Differential Reactivity with RCA-1 Facilitates Epitope Localization

In a separate study, we have shown preliminarily that Tier I and II V$_H$Hs recognize recombinant non-glycosylated RTB by ELISA, indicating that the antibodies recognize protein epitopes not RTB's high mannose side chains (D. Vance, *unpublished results*). Therefore, as an additional strategy for epitope mapping within the supercluster II footprint, we examined the reactivity of the Tier I-III V$_H$Hs with *Ricinius communis* agglutinin 1 (RCA-1) (Figure 3) [20]. The B chain of RCA-1 shares 84% amino acid identity with RTB. Within RTB domain 2, in particular, there are well delineated islands of identity and dissimilarity (Figures 4A and 5). Thus, ricin or RCA-1 was immobilized onto microtiter plates and was then probed with the Tier I-III V$_H$Hs across a range of concentrations (0.2–2000 nM). Within the Tier I V$_H$Hs, JIZ-B7 reacted equally well with RCA-1 and ricin, whereas the other three V$_H$Hs (V5E4, V2C11, and V5G1) preferentially reacted with ricin (Figure 3A). Thus, JIZ-B7 recognizes an epitope that is conserved between the toxin and the agglutinin, whereas V5E4, V2C11, and V5G1 recognize ricin-specific epitopes. A similar pattern was observed within the Tier II V$_H$Hs in that V5D1 reacted equally well with RCA-1 and ricin, whereas the other three V$_H$Hs (V2D4, V4A1, and V8D12) preferentially reacted with ricin (Figure 3B). Finally, the Tier III V$_H$Hs displayed a range of reactivity profiles, which will be interpreted in the following section. It is important to note that while RCA-1 profiling experiments are approximate and qualitative in nature, they have proven extremely valuable in relative epitope positioning [20,27,30,31].

A. Tier I

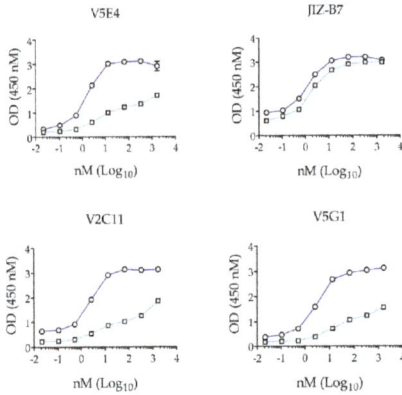

B. Tier II

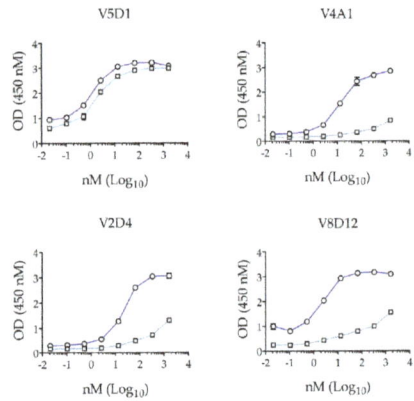

C. Tier III

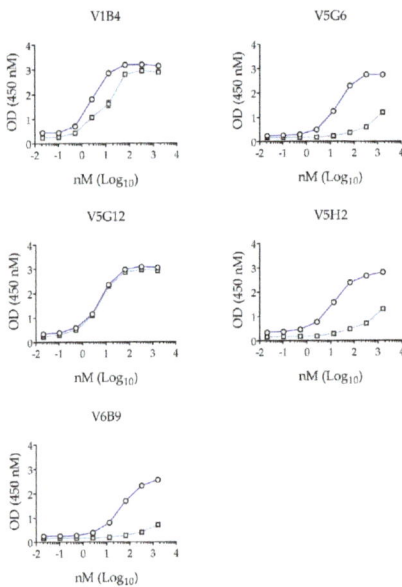

D.

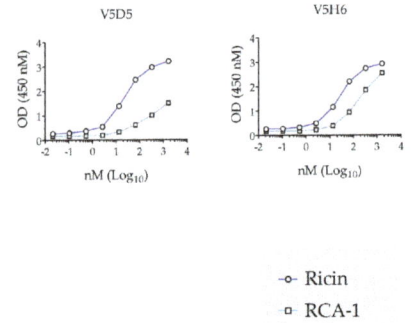

Figure 3. Reactivity profiles of Supercluster II V$_H$Hs with ricin and RCA-1. Ricin (circles) and RCA-1 (squares) were captured on microtiter plates using a mAb against RTB's domain 1 and then probed with indicated Tier I-III V$_H$Hs (Panels (**A–C**), respectively) or two control V$_H$Hs (Panel (**D**)) followed by anti-E-tag-HRP secondary antibody. All V$_H$Hs were tested and repeated at least three times.

2.5. Relative Epitope Positioning within Supercluster II

Based on V$_H$H competition results and RCA-1 reactivity profiles, we tentatively positioned the epitopes of the 13 Tier I-III V$_H$Hs on RTB's subdomain 2 (Figure 4). The localization studies also took into account previous competition profiles with asialofetuin (ASF) [17], which we have noted when appropriate.

Tier I: Based on the results presented in Table 1 and Figure 3, JIZ-B7's binding site was localized to a patch of identity on RTB's subdomain 2γ (238–240) that abuts V5E1's epitope on RTA and the Gal/GalNAc binding site on RTB. Based on Figure 3 and Table 2, the epitopes recognized by the remaining Tier I V$_H$Hs (V5E4, V2C11, V5G1) were positioned along a ridge of dissimilarity that forms a crescent shape around the Gal/GalNAc binding pocket (249–254). V2C11 was placed closer to the 2γ binding pocket based on a reduced binding activity when RTB is complexed with ASF, as described [17]. The two V2C11 family members, V2D4 (Tier II) and V5H2 (Tier III), were assigned the same epitope position as V2C11, based on homology modeling studies conducted with other V$_H$H family members [32].

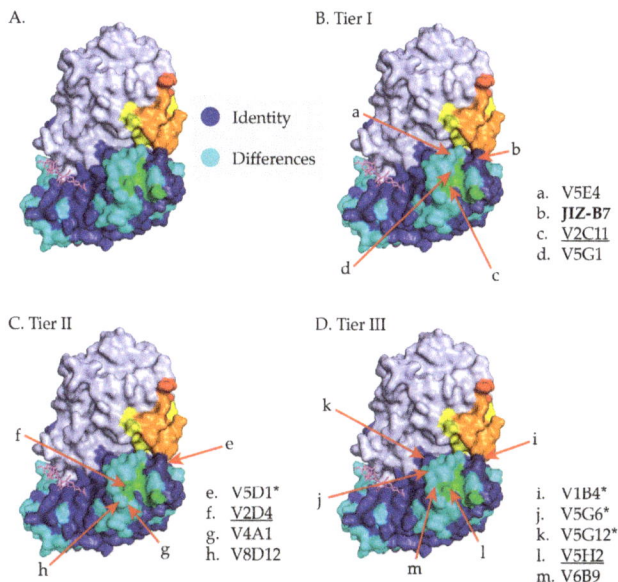

Figure 4. Supercluster II V$_H$H epitope positioning on RTB. (**A**) PyMol image of ricin holotoxin with the following landmarks highlighted: RTA (light blue), RTB's 2γ Gal/GalNAc binding pocket (green), SyH7's epitope (yellow), V5E1's epitope (red), and the overlap of SyH7 and V5E1 epitopes (orange). Residues on RTB that are identical on RCA-1 are colored dark blue, while non-identical residues are colored aqua blue. (**B–D**) Proposed locations of epitopes recognized by V$_H$Hs in Tiers I–III. JIZ-B7 is bolded for emphasis. Asterisks denote holotoxin-specific V$_H$Hs, while members of the V2C11 family are underlined.

```
        1         10        20        30        40        50        60
        |---------+---------+---------+---------+---------+---------|
RTB     ADVCMDPEPIVRIVGRNGLCVDVRDGRFHNGNAIQLWPCKSNTDANQLWTLKRDNTIRSN
RCB     ......................TGEE.FD..P...........W......RK.S.....


        61        70        80        90        100       110       120
        |---------+---------+---------+---------+---------+---------|
RTB     GKCLTTYGYSPGVYVMIYDCNTAATDATRWQIWDNGTIINPRSSLVLAATSGNSGTTLTV
RCB     .....ISKS..RQQ.V..N.S..TVG.........R.......G............K...


        121       130       140       150       160       170       180
        |---------+---------+---------+---------+---------+---------|
RTB     QTNIYAVSQGWLPTNNTQPFVTTIVGLYGLCLQANSGQVWIEDCSSEKAEQQWALYADGS
RCB     ..........................M.......K..L...T................


        181       190       200       210       220       230       240
        |---------+---------+---------+---------+---------+---------|
RTB     IRPQQNRDNCLTSDSNIRETVVKILSCGPASSGQRWMFKNGTILNLYSGLVLDVRASDPS
RCB     ............T.A..KG.........................N......R....


        241       250       261
        |---------+---------+|
RTB     LKQIILYPLHGDPNQIWLPLF
RCB     .....VH.F..NL........
```

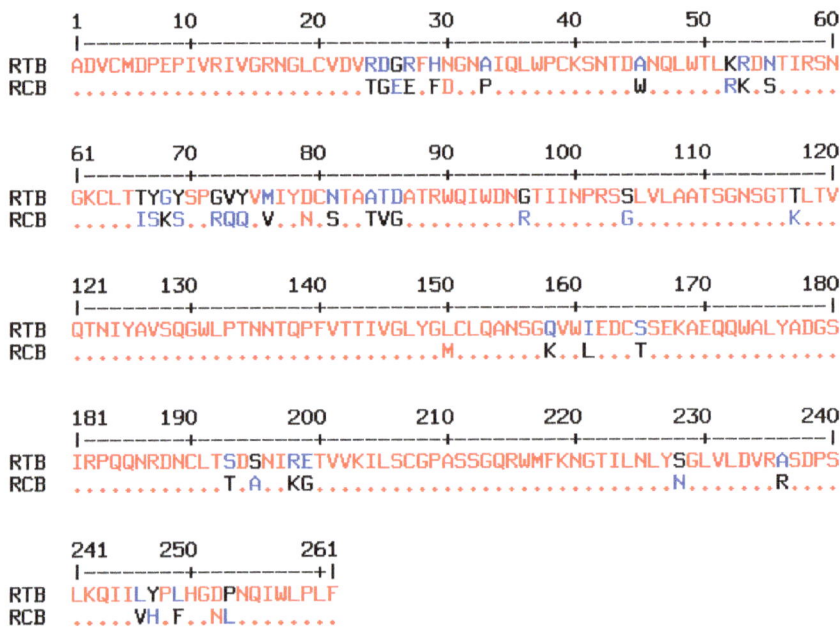

Figure 5. Amino acid sequence alignment of the binding subunits of ricin (RTB) and RCA-I (RCB). Red dots represent regions of sequence identity. Blue amino acids represent regions of low consensus value (less than 50%) and black amino acids represent neutral sequences.

Tier II: V2D4, V4A1, and V8D12's epitopes were positioned within a region of dissimilarity just below the RTB 2γ Gal/GalNAc binding pocket. The V_HHs were positioned further from the RTA interface, as compared to the Tier I V_HHs, since they were competed less by V5E1 and the Tier I V_HHs themselves. Finally, the epitope recognized by V5D1, a holotoxin-specific antibody, was localized to region close to both JIZ-B7's putative binding site and V5E1's epitope on RTA, as defined by X-ray crystallography [24].

Tier III: Tier III V_HHs were positioned around the periphery of the Gal/GalNAc binding pocket and at a distance from V5E1's epitope, with the exception of V1B4, which was shown to compete with SyH7 and V5E1 and demonstrated a near similar reactivity with ricin and RCA-1. V5G6 and V5G12 are holotoxin binders and are therefore positioned close to the interface of RTA and RTB, away from V5E1 and SyH7's epitope. Lastly, V6B9's competition profile with Tier I V_HHs and its ricin specific reactivity placed its binding site just outside of Tier I V_HHs' epitopes.

2.6. TNA Does Not Correlate with V_HHs' Ability to Block Ricin Attachment to Receptors

It is intriguing that most of the V_HHs that are characterized in this study are proposed to have binding sites in very close proximity to RTB's Gal/GalNAc binding pocket, which is known to play a role in the attachment and entry of ricin into host cells [13,33]. It was therefore imperative that we examine what impact, if any, that the V_HHs have on RTB-receptor interactions, and determine whether there was a correlation with TNA. We used both cell-based and cell-free assays to test this experimentally. In the cell-based assay, V_HHs were mixed with FITC-labeled ricin and then applied to THP-1 cells, a human monocyte cell line, maintained at 4 °C to permit attachment but not toxin internalization and then examined by flow cytometry. In the cell free assay, biotinylated ricin and individual V_HHs were mixed together and then applied to microtiter plates that were coated with ASF. ASF is routinely used as a receptor for RTB because it displays three Asn-linked triantennary complex carbohydrate chains with terminal Gal and GalNac moieties [15,33,34]. The results of these assays are presented in Table 3. In essence, there was no correlation between a V$_H$H's IC$_{50}$ and inhibition of ricin attachment to THP-1 cells (Pearson correlation coefficient = −0.004, p = 0.9918) or ASF (Pearson correlation coefficient = 0.085, p = 0.8413) (Figure 6). In fact, V5D5 is a potent inhibitor of ricin attachment but lacks TNA and is postulated to recognize an epitope outside of Supercluster II (Table 3). JIZ-B7, in contrast, has potent TNA but only affected ricin binding to target cells by ~30%. These results support a model in which the Tier I and Tier II supercluster II antibodies neutralize ricin by a mechanism other than the inhibition of attachment, most likely through the interference with ricin retrograde transport from the plasma membrane to the TGN [35,36]. Further speculation of supercluster II antibodies mechanism of action will be described in the discussion.

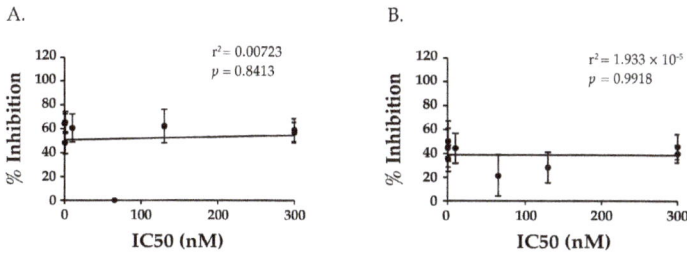

Figure 6. Toxin neutralizing activity of Supercluster II antibodies does not correlate with inhibition of cell attachment. The IC$_{50}$s of the Supercluster II V$_H$Hs (from Table 2) and their abilities to inhibit ricin attachment to (**A**) asialofetuin (ASF) and (**B**) THP-1 cells (from Table 3) were subjected to linear regression analysis, as described in the Materials and Methods. There was no correlation in either assay between a V$_H$H's ability to inhibit ricin attachment to receptors and toxin-neutralizing activity.

Table 3. V$_H$H interference of ricin attachment to host cell receptors.

Tier	V$_H$H	% Inhibition	
		ASF [a]	THP-1 [b]
I	V5E4	64 ± 8	45 ± 16
	JIZ-B7	48 ± 9	36 ± 11
	V2C11	65 ± 11	50 ± 16
	V5G1	60 ± 12	44 ± 12
II	V5D1 *	0	22 ± 17
	V2D4	62 ± 14	28 ± 12
	V4A1	59 ± 9	40 ± 7
	V8D12	57 ± 8	46 ± 10
III	V1B4 *	19 ± 18	27 ± 18
	V5G6 *	38 ± 13	16 ± 14
	V5G12 *	0	3 ± 14
	V5H2	55 ± 2	40 ± 6
	V6B9	57 ± 3	38 ± 8
Controls	V5D5	84 ± 10	90 ± 1
	V5H6	6 ± 11	18 ± 8

Binding assays were performed as described in Materials and Methods. Listed V$_H$Hs were [a] premixed with biotinylated ricin (0.25 µg/mL) and was applied to ASF coated plates or [b] premixed with FITC-labeled ricin (1 µg/mL), applied to THP-1 cells, and subjected to flow cytometry. Numbers listed represent % ricin binding inhibition ± standard deviation for three replicates. The asterisks (*) denote V$_H$Hs whose epitopes are holotoxin-specific. JIZ-B7 is bolded for emphasis. Underlines indicate members of the V2C11 family of V$_H$Hs.

3. Conclusions and Discussion

We have tentatively identified an aggregate of B cell epitopes, which we refer to as "supercluster II", which are located at the RTA-RTB interface. SCII is roughly delineated by SyH7's binding site on RTA and the Gal/GalNAc CRD on RTB domain 2 (Figure 7). At the core of SCII is a toxin-neutralizing hotspot that is defined by competition ELISAs that includes V5E1 and four RTB-specific Tier I V_HHs, JIZ-B7, V5E4, V2C11, and V5G1. The recently available X-ray crystal structure of V5E1 in complex with RTA affords insight into the structural nature of this neutralizing hotspot [24]. V5E1 forms a total of 10 hydrogen bonds and two salt bridges with RTA; contact with loop F–G (residues 192–196) is proposed to be particularly important for neutralizing activity. However, V5E1 also interacts with RTB, as was evident when the RTA-V5E1 complex was superimposed on the structure of ricin holotoxin (Figure 2A) [24]. More specifically, V5E1's CDR3 likely forms an H-bond with RTB residue Ala237, resulting in 237 $Å^2$ of buried surface area. The degree of competition (~95%) between V5E1 and JIZ-B7 for binding to ricin toxin, as shown in Table 1, suggests that the two antibodies engage the same (or closely spaced residues) on RTB. This information was taken into account when positioning JIZ-B7's epitope on RTB to a patch of RCA-1 amino acid identity located just below V5E1's binding site on RTA (Figure 4). The epitopes recognized by the remaining three Tier I V_HHs (V5E4, V2C11, and V5G1), as well as the Tier II V_HHs, were positioned concentrically away from V5E1's contact point based on values obtained in competition ELISAs (Figures 4 and 6). Within the Tier I and II V_HHs, toxin-neutralizing activity was associated with proximity to V5E1's epitope, not the RTB's Gal/GalNAc CRD, further supporting the notion of there being a core neutralizing hotspot that is situated at the RTA-RTB interface. That said, we recognize that the competition ELISAs and differential reactivity profiles with RCA-1 presented in this manuscript enable us to only approximate the location of the epitopes that are recognized by Tier I-III V_HHs. More definitive epitope positioning awaits the structural analysis of a Tier I V_HHs in complex with ricin holotoxin, as well as mutagenesis of residues on ricin toxin proposed to be involved in antibody contact. Both these tracks of investigation are currently ongoing.

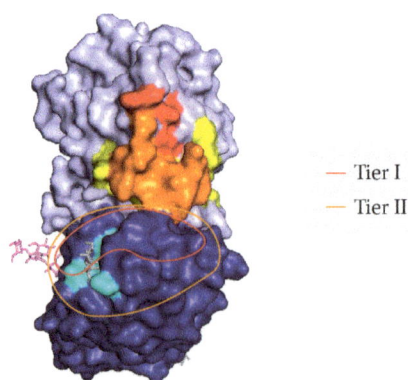

Figure 7. Proposed Supercluster II toxin-neutralizing hotspot at the RTA-RTB interface. PyMol image of ricin holotoxin (as per Figure 1) with the following landmarks highlighted: RTA (light blue), RTB (dark blue), RTB's 2γ Gal/GalNAc binding pocket (aqua blue), SyH7's epitope (yellow), V5E1's epitope (red), and the overlap of SyH7 and V5E1 epitopes (orange). The Tier I V_HH epitopes are proposed to be confined within the red trace, while the Tier II V_HH epitopes are proposed to be within the orange trace.

RTB should be an "easy target" for neutralizing antibodies, yet very few have been identified, and of those that have been described, their mechanisms of action remain incompletely defined. As a rule, antibodies against RTB have been proposed to function by one of two mechanisms: blocking ricin attachment to cell surfaces (so-called type I) or interrupting ricin intracellular trafficking (so called type II) [36,37]. The results presented in Table 3 and Figure 6 argue against the Tier I V$_H$Hs (JIZ-B7, V5E4, V2C11, and V5G1) interfering with RTB's ability to engage with host cell receptors. Therefore, we speculate that JIZ-B7, as well as V5E4, V2C11, and V5G1, likely influence toxin endocytosis and/or retrograde transport. This would be consistent with the proposed mode of action of cluster II antibodies like SyH7, which delayed ricin's egress from EEA-1(+) and Rab7(+) vesicles, and enhanced toxin accumulation in LAMP-1(+) vesicles, suggesting a degradation in lysosomes [37]. Going forward, it will be extremely interesting to compare the effects of V5E1 and JIZ-B7 on ricin endocytosis, retrograde transport, and even the efficiency of RTA retrotranslocation. Legler and colleagues demonstrated recently that RTA-specific V$_H$Hs with toxin-neutralizing activity limit the RTA thermal unfolding and/or conformational changes that would likely be encountered during toxin endocytosis (e.g., low pH). It is possible that the Tier I and II V$_H$Hs described in our current study may similarly influence the stability of RTB. However, when compared to RTA, very little is known about RTB's unfolding dynamics. Even the issue of cooperativity between RTB's two carbohydrate recognition domains remains unresolved [38]. We have only just initiated experiments to define the influence of pH on RTB association/dissociation with Gal/GalNAc residues and how single chain antibodies like JIZ-B7 might influence those dynamics.

On a final note, it is interesting to speculate that "SCII-like" mAbs may have been first identified several decades ago. For example, Chanh and colleagues described a toxin-neutralizing mAb BG11-G2, which did not bind to either purified ricin chain A or chain B, but recognized an antigenic determinant whose conformation requires the combination of the two chains [39]. Others have also described holotoxin-specific mAbs with toxin-neutralizing activity [40,41]. While there are likely many quaternary B cell epitopes along the RTA-RTB interface, it is clear from our current study that supercluster II is a region of the toxin that is vulnerable to neutralization.

4. Methods

4.1. Chemicals, Biological Reagents and Cell Lines

Labeled and unlabeled ricin toxin (*Ricinus communis* agglutinin II) and *Ricinus communis* agglutinin I (RCA-I) were purchased from Vector Laboratories (Burlingame, CA, USA). Ricin was dialyzed against phosphate buffered saline (PBS) at 4 °C in 10,000 MW cutoff Slide-A-Lyzer dialysis cassettes (ThermoFisher Pierce Protein Biology, Rockford, IL, USA) prior to use in cytotoxicity studies. Unless noted otherwise, all of the other chemicals were obtained from Sigma-Aldrich (St. Louis, MO, USA). Vero cells and THP-1 cells were obtained from the American Type Culture Collection (ATCC, Manassas, VA, USA), and propagated as recommended by ATCC. Cell culture media were provided by the Wadsworth Center's tissue culture facility.

4.2. V$_H$Hs and mAbs

Expression and purification of the ricin-specific single domain antibodies (V$_H$H) was done as described previously [16,24,25]. The murine mAb SyH7, as described previously [22], was purified from hybridoma supernatants by Protein A chromatography at the Dana Farber Cancer Institute's Monoclonal Antibody Core facility (DFCI, Boston, MA, USA).

4.3. ELISA

Epitope profiling immunocompetition capture ELISAs ("EPICC") were performed as follows: Nunc Maxisorb F96 microtiter plates (ThermoFisher Scientific, Pittsburgh, PA, USA) were coated overnight with 1 μg/mL capture mAb (SyH7) or V$_H$Hs (V5E1, V5E4, JIZ-B7, V2C11, or V5G1). Plates were blocked with 2% goat serum, washed, and incubated with analyte V$_H$Hs (330 nM) that had been mixed with biotinylated-ricin (b-R). The amount of b-R that was used in the competition ELISA was adjusted to achieve the EC$_{90}$ of each capture antibody. After 1 h, the plates were washed and developed with streptavidin-HRP antibody (1:500; SouthernBiotech, Birmingham, AL, USA) and

3,3′,5,5′-tetramethylbenzidine (TMB; Kirkegaard & Perry Labs, Gaithersburg, MD, USA). The plates were analyzed with a SpectraMax 250 spectrophotometer equipped with Softmax Pro 7 software (Molecular Devices, Sunnyvale, CA, USA). The percent (%) inhibition of ricin binding to capture antibody in the presence of competitor $V_H H$ was calculated from the optical density (OD) values as follows: 1-value OD_{450} (biotin-ricin + analyte $V_H H$)/value OD_{450} (biotin-ricin without analyte $V_H H$) × 100. For RCA-1 ELISAs, microtiter pates were coated with 24B11 (1 µg/mL) for 1 h and then incubated with either ricin or RCA-1 (1 µg/mL) for 1 h and then blocked overnight. The query $V_H Hs$ were serially diluted (1:5), starting at 1.65 µM and detected using goat anti-E tag-HRP secondary antibody [16].

4.4. Vero Cell Cytotoxicity Assays

Vero cell cytotoxicity assays were performed as previously described [42]. Vero cells were detached from culture dishes with trypsin, adjusted to ~5 × 10^4 cells per ml, and seeded (100 µL/well) into white 96-well plates (Corning Life Sciences, Corning, NY, USA), and allowed to adhere overnight. Cells were then treated with ricin (0.01 µg/mL; 154 pM), ricin:$V_H H$ mixtures, or medium alone (negative control) for 2 h at 37 °C. Cells were then washed and incubated with 10% FBS in medium for 48 h. Cell viability was assessed using CellTiter-GLO (Promega, Madison, WI, USA). All of the treatments were performed in triplicate, and 100% viability was defined as the average value obtained from wells in which cells were treated with medium only. All of the experiments were repeated at least three times independently.

4.5. THP-1 Cell Attachment Assay

THP-1 cell attachment assays were performed as previously described [20]. THP-1 cells were collected by gentle, low speed centrifugation (400× *g* for 5 min), adjusted to 5 × 10^6 cells per mL and seeded (200 µL/well) into clear U-bottom 96-well plates (BD Bioscience, San Jose, CA, USA). Cells were kept on ice at 4 °C for 20 min before treatment and throughout the experiment to prevent toxin internalization. Cells were then incubated with either FITC-labeled ricin (200 ng/well), FITC-ricin:$V_H H$ (330 nM) or with serum-free medium as a negative control for 30 min. Cells were then washed with ice cold PHEM buffer three times to remove unbound $V_H H$:ricin complexes and were fixed in 4% formaldehyde in PHEM. Fluorescence was measured using the FACS Calibur flow cytometer (BD Bioscience, San Jose, CA, USA). A minimum of 10,000 events was analyzed per sample; experiments were repeated three independent times.

4.6. Statistical Analyses

Statistical analyses were performed in GraphPad Prism version 7.03 (GraphPad Software, San Diego, CA, USA). The relationship between IC_{50} values and percent inhibition of Tier I and II $V_H Hs$, as described in Figure 2, was done using linear regression analysis. Tier III $V_H Hs$ were excluded due to the lack of a quantifiable IC_{50} value. To compare the levels of V5E1 competition of individual $V_H Hs$, as shown in Table 2, we performed a one-way ANOVA with Dunnett's multiple comparisons test. *p* values < 0.05 were considered significant.

4.7. Modeling of Ricin Toxin

Images of ricin holotoxin (PBD ID 2AAI) or RTA-$V_H H$ complexes were generated using PyMOL (The PyMOL Molecular Graphics System, Schrodinger LLC, San Diego, CA, USA).

Supplementary Materials: The following are available online at www.mdpi.com/2072-6651/9/12/378/s1, Table S1: VHH-SyH7 Competition Results, Figure S1: Sequence alignments of four Tier I VHHs and V2C11 family VHHs.

Acknowledgments: We gratefully acknowledge Renjie Song in the Wadsworth Center's Immunology Core facility for assisting with SPR. We thank Greta Van Slyke (Wadsworth Center) and Michael Rudolph (New York Structural Biology Center) for technical assistance with epitope mapping studies. We also thank Timothy Czajka for helping with manuscript revisions. Research reported in this work was supported by Contract No. HHSN272201400021C from the National Institutes of Allergy and Infectious Diseases, National Institutes of Health. The content is solely the responsibility of the authors and does not necessarily represent the official views of the National Institutes of Health. The funders had no role in study design, data collection and analysis, decision to publish, or preparation of the manuscript.

Author Contributions: A.Y.P., D.J.V., Y.R. and N.J.M. conceived, designed, and performed the experiments; A.Y.P., D.J.V., Y.R., D.E., N.J.M. analyzed the data; A.Y.P., D.J.V. and N.J.M. wrote the paper.

Conflicts of Interest: The authors declare no conflict of interest. The funding sponsors had no role in the design of the study; in the collection, analyses, or interpretation of data; in the writing of the manuscript, and in the decision to publish the results.

References

1. Schrot, J.; Weng, A.; Melzig, M.F. Ribosome-inactivating and related proteins. *Toxins (Basel)* **2015**, *7*, 1556–1615. [CrossRef] [PubMed]
2. Lamb, F.I.; Roberts, L.M.; Lord, J.M. Nucleotide sequence of cloned cdna coding for preproricin. *Eur. J. Biochem.* **1985**, *148*, 265–270. [CrossRef] [PubMed]
3. Lord, J.M. Precursors of ricin and ricinus communis agglutinin. Glycosylation and processing during synthesis and intracellular transport. *Eur. J. Biochem.* **1985**, *146*, 411–416. [CrossRef] [PubMed]
4. Jolliffe, N.A.; Craddock, C.P.; Frigerio, L. Pathways for protein transport to seed storage vacuoles. *Biochem. Soc. Trans.* **2005**, *33*, 1016–1018. [CrossRef] [PubMed]
5. Endo, Y.; Mitsui, K.; Motizuki, M.; Tsurugi, K. The mechanism of action of ricin and related toxic lectins on eukaryotic ribosomes. The site and the characteristics of the modification in 28 s ribosomal rna caused by the toxins. *J. Biol. Chem.* **1987**, *262*, 5908–5912. [PubMed]
6. Endo, Y.; Tsurugi, K. Rna n-glycosidase activity of ricin a-chain. Mechanism of action of the toxic lectin ricin on eukaryotic ribosomes. *J. Biol. Chem.* **1987**, *262*, 8128–8130. [PubMed]
7. Tesh, V.L. The induction of apoptosis by shiga toxins and ricin. *Curr. Top. Microbiol. Immunol.* **2012**, *357*, 137–178. [PubMed]
8. Montfort, W.; Villafranca, J.E.; Monzingo, A.F.; Ernst, S.R.; Katzin, B.; Rutenber, E.; Xuong, N.H.; Hamlin, R.; Robertus, J.D. The three-dimensional structure of ricin at 2.8 a. *J. Biol. Chem.* **1987**, *262*, 5398–5403. [PubMed]
9. Rutenber, E.; Katzin, B.J.; Ernst, S.; Collins, E.J.; Mlsna, D.; Ready, M.P.; Robertus, J.D. Crystallographic refinement of ricin to 2.5 a. *Proteins* **1991**, *10*, 240–250. [CrossRef] [PubMed]
10. Rutenber, E.; Ready, M.; Robertus, J.D. Structure and evolution of ricin b chain. *Nature* **1987**, *326*, 624–626. [CrossRef] [PubMed]
11. Sandvig, K.; Olsnes, S.; Pihl, A. Kinetics of binding of the toxic lectins abrin and ricin to surface receptors of human cells. *J. Biol. Chem.* **1976**, *251*, 3977–3984. [PubMed]
12. Cummings, R.; Etzler, M. R-type lectins. In *Essentials of Glycobiology*; Varki, A., Cummings, R., Esko, J., Freeze, H., Stanley, P., Bertozzi, C., Hart, G., Etzler, M., Eds.; Cold Spring Harbor Laboratory Press: Cold Spring Harbor, NY, USA, 2009.
13. Newton, D.L.; Wales, R.; Richardson, P.T.; Walbridge, S.; Saxena, S.K.; Ackerman, E.J.; Roberts, L.M.; Lord, J.M.; Youle, R.J. Cell surface and intracellular functions for ricin galactose binding. *J. Biol. Chem.* **1992**, *267*, 11917–11922. [PubMed]
14. Zentz, C.; Frenoy, J.P.; Bourrillon, R. Binding of galactose and lactose to ricin. Equilibrium studies. *Biochim. Biophys. Acta* **1978**, *536*, 18–26. [CrossRef]
15. Baenziger, J.U.; Fiete, D. Structural determinants of ricinus communis agglutinin and toxin specificity for oligosaccharides. *J. Biol. Chem.* **1979**, *254*, 9795–9799. [PubMed]
16. Vance, D.J.; Tremblay, J.M.; Mantis, N.J.; Shoemaker, C.B. Stepwise engineering of heterodimeric single domain camelid vhh antibodies that passively protect mice from ricin toxin. *J. Biol. Chem.* **2013**, *288*, 36538–36547. [CrossRef] [PubMed]
17. Vance, D.J.; Tremblay, J.M.; Rong, Y.; Angalakurthi, S.K.; Volkin, D.B.; Middaugh, C.R.; Weis, D.D.; Shoemaker, C.B.; Mantis, N.J. High-resolution epitope positioning of a large collection of neutralizing and non-neutralizing single domain antibodies on ricin toxin's enzymatic and binding subunits. *Clin. Vaccine Immunol.* **2017**. [CrossRef] [PubMed]

18. O'Hara, J.M.; Kasten-Jolly, J.C.; Reynolds, C.E.; Mantis, N.J. Localization of non-linear neutralizing b cell epitopes on ricin toxin's enzymatic subunit (rta). *Immunol. Lett.* **2014**, *158*, 7–13. [CrossRef] [PubMed]

19. Toth, R.T.; Angalakurthi, S.K.; Van Slyke, G.; Vance, D.J.; Hickey, J.M.; Joshi, S.B.; Middaugh, C.R.; Volkin, D.B.; Weis, D.D.; Mantis, N.J. High-definition mapping of four spatially distinct neutralizing epitope clusters on rivax, a candidate ricin toxin subunit vaccine. *Clin. Vaccine Immunol.* **2017**. [CrossRef] [PubMed]

20. Yermakova, A.; Vance, D.J.; Mantis, N.J. Sub-domains of ricin's b subunit as targets of toxin neutralizing and non-neutralizing monoclonal antibodies. *PLoS ONE* **2012**, *7*, e44317. [CrossRef] [PubMed]

21. Lebeda, F.J.; Olson, M.A. Prediction of a conserved, neutralizing epitope in ribosome-inactivating proteins. *Int. J. Biol. Macromol.* **1999**, *24*, 19–26. [CrossRef]

22. O'Hara, J.M.; Neal, L.M.; McCarthy, E.A.; Kasten-Jolly, J.A.; Brey, R.N., 3rd; Mantis, N.J. Folding domains within the ricin toxin a subunit as targets of protective antibodies. *Vaccine* **2010**, *28*, 7035–7046.

23. Vance, D.J.; Mantis, N.J. Resolution of two overlapping neutralizing b cell epitopes within a solvent exposed, immunodominant alpha-helix in ricin toxin's enzymatic subunit. *Toxicon* **2012**, *60*, 874–877. [CrossRef] [PubMed]

24. Rudolph, M.J.; Vance, D.J.; Cassidy, M.S.; Rong, Y.; Mantis, N.J. Structural analysis of single domain antibodies bound to a second neutralizing hot spot on ricin toxin's enzymatic subunit. *J. Biol. Chem.* **2017**, *292*, 872–883. [CrossRef] [PubMed]

25. Herrera, C.; Vance, D.J.; Eisele, L.E.; Shoemaker, C.B.; Mantis, N.J. Differential neutralizing activities of a single domain camelid antibody (vhh) specific for ricin toxin's binding subunit (rtb). *PLoS ONE* **2014**, *9*, e99788. [CrossRef] [PubMed]

26. Herrera, C.; Tremblay, J.M.; Shoemaker, C.B.; Mantis, N.J. Mechanisms of ricin toxin neutralization revealed through engineered homodimeric and heterodimeric camelid antibodies. *J. Biol. Chem.* **2015**, *290*, 27880–27889. [CrossRef] [PubMed]

27. Herrera, C.; Klokk, T.I.; Cole, R.; Sandvig, K.; Mantis, N.J. A bispecific antibody promotes aggregation of ricin toxin on cell surfaces and alters dynamics of toxin internalization and trafficking. *PLoS ONE* **2016**, *11*, e0156893. [CrossRef] [PubMed]

28. Rosenfeld, R.; Alcalay, R.; Mechaly, A.; Lapidoth, G.; Epstein, E.; Kronman, C.; Fleishman, S.J.; Mazor, O. Improved antibody-based ricin neutralization by affinity maturation is correlated with slower off-rate values. *Protein Eng. Des. Sel. PEDS* **2017**, *30*, 611–617. [CrossRef] [PubMed]

29. Derking, R.; Ozorowski, G.; Sliepen, K.; Yasmeen, A.; Cupo, A.; Torres, J.L.; Julien, J.P.; Lee, J.H.; van Montfort, T.; de Taeye, S.W.; et al. Comprehensive antigenic map of a cleaved soluble hiv-1 envelope trimer. *PLoS Pathog.* **2015**, *11*, e1004767. [CrossRef] [PubMed]

30. Rong, Y.; Van Slyke, G.; Vance, D.J.; Westfall, J.; Ehrbar, D.; Mantis, N.J. Spatial location of neutralizing and non-neutralizing b cell epitopes on domain 1 of ricin toxin's binding subunit. *PLoS ONE* **2017**, *12*, e0180999. [CrossRef] [PubMed]

31. Yermakova, A.; Mantis, N.J. Protective immunity to ricin toxin conferred by antibodies against the toxin's binding subunit (rtb). *Vaccine* **2011**, *29*, 7925–7935. [CrossRef] [PubMed]

32. Bazzoli, A.; Vance, D.J.; Rudolph, M.J.; Rong, Y.; Angalakurthi, S.K.; Toth, R.T.T.; Middaugh, C.R.; Volkin, D.B.; Weis, D.D.; Karanicolas, J.; et al. Using homology modeling to interrogate binding affinity in neutralization of ricin toxin by a family of single domain antibodies. *Proteins* **2017**, *85*, 1994–2008. [CrossRef] [PubMed]

33. Wales, R.; Richardson, P.T.; Roberts, L.M.; Woodland, H.R.; Lord, J.M. Mutational analysis of the galactose binding ability of recombinant ricin b chain. *J. Biol. Chem.* **1991**, *266*, 19172–19179. [PubMed]

34. Green, E.D.; Adelt, G.; Baenziger, J.U.; Wilson, S.; Van Halbeek, H. The asparagine-linked oligosaccharides on bovine fetuin. Structural analysis of n-glycanase-released oligosaccharides by 500-megahertz 1 h nmr spectroscopy. *J. Biol. Chem.* **1988**, *263*, 18253–18268. [PubMed]

35. Song, K.; Mize, R.R.; Marrero, L.; Corti, M.; Kirk, J.M.; Pincus, S.H. Antibody to ricin a chain hinders intracellular routing of toxin and protects cells even after toxin has been internalized. *PLoS ONE* **2013**, *8*, e62417. [CrossRef] [PubMed]

36. Yermakova, A.; Klokk, T.I.; Cole, R.; Sandvig, K.; Mantis, N.J. Antibody-mediated inhibition of ricin toxin retrograde transport. *MBio* **2014**, *5*, e00995. [CrossRef] [PubMed]

37. Yermakova, A.; Klokk, T.I.; O'Hara, J.M.; Cole, R.; Sandvig, K.; Mantis, N.J. Neutralizing monoclonal antibodies against disparate epitopes on ricin toxin's enzymatic subunit interfere with intracellular toxin transport. *Sci. Rep.* **2016**, *6*, 22721. [CrossRef] [PubMed]

38. Yao, J.; Nellas, R.B.; Glover, M.M.; Shen, T. Stability and sugar recognition ability of ricin-like carbohydrate binding domains. *Biochemistry* **2011**, *50*, 4097–4104. [CrossRef] [PubMed]

39. Chanh, T.C.; Romanowski, M.J.; Hewetson, J.F. Monoclonal antibody prophylaxis against the in vivo toxicity of ricin in mice. *Immunol. Investig.* **1993**, *22*, 63–72. [CrossRef]

40. Colombatti, M.; Pezzini, A.; Colombatti, A. Monoclonal antibodies against ricin: Effects on toxin function. *Hybridoma* **1986**, *5*, 9–19. [CrossRef] [PubMed]

41. Maddaloni, M.; Cooke, C.; Wilkinson, R.; Stout, A.V.; Eng, L.; Pincus, S.H. Immunological characteristics associated with the protective efficacy of antibodies to ricin. *J. Immunol.* **2004**, *172*, 6221–6228. [CrossRef] [PubMed]

42. Wahome, P.G.; Mantis, N.J. High-throughput, cell-based screens to identify small-molecule inhibitors of ricin toxin and related category b ribosome inactivating proteins (rips). *Curr. Protoc. Toxicol.* **2013**. [CrossRef]

Review

Shiga Toxin Therapeutics: Beyond Neutralization

Gregory Hall, Shinichiro Kurosawa and Deborah J. Stearns-Kurosawa *

Department of Pathology and Laboratory Medicine, Boston University School of Medicine, Boston, MA 02118, USA; grhall@bu.edu (G.H.); kurosawa@bu.edu (S.K.)
* Correspondence: dstearns@bu.edu; Tel.: +1-617-414-7092

Academic Editors: Tomas Girbes, Daniel Gillet and Julien Barbier
Received: 25 August 2017; Accepted: 15 September 2017; Published: 19 September 2017

Abstract: Ribotoxic Shiga toxins are the primary cause of hemolytic uremic syndrome (HUS) in patients infected with Shiga toxin-producing enterohemorrhagic *Escherichia coli* (STEC), a pathogen class responsible for epidemic outbreaks of gastrointestinal disease around the globe. HUS is a leading cause of pediatric renal failure in otherwise healthy children, resulting in a mortality rate of 10% and a chronic morbidity rate near 25%. There are currently no available therapeutics to prevent or treat HUS in STEC patients despite decades of work elucidating the mechanisms of Shiga toxicity in sensitive cells. The preclinical development of toxin-targeted HUS therapies has been hindered by the sporadic, geographically dispersed nature of STEC outbreaks with HUS cases and the limited financial incentive for the commercial development of therapies for an acute disease with an inconsistent patient population. The following review considers potential therapeutic targeting of the downstream cellular impacts of Shiga toxicity, which include the unfolded protein response (UPR) and the ribotoxic stress response (RSR). Outcomes of the UPR and RSR are relevant to other diseases with large global incidence and prevalence rates, thus reducing barriers to the development of commercial drugs that could improve STEC and HUS patient outcomes.

Keywords: Shiga toxin; Shiga-like toxins; STX1; STX2; Shiga toxin-producing *E. coli*; STEC; hemolytic uremic syndrome; unfolded protein response; ribotoxic stress response; ribotoxin

---

1. Introduction

Shiga toxins are ribotoxic proteins produced by several species of bacteria responsible for epidemic outbreaks of human gastrointestinal disease [1,2]. The prototypical toxin of this group is Shiga toxin produced by *Shigella dysenteriae* Type 1, an etiologic cause of bacterial dysentery associated with contaminated water supplies [3,4]. The related proteins Shiga-like toxin 1 (STX1) and Shiga-like toxin 2 (STX2) are produced by various pathogenic strains of Shiga toxin-producing *Escherichia coli* (STEC) responsible for food-borne illnesses globally, including numerous outbreaks in the United States, Europe, South America, and Japan [5–7]. STX1 and STX2 are encoded within the genome of lysogenized bacteriophages that can be transferred between related bacteria, creating a diverse array of bacterial strains secreting one or more toxin subtypes [1,8].

Shiga toxins are the etiologic cause of post-diarrheal hemolytic uremic syndrome (HUS), a thrombotic microangiopathy characterized by thrombocytopenia, hemolytic anemia, and acute renal failure following a course of bacterially induced hemorrhagic diarrhea [9–12]. Neurologic disease is a frequent complication of STEC infection via imprecisely defined mechanistic causes [12–14]. Approximately 5–30% of patients suffer long term morbidity from chronic renal insufficiency, hypertension, or neurological deficits following the resolution of active HUS [15]. Children younger than 2 years of age are particularly susceptible to Shiga toxin-induced HUS, and the overall HUS rates vary between 5–15% of confirmed STEC cases depending on the infecting bacterial strain. The recent European outbreak involving an atypical STEC O104:H4 strain showed substantially higher rates of adult HUS in part due to its enteroaggregative properties, and future emerging Shiga toxin-producing pathogens may have variant epidemiological profiles [6,16,17]. STEC strains are susceptible to antibiotics, but antibiotic therapy is generally contraindicated due to an association of antibiotic treatment with increased toxin production and risk of HUS development [18,19]. However, antibiotic treatment appeared to be effective during the European O104:H4 outbreak, and this was later confirmed by in vitro evaluation of patient isolates [20]. This highlights a need for rapid and specific clinical laboratory serotyping coupled with toxin detection, a technology that is not yet available commercially. As a result, the standard of care remains supportive and avoids antibiotics. The clinical management of STEC cases is complicated further by the lack of validated clinical biomarkers capable of predicting HUS onset prior to the development of thrombocytopenia and renal damage. There are no commercially approved therapeutics that specifically treat or prevent HUS caused by Shiga toxin-producing pathogens, and supportive care with careful fluid management is the recommended treatment following diagnosis [21]. Plasmapheresis and treatment with the C5 complement inhibitor Eculizumab® have not shown consistent clinical benefits in human patients [22,23]. Due to the diversity of *E. coli* serotypes capable of causing Shiga toxin-mediated disease and the potential of new emerging Shiga toxin-producing pathogens, treatments that target the activity of the toxin are currently being sought to prevent the development of HUS and to improve HUS patient outcomes.

The focus of therapeutic development for Shiga toxicosis and HUS has been the blockade of toxin activity or intracellular trafficking. Thus far, no Shiga toxin-specific therapeutic has advanced past Phase II clinical trials in the United States, partially due to the difficulties in drug development for a sporadic acute disease [24]. In this review, an alternate strategy of therapeutic development is explored that proposes to target the downstream signaling and outcomes of Shiga toxin activity. The overlap of Shiga toxin-induced stress pathways with common diseases may lead to a more rapid development and approval of commercially available therapeutics to improve patient outcomes compared to the direct targeting of the toxin itself.

2. Shiga Toxin Structure and Activity

Shiga toxins are AB5 toxins composed of a single A subunit and a pentameric B subunit [2,25]. Shiga toxins bind to the cell membrane glycolipid globotriaocylceramide (Gb3) via three binding sites on the B subunit to initiate endocytosis and gain cellular entry [26,27]. Retrograde trafficking machinery shuttle the toxin from the early endosome through the Golgi and endoplasmic reticulum (ER) [28]. During this process, furin-like proteases cleave a target site within the A subunit to create a catalytically active A subunit [29,30]. The active toxin is then transported from the ER into the cytosol to reach its target, the 28S rRNA of ribosomes [31,32]. Shiga toxins depurinate the conserved adenine residue 2260 within the sarcin–ricin loop of 28S eukaryotic rRNA via N-glycosidase activity to inhibit binding by the elongation factor EIF2a, thus terminating peptide elongation [31,33]. Other ribosome-inactivating toxins act on the sarcin–ricin loop of 28S eukaryotic rRNA, and the specific mechanism of toxicity via depurination is shared between Shiga toxins and the plant-derived toxin ricin [32]. Shiga toxins are highly toxic to Gb3-positive cells in culture, though their sensitivity is known to vary widely between cell lines and toxin subtypes [34–36]. Shiga toxicity in rodents, pigs, rabbits, and non-human primates differs in the distribution of tissues affected, presumably due to species differences in cellular Gb3 expression and localization. Rodents and rabbits develop gastrointestinal and renal tubular epithelial lesions when challenged with toxin, but fail to develop the glomerular endothelial damage and clinical HUS seen in intoxicated human patients and non-human primates [37–40]. Toxin-induced alterations in ribosomal structure and activity initiate cell stress pathways known as the unfolded protein response (UPR) and the ribotoxic stress response (RSR), which activate a variety of pro-inflammatory and pro-apoptotic cellular effector proteins that contribute to cellular dysfunction and disease [41–46]. The outcomes of these stress responses are inflammatory cytokine secretion, cellular apoptosis, and endothelial dysfunction, all of which are potential contributors to Shiga toxin-induced disease in vivo.

3. Current State of Shiga Toxin-Targeted Therapeutic Development

The field of Shiga toxin therapeutic development has focused on inhibiting the toxins at various points in the pathway between entry into the cell and ribosomal depurination by activated toxin A subunits (Figure 1). The therapeutic targeting of the toxin is conceptually justified due to a direct causal link between Shiga toxin and the development of HUS [10,12]. While this review will briefly summarize the current state of toxin-directed therapeutics, a more complete review was recently provided by Melton-Celsa and O'Brien [24].

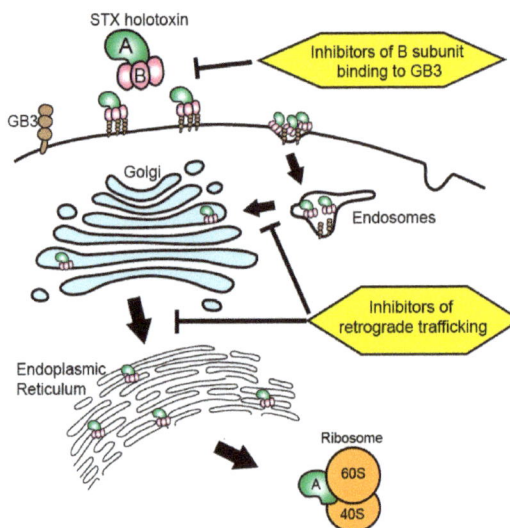

Figure 1. Therapeutic targeting of Shiga toxin internalization and retrograde trafficking. Shiga toxins bind to cell membrane globotriaosylceramide (Gb3) to initiate internalization via endocytosis. The toxin then undergoes retrograde trafficking from endosomes through the Golgi apparatus and endoplasmic reticulum. The holotoxin is processed during trafficking to release active toxin A subunits into the cytosol to interact with 60S ribosomal components, resulting in ribotoxicity through depurination of rRNA adenine residues within the sarcin–ricin loop. Therapies inhibiting the interaction of Shiga toxin B subunits with Gb3 at the cell surface or inhibiting aspects of the retrograde trafficking system seek to ameliorate Shiga toxicity by preventing the toxin from reaching cytosolic ribosomes to initiate ribotoxicity. STX: Shiga-like toxin.

Toxin-neutralizing therapeutics and vaccines for other toxin-based diseases, such as tetanus and anthrax, have yielded excellent clinical outcomes, thus providing inspiration for Shiga-like toxin (STX)-neutralizing therapeutic strategies [47,48] (Table 1). Trials using polyclonal anti-sera or monoclonal anti-Shiga toxin antibodies in animal models of Shiga toxicosis successfully rescue from mortality if treatment is given within 48 h of toxin exposure [49–52]. Monoclonal humanized murine anti-Shiga toxin antibodies have reached Phase II clinical trials within the U.S., but none have completed Phase III trials [53,54]. Camelid heavy-chain-only antibody constructs containing heavy-chain oligomers specific for the toxin have been developed and are effective in murine and porcine models, but have not been used in humans to date [55]. The heavy-chain antibody constructs have been administered as a bolus injection or introduced via a replication-incompetent adenoviral construct, with both routes of administration conferring protection against STXs [56]. One of the challenges facing antibody-based Shiga toxin neutralizers is the variety of toxin subtypes present in various STEC strains. A monoclonal antibody must neutralize STX1 and multiple STX2 subtypes to be universally effective in STEC patients, since each STEC strain can secrete one or more subtypes.

Specific multivalent binding of the Shiga toxin B subunit to Gb3 in all STX subtypes relevant to the human disease has led to the generation of synthetic Gb3 analog constructs capable of inhibitory toxin binding. Synsorb-Pk® is a silicon dioxide backbone containing bound Gb3 capable of neutralizing Shiga toxins in vitro [57]. A Phase II trial in pediatric HUS patients failed to show any clinical improvement following treatment with Synsorb-Pk®, possibly due to the inability of the orally administered drug to adsorb toxin produced at the mucosal surface or injected directly into epithelial cells by STEC [58]. The Nishikawa group has generated metallic backbone tetravalent peptides capable of binding to the Gb3 binding sites of STX1 and STX2 B subunits [59,60]. Variations of the construct have been found to be protective when administered systemically in murine and non-human primate models of Shiga toxicosis, but no human trials have been performed [61,62]. It remains to be seen if immunologic or Gb3-analog toxin neutralizers can be effective in the clinical environment where STEC-infected patients have likely been exposed to Shiga toxins for several days prior to presentation, or if preventing further toxin internalization will improve patient outcomes.

The blockade or alteration of the retrograde trafficking system is an alternate strategy for the inhibition of Shiga toxin activity. Stechmann et al. characterized two small molecule re-localizers of the vesicular transport SNARE protein Syntaxin 5 that were capable of rescuing protein synthesis in cells incubated with ricin and Shiga toxins for 4 h [63]. They named the compounds Retro-1 and 2, and found that the drugs did not impact the localization of several other cargo transport proteins in vitro. Retro-2 was found to be partially protective in a murine model of STEC infection when given orally prior to bacterial toxin induction with mitomycin C and completely protective against nasal ricin challenge if given prophylactically [63,64]. The metallic cofactor manganese has also been found to protect mice from STX1 toxicity by stimulating the degradation of the endosome-to-Golgi transport protein Gpp130, but doses were given prophylactically every 24 h for 5 days prior to a single bolus injection of toxin [65]. Manganese failed to protect mice from STX2 toxicity due to STX2 transport through proteins other than Gpp130, limiting its usefulness clinically [66]. To date, there have been no published accounts of reduced clinical toxicity when retrograde trafficking inhibitors are given after toxin exposure in vivo, a condition that more closely replicates the likely clinical scenario in diagnosed STEC patients.

Table 1. Shiga toxin-directed therapeutics in pre-clinical and early clinical development.

Therapeutic	Drug Class	Target	Mechanism of Action	Animal Models Tested	Clinical Trials Completed	References
Anti-sera	Polyclonal antibodies	STX, STX2	Circulating toxin neutralization	Pig, rabbit	None	[49,67]
Urtoxezumab®	Humanized murine monoclonal antibody	STX2	Circulating toxin neutralization	Rodent, NHP	Phase II	[50,53]
cαSTX1 and cαSTX2	Humanized murine monoclonal antibody	STX1, STX2	Circulating toxin neutralization	Rodent	Phase I	[54,68]
Murine anti-STX2	Murine monoclonal antibody	STX2	Circulating toxin neutralization	Rodent	None	[52]
Anti-STX antibodies (various clones)	Human monoclonal antibody	STX1, STX2	Circulating toxin neutralization	Rodent, pig	None	[51,69,70]
Camelid anti-STX oligomers	VHH-based neutralizing agent	STX1, STX2	Circulating toxin neutralization	Rodent	None	[55,71]
Adenoviral anti-STX2 construct	VHH-based neutralizing agent	STX2	Circulating toxin neutralization	Rodent, pig	None	[56]
Tetravalent peptides	Gb3 analogs	STX1, STX2	Circulating toxin neutralization	Rodent, non-human primate	None	[59–62]
Synsorb-Pk®	Silicon dioxide-Gb3 construct	STX1, STX2	Gastrointestinal toxin neutralization	None	Phase II (failed)	[57,58]
Retro 1 and Retro 2	Small molecule inhibitors	STX1, STX2	Retrograde trafficking inhibitor	Rodent	None	[63,64]
Manganese	Enzyme cofactor	STX1	Retrograde trafficking inhibitor	Rodent	None	[65]

4. Beyond Toxin Neutralization

While many toxin neutralizers and inhibitors have shown promise in pre-clinical settings and limited Phase II clinical trials, a direct inhibitor of Shiga toxins has not been successfully brought to market to prevent or ameliorate HUS in clinical STEC patients. Commercial barriers to treatments that specifically target sporadic acute epidemic diseases include difficulties in distribution and storage of the therapeutic as well as a reduced financial incentive to develop and produce therapeutics for an inconstant patient cohort. In addition, the usual HUS rate during an outbreak makes the statistics and finances for a clinical trial almost insurmountable. A simple design of two study groups (STEC versus STEC + HUS; α = 0.05, power 80%) and a 5% HUS rate will need over 600 patients to show effective drug prevention of HUS, which has been the criteria required by the FDA. Most outbreaks are less than 100 patients [5,72]. It is with these barriers in mind that an alternate strategy for STEC and HUS therapy should be considered that does not directly inhibit toxin activity but instead attempts to modulate the downstream stress responses to the catalytic activity of the toxin. Shiga toxins induce cellular stress pathways in sensitive cells following ribotoxicity and translational inhibition. Unfolded or incomplete proteins are detected in the ER by sensor proteins to initiate signaling cascades termed the unfolded protein response (UPR), and changes in ribosome conformation caused by depurination of the sarcin–ricin loop initiate separate signaling pathways termed the ribotoxic stress response (RSR). The UPR and RSR are relevant to other diseases, thus increasing the number of commercially viable drug candidates for the treatment of STEC and HUS patients. It is also possible that by the time of diagnosis, Shiga toxin internalization, processing, and ribosomal inactivation have already progressed to the point where toxin neutralizers and inhibitors alone will not be the most effective way of reducing cellular damage and HUS development. Targeting the UPR and RSR in STEC patients may be successful clinically, either alone or in combination with anti-toxin therapy.

5. The Unfolded Protein Response (UPR)

The UPR is a eukaryotic cellular stress response triggered by the accumulation of unfolded or improperly folded peptides within the lumen of ER, also known as ER stress. The UPR is initiated by several proteins located within the ER lumen that sense unfolded proteins and are then activated to initiate signaling cascades that attempt to restore protein homeostasis. The best characterized sensors of ER stress are protein kinase R-like endoplasmic reticulum kinase (PERK), inositol-requiring protein 1α (IRE1α), and activating transcription factor 6α (ATF6α) [73].

IRE1α is an ER membrane-bound protein that activates its endoribonuclease activity via the dissociation of the inhibitory chaperone protein GRP78 during ER stress. IRE1α cleaves a 26 nucleotide segment from xbox binding protein 1 (XBP1) mRNA. The frameshift created by this cleavage leads to the translation of the active form of XBP1, which acts as a transcription factor for protein folding, protein transport, protein degradation, and mRNA degradation-associated genes. Concurrent with IRE1α activation, the ER stress sensor protein ATF6α moves from the ER to the Golgi for processing into its active form. Active ATF6α travels to the nucleus to upregulate the transcription of chaperone proteins. The combination of ATF6α and IRE1α activation leads to an upregulation of protein-folding and degradation machinery to restore protein homeostasis [73,74].

The ER membrane-bound kinase PERK is activated to initiate a separate signaling pathway during ER stress via phosphorylation and the inactivation of eukaryotic translation initiation factor 2a (eIF2a). The inactivation of eIF2a inhibits translation to allow cellular machinery time to properly fold or degrade misfolded proteins. Phosphorylated eIF2a simultaneously upregulates several transcripts coding for proteins involved in protein degradation, mRNA degradation, chaperone proteins to aid in peptide folding, and protein transporters. A key known transcriptional target of phosphorylated eIF2a is activating transcription factor 4 (ATF4). ATF4 is a transcription factor for antioxidant and amino acid biosynthesis genes as well as CCAAT/enhancer-binding protein (C/EBP) homologous protein (CHOP). CHOP forms heterodimers with ATF4 to upregulate the transcription of additional UPR targets, including growth arrest and DNA-damage-inducible protein 34 (GADD34). GADD34 is a phosphatase that dephosphorylates eIF2a to reinitiate ribosomal translation [73–75]. The activation of ATF6α and IRE1α occurs transiently during ER stress, but the activation of PERK persists until ER stress is resolved [76].

While a short term activation of the UPR promotes survival of the cell, chronic UPR activation eventually leads to cellular apoptosis. The transcription factor CHOP concurrently suppresses anti-apoptotic Bcl-2 expression and enhances pro-apoptotic factor expression if chronically present in the cell. CHOP activates ER oxidase 1α (ERO1α), a protein that facilitates disulfide bond formation in the ER [77]. In the context of unresolved ER stress, the reactive oxygen species generated by disulfide bond formation lead to increased Ca^{+2} efflux from the ER, which in turn causes mitochondrial stress. Resuming ribosomal translation via GADD34 phosphatase activity on eIF2a in the context of unresolved ER stress leads to further generation of reactive oxygen species and ATP depletion [75]. CHOP appears to be the key factor in the progression of the UPR to apoptosis, as the knockout of CHOP in cultured cells and mice protects from apoptosis in various models of chronic UPR [77,78]. A chronic upregulation of CHOP leads to an upregulation of Bcl-2 interacting mediator of cell death (BIM) protein expression with a concurrent downregulation of the anti-apoptotic protein Bcl-2. BIM activates caspase 3, which in turn activates the caspase 8 executioner complex to initiate apoptosis [73].

6. The UPR in Health, Disease, and Shiga Toxicosis

Although the UPR is considered a stress response, it is a critical process involved in the normal function of eukaryotic organisms during development, cellular differentiation, and during times of intense cellular metabolism. During the differentiation of cells, large scale shifts in protein synthesis occur that can transiently lead to excess amounts of unfolded proteins. B cell development is a known example where abrogation of the UPR hinders cellular functional differentiation [79–81]. Cells that acutely produce and secrete proteins in response to stimulus also rely on the UPR to function normally and avoid cell death. Insulin-producing beta cells of the pancreas and activated inflammatory cells secreting cytokine and chemokine proteins are known to activate the UPR during stimulation [82–84].

Induction of the UPR by Shiga toxins has been documented in various susceptible cell types in culture as well as the renal tissue of in vivo mouse models of Shiga intoxication. While apoptosis occurs in susceptible cells exposed to Shiga toxins, it remains unclear if the UPR is the key driver of cellular apoptosis or if other cell stress response pathways, such as the RSR and TNFα death signaling, are involved [85,86]. STXs are known to associate with the ER chaperone proteins HEDJ and BiP during retrograde transport, possibly contributing to initiation of the UPR in addition to ribosomal inactivation [87,88]. Human monocytic leukemia cells upregulate IRE1α, p-PERK, and CHOP when exposed to STX1, and subsequently generate active forms of XBP1 and caspase 8 [44]. The knockdown of CHOP in THP-1 cells prevents apoptosis and caspase 8 activation [89]. Human brain microvascular endothelial cells showed a similar upregulation of CHOP and generation of caspase 8 when exposed to STX2 [90]. Mouse models injected with purified STX2 or infected with the intestinal pathogen *Citrobacter rodentium* carrying a STX2-containing plasmid showed increased renal expression of CHOP and spliced XBP1 transcripts as well as reduced Bcl-2 transcripts [46].

The UPR is also induced in a variety of diseases associated with chronically increased cellular protein production and/or secretion (Table 2). Insulin resistance and diabetes chronically induce the UPR in insulin-producing beta cells, leading to beta cell loss over time [84,91]. In diabetes-prone mice, the deletion of CHOP and other late-UPR factors rescues mice from the loss of beta cells. Non-alcoholic fatty liver disease leads to chronic UPR in hepatocytes due to increased fatty acid metabolic pathway activity, eventually contributing to intracellular lipid accumulation and hepatocyte dysfunction [92]. Neurons upregulate the UPR during neurodegenerative diseases associated with protein aggregates, such as Parkinson's disease and Alzheimer's disease [93]. Defective UPR responses in gastrointestinal Paneth cells have been suggested to contribute to inflammatory bowel disease, with unresolved ER stress leading to inflammation and a loss of microbiome homeostasis [94,95]. ER stress responses have also been found to contribute to cardiac dysfunction following hypoxia, and are associated with aberrant angiogenesis during vascular retinopathies and neoplastic growth [96–100]. Therapeutic development targeting the UPR is relevant for common chronic diseases, and could also be promising for the treatment of Shiga-intoxicated patients, thus reducing economic barriers for the development of treatments for STEC and HUS.

Table 2. Diseases associated with Unfolded Protein Response (UPR) Activation.

Disease	Cells Type Affected	Outcome	Model System(s)	Reference
Shiga toxicosis following STEC infection	Leukocytes, endothelial cells, renal epithelium, gastrointestinal epithelium	Hemolytic uremic syndrome? Inflammatory cytokine secretion?	Rodent, human monocyte, renal epithelial, and endothelial cells in vitro	[34,44,46,89,90]
Diabetes mellitus	Pancreatic beta cells	Loss of insulin production	Rodent	[91,92]
Obesity	Hepatocytes	Hepatic lipidosis, insulin resistance	Rodent, various hepatocyte cell lines in vitro	[91,92]
Inflammatory Bowel Disease	Intestinal Paneth and goblet cells	Loss of Paneth cells, gastrointestinal inflammation	Rodent	[94,95]
Neurodegenerative Diseases	Neurons	Neuron dysfunction and degeneration	Rodent	[93]
Vascular retinopathies	Retinal endothelial and pigmented epithelial cells	Aberrant angiogenesis	Rodent, human retinal endothelial cells and pigmented retinal epithelial cells in vitro	[96]
Cardiac disease	Cardiomyocytes	Cardiac hypertrophy, arrhythmias, cardiac fibrosis	Rodent, rabbit, human cardiomyocytes in vitro	[97]
Neoplasia	Malignant cells	Inflammatory cytokine secretion, angiogenesis, tumor survival	Human-mouse xenografts, neoplastic cells in vitro	[98–100]

STEC: Shiga toxin-producing enterohemorrhagic *Escherichia coli*.

7. Targeting the UPR

The most promising approach to the treatment of UPR-related diseases is to target the downstream effectors of chronic CHOP–ATF6 activity leading to the upregulation of apoptotic factors, the downregulation of anti-apoptotic factors, and the eventual activation of caspase 3 and caspase 8 to initiate apoptosis (Figure 2). A high-throughput therapeutic screening technique validated this theory in a variety of bacterial toxins including ricin, anthrax toxin, and diphtheria toxin by identifying a universally protective compound called biothionol. Biothionol is a small molecule caspase 3, 6, and 7 inhibitor found to prevent cytotoxicity by ricin in RAW264.7 and C32 cells in the face of ribosomal inactivation, and did not show signs of toxicity in mice [101]. Bithionol has not been used in STX models to date. Ouabain is a cardiotonic steroid that decreases Ca^{+2} fluxes intracellularly and increases Bcl-XL expression via the activation of Nf-KB p65 subunits, thus blunting the outcomes of chronic UPR activation. Treatment of rat proximal renal tubular cells exposed to STX2 with Ouabain in vitro prevented caspase 3 activation and cellular apoptosis. Ouabain was also found to protect human renal tubular epithelial cells and human glomerular endothelial cells from STX2-induced apoptosis in vitro [101]. Treatment of mice with continuous subcutaneous infusion of Ouabain prevented renal tubular epithelial apoptosis and a loss of podocytes 48 h after injection with STX2.

An alternate protective strategy is to increase the cellular capacity to resolve the UPR in an attempt to avoid chronic ER stress (Figure 2). Small molecule protein chaperones that improve protein folding in cells undergoing ER stress improve outcomes in mouse models of diabetes and inflammatory bowel disease, and have shown positive results in human patients with insulin resistance or cirrhosis [102–105]. Extendin-4 is an agonist of glucagon-like peptide agonist 1 that increases the expression of ATF4 in pancreatic beta cells to increase the protective effects of the early UPR [106]. Extendin-4 has been approved for use in human diabetics in the United States and Europe. To date, therapeutics that enhance UPR resolution have not been studied in the context of Shiga toxin-mediated disease in vitro or in vivo.

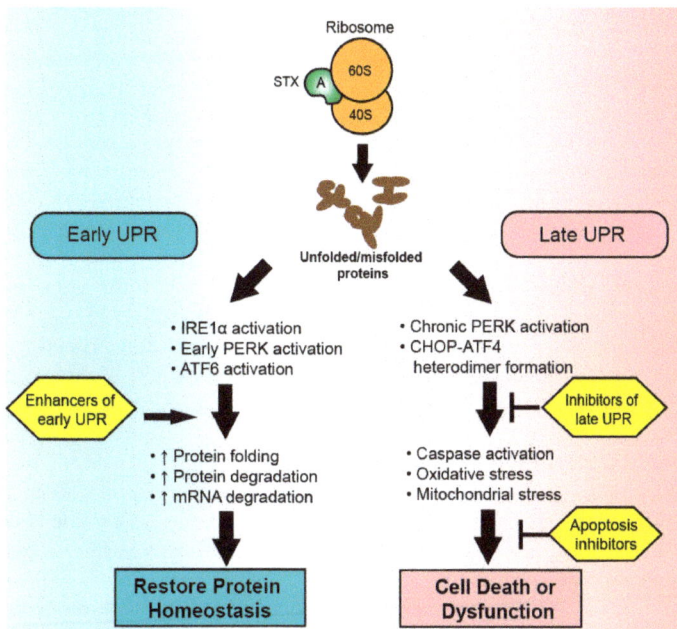

Figure 2. **Therapeutic targeting of the unfolded protein response (UPR) during Shiga toxicity.** Following ribosomal inhibition by Shiga toxin A subunits, an accumulation of unfolded and misfolded proteins is detected by sensor proteins to initiate the UPR. Therapeutics enhancing the early UPR seek to increase cellular capacity to resolve endoplasmic reticulum stress via the restoration of protein homeostasis. A chronic activation of UPR results in apoptosis or cellular dysfunction via the activity of CHOP-ATF4 heterodimers. Therapies targeting the late UPR seek to inhibit the activity or formation of CHOP–ATF4 heterodimers or inhibit initiators of apoptosis to preserve cellular function.

8. The Ribotoxic Response

In addition to the UPR, damage to domains V or VI of the ribosomal 28S rRNA has been found to initiate a stress response signaling cascade termed the ribotoxic stress response (RSR) [107]. The activity of ribotoxins such as ricin, sarcin, anisomycin, and deoxynivalenol (DON) leads to the activation of double-stranded-RNA-activated kinase R (PKR), likely through the homodimerization and transphosphorylation of ribosome-associated PKR [108]. PKR phosphorylates the translation factor eIF2a to inactivate it and initiate a signaling cascade leading to variable activation of classical p38 MAP kinases, ERK, and JNK depending on the cell type. It should be noted that both the UPR and the ribotoxic response involve the phosphorylation of eIF2a, but in cell culture experiments the ribotoxic stress response could be elicited at low levels of translational inhibition by ribotoxins. Furthermore, ribotoxins, such as emetine and pactamycin, that act on ribosomal regions other than domain V or VI fail to elicit the activation of PKR with downstream kinase activation despite their ability to inhibit protein synthesis [107]. These findings suggest that the RSR is an independent stress pathway that is distinct from the UPR and is not reliant on translational inhibition. Hematopoetic cell kinase (Hck) was found to associate with PKR and the 40S ribosome in human monocytic U937 cells, and the pharmacological inhibition of Hck activity prevented p38 MAPK phosphorylation and IL-8 secretion in response to DON, suggesting that Hck is also a key sensor of ribosomal damage necessary for downstream effector kinase signaling [109]. While the details of the signaling pathway between PKR and downstream effector kinase activity remain unclear, the zipper sterile alpha motif kinase (ZAK) has been identified as a critical component to signaling via pharmacological and siRNA knockdown studies in vitro. An abrogation of ZAK activity prevents the activation of p38 MAPK, JNK, and ERK in Hct-8 and Vero cells incubated with STX1 [81]. The ribotoxic stress response may also lead to cell death if chronic stimulation occurs, with NLRP3 inflammasome activation and pyroptosis via caspase 1 activation documented in THP-1 cells incubated with active STX1 in a concentration-dependent manner [41].

The in vitro impact of the RSR varies depending on the cell type, but is generally characterized by an upregulation of inflammatory cytokine and chemokine transcription and translation. Differentiated macrophage-like THP-1 cells secrete TNFα, IL-1β, IL-8, and MIP-1α in response to stimulation with STXs, and the response could be blunted through an inhibition of JNK, p38 MAPK, and ERK with variance in effect based on the kinase inhibited [43,110–112]. Human intestinal epithelial Hct-8 cells secreted the chemokine IL-8 when incubated with STX1 or the ribotoxic antibiotic anisomycin, and concurrent incubation with *E. coli* flagella and STX2 led to a superinduction of IL-8 secretion compared to flagella alone [113,114]. Interleukin 8 secretion was reduced through the inhibition of ERK, p38 MAPK, or JNK signaling or through use of a ZAK inhibitor. The amplification of MAPK, JNK, and ERK phosphorylation has also been documented in human dendritic cells co-stimulated with lipopolysaccharide (LPS) and anisomycin, with increased TNFα and IL12 secretion compared to stimulation with LPS alone [115]. Human brain endothelial cells secreted IL-6 and IL-8 in response to STX1, and the amount of protein secreted was amplified by costimulation with TNFα and STX1 concurrently [116,117]. Human microvascular endothelial cells were found to upregulate the chemokines IL-8, CXCL4, CXCR7, and CXCL12 in response to challenge with STX holotoxins, with delayed degradation of chemokine transcripts [118].

In vivo experiments in mice and non-human primates (NHPs) have documented acute tissue inflammation in the kidneys as well as increased circulating acute inflammatory proteins following exposure to Shiga toxins via injection or gastrointestinal infection with STEC [40,52,119]. In both mice and NHPs, increased circulating and renal TNFα, IL-6, IL-1β, and CXC chemokines are present following injection with purified STX1 and STX2. Mice lacking the gene encoding ZAK were protected from gastrointestinal ricin toxicity, with reduced CXCL1 production following depurination of the sarcin–ricin loop [120]. The treatment of rabbits with the ZAK kinase inhibitor imatinib reduced the number of neutrophils infiltrating colonic tissue infected with STEC [121]. In human HUS patients, serum and urine IL-6 levels correlated with severity of disease, and HUS patients suffering from neurologic complications had detectable increases in brain IL-1β content [122–124].

9. Targeting the RSR and Inflammation during Shiga Toxicosis

Limited research has been performed to evaluate the impact of targeted anti-inflammatory therapy on susceptibility to STX-mediated tissue injury, morbidity, and mortality. Alves-Rosa et al. hypothesized that the enhanced inflammatory response documented in vitro in cells costimulated with STXs and LPS could be blunted with anti-LPS antibodies in vivo. Mice immunized with *E. coli* O111:B4 were challenged with LPS and STXs following the confirmation of circulating anti-LPS IgG antibodies. There was no difference in mortality in immunized mice following STX + LPS challenge; however, the immunized mice failed to suppress circulating TNFα following intravenous LPS challenge compared to naïve mice, suggesting that the immunization did not prevent pattern recognition receptor activation by circulating LPS [125]. Mice pretreated with immunosuppressive doses of dexamethasone exhibited greater survival following STX2 challenge, reduced numbers of Gb3-positive CNS cells, reduced damage to the blood brain barrier, and a reduction in the number of activated astrocytes compared to controls [126]. The vasoactive drug anisodamine, an inhibitor of TNFα secretion, improved the survival of mice injected with lethal doses of STX. The improved survival was reversed via a concurrent injection of recombinant TNFα with STX, suggesting that the suppression of TNFα by anisodamine provided significant protection from STX-induced disease [127]. The pretreatment of mice with MCP-1-,

MIP-1α-, or RANTES-neutralizing antibodies protected against renal fibrin deposition following STX + LPS injections [128]. Isogai et al. found that germ-free mice pretreated with anti-TNF antibodies and infected with STEC were protected from clinical signs of morbidity, reduced renal pathology on histology, and reduced renal IL-1β, TNFα, and IL-6 protein concentrations despite similar levels of intestinal colonization and fecal STX content compared to controls [129].

A broad range of diseases are either driven through inflammatory responses or are complicated by inflammatory dysfunction. Anti-cytokine and chemokine therapeutics are being developed for use in diseases ranging from rheumatoid arthritis to inflammatory bowel disease, with multiple commercially approved drugs already available in the United States [130–133]. General anti-inflammatory drugs, such as glucocorticoids and NSAIDs, have been available for decades to treat acute inflammatory diseases. Further preclinical studies are necessary to characterize the role of inflammation in the pathogenesis of STX-induced HUS, as the role of the host response to STEC infection in the pathogenesis of HUS remains unclear [10–12]. Further work is necessary to determine if specific host cytokine responses are necessary for the development of HUS to determine the suitability of RSR elements as therapeutic targets for clinical study (Figure 3).

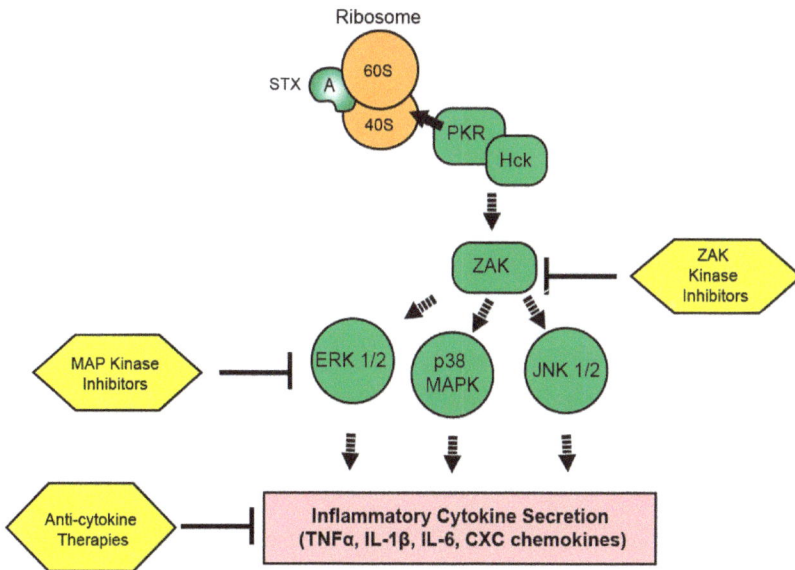

Figure 3. Therapeutic targeting of the ribotoxic stress response (RSR) during Shiga toxicity. Depurinated ribosomes are detected by PKR and Hck complexes to initiate the RSR. Signaling cascades involving ZAK and MAP kinases result in the upregulation of inflammatory transcripts and cytokine secretion depending on the intoxicated cell type. Kinase inhibitors targeting the RSR pathway or inhibitors of secreted cytokine activity could modulate Shiga toxicity through reduction of inflammatory cytokine-driven components of disease progression.

10. Future Directions

Research focusing on traditional toxin neutralizers has yielded several candidate molecules successful in animal models, but none have achieved FDA approval for use in patients. The epidemiologic profile of STEC outbreaks complicates pharmaceutical development due to its acute, sporadic, and geographically dispersed distribution of cases. Focusing on the downstream molecular impacts of STX, such as the UPR and RSR, have the advantage of utilizing therapeutics developed for use in common chronic diseases that have a greater financial incentive for pharmaceutical development.

Further preclinical study is required to determine the roles of the UPR and RSR during development of HUS in order to identify and validate potential novel therapeutic targets to improve STEC and HUS patient outcomes.

Acknowledgments: This work was supported by funding from NIH RO1 R01 AI 102931 (S.K.). G.H. was supported by a fellowship provided by the NIH Research Training in Immunology Grant T32 AI007309.

Author Contributions: G.H., S.K. and D.J.S.-K conceived and structured the review; G.H. wrote the first draft and all authors reviewed and revised the manuscript.

Conflicts of Interest: The authors declare no conflict of interest.

References

1. Krüger, A.; Lucchesi, P.M.A. Shiga toxins and stx phages: highly diverse entities. *Microbiology* **2015**, *161*, 451–462. [CrossRef] [PubMed]
2. Melton-Celsa, A.R. Shiga Toxin (Stx) Classification, Structure, and Function. *Microbiol. Spectr.* **2014**, *2*. [CrossRef] [PubMed]
3. Njamkepo, E.; Fawal, N.; Tran-Dien, A.; Hawkey, J.; Strockbine, N.; Jenkins, C.; Talukder, K.A.; Bercion, R.; Kuleshov, K.; Kolínská, R.; et al. Global phylogeography and evolutionary history of Shigella dysenteriae type 1. *Nat. Microbiol.* **2016**, *1*, 16027. [CrossRef] [PubMed]
4. Zaidi, M.B.; Estrada-García, T. Shigella: A Highly Virulent and Elusive Pathogen. *Curr. Trop. Med. Rep.* **2014**, *1*, 81–87. [CrossRef] [PubMed]
5. Rangel, J.M.; Sparling, P.H.; Crowe, C.; Griffin, P.M.; Swerdlow, D.L. Epidemiology of *Escherichia coli* O157:H7 Outbreaks, United States, 1982–2002. *Emerg. Infect. Dis.* **2005**, *11*, 603–609. [CrossRef] [PubMed]
6. Frank, C.; Werber, D.; Cramer, J.P.; Askar, M.; Faber, M.; an der Heiden, M.; Bernard, H.; Fruth, A.; Prager, R.; Spode, A.; et al. Epidemic Profile of Shiga-Toxin–Producing *Escherichia coli* O104:H4 Outbreak in Germany. *N. Engl. J. Med.* **2011**, *365*, 1771–1780. [CrossRef] [PubMed]
7. Terajima, J.; Iyoda, S.; Ohnishi, M.; Watanabe, H. Shiga Toxin (Verotoxin)-Producing *Escherichia coli* in Japan. *Microbiol. Spectr.* **2014**, *2*. [CrossRef] [PubMed]
8. O'Brien, A.D.; Newland, J.W.; Miller, S.F.; Holmes, R.K.; Smith, H.W.; Formal, S.B. Shiga-like toxin-converting phages from *Escherichia coli* strains that cause hemorrhagic colitis or infantile diarrhea. *Science* **1984**, *226*, 694–696. [CrossRef] [PubMed]
9. Karmali, M.A.; Steele, B.T.; Petric, M.; Lim, C. Sporadic cases of haemolytic-uraemic syndrome associated with faecal cytotoxin and cytotoxin-producing *Escherichia coli* in stools. *Lancet Lond. Engl.* **1983**, *1*, 619–620. [CrossRef]
10. Mayer, C.L.; Leibowitz, C.S.; Kurosawa, S.; Stearns-Kurosawa, D.J. Shiga toxins and the pathophysiology of hemolytic uremic syndrome in humans and animals. *Toxins* **2012**, *4*, 1261–1287. [CrossRef] [PubMed]
11. Walker, C.L.F.; Applegate, J.A.; Black, R.E. Haemolytic-Uraemic Syndrome as a Sequela of Diarrhoeal Disease. *J. Health Popul. Nutr.* **2012**, *30*, 257–261. [CrossRef] [PubMed]
12. Karpman, D.; Loos, S.; Tati, R.; Arvidsson, I. Haemolytic uraemic syndrome. *J. Intern. Med.* **2017**, *281*, 123–148. [CrossRef] [PubMed]
13. Weissenborn, K.; Donnerstag, F.; Kielstein, J.T.; Heeren, M.; Worthmann, H.; Hecker, H.; Schmitt, R.; Schiffer, M.; Pasedag, T.; Schuppner, R.; et al. Neurologic manifestations of E coli infection-induced hemolytic-uremic syndrome in adults. *Neurology* **2012**, *79*, 1466–1473. [CrossRef] [PubMed]
14. Takanashi, J.; Taneichi, H.; Misaki, T.; Yahata, Y.; Okumura, A.; Ishida, Y.; Miyawaki, T.; Okabe, N.; Sata, T.; Mizuguchi, M. Clinical and radiologic features of encephalopathy during 2011 E coli O111 outbreak in Japan. *Neurology* **2014**, *82*, 564–572. [CrossRef] [PubMed]
15. Rosales, A.; Hofer, J.; Zimmerhackl, L.-B.; Jungraithmayr, T.C.; Riedl, M.; Giner, T.; Strasak, A.; Orth-Höller, D.; Würzner, R.; Karch, H.; et al. Need for Long-term Follow-up in Enterohemorrhagic *Escherichia coli*–Associated Hemolytic Uremic Syndrome Due to Late-Emerging Sequelae. *Clin. Infect. Dis.* **2012**, *54*, 1413–1421. [CrossRef] [PubMed]
16. Scallan, E.; Mahon, B.E.; Hoekstra, R.M.; Griffin, P.M. Estimates of Illnesses, Hospitalizations, and Deaths Caused By Major Bacterial Enteric Pathogens in Young Children in the United States. *Pediatr. Infect. Dis. J.* **2012**, *32*, 217–221. [CrossRef] [PubMed]

17. Bielaszewska, M.; Mellmann, A.; Zhang, W.; Köck, R.; Fruth, A.; Bauwens, A.; Peters, G.; Karch, H. Characterisation of the *Escherichia coli* strain associated with an outbreak of haemolytic uraemic syndrome in Germany, 2011: a microbiological study. *Lancet Infect. Dis.* **2011**, *11*, 671–676. [CrossRef]

18. Wong, C.S.; Jelacic, S.; Habeeb, R.L.; Watkins, S.L.; Tarr, P.I. The risk of the hemolytic–uremic syndrome after antibiotic treatment of escherichia coli o157:h7 infections. *N. Engl. J. Med.* **2000**, *342*, 1930–1936. [CrossRef] [PubMed]

19. Freedman, S.B.; Xie, J.; Neufeld, M.S.; Hamilton, W.L.; Hartling, L.; Tarr, P.I.; Nettel-Aguirre, A.; Chuck, A.; Lee, B.; Johnson, D.; et al. Shiga Toxin–Producing *Escherichia coli* Infection, Antibiotics, and Risk of Developing Hemolytic Uremic Syndrome: A Meta-analysis. *Clin. Infect. Dis.* **2016**, *62*, 1251–1258. [CrossRef] [PubMed]

20. Corogeanu, D.; Willmes, R.; Wolke, M.; Plum, G.; Utermöhlen, O.; Krönke, M. Therapeutic concentrations of antibiotics inhibit Shiga toxin release from enterohemorrhagic *E. coli* O104:H4 from the 2011 German outbreak. *BMC Microbiol.* **2012**, *12*, 160. [CrossRef] [PubMed]

21. Ardissino, G.; Tel, F.; Possenti, I.; Testa, S.; Consonni, D.; Paglialonga, F.; Salardi, S.; Borsa-Ghiringhelli, N.; Salice, P.; Tedeschi, S.; et al. Early Volume Expansion and Outcomes of Hemolytic Uremic Syndrome. *Pediatrics* **2016**, *137*, e20152153. [CrossRef] [PubMed]

22. Delmas, Y.; Vendrely, B.; Clouzeau, B.; Bachir, H.; Bui, H.-N.; Lacraz, A.; Hélou, S.; Bordes, C.; Reffet, A.; Llanas, B.; et al. Outbreak of *Escherichia coli* O104:H4 haemolytic uraemic syndrome in France: Outcome with eculizumab. *Nephrol. Dial. Transplant.* **2014**, *29*, 565–572. [CrossRef] [PubMed]

23. Kielstein, J.T.; Beutel, G.; Fleig, S.; Steinhoff, J.; Meyer, T.N.; Hafer, C.; Kuhlmann, U.; Bramstedt, J.; Panzer, U.; Vischedyk, M.; et al. Collaborators of the DGfN STEC-HUS registry Best supportive care and therapeutic plasma exchange with or without eculizumab in Shiga-toxin-producing *E. coli* O104:H4 induced haemolytic-uraemic syndrome: An analysis of the German STEC-HUS registry. *Nephrol. Dial. Transplant.* **2012**, *27*, 3807–3815. [CrossRef] [PubMed]

24. O'Brien, A.D.; Melton-Celsa, A.R. New Therapeutic Developments against Shiga Toxin-Producing *Escherichia coli*. *Microbiol. Spectr.* **2014**, *2*. [CrossRef]

25. Fraser, M.E.; Fujinaga, M.; Cherney, M.M.; Melton-Celsa, A.R.; Twiddy, E.M.; O'Brien, A.D.; James, M.N.G. Structure of Shiga Toxin Type 2 (Stx2) from *Escherichia coli* O157:H7. *J. Biol. Chem.* **2004**, *279*, 27511–27517. [CrossRef] [PubMed]

26. Ling, H.; Boodhoo, A.; Hazes, B.; Cummings, M.D.; Armstrong, G.D.; Brunton, J.L.; Read, R.J. Structure of the Shiga-like Toxin I B-Pentamer Complexed with an Analogue of Its Receptor Gb3. *Biochemistry (Mosc.)* **1998**, *37*, 1777–1788. [CrossRef] [PubMed]

27. Lauvrak, S.U.; Torgersen, M.L.; Sandvig, K. Efficient endosome-to-Golgi transport of Shiga toxin is dependent on dynamin and clathrin. *J. Cell Sci.* **2004**, *117*, 2321–2331. [CrossRef] [PubMed]

28. Sandvig, K.; Bergan, J.; Dyve, A.-B.; Skotland, T.; Torgersen, M.L. Endocytosis and retrograde transport of Shiga toxin. *Toxicon* **2010**, *56*, 1181–1185. [CrossRef] [PubMed]

29. Garred, O.; van Deurs, B.; Sandvig, K. Furin-induced Cleavage and Activation of Shiga Toxin. *J. Biol. Chem.* **1995**, *270*, 10817–10821. [CrossRef] [PubMed]

30. Garred, Ø.; Dubinina, E.; Holm, P.K.; Olsnes, S.; Van Deurs, B.; Kozlov, J.V.; Sandvig, K. Role of Processing and Intracellular Transport for Optimal Toxicity of Shiga Toxin and Toxin Mutants. *Exp. Cell Res.* **1995**, *218*, 39–49. [CrossRef] [PubMed]

31. Endo, Y.; Tsurugi, K.; Yutsudo, T.; Takeda, Y.; Ogasawara, T.; Igarashi, K. Site of action of a Vero toxin (VT2) from *Escherichia coli* O157:H7 and of Shiga toxin on eukaryotic ribosomes. *Eur. J. Biochem.* **1988**, *171*, 45–50. [CrossRef] [PubMed]

32. Shi, W.-W.; Mak, A.N.-S.; Wong, K.-B.; Shaw, P.-C. Structures and Ribosomal Interaction of Ribosome-Inactivating Proteins. *Molecules* **2016**, *21*, 1588. [CrossRef] [PubMed]

33. Furutani, M.; Kashiwagi, K.; Ito, K.; Endo, Y.; Igarashi, K. Comparison of the modes of action of a vero toxin (a Shiga-like toxin) from *Escherichia coli*, of ricin, and of α-sarcin. *Arch. Biochem. Biophys.* **1992**, *293*, 140–146. [CrossRef]

34. Lentz, E.K.; Leyva-Illades, D.; Lee, M.-S.; Cherla, R.P.; Tesh, V.L. Differential Response of the Human Renal Proximal Tubular Epithelial Cell Line HK-2 to Shiga Toxin Types 1 and 2. *Infect. Immun.* **2011**, *79*, 3527–3540. [CrossRef] [PubMed]

35. Lee, S.-Y.; Cherla, R.P.; Caliskan, I.; Tesh, V.L. Shiga Toxin 1 Induces Apoptosis in the Human Myelogenous Leukemia Cell Line THP-1 by a Caspase-8-Dependent, Tumor Necrosis Factor Receptor-Independent Mechanism. *Infect. Immun.* **2005**, *73*, 5115–5126. [CrossRef] [PubMed]

36. Lindgren, S.W.; Samuel, J.E.; Schmitt, C.K.; O'Brien, A.D. The specific activities of Shiga-like toxin type II (SLT-II) and SLT-II-related toxins of enterohemorrhagic *Escherichia coli* differ when measured by Vero cell cytotoxicity but not by mouse lethality. *Infect. Immun.* **1994**, *62*, 623–631. [PubMed]

37. García, A.; Bosques, C.J.; Wishnok, J.S.; Feng, Y.; Karalius, B.J.; Butterton, J.R.; Schauer, D.B.; Rogers, A.B.; Fox, J.G. Renal Injury Is a Consistent Finding in Dutch Belted Rabbits Experimentally Infected with Enterohemorrhagic *Escherichia coli. J. Infect. Dis.* **2006**, *193*, 1125–1134. [CrossRef] [PubMed]

38. Richardson, S.E.; Rotman, T.A.; Jay, V.; Smith, C.R.; Becker, L.E.; Petric, M.; Olivieri, N.F.; Karmali, M.A. Experimental verocytotoxemia in rabbits. *Infect. Immun.* **1992**, *60*, 4154–4167. [PubMed]

39. Tesh, V.L.; Burris, J.A.; Owens, J.W.; Gordon, V.M.; Wadolkowski, E.A.; O'Brien, A.D.; Samuel, J.E. Comparison of the relative toxicities of Shiga-like toxins type I and type II for mice. *Infect. Immun.* **1993**, *61*, 3392–3402. [PubMed]

40. Stearns-Kurosawa, D.J.; Oh, S.-Y.; Cherla, R.P.; Lee, M.-S.; Tesh, V.L.; Papin, J.; Henderson, J.; Kurosawa, S. Distinct Renal Pathology and a Chemotactic Phenotype after Enterohemorrhagic *Escherichia coli* Shiga Toxins in Non-Human Primate Models of Hemolytic Uremic Syndrome. *Am. J. Pathol.* **2013**, *182*, 1227–1238. [CrossRef] [PubMed]

41. Lee, M.-S.; Kwon, H.; Lee, E.-Y.; Kim, D.-J.; Park, J.-H.; Tesh, V.L.; Oh, T.-K.; Kim, M.H. Shiga Toxins Activate the NLRP3 Inflammasome Pathway To Promote Both Production of the Proinflammatory Cytokine Interleukin-1β and Apoptotic Cell Death. *Infect. Immun.* **2016**, *84*, 172–186. [CrossRef] [PubMed]

42. Harrison, L.M.; van Haaften, W.C.E.; Tesh, V.L. Regulation of Proinflammatory Cytokine Expression by Shiga Toxin 1 and/or Lipopolysaccharides in the Human Monocytic Cell Line THP-1. *Infect. Immun.* **2004**, *72*, 2618–2627. [CrossRef] [PubMed]

43. Leyva-Illades, D.; Cherla, R.P.; Lee, M.-S.; Tesh, V.L. Regulation of Cytokine and Chemokine Expression by the Ribotoxic Stress Response Elicited by Shiga Toxin Type 1 in Human Macrophage-Like THP-1 Cells. *Infect. Immun.* **2012**, *80*, 2109–2120. [CrossRef] [PubMed]

44. Lee, S.-Y.; Lee, M.-S.; Cherla, R.P.; Tesh, V.L. Shiga toxin 1 induces apoptosis through the endoplasmic reticulum stress response in human monocytic cells. *Cell. Microbiol.* **2008**, *10*, 770–780. [CrossRef] [PubMed]

45. Jandhyala, D.M.; Ahluwalia, A.; Schimmel, J.J.; Rogers, A.B.; Leong, J.M.; Thorpe, C.M. Activation of the Classical Mitogen-Activated Protein Kinases Is Part of the Shiga Toxin-Induced Ribotoxic Stress Response and May Contribute to Shiga Toxin-Induced Inflammation. *Infect. Immun.* **2016**, *84*, 138–148. [CrossRef] [PubMed]

46. Parello, C.; Mayer, C.; Lee, B.; Motomochi, A.; Kurosawa, S.; Stearns-Kurosawa, D. Shiga Toxin 2-Induced Endoplasmic Reticulum Stress Is Minimized by Activated Protein C but Does Not Correlate with Lethal Kidney Injury. *Toxins* **2015**, *7*, 170–186. [CrossRef] [PubMed]

47. Huang, E.; Pillai, S.K.; Bower, W.A.; Hendricks, K.A.; Guarnizo, J.T.; Hoyle, J.D.; Gorman, S.E.; Boyer, A.E.; Quinn, C.P.; Meaney-Delman, D. Antitoxin Treatment of Inhalation Anthrax: A Systematic Review. *Health Secur.* **2015**, *13*, 365. [CrossRef] [PubMed]

48. Lukić, I.; Marinković, E.; Filipović, A.; Krnjaja, O.; Kosanović, D.; Inić-Kanada, A.; Stojanović, M. Key protection factors against tetanus: Anti-tetanus toxin antibody affinity and its ability to prevent tetanus toxin—Ganglioside interaction. *Toxicon* **2015**, *103*, 135–144. [CrossRef] [PubMed]

49. Donohue-Rolfe, A.; Kondova, I.; Mukherjee, J.; Chios, K.; Hutto, D.; Tzipori, S. Antibody-Based Protection of Gnotobiotic Piglets Infected with *Escherichia coli* O157:H7 against Systemic Complications Associated with Shiga Toxin 2. *Infect. Immun.* **1999**, *67*, 3645–3648. [PubMed]

50. Yamagami, S.; Motoki, M.; Kimura, T.; Izumi, H.; Takeda, T.; Katsuura, Y.; Matsumoto, Y. Efficacy of Postinfection Treatment with Anti-Shiga Toxin (Stx) 2 Humanized Monoclonal Antibody TMA-15 in Mice Lethally Challenged with Stx-Producing *Escherichia coli. J. Infect. Dis.* **2001**, *184*, 738–742. [CrossRef] [PubMed]

51. Sheoran, A.S.; Chapman, S.; Singh, P.; Donohue-Rolfe, A.; Tzipori, S. Stx2-Specific Human Monoclonal Antibodies Protect Mice against Lethal Infection with *Escherichia coli* Expressing Stx2 Variants. *Infect. Immun.* **2003**, *71*, 3125–3130. [CrossRef] [PubMed]

52. Sauter, K.A.D.; Melton-Celsa, A.R.; Larkin, K.; Troxell, M.L.; O'Brien, A.D.; Magun, B.E. Mouse Model of Hemolytic-Uremic Syndrome Caused by Endotoxin-Free Shiga Toxin 2 (Stx2) and Protection from Lethal Outcome by Anti-Stx2 Antibody. *Infect. Immun.* **2008**, *76*, 4469–4478. [CrossRef] [PubMed]

53. López, E.L.; Contrini, M.M.; Glatstein, E.; Ayala, S.G.; Santoro, R.; Allende, D.; Ezcurra, G.; Teplitz, E.; Koyama, T.; Matsumoto, Y.; et al. Safety and Pharmacokinetics of Urtoxazumab, a Humanized Monoclonal Antibody, against Shiga-Like Toxin 2 in Healthy Adults and in Pediatric Patients Infected with Shiga-Like Toxin-Producing *Escherichia coli*. *Antimicrob. Agents Chemother.* **2010**, *54*, 239–243. [CrossRef] [PubMed]

54. Dowling, T.C.; Chavaillaz, P.A.; Young, D.G.; Melton-Celsa, A.; O'Brien, A.; Thuning-Roberson, C.; Edelman, R.; Tacket, C.O. Phase 1 Safety and Pharmacokinetic Study of Chimeric Murine-Human Monoclonal Antibody cαStx2 Administered Intravenously to Healthy Adult Volunteers. *Antimicrob. Agents Chemother.* **2005**, *49*, 1808–1812. [CrossRef] [PubMed]

55. Mejías, M.P.; Hiriart, Y.; Lauché, C.; Fernández-Brando, R.J.; Pardo, R.; Bruballa, A.; Ramos, M.V.; Goldbaum, F.A.; Palermo, M.S.; Zylberman, V. Development of camelid single chain antibodies against Shiga toxin type 2 (Stx2) with therapeutic potential against Hemolytic Uremic Syndrome (HUS). *Sci. Rep.* **2016**, *6*, 24913. [CrossRef] [PubMed]

56. Sheoran, A.S.; Dmitriev, I.P.; Kashentseva, E.A.; Cohen, O.; Mukherjee, J.; Debatis, M.; Shearer, J.; Tremblay, J.M.; Beamer, G.; Curiel, D.T.; et al. Adenovirus Vector Expressing Stx1/Stx2-Neutralizing Agent Protects Piglets Infected with *Escherichia coli* O157:H7 against Fatal Systemic Intoxication. *Infect. Immun.* **2015**, *83*, 286–291. [CrossRef] [PubMed]

57. Armstrong, G.D.; Rowe, P.C.; Goodyer, P.; Orrbine, E.; Klassen, T.P.; Wells, G.; MacKenzie, A.; Lior, H.; Blanchard, C.; Auclair, F.; et al. A Phase I Study of Chemically Synthesized Verotoxin (Shiga-like Toxin) Pk-Trisaccharide Receptors Attached to Chromosorb for Preventing Hemolytic-Uremic Syndrome. *J. Infect. Dis.* **1995**, *171*, 1042–1045. [CrossRef] [PubMed]

58. Trachtman, H.; Cnaan, A.; Christen, E.; Gibbs, K.; Zhao, S.; Acheson, D.W.K.; Weiss, R.; Kaskel, F.J.; Spitzer, A.; Hirschman, G.H. Investigators of the HUS-SYNSORB Pk Multicenter Clinical Trial Effect of an oral Shiga toxin-binding agent on diarrhea-associated hemolytic uremic syndrome in children: a randomized controlled trial. *JAMA* **2003**, *290*, 1337–1344. [CrossRef] [PubMed]

59. Kato, M.; Watanabe-Takahashi, M.; Shimizu, E.; Nishikawa, K. Identification of a Wide Range of Motifs Inhibitory to Shiga Toxin by Affinity-Driven Screening of Customized Divalent Peptides Synthesized on a Membrane. *Appl. Environ. Microbiol.* **2015**, *81*, 1092–1100. [CrossRef] [PubMed]

60. Mitsui, T.; Watanabe-Takahashi, M.; Shimizu, E.; Zhang, B.; Funamoto, S.; Yamasaki, S.; Nishikawa, K. Affinity-based screening of tetravalent peptides identifies subtype-selective neutralizers of Shiga toxin 2d, a highly virulent subtype, by targeting a unique amino acid involved in its receptor recognition. *Infect. Immun.* **2016**, *84*, 2653–2661, IAI.00149-16. [CrossRef] [PubMed]

61. Stearns-Kurosawa, D.J.; Collins, V.; Freeman, S.; Debord, D.; Nishikawa, K.; Oh, S.-Y.; Leibowitz, C.S.; Kurosawa, S. Rescue from lethal Shiga toxin 2-induced renal failure with a cell-permeable peptide. *Pediatr. Nephrol.* **2011**, *26*, 2031–2039. [CrossRef] [PubMed]

62. Tsutsuki, K.; Watanabe-Takahashi, M.; Takenaka, Y.; Kita, E.; Nishikawa, K. Identification of a Peptide-Based Neutralizer That Potently Inhibits Both Shiga Toxins 1 and 2 by Targeting Specific Receptor-Binding Regions. *Infect. Immun.* **2013**, *81*, 2133–2138. [CrossRef] [PubMed]

63. Stechmann, B.; Bai, S.-K.; Gobbo, E.; Lopez, R.; Merer, G.; Pinchard, S.; Panigai, L.; Tenza, D.; Raposo, G.; Beaumelle, B.; et al. Inhibition of Retrograde Transport Protects Mice from Lethal Ricin Challenge. *Cell* **2010**, *141*, 231–242. [CrossRef] [PubMed]

64. Secher, T.; Shima, A.; Hinsinger, K.; Cintrat, J.C.; Johannes, L.; Barbier, J.; Gillet, D.; Oswald, E. Retrograde Trafficking Inhibitor of Shiga Toxins Reduces Morbidity and Mortality of Mice Infected with Enterohemorrhagic *Escherichia coli*. *Antimicrob. Agents Chemother.* **2015**, *59*, 5010–5013. [CrossRef] [PubMed]

65. Mukhopadhyay, S.; Linstedt, A.D. Manganese Blocks Intracellular Trafficking of Shiga Toxin and Protects Against Shiga Toxicosis. *Science* **2012**, *335*, 332–335. [CrossRef] [PubMed]

66. Mukhopadhyay, S.; Redler, B.; Linstedt, A.D. Shiga toxin–binding site for host cell receptor GPP130 reveals unexpected divergence in toxin-trafficking mechanisms. *Mol. Biol. Cell* **2013**, *24*, 2311–2318. [CrossRef] [PubMed]

67. Keren, D.F.; Brown, J.E.; McDonald, R.A.; Wassef, J.S. Secretory immunoglobulin A response to Shiga toxin in rabbits: kinetics of the initial mucosal immune response and inhibition of toxicity in vitro and in vivo. *Infect. Immun.* **1989**, *57*, 1885–1889. [PubMed]

68. Bitzan, M.; Poole, R.; Mehran, M.; Sicard, E.; Brockus, C.; Thuning-Roberson, C.; Rivière, M. Safety and Pharmacokinetics of Chimeric Anti-Shiga Toxin 1 and Anti-Shiga Toxin 2 Monoclonal Antibodies in Healthy Volunteers. *Antimicrob. Agents Chemother.* **2009**, *53*, 3081–3087. [CrossRef] [PubMed]

69. Jeong, K.; Tzipori, S.; Sheoran, A.S. Shiga Toxin 2-Specific but Not Shiga Toxin 1-Specific Human Monoclonal Antibody Protects Piglets Challenged with Enterohemorrhagic *Escherichia coli* Producing Shiga Toxin 1 and Shiga Toxin 2. *J. Infect. Dis.* **2010**, *201*, 1081–1083. [CrossRef] [PubMed]

70. Sheoran, A.S.; Chapman-Bonofiglio, S.; Harvey, B.R.; Mukherjee, J.; Georgiou, G.; Donohue-Rolfe, A.; Tzipori, S. Human Antibody against Shiga Toxin 2 Administered to Piglets after the Onset of Diarrhea Due to *Escherichia coli* O157:H7 Prevents Fatal Systemic Complications. *Infect. Immun.* **2005**, *73*, 4607–4613. [CrossRef] [PubMed]

71. Tremblay, J.M.; Mukherjee, J.; Leysath, C.E.; Debatis, M.; Ofori, K.; Baldwin, K.; Boucher, C.; Peters, R.; Beamer, G.; Sheoran, A.; et al. A Single VHH-Based Toxin-Neutralizing Agent and an Effector Antibody Protect Mice against Challenge with Shiga Toxins 1 and 2. *Infect. Immun.* **2013**, *81*, 4592–4603. [CrossRef] [PubMed]

72. Luna-Gierke, R.E.; Griffin, P.M.; Gould, L.H.; Herman, K.; Bopp, C.A.; Strockbine, N.; Mody, R.K. Outbreaks of non-O157 Shiga toxin-producing *Escherichia coli* infection: USA. *Epidemiol. Infect.* **2014**, *142*, 2270–2280. [CrossRef] [PubMed]

73. Hetz, C. The unfolded protein response: controlling cell fate decisions under ER stress and beyond. *Nat. Rev. Mol. Cell Biol.* **2012**, *13*, 89–102. [CrossRef] [PubMed]

74. Wang, M.; Kaufman, R.J. Protein misfolding in the endoplasmic reticulum as a conduit to human disease. *Nature* **2016**, *529*, 326–335. [CrossRef] [PubMed]

75. Han, J.; Back, S.H.; Hur, J.; Lin, Y.-H.; Gildersleeve, R.; Shan, J.; Yuan, C.L.; Krokowski, D.; Wang, S.; Hatzoglou, M.; et al. ER-stress-induced transcriptional regulation increases protein synthesis leading to cell death. *Nat. Cell Biol.* **2013**, *15*, 481–490. [CrossRef] [PubMed]

76. Lin, J.H.; Li, H.; Yasumura, D.; Cohen, H.R.; Zhang, C.; Panning, B.; Shokat, K.M.; LaVail, M.M.; Walter, P. IRE1 Signaling Affects Cell Fate During the Unfolded Protein Response. *Science* **2007**, *318*, 944–949. [CrossRef] [PubMed]

77. Marciniak, S.J.; Yun, C.Y.; Oyadomari, S.; Novoa, I.; Zhang, Y.; Jungreis, R.; Nagata, K.; Harding, H.P.; Ron, D. CHOP induces death by promoting protein synthesis and oxidation in the stressed endoplasmic reticulum. *Genes Dev.* **2004**, *18*, 3066–3077. [CrossRef] [PubMed]

78. Song, B.; Scheuner, D.; Ron, D.; Pennathur, S.; Kaufman, R.J. *Chop* deletion reduces oxidative stress, improves β cell function, and promotes cell survival in multiple mouse models of diabetes. *J. Clin. Investig.* **2008**, *118*, 3378–3389. [CrossRef] [PubMed]

79. Reimold, A.M.; Iwakoshi, N.N.; Manis, J.; Vallabhajosyula, P.; Szomolanyi-Tsuda, E.; Gravallese, E.M.; Friend, D.; Grusby, M.J.; Alt, F.; Glimcher, L.H. Plasma cell differentiation requires the transcription factor XBP-1. *Nature* **2001**, *412*, 300–307. [CrossRef] [PubMed]

80. Zhang, K.; Wong, H.N.; Song, B.; Miller, C.N.; Scheuner, D.; Kaufman, R.J. The unfolded protein response sensor IRE1α is required at 2 distinct steps in B cell lymphopoiesis. *J. Clin. Investig.* **2005**, *115*, 268. [CrossRef] [PubMed]

81. Shaffer, A.L.; Shapiro-Shelef, M.; Iwakoshi, N.N.; Lee, A.-H.; Qian, S.-B.; Zhao, H.; Yu, X.; Yang, L.; Tan, B.K.; Rosenwald, A.; et al. XBP1, Downstream of Blimp-1, Expands the Secretory Apparatus and Other Organelles, and Increases Protein Synthesis in Plasma Cell Differentiation. *Immunity* **2004**, *21*, 81–93. [CrossRef] [PubMed]

82. Iwakoshi, N.N.; Pypaert, M.; Glimcher, L.H. The transcription factor XBP-1 is essential for the development and survival of dendritic cells. *J. Exp. Med.* **2007**, *204*, 2267–2275. [CrossRef] [PubMed]

83. Kamimura, D.; Bevan, M.J. Endoplasmic Reticulum Stress Regulator XBP-1 Contributes to Effector CD8+ T Cell Differentiation during Acute Infection. *J. Immunol. Baltim. Md 1950* **2008**, *181*, 5433. [CrossRef]

84. Lee, A.-H.; Heidtman, K.; Hotamisligil, G.S.; Glimcher, L.H. Dual and opposing roles of the unfolded protein response regulated by IRE1α and XBP1 in proinsulin processing and insulin secretion. *Proc. Natl. Acad. Sci. USA* **2011**, *108*, 8885–8890. [CrossRef] [PubMed]

85. Lee, M.-S.; Koo, S.; Jeong, D.; Tesh, V. Shiga Toxins as Multi-Functional Proteins: Induction of Host Cellular Stress Responses, Role in Pathogenesis and Therapeutic Applications. *Toxins* **2016**, *8*, 77. [CrossRef] [PubMed]

86. Tesh, V.L. The induction of apoptosis by Shiga toxins and ricin. *Curr. Top. Microbiol. Immunol.* **2012**, *357*, 137–178. [CrossRef] [PubMed]

87. Falguières, T.; Johannes, L. Shiga toxin B-subunit binds to the chaperone BiP and the nucleolar protein B23. *Biol. Cell* **2006**, *98*, 125–134. [CrossRef] [PubMed]

88. Yu, M.; Haslam, D.B. Shiga Toxin Is Transported from the Endoplasmic Reticulum following Interaction with the Luminal Chaperone HEDJ/ERdj3. *Infect. Immun.* **2005**, *73*, 2524–2532. [CrossRef] [PubMed]

89. Lee, M.-S.; Cherla, R.P.; Lentz, E.K.; Leyva-Illades, D.; Tesh, V.L. Signaling through C/EBP Homologous Protein and Death Receptor 5 and Calpain Activation Differentially Regulate THP-1 Cell Maturation-Dependent Apoptosis Induced by Shiga Toxin Type 1. *Infect. Immun.* **2010**, *78*, 3378–3391. [CrossRef] [PubMed]

90. Fujii, J.; Wood, K.; Matsuda, F.; Carneiro-Filho, B.A.; Schlegel, K.H.; Yutsudo, T.; Binnington-Boyd, B.; Lingwood, C.A.; Obata, F.; Kim, K.S.; et al. Shiga Toxin 2 Causes Apoptosis in Human Brain Microvascular Endothelial Cells via C/EBP Homologous Protein. *Infect. Immun.* **2008**, *76*, 3679–3689. [CrossRef] [PubMed]

91. Özcan, U.; Cao, Q.; Yilmaz, E.; Lee, A.-H.; Iwakoshi, N.N.; Özdelen, E.; Tuncman, G.; Görgün, C.; Glimcher, L.H.; Hotamisligil, G.S. Endoplasmic Reticulum Stress Links Obesity, Insulin Action, and Type 2 Diabetes. *Science* **2004**, *306*, 457–461. [CrossRef] [PubMed]

92. Fu, S.; Yang, L.; Li, P.; Hofmann, O.; Dicker, L.; Hide, W.; Lin, X.; Watkins, S.M.; Ivanov, A.R.; Hotamisligil, G.S. Aberrant lipid metabolism disrupts calcium homeostasis causing liver endoplasmic reticulum stress in obesity. *Nature* **2011**, *473*, 528–531. [CrossRef] [PubMed]

93. Hetz, C.; Mollereau, B. Disturbance of endoplasmic reticulum proteostasis in neurodegenerative diseases. *Nat. Rev. Neurosci.* **2014**, *15*, 233–249. [CrossRef] [PubMed]

94. Kaser, A.; Lee, A.-H.; Franke, A.; Glickman, J.N.; Zeissig, S.; Tilg, H.; Nieuwenhuis, E.E.S.; Higgins, D.E.; Schreiber, S.; Glimcher, L.H.; et al. XBP1 Links ER Stress to Intestinal Inflammation and Confers Genetic Risk for Human Inflammatory Bowel Disease. *Cell* **2008**, *134*, 743–756. [CrossRef] [PubMed]

95. Grootjans, J.; Kaser, A.; Kaufman, R.J.; Blumberg, R.S. The unfolded protein response in immunity and inflammation. *Nat. Rev. Immunol.* **2016**, *16*, 469–484. [CrossRef] [PubMed]

96. Zhang, S.X.; Ma, J.H.; Bhatta, M.; Fliesler, S.J.; Wang, J.J. The unfolded protein response in retinal vascular diseases: Implications and therapeutic potential beyond protein folding. *Prog. Retin. Eye Res.* **2015**, *45*, 111–131. [CrossRef] [PubMed]

97. Wang, S.; Binder, P.; Fang, Q.; Wang, Z.; Xiao, W.; Liu, W.; Wang, X. Endoplasmic reticulum stress in the heart: insights into mechanisms and drug targets. *Br. J. Pharmacol.* **2017**. [CrossRef] [PubMed]

98. Wang, Y.; Alam, G.N.; Ning, Y.; Visioli, F.; Dong, Z.; Nör, J.E.; Polverini, P.J. The Unfolded Protein Response Induces the Angiogenic Switch in Human Tumor Cells through the PERK/ATF4 Pathway. *Cancer Res.* **2012**, *72*, 5396–5406. [CrossRef] [PubMed]

99. Rodvold, J.J.; Chiu, K.T.; Hiramatsu, N.; Nussbacher, J.K.; Galimberti, V.; Mahadevan, N.R.; Willert, K.; Lin, J.H.; Zanetti, M. Intercellular transmission of the unfolded protein response promotes survival and drug resistance in cancer cells. *Sci. Signal* **2017**, *10*, 1–12. [CrossRef] [PubMed]

100. Blais, J.D.; Addison, C.L.; Edge, R.; Falls, T.; Zhao, H.; Wary, K.; Koumenis, C.; Harding, H.P.; Ron, D.; Holcik, M.; et al. Perk-Dependent Translational Regulation Promotes Tumor Cell Adaptation and Angiogenesis in Response to Hypoxic Stress. *Mol. Cell. Biol.* **2006**, *26*, 9517–9532. [CrossRef] [PubMed]

101. Amaral, M.M.; Girard, M.C.; Álvarez, R.S.; Paton, A.W.; Paton, J.C.; Repetto, H.A.; Sacerdoti, F.; Ibarra, C.A. Ouabain Protects Human Renal Cells against the Cytotoxic Effects of Shiga Toxin Type 2 and Subtilase Cytotoxin. *Toxins* **2017**, *9*. [CrossRef] [PubMed]

102. Engin, F.; Yermalovich, A.; Nguyen, T.; Hummasti, S.; Fu, W.; Eizirik, D.L.; Mathis, D.; Hotamisligil, G.S. Restoration of the Unfolded Protein Response in Pancreatic β Cells Protects Mice against Type 1 Diabetes. *Sci. Transl. Med.* **2013**, *5*, 1–15. [CrossRef] [PubMed]

103. Cao, S.S.; Zimmermann, E.M.; Chuang, B.; Song, B.; Nwokoye, A.; Wilkinson, J.E.; Eaton, K.A.; Kaufman, R.J. The Unfolded Protein Response and Chemical Chaperones Reduce Protein Misfolding and Colitis in Mice. *Gastroenterology* **2013**, *144*, 989–1000. [CrossRef] [PubMed]

104. Kars, M.; Yang, L.; Gregor, M.F.; Mohammed, B.S.; Pietka, T.A.; Finck, B.N.; Patterson, B.W.; Horton, J.D.; Mittendorfer, B.; Hotamisligil, G.S.; et al. Tauroursodeoxycholic Acid May Improve Liver and Muscle but Not Adipose Tissue Insulin Sensitivity in Obese Men and Women. *Diabetes* **2010**, *59*, 1899–1905. [CrossRef] [PubMed]

105. Xiao, C.; Giacca, A.; Lewis, G.F. Sodium Phenylbutyrate, a Drug With Known Capacity to Reduce Endoplasmic Reticulum Stress, Partially Alleviates Lipid-Induced Insulin Resistance and β-Cell Dysfunction in Humans. *Diabetes* **2011**, *60*, 918–924. [CrossRef] [PubMed]

106. Yusta, B.; Baggio, L.L.; Estall, J.L.; Koehler, J.A.; Holland, D.P.; Li, H.; Pipeleers, D.; Ling, Z.; Drucker, D.J. GLP-1 receptor activation improves β cell function and survival following induction of endoplasmic reticulum stress. *Cell Metab.* **2006**, *4*, 391–406. [CrossRef] [PubMed]

107. Iordanov, M.S.; Pribnow, D.; Magun, J.L.; Dinh, T.H.; Pearson, J.A.; Chen, S.L.; Magun, B.E. Ribotoxic stress response: activation of the stress-activated protein kinase JNK1 by inhibitors of the peptidyl transferase reaction and by sequence-specific RNA damage to the alpha-sarcin/ricin loop in the 28S rRNA. *Mol. Cell. Biol.* **1997**, *17*, 3373. [CrossRef] [PubMed]

108. Zhou, H.-R.; Lau, A.S.; Pestka, J.J. Role of Double-Stranded RNA-Activated Protein Kinase R (PKR) in Deoxynivalenol-Induced Ribotoxic Stress Response. *Toxicol. Sci.* **2003**, *74*, 335–344. [CrossRef] [PubMed]

109. Bae, H.; Gray, J.S.; Li, M.; Vines, L.; Kim, J.; Pestka, J.J. Hematopoietic Cell Kinase Associates with the 40S Ribosomal Subunit and Mediates the Ribotoxic Stress Response to Deoxynivalenol in Mononuclear Phagocytes. *Toxicol. Sci.* **2010**, *115*, 444–452. [CrossRef] [PubMed]

110. Foster, G.H.; Tesh, V.L. Shiga toxin 1-induced activation of c-Jun NH2-terminal kinase and p38 in the human monocytic cell line THP-1: possible involvement in the production of TNF-α. *J. Leukoc. Biol.* **2002**, *71*, 107–114. [PubMed]

111. Brandelli, J.R.; Griener, T.P.; Laing, A.; Mulvey, G.; Armstrong, G.D. The Effects of Shiga Toxin 1, 2 and Their Subunits on Cytokine and Chemokine Expression by Human Macrophage-Like THP-1 Cells. *Toxins* **2015**, *7*, 4054–4066. [CrossRef] [PubMed]

112. Moazzezy, N.; Oloomi, M.; Bouzari, S. Effect of Shiga Toxin and Its Subunits on Cytokine Induction in Different Cell Lines. *Int. J. Mol. Cell. Med.* **2014**, *3*, 108. [PubMed]

113. Smith, W.E.; Kane, A.V.; Campbell, S.T.; Acheson, D.W.K.; Cochran, B.H.; Thorpe, C.M. Shiga Toxin 1 Triggers a Ribotoxic Stress Response Leading to p38 and JNK Activation and Induction of Apoptosis in Intestinal Epithelial Cells. *Infect. Immun.* **2003**, *71*, 1497–1504. [CrossRef] [PubMed]

114. Jandhyala, D.M.; Rogers, T.J.; Kane, A.; Paton, A.W.; Paton, J.C.; Thorpe, C.M. Shiga Toxin 2 and Flagellin from Shiga-Toxigenic *Escherichia coli* Superinduce Interleukin-8 through Synergistic Effects on Host Stress-Activated Protein Kinase Activation. *Infect. Immun.* **2010**, *78*, 2984–2994. [CrossRef] [PubMed]

115. Bunyard, P.; Handley, M.; Pollara, G.; Rutault, K.; Wood, I.; Chaudry, M.; Alderman, C.; Foreman, J.; Katz, D.R.; Chain, B.M. Ribotoxic stress activates p38 and JNK kinases and modulates the antigen-presenting activity of dendritic cells. *Mol. Immunol.* **2003**, *39*, 815–827. [CrossRef]

116. Eisenhauer, P.B.; Jacewicz, M.S.; Conn, K.J.; Koul, O.; Wells, J.M.; Fine, R.E.; Newburg, D.S. *Escherichia coli* Shiga toxin 1 and TNF-α induce cytokine release by human cerebral microvascular endothelial cells. *Microb. Pathog.* **2004**, *36*, 189–196. [CrossRef] [PubMed]

117. Ramegowda, B.; Samuel, J.E.; Tesh, V.L. Interaction of Shiga Toxins with Human Brain Microvascular Endothelial Cells: Cytokines as Sensitizing Agents. *J. Infect. Dis.* **1999**, *180*, 1205–1213. [CrossRef] [PubMed]

118. Petruzziello-Pellegrini, T.N.; Yuen, D.A.; Page, A.V.; Patel, S.; Soltyk, A.M.; Matouk, C.C.; Wong, D.K.; Turgeon, P.J.; Fish, J.E.; Ho, J.J.D.; et al. The CXCR4/CXCR7/SDF-1 pathway contributes to the pathogenesis of Shiga toxin–associated hemolytic uremic syndrome in humans and mice. *J. Clin. Investig.* **2012**, *122*, 759. [CrossRef] [PubMed]

119. Stearns-Kurosawa, D.J.; Collins, V.; Freeman, S.; Tesh, V.L.; Kurosawa, S. Distinct physiologic and inflammatory responses elicited in baboons after challenge with Shiga toxin type 1 or 2 from enterohemorrhagic *Escherichia coli*. *Infect. Immun.* **2010**, *78*, 2497–2504. [CrossRef] [PubMed]

120. Jandhyala, D.M.; Wong, J.; Mantis, N.J.; Magun, B.E.; Leong, J.M.; Thorpe, C.M. A Novel Zak Knockout Mouse with a Defective Ribotoxic Stress Response. *Toxins* **2016**, *8*. [CrossRef] [PubMed]

121. Stone, S.M.; Thorpe, C.M.; Ahluwalia, A.; Rogers, A.B.; Obata, F.; Vozenilek, A.; Kolling, G.L.; Kane, A.V.; Magun, B.E.; Jandhyala, D.M. Shiga toxin 2-induced intestinal pathology in infant rabbits is A-subunit dependent and responsive to the tyrosine kinase and potential ZAK inhibitor imatinib. *Front. Cell. Infect. Microbiol.* **2012**, *2*. [CrossRef] [PubMed]

122. Karpman, D.; Andreasson, A.; Thysell, H.; Kaplan, B.S.; Svanborg, C. Cytokines in childhood hemolytic uremic syndrome and thrombotic thrombocytopenic purpura. *Pediatr. Nephrol. Berl. Ger.* **1995**, *9*, 694–699. [CrossRef]

123. Hagel, C.; Krasemann, S.; Löffler, J.; Püschel, K.; Magnus, T.; Glatzel, M. Upregulation of Shiga Toxin Receptor CD77/Gb3 and Interleukin-1β Expression in the Brain of EHEC Patients with Hemolytic Uremic Syndrome and Neurologic Symptoms. *Brain Pathol.* **2015**, *25*, 146–156. [CrossRef] [PubMed]

124. Lopez, E.L.; Contrini, M.M.; Devoto, S.; De Rosa, M.F.P.; Grana, M.G.; Genero, M.H.; Canepa, C.; Gomez, H.F.; Cleary, T.G. Tumor necrosis factor concentrations in hemolytic uremic syndrome patients and children with bloody diarrhea in Argentina. *Pediatr. Infect. Dis. J.* **1995**, *14*, 594–598. [CrossRef] [PubMed]

125. Alves-Rosa, F.; Beigier-Bompadre, M.; Fernández, G.; Barrionuevo, P.; Mari, L.; Palermo, M.; Isturiz, M. Tolerance to lipopolysaccharide (LPS) regulates the endotoxin effects on Shiga toxin-2 lethality. *Immunol. Lett.* **2001**, *76*, 125–131. [CrossRef]

126. Pinto, A.; Cangelosi, A.; Geoghegan, P.A.; Goldstein, J. Dexamethasone prevents motor deficits and neurovascular damage produced by shiga toxin 2 and lipopolysaccharide in the mouse striatum. *Neuroscience* **2017**, *344*, 25–38. [CrossRef] [PubMed]

127. Zhang, H.-M.; Ou, Z.L.; Gondaira, F.; Ohmura, M.; Kojio, S.; Yamamoto, T. Protective effect of anisodamine against Shiga toxin-1: Inhibition of cytokine production and increase in the survival of mice. *J. Lab. Clin. Med.* **2001**, *137*, 93–100. [CrossRef] [PubMed]

128. Keepers, T.R.; Gross, L.K.; Obrig, T.G. Monocyte Chemoattractant Protein 1, Macrophage Inflammatory Protein 1α, and RANTES Recruit Macrophages to the Kidney in a Mouse Model of Hemolytic-Uremic Syndrome. *Infect. Immun.* **2007**, *75*, 1229–1236. [CrossRef] [PubMed]

129. Isogai, E.; Isogai, H.; Hirose, K.; Kubota, T.; Kimura, K.; Fujii, N.; Hayashi, S.; Takeshi, K.; Oguma, K. Therapeutic effect of anti-TNF-α antibody and levofloxacin (LVFX) in a mouse model of enterohemorrhagic *Escherichia coli* O157 infection. *Comp. Immunol. Microbiol. Infect. Dis.* **2001**, *24*, 217–231. [CrossRef]

130. Argollo, M.; Fiorino, G.; Hindryck, P.; Peyrin-Biroulet, L.; Danese, S. Novel therapeutic targets for inflammatory bowel disease. *J. Autoimmun.* **2017**. [CrossRef] [PubMed]

131. Mateen, S.; Zafar, A.; Moin, S.; Khan, A.Q.; Zubair, S. Understanding the role of cytokines in the pathogenesis of rheumatoid arthritis. *Clin. Chim. Acta* **2016**, *455*, 161–171. [CrossRef] [PubMed]

132. Marcuzzi, A.; Piscianz, E.; Valencic, E.; Monasta, L.; Vecchi Brumatti, L.; Tommasini, A. To Extinguish the Fire from Outside the Cell or to Shutdown the Gas Valve Inside? Novel Trends in Anti-Inflammatory Therapies. *Int. J. Mol. Sci.* **2015**, *16*, 21277–21293. [CrossRef] [PubMed]

133. Mandrup-Poulsen, T. Interleukin-1 Antagonists and Other Cytokine Blockade Strategies for Type 1 Diabetes. *Rev. Diabet. Stud. RDS* **2012**, *9*, 338–347. [CrossRef] [PubMed]

Review

The Use of Plant-Derived Ribosome Inactivating Proteins in Immunotoxin Development: Past, Present and Future Generations

Aleksander Rust [1],*, Lynda J. Partridge [2], Bazbek Davletov [3] and Guillaume M. Hautbergue [4],*

[1] Structural and Molecular Biology, Division of Biosciences, Faculty of Life Sciences,
University College London, London WC1E 6BT, UK

[2] Department of Molecular Biology and Biotechnology, University of Sheffield, Firth Court, Western Bank,
Sheffield S10 2TN, UK; l.partridge@sheffield.ac.uk

[3] Department of Biomedical Science, University of Sheffield, Firth Court, Western Bank,
Sheffield S10 2TN, UK; b.davletov@sheffield.ac.uk

[4] Sheffield Institute for Translational Neuroscience, Department of Neuroscience, University of Sheffield,
385a Glossop Road, Sheffield S10 2HQ, UK

* Correspondence: a.rust@ucl.ac.uk (A.R.); g.hautbergue@sheffield.ac.uk (G.M.H.);
Tel.: +44-207-679-2021 (A.R.); +44-114-222-2252 (G.M.H.)

Academic Editors: Julien Barbier and Daniel Gillet
Received: 29 September 2017; Accepted: 24 October 2017; Published: 27 October 2017

Abstract: Ribosome inactivating proteins (RIPs) form a class of toxins that was identified over a century ago. They continue to fascinate scientists and the public due to their very high activity and long-term stability which might find useful applications in the therapeutic killing of unwanted cells but can also be used in acts of terror. We will focus our review on the canonical plant-derived RIPs which display ribosomal RNA N-glycosidase activity and irreversibly inhibit protein synthesis by cleaving the 28S ribosomal RNA of the large 60S subunit of eukaryotic ribosomes. We will place particular emphasis on therapeutic applications and the generation of immunotoxins by coupling antibodies to RIPs in an attempt to target specific cells. Several generations of immunotoxins have been developed and we will review their optimisation as well as their use and limitations in pre-clinical and clinical trials. Finally, we endeavour to provide a perspective on potential future developments for the therapeutic use of immunotoxins.

Keywords: ribosome inactivating proteins; immunotoxins; therapeutic applications

---

1. Introduction

Ribosome inactivating proteins (RIPs) are a family of toxins that irreversibly inhibit eukaryotic protein synthesis [1–3]. Although most commonly found in plants, RIPs have also been identified in bacteria, fungi and even in two mosquito species [4]. RIPs form a heterogeneous group of proteins with varied enzyme functionalities and several classifications have been proposed [1,5]. This review will focus on the canonical plant-derived RIPs as they are the most characterised and commonly used toxins in immunotoxin development. In particular, we will review how these RIPs have been utilised in the potential treatment of cancer which currently holds the most promising applications for immunotoxin therapy.

Purification of the RIP ricin and subsequent immunological experiments carried out by Paul Ehrlich in the early 19th century led to the concept of the 'magic bullet' [2]. This is the idea of an exceptionally toxic molecule that can specifically attack and kill target cells [6]. It became a highly attractive concept for the treatment of cancer and was built upon in the 1970s when the first immunotoxins—a protein synthesis inhibiting toxin conjugated to a targeting antibody—were generated [7]. As recombinant technology has started to dominate immunotoxin design incorporating bacterial toxic enzymes, the use of plant RIPs has recently decreased; however, these toxins still display a number of characteristics that make them attractive in immunotoxin technology. Moreover, pre-clinical studies using plant RIPs are showing promise for their use in future therapy.

2. Ribosome Inactivating Proteins with Ribosomal RNA *N*-Glycosidase Activity

RIPs are found almost ubiquitously among plants and are thought to act as a form of immune defence, as upregulation of expression can be seen following viral infection and contamination with microorganisms [8] as well as abiotic stress [9]. RIPs are renowned active substances that have been used in traditional Chinese medicine for centuries. Trichosanthin is one typical example that shows promising outcomes in the killing of cancer cells, particularly hepatocellular carcinoma, both in vitro and in vivo in a murine xenograft model [10]. There are currently almost 250 proteins that irreversibly inactivate protein synthesis. They can be divided into several broad groups with a two-category classification (type 1 and 2) prevailing (reviewed in [1,2,5]). Type 1 RIPs are monomeric proteins of approximately 30 kDa with enzymatic activity, and type 2 RIPs are heterodimeric proteins that contain an enzymatic domain of approximately 30 kDa (A-chain) linked via a disulphide bond to a second lectin-like domain of approximately 35 kDa (B-chain) which is able to bind to cells and facilitate internalisation [11].

Type 2 RIPs such as ricin are able to bind to sugars on the cell surface via their lectin-like B-chain. Although this means that type 2 RIPs are generally more potent than type 1 RIPs, cell binding alone is not sufficient to confer potency, as toxicity can vary greatly between type 2 RIPs—some RIPs being considered non-toxic. For instance, Ricinus agglutinin (RCA) and ricin are both type 2 RIPs found in castor beans, but ricin shows around 68-fold higher potency than RCA in cells, likely due to a decreased ability of RCA to translocate into the cytoplasm [11]. Similarly, some type 2 RIPs isolated from some *Sambucus* species, such as nigrins or ebulins, exhibit high ribosome-inactivating activities in cell-free systems but lack toxicity both in vitro and in vivo. This is because these RIPs follow a different intracellular trafficking route with the majority of molecules being either recycled to the cell surface or degraded [12,13]. These examples highlight the importance of intracellular trafficking following binding for mediating cytotoxicity. Studies using ricin show that, following binding, the toxin is taken up by both clathrin-dependent and -independent endocytosis and a small percentage localises with the trans-Golgi network, followed by retrograde transport to the ER [14]. Once in the lumen of the ER, it is thought that the A-chain is cleaved from the B-chain by the protein disulphide isomerase and is then processed by the ER as a misfolded protein, meaning that it is exported to the cytosol for degradation [15,16]. Upon entering the cytosol, the A-chain is refolded by the sequential utilisation of the Hsc70 and Hsp90 chaperone systems and the correctly folded native A-chain is then able to carry out its catalytic activity at the ribosomes [17].

Type 1 RIPs such as saporin and gelonin lack the cell-binding B-chain of type 2 RIPs and are therefore much less cytotoxic than most type 2 RIPs. It is thought that uptake generally occurs through a passive manner, such as by fluid-phase pinocytosis [18]. It has also been proposed that saporin can enter cells in a receptor-dependent manner, via binding to α2-macroglobulin receptors [19]. However, similar sensitivities to saporin have been observed between α2-macroglobulin receptor expressing and non-expressing cell lines which would indicate that saporin internalisation does not occur via this receptor [20]. The mechanism of endocytosis of type 1 RIPs remains unclear, but studies with saporin appear to show an internalisation mechanism that is independent of the Golgi apparatus, suggesting that it follows a distinct pathway to ricin [21]. Nevertheless, upon reaching the cytosol, many type 1 RIPs display a highly active enzymatic action, and artificial delivery into the cell or attachment to a targeting ligand leads to cytotoxicity with high potency [11,22].

Types 1 and 2 RIPs display ribosomal RNA *N*-glycosidase activity (EC 3.2.2.22) and cleave the 28S rRNA by removal of a single adenine residue (A4324 in rat rRNA) from a GAGA sequence at the universally conserved α-sarcin/ricin loop [1,2,5]. This further prevents the recruitment of eukaryotic elongation translation factors eEF1/2 to the 60S ribosomal subunit and translocation of the ribosome and protein synthesis [23–25] thus causing a complete and irreversible block of protein synthesis [26]. It was initially thought that cells underwent apoptosis after exposure to RIPs solely due to the ribotoxic stress response after inhibition of protein synthesis. However, more recent data suggests that RIPs may also exhibit other activities independently of their targeting of protein synthesis. For instance, it has been shown that RIPs show adenine glycosidase activity in DNA, RNA and poly (A) [27]. Additionally, ricin has been shown to cause early nuclear DNA damage independently of protein synthesis inhibition, and saporin S6 was shown to induce apoptosis through mitochondrial cascade prior to the onset of protein synthesis inhibition [28,29]. It has therefore been proposed that RIPs may induce apoptosis by a number of different mechanisms, of which inhibition of protein synthesis plays an important but not always essential role [30].

3. The Development of RIP-Based Immunotoxins

A common feature of many RIPs is their extraordinarily high level of potency. As with other protein synthesis inhibiting toxins, such as diphtheria toxin, it is thought that only a few molecules are needed to enter the cytosol of a cell for cell death to occur. This potency makes them highly attractive as a possible cancer therapeutic. However, this is a double-edged sword as toxicity is achieved in both healthy and malignant cells, meaning that these toxins must be efficiently targeted to cancer cells to convey specific anticancer activity. The main methods by which toxins are targeted to cancer cells are either by conjugation to an antibody to make an immunotoxin, or to a targeting ligand such as a growth factor or cytokine [31]. Immunotoxins are becoming the predominant choice as they allow for greater selectivity and flexibility when choosing a target. Selecting an appropriate target is of high importance when generating a targeted toxin, as it has a large impact on specificity and potency. The chosen target must be highly expressed on the surface of the cancer cell, but have relatively restricted expression in healthy tissue, as this limits on-target off-tumour toxicity [32]. For immunotoxins, this generally means targeting to tumour-associated antigens, which are highly expressed on the cell surface as a result of transformation [33]. Since the generation of the first immunotoxins in the 1970s, more than 450 immunotoxins have been used considering RIPs alone [31]. The most common plant-based RIPs used in immunotoxin development are the chain A of the type 2 RIP ricin as well as type 1 RIPs saporin and gelonin which all exhibit high potency and stability. It has also been recently suggested to use type 2 RIPs with low in vitro and in vivo toxicity but potent ribosomal RNA *N*-glycosidase activities in cell-free systems, such as the chains A of ebulins and nigrins from the *Sambucus* species, to generate alternative immunotoxins with very high cytotoxicity [12,13].

As antibody therapy and recombinant technology has advanced, so too has immunotoxin design, progressing from simple chemical conjugation of a native toxin to a whole antibody, to the recombinant engineering of modified toxin domains fused with humanised antibody fragments. The progression of development can be broadly split into three generations.

3.1. First-Generation Immunotoxins

The first-generation immunotoxins were developed in the early 1970s and were usually made using a full toxin chemically linked to a whole monoclonal antibody (Figure 1a).

Initial studies with first-generation immunotoxins were primarily carried out with the diphtheria toxin: a bacterial toxin that is analogous to type 2 RIPs in that it has two distinct domains for targeting and enzymatic action to inhibit protein synthesis [7,34,35]. Although these often gave promising results in vitro, a number of issues were encountered upon testing in vivo. The major drawback of these immunotoxins was the presence of the targeting domain, which meant that the protein was able to bind to and enter a wide range of different cells, irrespective of target antigen expression. This caused a high level of non-specific toxicity and intolerable side effects. At around this time the type 2 RIPs ricin and abrin were gaining attention as anti-cancer agents as they were found to more efficiently inhibit protein synthesis in certain tumour models than in healthy cells. Additionally, these unmodified toxins were shown to have anti-tumour properties in Ehrlich ascites tumour mouse models [36]. However, non-specific toxicity was still an issue with only very low doses needed to cause lethality [37]. Re-targeting of RIPs to cancer cells was therefore necessary, and they were increasingly used in the design of second-generation immunotoxins.

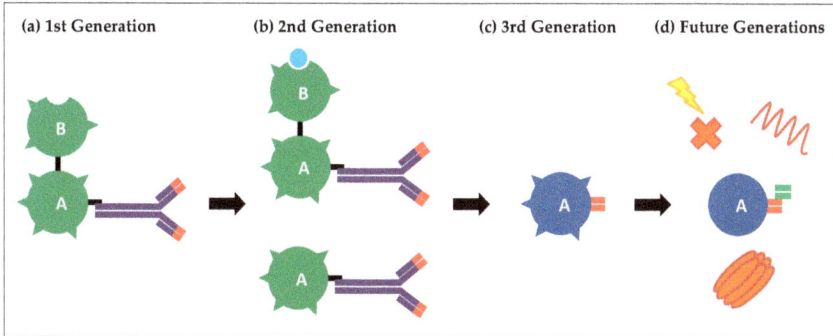

Figure 1. Diagram depicting the different generations of immunotoxins. (**a**) First-generation immunotoxins. Purified toxins were chemically linked to a targeting antibody; (**b**) Second-generation immunotoxins. Purified type 1 ribosome inactivating proteins (RIPs) or type 2 RIPs with B-chain either blocked or removed were chemically linked to a targeting antibody; (**c**) Third-generation immunotoxins. Recombinant purified toxins were fused to antibody targeting fragments; (**d**) Future generation immunotoxins. Toxins are modified to remove immunogenic epitopes and exhibit dual targeting abilities to improve specificity. They can be co-administered with endosome disruptive agents, such as pore-forming agents, endosome disruptive peptides, or photosensitisers, to increase intracellular delivery and potency.

Despite a large number of issues with these original toxins, impressive in vitro data displayed the potential of immunotoxins for cancer treatment and a large amount of effort was applied to increase effectiveness in vivo.

3.2. Second-Generation Immunotoxins

A greater understanding of toxin structure and function led to the second generation of immunotoxins in the mid-1970s. These usually consisted of a toxin lacking a cell-binding domain, which greatly reduced non-specific internalisation, allowing for the administration of higher doses and a greater therapeutic window (Figure 1b). The use of RIPs was, and still is, popular for this generation of immunotoxins as purification techniques are well established and they are highly stable proteins. Ricin-based immunotoxins of the second generation were produced in 1976 and consisted of the castor bean ricin A-chain chemically linked to cancer-targeting antibodies [38]. By the mid-1980s, a number of ricin A-chain conjugates had been developed which demonstrated efficacy against various cancer models in vitro including leukaemia and breast cancer, as well as showing efficacy in vivo [39].

In these second-generation toxins, the A-chain was isolated by purification of the complete protein followed by removal of the B-chain by reduction of the disulphide bond that joins the two domains. This made purification of utmost importance as, due to the high potency of native ricin, any residual full toxin would cause unwanted side effects. A second issue with the A-chain is that it is glycosylated, meaning that it can bind to and be internalised by cells expressing appropriate mannose receptors, leading to reduced stability in blood circulation and residual non-specific toxicity [40]. The A-chain therefore needed to be deglycosylated before use in immunotoxins, further complicating production. An alternative method was to use the entire ricin toxin with the B-chain blocked by chemical modification or the addition of lactose (Figure 1b). These conjugates often showed higher in vitro potency than A-chain only conjugates, and it was postulated that this was due to the B-chain facilitating entry into the cell cytosol [41], or unblocking of the B-chain upon cell binding, allowing for increased cellular interaction with galactose binding sites [42]. However, these blocked-ricin immunotoxins also demonstrated higher non-specific toxicity than their "A-chain only" counterparts.

The first type 1 RIP-based immunotoxin was generated in 1981 and consisted of gelonin conjugated to an anti-Thy1.1 antibody [43]. This showed high cytotoxicity in vitro in lymphoma cell lines and significantly prolonged the life of mice bearing lymphoma allografts. Pokeweed antiviral protein (PAP), another type 1 RIP, was also used in immunotoxin generation and showed similar potency against Thy1.1 positive cells when directly compared to a ricin A-chain immunotoxin [44]. Type 1 RIPs have an advantage over type 2 RIPs in that they natively lack a cell-binding domain, so further modification of the toxin is not necessary before antibody conjugation. Since then, saporin has become the more popular type 1 RIP for immunotoxin development due to its high thermodynamic stability [2].

The second generation of immunotoxins also saw the use of recombinant technology for toxin production and the ricin A-chain was expressed in *Escherichia coli* in 1987 [45]. Despite issues expressing plant-based toxins, recombinant expression holds a number of advantages over traditional purification methods. Firstly, it is a much more straightforward process and does not require complex and lengthy procedures to eliminate the residual B-chain. Secondly it is much safer, as no highly potent full toxin needs to be handled. Finally, prokaryotic expression means that the protein is not glycosylated, helping to prolong toxin stability and reduce toxicity in vivo. A number of type 1 RIPs, including saporin and gelonin, have also been generated recombinantly [46,47].

More promising in vivo data obtained from second-generation immunotoxins led to the first clinical trial in 1986 [48]. This trial used the ricin A-chain linked to a pan-T-lymphocyte antibody (T101) in the treatment of two leukaemia patients, and showed promising results with large drops in target cell numbers. Follow-up phase I and II trials also demonstrated efficacy, but revealed a number of dose-limiting side effects including a drop in serum albumin levels, weight gain, and oedema, caused by off-target toxicity of the immunotoxin [49]. The first clinical trial with a type 1 RIP-based immunotoxin was carried out in 1992 using a saporin-anti-CD30 conjugate for the treatment of Hodgkin's disease. It successfully led to a 50 to 75% reduction of the tumour mass in 3 out of 4 patients, but also caused side effects such as oedema [50]. Similar dose-limiting side effects were seen in a number of other immunotoxin clinical trials [51,52] and this was later attributed to vascular leak syndrome (VLS) [53]. VLS occurs when non-specific uptake of the toxin into endothelial cells causes cell death and leakage of fluid from the circulatory system into the interstitial space, leading to oedema, weight gain, hypoalbuminemia, pulmonary infiltrates, and hypotension [54].

As well as dose-limiting toxicity, further problems persisted in second-generation immunotoxins. As with the first generation, second-generation immunotoxins were chemically linked to targeting antibodies through a disulphide bond so that, upon entry into endosomes, an increase in reducing conditions breaks the bond and frees the toxin. However, premature nucleophilic attack of the disulphide bond, particularly in the anaerobic reducing environment of tumours, caused low stability in vivo [55]. Inefficient chemical linking also led to heterogeneous compositions of bound and unbound components which could interfere with antigen binding [56]. As well as instability and non-specific binding, these conjugates were also very large (180–200 kDa) which affected tumour penetration [57]. This is why the most promising results were seen in haematological cancers rather than solid tumours. Another factor affecting tumour penetration was the short half-life, which was around 30 min in native ricin A-chain immunotoxins. This issue was addressed by de-glycosylation of the toxin which helped to increase circulating half-life to 4–6 h [57]. Longer treatment periods and higher concentrations unveiled a further problem with immunotoxins, which is immunogenicity [58]. Both the antibody and toxin were obtained from non-human sources—mice and plants respectively—meaning that they contained many immunogenic epitopes. Repeated exposure would therefore elicit an immune response, negating the anti-tumour effect of the immunotoxin.

The second generation of immunotoxins showed large improvements over the first generation, allowing higher tolerance in animals and the first clinical trials. However, many problems persisted that limited the therapeutic benefit of this treatment. Advances in recombinant technology allowed for greater control over design and production of immunotoxins, so that a number of these issues could be addressed.

3.3. Third-Generation Immunotoxins

Third-generation immunotoxins utilise advanced genetic engineering techniques to generate fully recombinant toxins fused to the targeting domain of a monoclonal antibody [59] (Figure 1c). These are commonly known as recombinant immunotoxins (RITs). Most RITs are generated from bacterial toxins such as diphtheria toxin or pseudomonas exotoxin A, as these are easier to express in bacteria. A RIT consisting of the enzymatic and translocation domains of pseudomonas exotoxin A fused to the heavy and light chain portions of a monoclonal antibody targeting B3 was reported in 1995 [60]. This RIT demonstrated selective cytotoxicity in vitro and in vivo that was several-fold higher than an equivalent, chemically linked immunotoxin. The main advantage of this approach is that a homogenous population is produced which gives increased stability and reduced interference of antigen binding. This approach also allows the incorporation of more stable linkers that are cleavable only by intracellular enzymes. For instance, furin-sensitive linker sequences between the targeting and enzymatic domains have shown significant improvements in toxicity of pseudomonas exotoxin A immunotoxins [61].

The use of antibody fragments allowed the generation of immunotoxins of much smaller sizes. The smallest antibody fragments that retain targeting ability consist of one variable heavy chain and one variable light chain (scFv) and allow for the generation of immunotoxins as small as 60 kDa [56]. Although these showed increased tumour penetration, they were rapidly cleared from the blood (circulating half-life of approximately 20 min) leading to decreased therapeutic efficacy. Another issue with smaller RITs is that they are cleared by renal filtration and can accumulate in the kidney, leading to renal toxicity [62]. It is therefore thought that RITs should be over 65 kDa in size as this is the cut-off limit for macromolecules cleared by renal filtration [63]. Thus, a balance needs to be found between decreasing size to increase tumour penetration, without affecting plasma half-life. Another issue with the use of a single scFv antibody portion is instability, as this lacks the stabilising disulphide bond found in the absent portion [64]. As a result of this, the heavy and light chains can dissociate and bind to other dissociated chains leading to aggregation and loss of activity. However, further engineering introduced a disulphide bond into the Fv domain which prevented dissociation without affecting binding affinity [65]. Another alternative is the use of nanobodies, monomeric and variable heavy chain antibodies that occur naturally in the *Camelidae* family [66]. These offer a number of advantages over conventional antibodies, including their small size, low immunogenicity, high stability, and ease of production. They have been recently used successfully in the development of a number of immunotoxins [67–69].

Bacterial toxins are much more common in fusion immunotoxins as they are easier to express in bacterial systems. Type 1 RIP-based fusion proteins have been successfully produced in bacteria [70,71], but are less common as they often suffer from low stability and activity due to incorrect folding. However, it has been demonstrated that the yeast *Pichia pastoris* may be used for RIT generation [72], and a recent study has shown that saporin-antibody fusions generated using this system demonstrate comparable in vitro potency to bacterial-based immunotoxins [73]. Yeast may be a more suitable system for RIT production as they are eukaryotic and therefore capable of the complex folding and post-translational modifications necessary to generate functional RITs. A drawback is that they are eukaryotic and therefore susceptible to the cytotoxic mechanism of the toxin. However, certain strains of *Pichia pastoris* (e.g., GS115) have displayed a high level of resistance to toxic action which is attributed to the rapid secretion of the protein into the culture medium [72].

Due to the issues with generating recombinant RIP-based immunotoxins, the vast majority of clinical trials have been, or are being, carried out with bacterial-based immunotoxins, particularly pseudomonas exotoxin A. Indeed, most recent trials using ricin are based on deglycosylated A-chain or blocked-ricin second-generation immunotoxins [31,74–76]. Two earlier phase I trials using recombinant ricin were carried out in breast cancer patients, but significant off-target toxicity was still observed in both variants [77,78]. Phase I clinical trials have also been carried out using the recombinantly generated type 1 RIPs gelonin and bouganin. VB6-845 is a recombinant fusion protein consisting of modified bouganin and an anti-epithelial cell adhesion molecule (epCAM) Fab moiety which was tested in a phase I trial for treating epithelial tumours (Clinicaltrials.gov identifier: NCT00481936). However, results from this trial remain unpublished. Recombinant gelonin conjugated to a humanised anti-CD33 antibody (M195) has also been tested in a phase I trial for the treatment of refractory myeloid leukaemia [79]. Although low response rates were reported, this conjugate nevertheless exhibited promising characteristics when compared to bacterial immunotoxins, including low immunogenicity and an absence of VLS as a side effect.

4. Current Limitations and Future Generations of Immunotoxins

Advances in antibody therapy and recombinant technology have sparked a resurgence in interest for targeted toxins as a possible therapeutic approach, and a number of promising results have been obtained for the treatment of haematological malignancies. However, issues still need to be addressed to make these therapies more effective, particularly in solid tumours, including increasing bioavailability and tumour penetration, reducing immunogenicity and reducing off-target toxicity. Indeed, due to these issues, only one targeted toxin, denileukin diftitox, has FDA approval for cancer therapy in humans. Denileukin diftitox (also known as Ontak) is a targeted toxin made up of the enzymatic and translocation domains of *Diphtheria* toxin recombinantly fused to IL-2 used in the treatment of lymphoma [80]. The most successful application of a type 1 RIP-based targeted toxin is the use of recombinant saporin linked to Substance P for the treatment of bone cancer pain in old dogs [81,82]. The FDA has already approved Minor Use/Minor Species (MUMS) designation for this drug providing extended market exclusivity to treat the >10,000 annual cases of canine bone cancer-related pain.

Numerous groups are working on the inherent problems, and incorporation of their findings into immunotoxin design may drastically improve the efficacy of future generations. For example, an antibody can target the toxins only to a specific subset of cells, meaning that any heterogeneity within the tumour with regards to antigen expression will lead to resistant cancer cells. To overcome this, immunotoxins that can simultaneously target different antigens are being tested (Figure 1d). Combotox is the co-administration of ricin A-chain-based immunotoxins such as those that target CD19 and CD22 to treat B-lineage lymphoblastic leukaemia. This treatment has shown efficacy in phase I clinical trials [74]. A recent study demonstrated that the use of antibodies targeting the Neuron-glia 2 (NG2) or Ganglioside D3 (GD3) antigens can greatly increase the efficacy of saporin immunotoxins in in vitro models of glioblastoma [83]. NG2 and GD3 are respectively associated with two distinct cell sub-populations, fast dividing NG2 positive cells and GD3 positive cells that are involved in survival and migration. Targeting known sub-populations within a cancer will help to improve response rates and decrease relapse. As well as the co-administration of immunotoxins, the use of bi-specific antibodies allows one molecule to target multiple antigens. For instance, tandem scFv segments have been successfully used to target CD19 and CD22 in both pseudomonas exotoxin A and diphtheria toxin immunotoxins. These fusions have shown increased efficacy in B-cell lymphoma mice models when compared with immunotoxins targeting just one antigen [84,85]. As well as helping to overcome tumour-heterogeneity, dual-antigen targeting may help to decrease on-target, off-tumour toxicity caused by low-level antigen expression on healthy cells. Antigens found exclusively on cancer cells are rare which severely limits targeting choices [63]. Indeed, unexpected, low-level expression of a target antigen in certain tissues has limited the use of a number of immunotoxins [32]. The selection of two TAAs expressed on a cancer cell surface will help to increase specific tumour cell binding and reduce internalisation into healthy cells.

Off-target toxicity due to non-specific uptake of toxin leads to dose-limiting side effects such as VLS, hepatotoxicity, renal toxicity, and cardiac dysfunction. As mentioned previously, VLS was observed as a side effect in the first clinical trials and has continued to affect all immunotoxin treatments tested to date. It is particularly severe in ricin-based therapies, even resulting in fatalities in some trials [63,86]. One study identified short amino-acid motifs (x)-D-(y) (where x is L, I, G or V and y is V, L or S) which may bind to endothelial cells resulting in internalisation of the toxin and cell death [87]. Modification of these motifs in ricin led to the generation of a toxin with reduced ability to cause vascular leak syndrome in mice [88]. Pre-treatment of patients with steroids to reduce inflammatory responses has also been shown to help combat VLS [89].

Improved cytosolic delivery of the toxin would also help to reduce side effects, as lower doses would need to be administered to observe the same therapeutic effect. Additionally, more efficient cytosolic entry may give increased efficacy in solid tumours where only a small number of molecules reach target cells. It is thought that only a small percentage of internalised toxin is able to evade lysosomal degradation and enter the cytosol. Indeed, a study using targeted gelonin found a near-universal requirement of 5 million molecules needed to be internalised for cell killing, despite different routes of binding and internalisation [90]. Considering the high potency of RIPs with in vitro turnover of 28S rRNA depurination of 700–800 molecules/min, this strongly suggests that escape from the endolysosomal compartment is the rate-limiting step that determines efficacy. Various methods are in development to improve endosome escape, including the co-administration of endosome disrupting agents (Figure 1d). For instance, one study found that co-administration of targeted gelonin with listeriolysin targeted to the same antigen led to a large increase in efficacy and was well tolerated in vivo [91]. Lysteriolysin is a cytolysin protein produced by the bacterium *Listeria monocytogenes* which can lyse endosomes in a pH-dependent manner. Co-administration of plant-based saponins, glycosides that can form pores in membranes, has also been found to increase potency whilst maintaining target specificity of targeted toxins [92]. A range of endosome escape peptides are available which are able to disrupt membranes in a pH-dependent manner. For instance, the GALA peptide is a short, 30-residue, synthetic peptide with a repeating Glu-Ala-Leu-Ala sequence [93]. GALA mimics the function of viral fusion protein segments that mediate the escape of viral genes from endosomes into the cytosol. Endosomal acidification causes a rearrangement of the peptide structure from random to helical, giving it a high affinity for neutral or negatively charged membranes, leading to the formation of pores and destabilisation of the endosomal membrane [93]. An alternative method is the use of photochemical internalisation which utilises an endocytic vesicle-localising photosensitiser that generates reactive oxygen species upon exposure to light, triggering vesicle rupture [94]. Preferential retention of the sensitiser in tumour cells and focused light application using a laser adds a degree of selectivity to this technique which, combined with specific targeting by the toxin, can reduce side effects and increase the therapeutic window [95]. Photochemical internalisation has been used successfully in vivo to enhance the efficacy of a saporin-based immunotoxin that targets the cancer stem cell marker CD133 [96].

Development of humanised antibodies by combining the recognition domain with human framework regions has largely stopped immune reactions to the targeting moiety of immunotoxins [97], but problems persist with the toxins themselves. Methods employed to reduce immunogenicity of the toxins include chemical modification and removal of immunogenic epitopes. Chemical modification with polyethene glycol (PEGylation) is a common technique used to increase the plasma half-life of therapeutic proteins [98]. Moreover, site-specific PEGylation of an IL-2 targeting pseudomonas exotoxin A based immunotoxin was found to dramatically reduce immunogenicity in mice, and was thought to act by reducing protein degradation in antigen-presenting cells as well as shielding some epitopes following degradation [99]. Immunogenicity of toxins can be greatly reduced by the removal of immunogenic epitopes. In one study, the antigenic domains of gelonin were successfully mapped and deleted to create smaller, modified toxins that retained enzymatic activity but exhibited reduced antigenicity in vitro [100]. T-cell recognition epitopes were identified and mutated in the type I RIP bouganin to create an epitope-depleted mutant de-bouganin [101]. An immunotoxin based on de-bouganin has been shown to be well tolerated in vivo with minimal side effects and low immunogenicity [102]. B-cell epitopes have been successfully removed from pseudomonas exotoxin A by isolating antibodies from patients with immune resistance to this toxin and constructing a phage display library [103]. Alanine scanning mutagenesis was then used to locate the epitopes and an alternative toxin was generated (LR-O10) which showed low reactivity with human antisera but maintained high cytotoxic and anti-tumour activity. Another possibility is to utilise the wide range of plant-based RIPs that are available. These RIPs are often immunogenically distinct and so development of a treatment regimen that utilises a combination of immunotoxins containing different RIPs may help to prolong the number of treatment rounds whilst avoiding an immune response.

An additional way to improve the success of immunotoxin therapy may be to alter how these proteins are utilised. The majority of trials for solid cancers are carried out in patients with advanced disease and high tumour burden. A more effective strategy may be the use of immunotoxins following traditional chemotherapy to clear up minimal residual disease. This is the small number of cells that are often resistant to chemotherapy and remain circulating in the body with the ability to cause relapse [104]. This would be a more suitable target for immunotoxins as issues with tumour penetration would be bypassed. Immunotoxins have been thought to be particularly applicable to bladder cancer as this organ could be thought of as an "external" environment that would restrict the immune reactions and side effects caused by immunotoxins [105,106]. Moreover, it was recently reported that a complete and at least a 3 year long-lasting elimination of bladder cancer was observed following treatment with an immunotoxin prepared with the ricin A-chain [107].

5. Perspectives

A number of advances have been made in immunotoxin design which have taken them closer to clinical use as a therapy for a variety of cancers. RIPs have been instrumental in this advancement, from the original concept of the 'magic bullet' in the late 1800s to modern-day clinical trials. Recombinant technology has seen a rise in the use of bacterial toxins over RIPs. More recently, human cytotoxic enzymes such as granzyme B and ribonuclease have also been utilised in what has been called the fourth generation of immunotoxins [108–110]. However, numerous groups are still utilising plant-based RIPs to tackle current problems with these treatments including production [73], cell delivery [91,92] and immunogenicity [100,101]. These toxins are therefore useful tools for immunotoxin design and may yet be seen in future clinical trials.

Acknowledgments: G.M.H. acknowledges support from the Motor Neurone Disease Association (grants Apr16/846-791 and Apr17/854-791), the Medical Research Council (MRC grant MR/M010864/1) and the Thierry Latran Foundation (grant FTLAAP2016). Open access costs are covered by grant MRC MR/M010864/1.

Author Contributions: A.R. wrote the first draft of the manuscript. G.M.H. supervised the writing, edited and revised the manuscript with input from all authors.

Conflicts of Interest: The authors declare no conflict of interest.

References

1. Schrot, J.; Weng, A.; Melzig, M.F. Ribosome-inactivating and related proteins. *Toxins* **2015**, *7*, 1556–1615. [CrossRef] [PubMed]
2. Bolognesi, A.; Bortolotti, M.; Maiello, S.; Battelli, M.; Polito, L. Ribosome-inactivating proteins from plants: A historical overview. *Molecules* **2016**, *21*, 1627. [CrossRef] [PubMed]
3. Fabbrini, M.S.; Katayama, M.; Nakase, I.; Vago, R. Plant ribosome-inactivating proteins: Progesses, challenges and biotechnological applications (and a few digressions). *Toxins* **2017**, *9*, 314. [CrossRef] [PubMed]
4. Lapadula, W.J.; Sanchez Puerta, M.V.; Juri Ayub, M. Revising the taxonomic distribution, origin and evolution of ribosome inactivating protein genes. *PLoS ONE* **2013**, *8*, e72825. [CrossRef] [PubMed]

5. Walsh, M.J.; Dodd, J.E.; Hautbergue, G.M. Ribosome-inactivating proteins: Potent poisons and molecular tools. In *Virulence*; Taylor & Francis Group: Philadelphia, PA, USA, 2013; Volume 4, pp. 774–784.

6. Antignani, A.; Fitzgerald, D. Immunotoxins: The role of the toxin. *Toxins* **2013**, *5*, 1486–1502. [CrossRef] [PubMed]

7. Moolten, F.L.; Cooperband, S.R. Selective destruction of target cells by diphtheria toxin conjugated to antibody directed against antigens on the cells. *Science* **1970**, *169*, 68–70. [CrossRef] [PubMed]

8. Wong, R.N.S.; Mak, N.K.; Choi, W.T.; Law, P.T.W. Increased accumulation of trichosanthin in trichosanthes kirilowii induced by microorganisms. *J. Exp. Bot.* **1995**, *46*, 355–358. [CrossRef]

9. Rippmann, J.F.; Michalowski, C.B.; Nelson, D.E.; Bohnert, H.J. Induction of a ribosome-inactivating protein upon environmental stress. *Plant Mol. Biol.* **1997**, *35*, 701–709. [CrossRef] [PubMed]

10. Zhang, Y.H.; Wang, Y.; Yusufali, A.H.; Ashby, F.; Zhang, D.; Yin, Z.F.; Aslanidi, G.V.; Srivastava, A.; Ling, C.Q.; Ling, C. Cytotoxic genes from traditional chinese medicine inhibit tumor growth both in vitro and in vivo. *J. Integr. Med.* **2014**, *12*, 483–494. [CrossRef]

11. Stirpe, F.; Battelli, M.G. Ribosome-inactivating proteins: Progress and problems. *Cell. Mol. Life Sci.* **2006**, *63*, 1850–1866. [CrossRef] [PubMed]

12. Ferreras, J.M.; Citores, L.; Iglesias, R.; Jimenez, P.; Girbes, T. Use of ribosome-inactivating proteins from sambucus for the construction of immunotoxins and conjugates for cancer therapy. *Toxins* **2011**, *3*, 420–441. [CrossRef] [PubMed]

13. Jimenez, P.; Tejero, J.; Cordoba-Diaz, D.; Quinto, E.J.; Garrosa, M.; Gayoso, M.J.; Girbes, T. Ebulin from dwarf elder (*sambucus ebulus* L.): A mini-review. *Toxins* **2015**, *7*, 648–658. [CrossRef] [PubMed]

14. Sandvig, K.; van Deurs, B. Endocytosis, intracellular transport, and cytotoxic action of shiga toxin and ricin. *Physiol. Rev.* **1996**, *76*, 949–966. [PubMed]

15. Roberts, L.M.; Lord, J.M. Ribosome-inactivating proteins: Entry into mammalian cells and intracellular routing. *Mini Rev. Med. Chem.* **2004**, *4*, 505–512. [CrossRef] [PubMed]

16. Spooner, R.A.; Watson, P.D.; Marsden, C.J.; Smith, D.C.; Moore, K.A.; Cook, J.P.; Lord, J.M.; Roberts, L.M. Protein disulphide-isomerase reduces ricin to its a and b chains in the endoplasmic reticulum. *Biochem. J.* **2004**, *383*, 285–293. [CrossRef] [PubMed]

17. Spooner, R.A.; Hart, P.J.; Cook, J.P.; Pietroni, P.; Rogon, C.; Hohfeld, J.; Roberts, L.M.; Lord, J.M. Cytosolic chaperones influence the fate of a toxin dislocated from the endoplasmic reticulum. *Proc. Natl. Acad. Sci. USA* **2008**, *105*, 17408–17413. [CrossRef] [PubMed]

18. Polito, L.; Bortolotti, M.; Mercatelli, D.; Battelli, M.G.; Bolognesi, A. Saporin-s6: A useful tool in cancer therapy. *Toxins* **2013**, *5*, 1698–1722. [CrossRef] [PubMed]

19. Cavallaro, U.; Nykjaer, A.; Nielsen, M.; Soria, M.R. Alpha 2-macroglobulin receptor mediates binding and cytotoxicity of plant ribosome-inactivating proteins. *Eur. J. Biochem.* **1995**, *232*, 165–171. [CrossRef] [PubMed]

20. Bagga, S.; Hosur, M.V.; Batra, J.K. Cytotoxicity of ribosome-inactivating protein saporin is not mediated through alpha2-macroglobulin receptor. *FEBS Lett.* **2003**, *541*, 16–20. [CrossRef]

21. Vago, R.; Marsden, C.J.; Lord, J.M.; Ippoliti, R.; Flavell, D.J.; Flavell, S.U.; Ceriotti, A.; Fabbrini, M.S. Saporin and ricin a chain follow different intracellular routes to enter the cytosol of intoxicated cells. *FEBS J.* **2005**, *272*, 4983–4995. [CrossRef] [PubMed]

22. Rust, A.; Hassan, H.H.; Sedelnikova, S.; Niranjan, D.; Hautbergue, G.; Abbas, S.A.; Partridge, L.; Rice, D.; Binz, T.; Davletov, B. Two complementary approaches for intracellular delivery of exogenous enzymes. *Sci. Rep.* **2015**, *5*, 12444. [CrossRef] [PubMed]

23. Endo, Y.; Tsurugi, K. Rna n-glycosidase activity of ricin a-chain. Mechanism of action of the toxic lectin ricin on eukaryotic ribosomes. *J. Biol. Chem.* **1987**, *262*, 8128–8130. [PubMed]

24. Endo, Y.; Mitsui, K.; Motizuki, M.; Tsurugi, K. The mechanism of action of ricin and related toxic lectins on eukaryotic ribosomes. The site and the characteristics of the modification in 28 s ribosomal rna caused by the toxins. *J. Biol. Chem.* **1987**, *262*, 5908–5912. [PubMed]

25. Endo, Y.; Tsurugi, K. The rna n-glycosidase activity of ricin a-chain. The characteristics of the enzymatic activity of ricin a-chain with ribosomes and with rrna. *J. Biol. Chem.* **1988**, *263*, 8735–8739. [PubMed]

26. Montanaro, L.; Sperti, S.; Mattioli, A.; Testoni, G.; Stirpe, F. Inhibition by ricin of protein synthesis in vitro. Inhibition of the binding of elongation factor 2 and of adenosine diphosphate-ribosylated elongation factor 2 to ribosomes. *Biochem. J.* **1975**, *146*, 127–131. [CrossRef] [PubMed]

27. Barbieri, L.; Valbonesi, P.; Bonora, E.; Gorini, P.; Bolognesi, A.; Stirpe, F. Polynucleotide:Adenosine glycosidase activity of ribosome-inactivating proteins: Effect on dna, rna and poly(a). *Nucleic Acids Res.* **1997**, *25*, 518–522. [CrossRef] [PubMed]

28. Brigotti, M.; Alfieri, R.; Sestili, P.; Bonelli, M.; Petronini, P.G.; Guidarelli, A.; Barbieri, L.; Stirpe, F.; Sperti, S. Damage to nuclear dna induced by shiga toxin 1 and ricin in human endothelial cells. *FASEB J.* **2002**, *16*, 365–372. [CrossRef] [PubMed]

29. Sikriwal, D.; Ghosh, P.; Batra, J.K. Ribosome inactivating protein saporin induces apoptosis through mitochondrial cascade, independent of translation inhibition. *Int. J. Biochem. Cell Biol.* **2008**, *40*, 2880–2888. [CrossRef] [PubMed]

30. Das, M.K.; Sharma, R.S.; Mishra, V. Induction of apoptosis by ribosome inactivating proteins: Importance of n-glycosidase activity. *Appl. Biochem. Biotechnol.* **2012**, *166*, 1552–1561. [CrossRef] [PubMed]

31. Gilabert-Oriol, R.; Weng, A.; Mallinckrodt, B.; Melzig, M.F.; Fuchs, H.; Thakur, M. Immunotoxins constructed with ribosome-inactivating proteins and their enhancers: A lethal cocktail with tumor specific efficacy. *Curr. Pharm. Des.* **2014**, *20*, 6584–6643. [CrossRef] [PubMed]

32. Alewine, C.; Hassan, R.; Pastan, I. Advances in anticancer immunotoxin therapy. *Oncologist* **2015**, *20*, 176–185. [CrossRef] [PubMed]

33. Madhumathi, J.; Verma, R.S. Therapeutic targets and recent advances in protein immunotoxins. *Curr. Opin. Microbiol.* **2012**, *15*, 300–309. [CrossRef] [PubMed]

34. Moolten, F.L.; Capparell, N.J.; Cooperband, S.R. Antitumor effects of antibody-diphtheria toxin conjugates: Use of hapten-coated tumor cells as an antigenic target. *J. Natl. Cancer Inst.* **1972**, *49*, 1057–1062. [PubMed]

35. Samagh, B.S.; Gregory, K.F. Antibody to lactate dehydrogenase. V. Use as a carrier for introducing diphtheria toxin into mouse tumor cells. *Biochim. Biophys. Acta* **1972**, *273*, 188–198. [CrossRef]

36. Lin, J.Y.; Tserng, K.Y.; Chen, C.C.; Lin, L.T.; Tung, T.C. Abrin and ricin: New anti-tumour substances. *Nature* **1970**, *227*, 292–293. [CrossRef] [PubMed]

37. Fodstad, O.; Olsnes, S.; Pihl, A. Toxicity, distribution and elimination of the cancerostatic lectins abrin and ricin after parenteral injection into mice. *Br. J. Cancer* **1976**, *34*, 418–425. [CrossRef] [PubMed]

38. Moolten, F.; Zajdel, S.; Cooperband, S. Immunotherapy of experimental animal tumors with antitumor antibodies conjugated to diphtheria toxin or ricin. *Ann. N. Y. Acad. Sci.* **1976**, *277*, 690–699. [CrossRef] [PubMed]

39. Spitler, L.E. Immunotoxin therapy of malignant melanoma. *Med. Oncol. Tumor Pharmacother.* **1986**, *3*, 147–152. [PubMed]

40. Polito, L.; Djemil, A.; Bortolotti, M. Plant toxin-based immunotoxins for cancer therapy: A short overview. *Biomedicines* **2016**, *4*, 12. [CrossRef] [PubMed]

41. Youle, R.J.; Neville, D.M., Jr. Anti-thy 1.2 monoclonal antibody linked to ricin is a potent cell-type-specific toxin. *Proc. Natl. Acad. Sci. USA* **1980**, *77*, 5483–5486. [CrossRef] [PubMed]

42. Wawrzynczak, E.J.; Watson, G.J.; Cumber, A.J.; Henry, R.V.; Parnell, G.D.; Rieber, E.P.; Thorpe, P.E. Blocked and non-blocked ricin immunotoxins against the cd4 antigen exhibit higher cytotoxic potency than a ricin a chain immunotoxin potentiated with ricin b chain or with a ricin b chain immunotoxin. *Cancer Immunol. Immunother.* **1991**, *32*, 289–295. [CrossRef] [PubMed]

43. Thorpe, P.E.; Brown, A.N.; Ross, W.C.; Cumber, A.J.; Detre, S.I.; Edwards, D.C.; Davies, A.J.; Stirpe, F. Cytotoxicity acquired by conjugation of an anti-thy1.1 monoclonal antibody and the ribosome-inactivating protein, gelonin. *Eur. J. Biochem.* **1981**, *116*, 447–454. [CrossRef] [PubMed]

44. Ramakrishnan, S.; Houston, L.L. Comparison of the selective cytotoxic effects of immunotoxins containing ricin a chain or pokeweed antiviral protein and anti-thy 1.1 monoclonal antibodies. *Cancer Res.* **1984**, *44*, 201–208. [PubMed]

45. O'Hare, M.; Roberts, L.M.; Thorpe, P.E.; Watson, G.J.; Prior, B.; Lord, J.M. Expression of ricin a chain in escherichia coli. *FEBS Lett.* **1987**, *216*, 73–78. [CrossRef]

46. Prieto, I.; Lappi, D.A.; Ong, M.; Matsunami, R.; Benatti, L.; Villares, R.; Soria, M.; Sarmientos, P.; Baird, A. Expression and characterization of a basic fibroblast growth factor-saporin fusion protein in escherichia coli. *Ann. N. Y. Acad. Sci.* **1991**, *638*, 434–437. [CrossRef] [PubMed]

47. Rosenblum, M.G.; Kohr, W.A.; Beattie, K.L.; Beattie, W.G.; Marks, W.; Toman, P.D.; Cheung, L. Amino acid sequence analysis, gene construction, cloning, and expression of gelonin, a toxin derived from gelonium multiflorum. *J. Interferon Cytokine Res.* **1995**, *15*, 547–555. [CrossRef] [PubMed]

48. Laurent, G.; Pris, J.; Farcet, J.P.; Carayon, P.; Blythman, H.; Casellas, P.; Poncelet, P.; Jansen, F.K. Effects of therapy with t101 ricin a-chain immunotoxin in two leukemia patients. *Blood* **1986**, *67*, 1680–1687. [PubMed]

49. Spitler, L.E.; del Rio, M.; Khentigan, A.; Wedel, N.I.; Brophy, N.A.; Miller, L.L.; Harkonen, W.S.; Rosendorf, L.L.; Lee, H.M.; Mischak, R.P.; et al. Therapy of patients with malignant melanoma using a monoclonal antimelanoma antibody-ricin a chain immunotoxin. *Cancer Res.* **1987**, *47*, 1717–1723. [PubMed]

50. Falini, B.; Bolognesi, A.; Flenghi, L.; Tazzari, P.L.; Broe, M.K.; Stein, H.; Durkop, H.; Aversa, F.; Corneli, P.; Pizzolo, G.; et al. Response of refractory hodgkin's disease to monoclonal anti-cd30 immunotoxin. *Lancet* **1992**, *339*, 1195–1196. [CrossRef]

51. LeMaistre, C.F.; Rosen, S.; Frankel, A.; Kornfeld, S.; Saria, E.; Meneghetti, C.; Drajesk, J.; Fishwild, D.; Scannon, P.; Byers, V. Phase i trial of h65-rta immunoconjugate in patients with cutaneous t-cell lymphoma. *Blood* **1991**, *78*, 1173–1182. [PubMed]

52. Selvaggi, K.; Saria, E.A.; Schwartz, R.; Vlock, D.R.; Ackerman, S.; Wedel, N.; Kirkwood, J.M.; Jones, H.; Ernstoff, M.S. Phase I/II study of murine monoclonal antibody-ricin a chain (xomazyme-mel) immunoconjugate plus cyclosporine a in patients with metastatic melanoma. *J. Immunother. Emphasis Tumor Immunol.* **1993**, *13*, 201–207. [CrossRef] [PubMed]

53. Baluna, R.; Vitetta, E.S. Vascular leak syndrome: A side effect of immunotherapy. *Immunopharmacology* **1997**, *37*, 117–132. [CrossRef]

54. Vitetta, E.S. Immunotoxins and vascular leak syndrome. *Cancer J.* **2000**, *6* (Suppl. 3), S218–S224. [PubMed]

55. Ahmad, A.; Law, K. Strategies for designing antibody-toxin conjugates. *Trends Biotechnol.* **1988**, *6*, 246–251. [CrossRef]

56. Shan, L.; Liu, Y.; Wang, P. Recombinant immunotoxin therapy of solid tumors: Challenges and strategies. *J. Basic Clin. Med.* **2013**, *2*, 1–6. [PubMed]

57. Hertler, A.A.; Frankel, A.E. Immunotoxins: A clinical review of their use in the treatment of malignancies. *J. Clin. Oncol.* **1989**, *7*, 1932–1942. [CrossRef] [PubMed]

58. Harkonen, S.; Stoudemire, J.; Mischak, R.; Spitler, L.E.; Lopez, H.; Scannon, P. Toxicity and immunogenicity of monoclonal antimelanoma antibody-ricin a chain immunotoxin in rats. *Cancer Res.* **1987**, *47*, 1377–1382. [PubMed]

59. Li, M.; Liu, Z.S.; Liu, X.L.; Hui, Q.; Lu, S.Y.; Qu, L.L.; Li, Y.S.; Zhou, Y.; Ren, H.L.; Hu, P. Clinical targeting recombinant immunotoxins for cancer therapy. *OncoTargets Ther.* **2017**, *10*, 3645–3665. [CrossRef] [PubMed]

60. Pastan, I.H.; Pai, L.H.; Brinkmann, U.; Fitzgerald, D.J. Recombinant toxins: New therapeutic agents for cancer. *Ann. N. Y. Acad. Sci.* **1995**, *758*, 345–354. [CrossRef] [PubMed]

61. Weldon, J.E.; Skarzynski, M.; Therres, J.A.; Ostovitz, J.R.; Zhou, H.; Kreitman, R.J.; Pastan, I. Designing the furin-cleavable linker in recombinant immunotoxins based on pseudomonas exotoxin a. *Bioconj. Chem.* **2015**, *26*, 1120–1128. [CrossRef] [PubMed]

62. Vallera, D.A.; Panoskaltsis-Mortari, A.; Blazar, B.R. Renal dysfunction accounts for the dose limiting toxicity of dt390anti-cd3sfv, a potential new recombinant anti-gvhd immunotoxin. *Protein Eng.* **1997**, *10*, 1071–1076. [CrossRef] [PubMed]

63. Pennell, C.A.; Erickson, H.A. Designing immunotoxins for cancer therapy. *Immunol. Res.* **2002**, *25*, 177–191. [CrossRef]

64. Pastan, I.; Hassan, R.; FitzGerald, D.J.; Kreitman, R.J. Immunotoxin treatment of cancer. *Annu. Rev. Med.* **2007**, *58*, 221–237. [CrossRef] [PubMed]

65. Brinkmann, U.; Pai, L.H.; FitzGerald, D.J.; Willingham, M.; Pastan, I. B3(fv)-pe38kdel, a single-chain immunotoxin that causes complete regression of a human carcinoma in mice. *Proc. Natl. Acad. Sci. USA* **1991**, *88*, 8616–8620. [CrossRef] [PubMed]

66. Hamers-Casterman, C.; Atarhouch, T.; Muyldermans, S.; Robinson, G.; Hamers, C.; Songa, E.B.; Bendahman, N.; Hamers, R. Naturally occurring antibodies devoid of light chains. *Nature* **1993**, *363*, 446–448. [CrossRef] [PubMed]

67. Li, T.; Qi, S.; Unger, M.; Hou, Y.N.; Deng, Q.W.; Liu, J.; Lam, C.M.; Wang, X.W.; Xin, D.; Zhang, P.; et al. Immuno-targeting the multifunctional cd38 using nanobody. *Sci. Rep.* **2016**, *6*, 27055. [CrossRef] [PubMed]

68. Yu, Y.; Li, J.; Zhu, X.; Tang, X.; Bao, Y.; Sun, X.; Huang, Y.; Tian, F.; Liu, X.; Yang, L. Humanized cd7 nanobody-based immunotoxins exhibit promising anti-t-cell acute lymphoblastic leukemia potential. *Int. J. Nanomed.* **2017**, *12*, 1969–1983. [CrossRef] [PubMed]

69. Tang, J.; Li, J.; Zhu, X.; Yu, Y.; Chen, D.; Yuan, L.; Gu, Z.; Zhang, X.; Qi, L.; Gong, Z.; et al. Novel cd7-specific nanobody-based immunotoxins potently enhanced apoptosis of cd7-positive malignant cells. *Oncotarget* **2016**, *7*, 34070–34083. [CrossRef] [PubMed]

70. Rosenblum, M.G.; Cheung, L.H.; Liu, Y.; Marks, J.W., 3rd. Design, expression, purification, and characterization, in vitro and in vivo, of an antimelanoma single-chain fv antibody fused to the toxin gelonin. *Cancer Res.* **2003**, *63*, 3995–4002. [PubMed]

71. Zhou, H.; Ekmekcioglu, S.; Marks, J.W.; Mohamedali, K.A.; Asrani, K.; Phillips, K.K.; Brown, S.A.; Cheng, E.; Weiss, M.B.; Hittelman, W.N.; et al. The tweak receptor fn14 is a novel therapeutic target in melanoma: Immunotoxins targeting fn14 receptor for malignant melanoma treatment. *J. Investig. Dermatol.* **2013**, *133*, 1052–1062. [CrossRef] [PubMed]

72. Woo, J.H.; Liu, Y.Y.; Mathias, A.; Stavrou, S.; Wang, Z.; Thompson, J.; Neville, D.M., Jr. Gene optimization is necessary to express a bivalent anti-human anti-t cell immunotoxin in pichia pastoris. *Protein Expr. Purif.* **2002**, *25*, 270–282. [CrossRef]

73. Della Cristina, P.; Castagna, M.; Lombardi, A.; Barison, E.; Tagliabue, G.; Ceriotti, A.; Koutris, I.; Di Leandro, L.; Giansanti, F.; Vago, R.; et al. Systematic comparison of single-chain fv antibody-fusion toxin constructs containing pseudomonas exotoxin a or saporin produced in different microbial expression systems. *Microb. Cell Fact.* **2015**, *14*, 19. [CrossRef] [PubMed]

74. Schindler, J.; Gajavelli, S.; Ravandi, F.; Shen, Y.; Parekh, S.; Braunchweig, I.; Barta, S.; Ghetie, V.; Vitetta, E.; Verma, A. A phase i study of a combination of anti-cd19 and anti-cd22 immunotoxins (combotox) in adult patients with refractory b-lineage acute lymphoblastic leukaemia. *Br. J. Haematol.* **2011**, *154*, 471–476. [CrossRef] [PubMed]

75. Herrera, L.; Bostrom, B.; Gore, L.; Sandler, E.; Lew, G.; Schlegel, P.G.; Aquino, V.; Ghetie, V.; Vitetta, E.S.; Schindler, J. A phase 1 study of combotox in pediatric patients with refractory b-lineage acute lymphoblastic leukemia. *J. Pediatr. Hematol. Oncol.* **2009**, *31*, 936–941. [CrossRef] [PubMed]

76. Furman, R.R.; Grossbard, M.L.; Johnson, J.L.; Pecora, A.L.; Cassileth, P.A.; Jung, S.H.; Peterson, B.A.; Nadler, L.M.; Freedman, A.; Bayer, R.L.; et al. A phase iii study of anti-b4-blocked ricin as adjuvant therapy post-autologous bone marrow transplant: Calgb 9254. *Leuk. Lymphoma* **2011**, *52*, 587–596. [CrossRef] [PubMed]

77. Gould, B.J.; Borowitz, M.J.; Groves, E.S.; Carter, P.W.; Anthony, D.; Weiner, L.M.; Frankel, A.E. Phase i study of an anti-breast cancer immunotoxin by continuous infusion: Report of a targeted toxic effect not predicted by animal studies. *J. Natl. Cancer Inst.* **1989**, *81*, 775–781. [CrossRef] [PubMed]

78. Weiner, L.M.; O'Dwyer, J.; Kitson, J.; Comis, R.L.; Frankel, A.E.; Bauer, R.J.; Konrad, M.S.; Groves, E.S. Phase I evaluation of an anti-breast carcinoma monoclonal antibody 260f9-recombinant ricin a chain immunoconjugate. *Cancer Res.* **1989**, *49*, 4062–4067. [PubMed]

79. Borthakur, G.; Rosenblum, M.G.; Talpaz, M.; Daver, N.; Ravandi, F.; Faderl, S.; Freireich, E.J.; Kadia, T.; Garcia-Manero, G.; Kantarjian, H.; et al. Phase 1 study of an anti-cd33 immunotoxin, humanized monoclonal antibody m195 conjugated to recombinant gelonin (hum-195/rgel), in patients with advanced myeloid malignancies. *Haematologica* **2013**, *98*, 217–221. [CrossRef] [PubMed]

80. Turturro, F. Denileukin diftitox: A biotherapeutic paradigm shift in the treatment of lymphoid-derived disorders. *Expert Rev. Anticancer Ther.* **2007**, *7*, 11–17. [CrossRef] [PubMed]

81. Hayashida, K. Substance p-saporin for bone cancer pain in dogs: Can man's best friend solve the lost in translation problem in analgesic development? *Anesthesiology* **2013**, *119*, 999–1000. [CrossRef] [PubMed]

82. Brown, D.C.; Agnello, K. Intrathecal substance p-saporin in the dog: Efficacy in bone cancer pain. *Anesthesiology* **2013**, *119*, 1178–1185. [CrossRef] [PubMed]

83. Higgins, S.C.; Fillmore, H.L.; Ashkan, K.; Butt, A.M.; Pilkington, G.J. Dual targeting ng2 and gd3a using mab-zap immunotoxin results in reduced glioma cell viability in vitro. *Anticancer Res.* **2015**, *35*, 77–84. [PubMed]

84. Vallera, D.A.; Todhunter, D.A.; Kuroki, D.W.; Shu, Y.; Sicheneder, A.; Chen, H. A bispecific recombinant immunotoxin, dt2219, targeting human cd19 and cd22 receptors in a mouse xenograft model of b-cell leukemia/lymphoma. *Clin. Cancer Res.* **2005**, *11*, 3879–3888. [CrossRef] [PubMed]

85. Vallera, D.A.; Oh, S.; Chen, H.; Shu, Y.; Frankel, A.E. Bioengineering a unique deimmunized bispecific targeted toxin that simultaneously recognizes human cd22 and cd19 receptors in a mouse model of b-cell metastases. *Mol. Cancer Ther.* **2010**, *9*, 1872–1883. [CrossRef] [PubMed]

86. Fidias, P.; Grossbard, M.; Lynch, T.J., Jr. A phase ii study of the immunotoxin n901-blocked ricin in small-cell lung cancer. *Clin. Lung Cancer* **2002**, *3*, 219–222. [CrossRef] [PubMed]

87. Baluna, R.; Rizo, J.; Gordon, B.E.; Ghetie, V.; Vitetta, E.S. Evidence for a structural motif in toxins and interleukin-2 that may be responsible for binding to endothelial cells and initiating vascular leak syndrome. *Proc. Natl. Acad. Sci. USA* **1999**, *96*, 3957–3962. [CrossRef] [PubMed]

88. Smallshaw, J.E.; Ghetie, V.; Rizo, J.; Fulmer, J.R.; Trahan, L.L.; Ghetie, M.A.; Vitetta, E.S. Genetic engineering of an immunotoxin to eliminate pulmonary vascular leak in mice. *Nat. Biotechnol.* **2003**, *21*, 387–391. [CrossRef] [PubMed]

89. Pastan, I.; Hassan, R.; Fitzgerald, D.J.; Kreitman, R.J. Immunotoxin therapy of cancer. *Nat. Rev. Cancer* **2006**, *6*, 559–565. [CrossRef] [PubMed]

90. Pirie, C.M.; Hackel, B.J.; Rosenblum, M.G.; Wittrup, K.D. Convergent potency of internalized gelonin immunotoxins across varied cell lines, antigens, and targeting moieties. *J. Biol. Chem.* **2011**, *286*, 4165–4172. [CrossRef] [PubMed]

91. Pirie, C.M.; Liu, D.V.; Wittrup, K.D. Targeted cytolysins synergistically potentiate cytoplasmic delivery of gelonin immunotoxin. *Mol. Cancer Ther.* **2013**, *12*, 1774–1782. [CrossRef] [PubMed]

92. Fuchs, H.; Bachran, D.; Panjideh, H.; Schellmann, N.; Weng, A.; Melzig, M.F.; Sutherland, M.; Bachran, C. Saponins as tool for improved targeted tumor therapies. *Curr. Drug Targets* **2009**, *10*, 140–151. [CrossRef] [PubMed]

93. Nakase, I.; Kobayashi, S.; Futaki, S. Endosome-disruptive peptides for improving cytosolic delivery of bioactive macromolecules. *Biopolymers* **2010**, *94*, 763–770. [CrossRef] [PubMed]

94. Berg, K.; Folini, M.; Prasmickaite, L.; Selbo, P.K.; Bonsted, A.; Engesaeter, B.O.; Zaffaroni, N.; Weyergang, A.; Dietze, A.; Maelandsmo, G.M.; et al. Photochemical internalization: A new tool for drug delivery. *Curr. Pharm. Biotechnol.* **2007**, *8*, 362–372. [CrossRef] [PubMed]

95. Weyergang, A.; Selbo, P.K.; Berstad, M.E.; Bostad, M.; Berg, K. Photochemical internalization of tumor-targeted protein toxins. *Lasers Surg. Med.* **2011**, *43*, 721–733. [CrossRef] [PubMed]

96. Bostad, M.; Olsen, C.E.; Peng, Q.; Berg, K.; Hogset, A.; Selbo, P.K. Light-controlled endosomal escape of the novel cd133-targeting immunotoxin ac133-saporin by photochemical internalization—A minimally invasive cancer stem cell-targeting strategy. *J. Control. Release* **2015**, *206*, 37–48. [CrossRef] [PubMed]

97. Queen, C.; Schneider, W.P.; Selick, H.E.; Payne, P.W.; Landolfi, N.F.; Duncan, J.F.; Avdalovic, N.M.; Levitt, M.; Junghans, R.P.; Waldmann, T.A. A humanized antibody that binds to the interleukin 2 receptor. *Proc. Natl. Acad. Sci. USA* **1989**, *86*, 10029–10033. [CrossRef] [PubMed]

98. Molineux, G. Pegylation: Engineering improved pharmaceuticals for enhanced therapy. *Cancer Treat. Rev.* **2002**, *28*, 13–16. [CrossRef]

99. Tsutsumi, Y.; Onda, M.; Nagata, S.; Lee, B.; Kreitman, R.J.; Pastan, I. Site-specific chemical modification with polyethylene glycol of recombinant immunotoxin anti-tac(fv)-pe38 (lmb-2) improves antitumor activity and reduces animal toxicity and immunogenicity. *Proc. Natl. Acad. Sci. USA* **2000**, *97*, 8548–8553. [CrossRef] [PubMed]

100. Cheung, L.H.; Marks, J.W.; Rosenblum, M.G. Development of "designer toxins" with reduced antigenicity and size. *Proc. Amer. Assoc. Cancer Res.* **2004**, *64*, 874–875.

101. Cizeau, J.; Grenkow, D.M.; Brown, J.G.; Entwistle, J.; MacDonald, G.C. Engineering and biological characterization of vb6-845, an anti-epcam immunotoxin containing a t-cell epitope-depleted variant of the plant toxin bouganin. *J. Immunother.* **2009**, *32*, 574–584. [CrossRef] [PubMed]

102. Entwistle, J.; Brown, J.G.; Chooniedass, S.; Cizeau, J.; MacDonald, G.C. Preclinical evaluation of vb6–845: An anti-epcam immunotoxin with reduced immunogenic potential. *Cancer Biother. Radiopharm.* **2012**, *27*, 582–592. [CrossRef] [PubMed]

103. Liu, W.; Onda, M.; Lee, B.; Kreitman, R.J.; Hassan, R.; Xiang, L.; Pastan, I. Recombinant immunotoxin engineered for low immunogenicity and antigenicity by identifying and silencing human b-cell epitopes. *Proc. Natl. Acad. Sci. USA* **2012**, *109*, 11782–11787. [CrossRef] [PubMed]

104. Tachtsidis, A.; McInnes, L.M.; Jacobsen, N.; Thompson, E.W.; Saunders, C.M. Minimal residual disease in breast cancer: An overview of circulating and disseminated tumour cells. *Clin. Exp. Metastasis* **2016**, *33*, 521–550. [CrossRef] [PubMed]

105. Sarosdy, M.F.; Hutzler, D.H.; Yee, D.; von Hoff, D.D. In vitro sensitivity testing of human bladder cancers and cell lines to tp-40, a hybrid protein with selective targeting and cytotoxicity. *J. Urol.* **1993**, *150*, 1950–1955. [CrossRef]

106. Battelli, M.G.; Polito, L.; Bolognesi, A.; Lafleur, L.; Fradet, Y.; Stirpe, F. Toxicity of ribosome-inactivating proteins-containing immunotoxins to a human bladder carcinoma cell line. *Int. J. Cancer* **1996**, *65*, 485–490. [CrossRef]

107. Li, C.; Yan, R.; Yang, Z.; Wang, H.; Zhang, R.; Chen, H.; Wang, J. Bcmab1-ra, a novel immunotoxin that bcmab1 antibody coupled to ricin a chain, can eliminate bladder tumor. *Oncotarget* **2017**, *8*, 46704–46705. [CrossRef] [PubMed]

108. Dalken, B.; Giesubel, U.; Knauer, S.K.; Wels, W.S. Targeted induction of apoptosis by chimeric granzyme b fusion proteins carrying antibody and growth factor domains for cell recognition. *Cell Death Differ.* **2006**, *13*, 576–585. [CrossRef] [PubMed]

109. Rybak, S.M.; Arndt, M.A.; Schirrmann, T.; Dubel, S.; Krauss, J. Ribonucleases and immunornases as anticancer drugs. *Curr. Pharm. Des.* **2009**, *15*, 2665–2675. [CrossRef] [PubMed]

110. Mathew, M.; Verma, R.S. Humanized immunotoxins: A new generation of immunotoxins for targeted cancer therapy. *Cancer Sci.* **2009**, *100*, 1359–1365. [CrossRef] [PubMed]

toxins

Article

Two Saporin-Containing Immunotoxins Specific for CD20 and CD22 Show Different Behavior in Killing Lymphoma Cells

Letizia Polito *,†, Daniele Mercatelli †, Massimo Bortolotti †, Stefania Maiello, Alice Djemil, Maria Giulia Battelli and Andrea Bolognesi

Department of Experimental, Diagnostic and Specialty Medicine—DIMES, General Pathology Section, Alma Mater Studiorum—University of Bologna, Via S. Giacomo 14, 40126 Bologna, Italy; danielemercatelli@gmail.com (D.M.); massimo.bortolotti2@unibo.it (M.B.); stefania.maiello2@unibo.it (S.M.); alice.djemil2@unibo.it (A.D.); mariagiulia.battelli@unibo.it (M.G.B.); andrea.bolognesi@unibo.it (A.B.)
* Correspondence: letizia.polito@unibo.it; Tel.: +39-051-209-4700
† These authors contributed equally to this work.

Academic Editors: Daniel Gillet and Julien Barbier
Received: 12 April 2017; Accepted: 26 May 2017; Published: 30 May 2017

Abstract: Immunotoxins (ITs) are hybrid proteins combining the binding specificity of antibodies with the cytocidal properties of toxins. They represent a promising approach to lymphoma therapy. The cytotoxicity of two immunotoxins obtained by chemical conjugation of the plant toxin saporin-S6 with the anti-CD20 chimeric antibody rituximab and the anti-CD22 murine antibody OM124 were evaluated on the CD20-/CD22-positive cell line Raji. Both ITs showed strong cytotoxicity for Raji cells, but the anti-CD22 IT was two logs more efficient in killing, probably because of its faster internalization. The anti-CD22 IT gave slower but greater caspase activation than the anti-CD20 IT. The cytotoxic effect of both immunotoxins can be partially prevented by either the pan-caspase inhibitor Z-VAD or the necroptosis inhibitor necrostatin-1. Oxidative stress seems to be involved in the cell killing activity of anti-CD20 IT, as demonstrated by the protective role of the H_2O_2 scavenger catalase, but not in that of anti-CD22 IT. Moreover, the IT toxicity can be augmented by the contemporary administration of other chemotherapeutic drugs, such as PS-341, MG-132, and fludarabine. These results contribute to the understanding of the immunotoxin mechanism of action that is required for their clinical use, either alone or in combination with other drugs.

Keywords: B-lymphoma; CD20; CD22; immunotherapy; immunotoxin; ribosome-inactivating proteins; saporin-S6

1. Introduction

Non-Hodgkin's lymphomas (NHLs) are a group of heterogeneous hematological malignancies with a wide range of aggressiveness. The majority of NHLs involve B-cells. Although NHLs are quite curable, approximately 50% of NHL patients either relapse or become refractory to conventional therapy. For these patients with poor outcomes, it is mandatory to find new therapeutic strategies, such as monoclonal antibody (mAb)-based immunotherapy. Several receptors are widely expressed on the surface of B-cell lymphomas. Two of them, CD20 and CD22, are expressed at high levels on normal mature B-cells and on a vast proportion of B-lymphoma cells but are absent from other normal tissues and hematopoietic stem cells. Efforts have been made toward developing mAbs that would specifically bind to CD20/CD22 as therapeutic agents for B-cell NHLs. The in vivo administration of these mAbs confirmed their high-affinity binding to B-lymphoma cells, while other hematological cells, such as T cells and stem cells, were not recognized. The positive outcome of clinical use of the anti-CD20 mAb rituximab represents a good example of the worldwide interest in anti-B-cell immunotherapy [1,2].

The cytotoxic effects of mAbs can be enhanced by their conjugation with drugs, radioisotopes, or toxic enzymes. The latter type of conjugates is classified as immunotoxins (ITs), which are chimeric proteins with specific cytotoxic effects [3,4]. Both plant and bacterial toxins can be utilized for the preparation of ITs, which are mostly designed for cancer immunotherapy and are particularly suitable for eliminating residual disease after chemo- or radiotherapy. To date, ITs have achieved excellent results in several different models in preclinical settings [5–7] and clinical trials [6,8,9]. The most promising results have been reported in the treatment of hematological malignancies [5,9].

Among plant toxins, ribosome-inactivating proteins (RIPs) have been widely used to produce ITs [10,11]. RIPs are largely distributed in the plant kingdom [12], and many plants producing RIPs have been used for centuries in traditional medicine [13]. RIPs are mainly divided into type 1, consisting of a single-chain enzymatic protein, and type 2, consisting of an A chain with the same enzymatic activity linked to a B chain with lectin properties [14]. The A chain cleaves one or more adenine molecules from ribosomal RNA, thus irreversibly damaging the ribosomes. The literature about type 1 RIP saporin is very abundant, comprising more than a thousand studies. Saporin-S6 has been well characterized, and because of its potency and stability, it has been largely employed in the production of conjugates and ITs for various purposes [9,15–20] and sometimes has been tested in clinical trials [9,11,21]. Saporin-S6, as many other type 1 and type 2 RIPs, can remove adenine from various biochemical targets in vitro, including ribosomal RNA, DNA, poly(A), and poly-ADP-ribosylated proteins [22–24]. Most likely, saporin-S6 is able to deadenylate different substrates also in vivo, and, as a consequence, once internalized in target cells, it can trigger several death pathways, including apoptosis, necroptosis, and to a lesser extent, necrosis [25]. This multipathway killing could also depend on RIP distribution inside the cell. Indeed, saporin-S6 has been shown to have a different intracellular routing path compared to other RIPs, such as ricin [26]. The accumulation of saporin-S6 in the endoplasmic reticulum and perinuclear cisternae has previously been reported [27] and could be in agreement with the unfolded protein response (UPR) evidenced for other RIPs [28]. Moreover, molecules of saporin-S6 were detected inside the nuclei of intoxicated cells together with DNA strand break formation, thus suggesting direct saporin-S6 action on nuclear DNA [27].

In previous studies, we demonstrated that the anti-tumor efficacy of both the anti-CD20 mAb rituximab and the anti-CD22 mAb OM124 can be strongly augmented by their conjugation to RIPs [15,16]. In this research, we compared the cytotoxic effects of two immunoconjugates obtained by the chemical linking of saporin-S6 with the two mAbs, rituximab and OM124, on the CD20-/CD22-positive Raji cell line. CD20 and CD22 were chosen as target antigens in our experiments because of their large expression on B-cell malignancies [1,2]. The cell death pathways involved were investigated both with direct methods and indirectly by using inhibitors of specific routes, such as the pan-caspase inhibitor Z-VAD, the necroptosis inhibitor necrostatin-1 (Nec-1) and the H_2O_2 scavenger catalase. Moreover, we explored the possibility of specifically augmenting the cytotoxic activity of the ITs with two proteasome inhibitors, PS-341 and MG-132, and with the chemotherapeutic drug fludarabine with the aim of increasing their efficiency in killing target cells.

2. Results

2.1. Immunotoxin Purification and Characterization

To obtain the anti-CD20 and anti-CD22 ITs, saporin-S6 was conjugated to the anti-CD20 chimeric mAb rituximab and to the anti-CD22 murine mAb OM124, respectively, through the insertion of an artificial disulfide bond, as described in the methods section. The derivatization procedures with 2-iminothiolane allowed the insertion of an average of about three and two thiol groups per molecule for rituximab and OM124, respectively. In the case of saporin-S6, about one thiol group was inserted for each molecule (Table 1). After conjugation, the composition of the two purified ITs was analyzed by sodium dodecyl sulfate-polyacrylamide gel electrophoresis (SDS-PAGE) under non-reducing conditions. Densitometric analysis revealed RIP/mAb molar ratios of 1.85 and 1.41 for the anti-CD20 and anti-CD22 ITs, respectively (Figure 1). The inhibitory activity of the immunoconjugates on cell-free protein synthesis was evaluated in vitro using a rabbit reticulocyte lysate system and compared to saporin-S6 activity. The ability of saporin-S6 to inhibit cell-free protein synthesis was retained almost intact after derivatization and conjugation to both antibodies. A similar concentration causing 50% protein synthesis inhibition (IC_{50}) was calculated for the two ITs (Figure 2a). The characteristics of the two immunoconjugates are summarized in Tables 1 and 2.

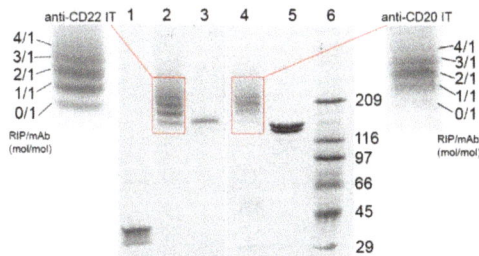

Figure 1. Analysis of rituximab/saporin-S6 (anti-CD20 IT) and OM124/saporin-S6 (anti-CD22 IT) immunotoxins by SDS–PAGE under non-reducing conditions. The gel was stained with Coomassie blue and subjected to densitometric analysis to establish the ribosome-inactivating protein (RIP)/mAb molar ratio. Lane 1: unconjugated saporin-S6, lane 2: anti-CD22 IT, lane 3: unconjugated OM124, lane 4: anti-CD20 IT, lane 5: unconjugated rituximab, lane 6: Mw standards expressed in kDa.

Table 1. Characteristics of the anti-CD20 IT and anti-CD22 IT.

Immunotoxin	Linker [#]/Ab (mol/mol)	Linker [#]/RIP (mol/mol)	RIP/Ab (mol/mol)
anti-CD20 IT	3.32	0.92	1.85
anti-CD22 IT	2.31	1.09	1.41

[#] The ribosome-inactivating protein (RIP) and the mAbs were modified using 2-iminothiolane linker, and the number of SH groups inserted per molecule is reported.

2.2. Effect of the Immunotoxins on Raji Protein Synthesis and Cell Viability

The ability of the ITs to inhibit cellular protein synthesis was assayed in CD20-/CD22-positive Raji cells after 96 h of treatment (Figure 2b) and was compared to the mixture of unconjugated mAbs and saporin-S6. The anti-CD20 IT showed excellent efficacy with IC_{50} values of 1.99 nM, and it was able to almost completely abolish protein synthesis at a 10 nM concentration, expressed as RIP content. The anti-CD22 IT showed a higher inhibitory activity in Raji cells than the anti-CD20 IT with an IC_{50} of 0.060 nM and a complete protein synthesis inhibition already at a 1 nM concentration. The mixture of free RIP and either anti-CD20 or anti-CD22 mAbs produced scarce effect on Raji cell protein synthesis, with IC_{50} values >100 nM (highest tested concentration).

The viability of Raji cells after IT exposure was measured in dose-response experiments after different incubation times (24, 48, 72, and 96 h) to evaluate the minimum time required to observe a cytotoxic effect (Figure 3). The IT cytotoxicity is related to the incubation time. The dose-response curves showed similar tendencies in both cases. Even in this test, the anti-CD22 IT confirmed its two log higher toxicity than the anti-CD20. The maximum cytotoxic effect, corresponding to the complete abolishment of cell viability, was observed after 96 h only for the highest tested doses: 100 nM and 1 nM for anti-CD20 and anti-CD22 ITs, respectively. The mixture of free RIP and either anti-CD20 or anti-CD22 mAbs produced no relevant effect on Raji cell viability in the tested concentration range (data not shown). The concentrations causing 50% cell viability reduction (EC_{50}) values for the anti-CD20 IT and anti-CD22 IT calculated after 96 h were 4.06 nM and 0.05 nM, respectively (Table 2).

The effect of immunotoxins was also evaluated on viability of CD20-/CD22-negative Jurkat cells. For both immunotoxins, after 96 h of incubation, the EC_{50} calculated was higher than the maximum tested concentration (100 nM) (Table 2).

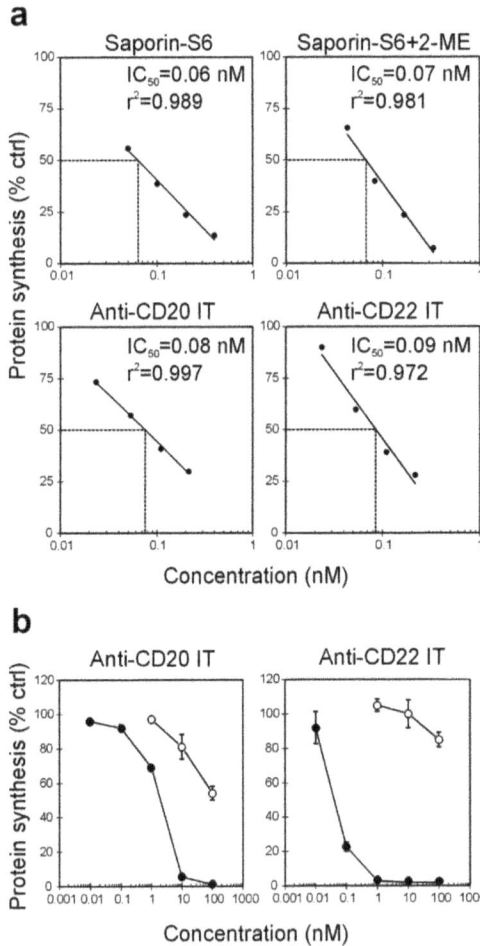

Figure 2. (**a**) The protein synthesis inhibitory activity of the two immunotoxins (ITs) and free RIP was evaluated using a cell-free rabbit reticulocyte lysate system. The IT rituximab/saporin-S6 (anti-CD20 IT) or OM124/saporin-S6 (anti-CD22 IT) and saporin-S6 were reduced with 20 mM 2-mercaptoethanol (2-ME) for 30 min at 37 °C and compared to native saporin-S6. The IC_{50} is the concentration of free or conjugated RIP causing 50% protein synthesis inhibition and was calculated by linear regression analysis. (**b**) Protein synthesis inhibition in Raji cells treated for 96 h with anti-CD20 or anti-CD22 ITs (black symbols) or the mixture of the respective unconjugated mAb and saporin-S6 (white symbols). Means ± S.D. of three independent experiments, each in triplicate, are given. SD never exceeded 10%.

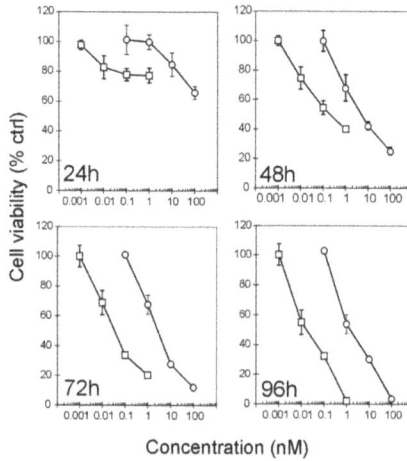

Figure 3. Effect of the immunotoxins on Raji cell viability. Raji cells were incubated for the indicated times with different doses of anti-CD20 IT (○) or anti-CD22 IT (□). Cell viability was measured by MTS salt reduction, as described in the methods section, and expressed as percentage of untreated cells. Means ± S.D. of three independent experiments, each in triplicate, are given. SD never exceeded 15%. At every tested time, the curves are significantly different from each other ($p < 0.0001$). MTS = 3-(4,5-dimethylthiazol-2-yl)-5-(3-carboxymethoxyphenyl)-2-(4-sulfophenyl)-2H-tetrazolium.

Table 2. Effect of the anti-CD20 IT and anti-CD22 IT on cell-free and cellular systems.

Immunotoxin	Cell-Free	Raji Cells [§]		Jurkat Cells [#]
	IC_{50} * (nM)	IC_{50} * (nM)	EC_{50} * (nM)	EC_{50} * (nM)
anti-CD20 IT	0.08	1.99	4.06	>100
anti-CD22 IT	0.09	0.06	0.05	>100

* The IC_{50} and EC_{50} of the immunotoxins refer to the RIP content. IC_{50}, concentration causing 50% protein synthesis inhibition; EC_{50}, concentration causing 50% cell viability reduction. [§] The IC_{50} and EC_{50} on Raji cells were calculated after 96 h of incubation with the immunotoxins. [#] The EC_{50} on Jurkat non target cells (CD20-/CD22-negative) was calculated after 96 h of incubation with the immunotoxins.

2.3. Evaluation of Internalization Time of the Immunotoxins

The binding of the ITs to the CD20 and CD22 membrane antigens in Raji cells was evaluated by cytofluorimetric analysis, after different incubation times with ITs. To allow binding and avoid the internalization of the complex, Raji cells were treated with ITs at a 10 nM concentration, for 30 min on ice. Cells were then incubated at 37 °C for different times ranging from 0 to 120 min. We considered as the maximum antigen binding the fluorescence intensity value obtained after 30 min incubation of cells with the ITs on ice, followed by 0 min exposure at 37 °C.

The two ITs have a similar binding intensity to Raji cells at 0′ (compare histograms in Figure 4a,b 0′). In the case of the anti-CD20 IT (Figure 4a,c), the positivity to FITC remained unchanged from 0 to 30 min at 37 °C. The IT bound to the membrane significantly decreased after 60 min and was almost completely absent after 120 min, indicating the partial and complete internalization of the CD20-IT complex, respectively.

The anti-CD22 IT showed a faster internalization of the antigen-IT complex in comparison to the CD20 one (Figure 4b,c). In fact, after 15 min of incubation at 37 °C, the observed binding was already significantly lower than that observed for cells incubated for 0 min at 37 °C ($p < 0.0001$). After 20 min the IT bound to membrane resulted strongly decreased, and after 30 min, the complex was completely internalized.

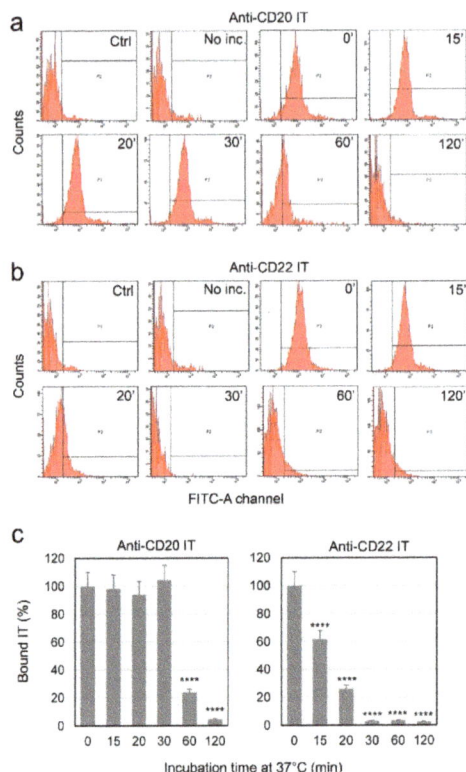

Figure 4. Evaluation of the internalization time of the antigen-immunotoxin complex by cytofluorimetric analysis in Raji cells. Samples were prepared by incubating cells with 10 nM anti-CD20 IT (**a**) or anti-CD22 IT (**b**) for 30 min on ice to allow the binding of the IT to the antigen, avoiding the internalization of the complex. After cell incubation for 0–120 min at 37 °C, the corresponding FITC-secondary antibody was added. Negative controls were carried out by incubating cells with complete medium alone (ctrl). A second series of controls were obtained without the 30 min pre-incubation at 0 °C and instead putting cells into contact with the IT for only an instant (No inc.). In Figure 4c, the percentage of cell membrane bound IT at the indicated times is reported. The bound IT is expressed as the percentage of mean fluorescence intensity values for each time point with respect to those of the 0 min samples, which was considered the maximum antigen binding. The values significantly lower than the 0 min samples are indicated by asterisks (**** $p < 0.0001$). The results are the means of three independent experiments.

2.4. Evaluation of Cell Death Pathways Induced by Immunotoxins in Raji Cells

The presence of membrane apoptotic and necrotic changes in Raji cells treated for 96 h with the ITs was evaluated by double staining with Annexin V-EGFP (AnnV) and propidium iodide (PI) at concentrations of 1 nM for anti-CD20 IT and 0.01 nM for anti-CD22 IT. As shown in Figure 5a, after exposure to ITs, approximately 50% (anti-CD20 IT) and 60% (anti-CD22 IT) of cells were positive for AnnV and PI double staining, indicating a late apoptosis stage. A very low percentage of necrotic cells (AnnV$^-$/PI$^+$) was evidenced for both ITs, 3.2% for anti-CD20 IT and 6.4% for anti-CD22 IT (Figure 5a), compared to approximately 0.5% in untreated cells.

The activation of effector caspases 3/7 was measured in Raji cells after 12, 24, 48, 72, and 96 h of treatment with the ITs at the same concentrations used for the AnnV/PI experiments. Both ITs caused a strong caspase 3/7 augmentation, but with different slopes of the activation curves. Raji cells treated with the anti-CD20 IT showed significant activation of caspase 3/7 already after 12 h ($p < 0.0001$). The caspase activation curve showed a fairly linear rise in the range of time tested, reaching approximately 900% of the value observed in untreated cells after 96 h (Figure 5b, left). In the case of the anti-CD22 IT, the caspase activation curve increased slowly over the first 48 h, becoming significant only after 24 h ($p < 0.001$). However, the caspases showed an exponential growth at 72 and 96 h, reaching approximately 2300% of untreated cells after 96 h (Figure 5b, right). In both cases, the mixture of unconjugated mAb and saporin-S6 at the same concentrations as the ITs produced no relevant activation of caspases 3/7 in Raji cells.

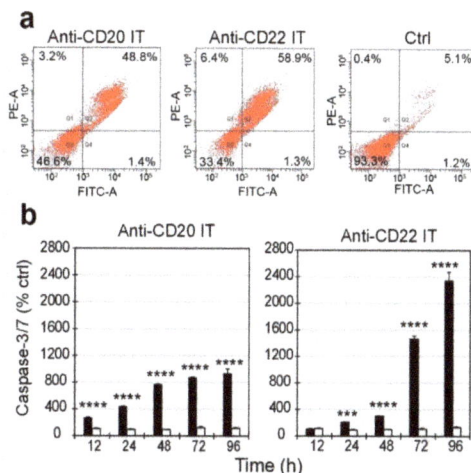

Figure 5. (**a**) Cytofluorimetric analysis of Annexin V/propidium iodide double staining of Raji cells treated for 96 h with 1 nM anti-CD20 IT or 0.01 nM anti-CD22 IT, i.e., the concentrations corresponding to their EC_{50} values. FITC-A channel (*x*-axis) is used for the detection of Annexin V-EGFP fluorescence. PE-A channel (*y*-axis) is used for the detection of propidium iodide fluorescence. (**b**) Caspase 3/7 activation in Raji cells exposed to 1 nM anti-CD20 IT or 0.01 nM anti-CD22 IT (black columns) or a mixture of unconjugated mAb and saporin-S6 (white columns). The expression of activated caspases was reported as percentage of untreated cell values. Means ± S.D. of three independent experiments, each in triplicate, are given. Statistical significance was determined by ANOVA/Bonferroni test. Asterisks indicate the significant difference in each experimental condition between IT and the mixture of mAb and RIP (**** $p < 0.0001$; *** $p < 0.001$).

2.5. Evaluation of the Protective Effect of Apoptosis and Necroptosis Inhibitors Z-VAD and Necrostatin-1 and the H_2O_2 Scavenger Catalase on Raji Cells

To determine the role of apoptosis and necroptosis in IT-induced cell death, we designed experiments that included the pan-caspase inhibitor Z-VAD and the necroptosis inhibitor Nec-1. Additionally, the involvement of oxidative stress was investigated by including the reactive oxygen species (ROS) scavenger catalase. Raji cells were treated with the anti-CD20 IT and anti-CD22 IT at 1 nM and 0.01 nM concentrations, respectively. The viability was measured after different amounts of time, ranging from 24 to 96 h of exposure to ITs in the presence or absence of Z-VAD (10 µM), Nec-1 (10 µM), or catalase (10 U/mL), added 3 h before the IT treatment (Figures 6 and 7).

The pan-caspase inhibitor Z-VAD was able to protect Raji cells from death triggered by ITs. As shown in Figure 6a, the protective effect of Z-VAD became significant from 48 h ($p < 0.0001$ for anti-CD20 IT, $p < 0.05$ for anti-CD22 IT) and increased over time, leading to a maximum protective effect on cell viability after 96 h. The necroptosis inhibitor Nec-1 protected Raji cells from IT-induced cell death similarly to Z-VAD (Figure 6b). The protective effect became significant from 48 h for the anti-CD20 IT ($p < 0.001$) and from 72 h for the anti-CD22 IT ($p < 0.0001$), leading to a maximum protective effect after 96 h. Surprisingly, different behavior was observed when cells were pretreated with catalase. In fact, catalase was able to protect cells from cytotoxicity induced by anti-CD20 IT with a similar trend than Z-VAD and Nec-1; its protective effect became significant after 48 h ($p < 0.001$) and reach its maximum at 96 h. By contrast, no protection was given by the scavenger in cells treated with anti-CD22 IT at any tested time (Figure 6c).

Visual inspection by phase-contrast microscopy of cells treated for 96 h with 1 nM anti-CD20 IT or with 0.01 nM anti-CD22 IT showed a strong cytotoxic effect with a marked reduction of cell viability compared to untreated cells (Figure 7). When pretreated with the inhibitors, Raji cells treated with the ITs appeared similar to untreated cells. Pretreatment with the scavenger catalase saved Raji cells from death triggered by anti-CD20 IT, but it was completely ineffective at rescuing cells treated with the anti-CD22 IT.

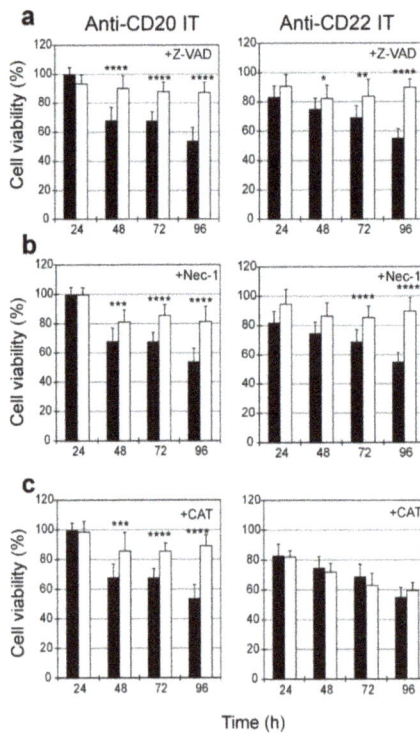

Figure 6. Viability of Raji cells treated for 24, 48, 72, and 96 h with 1 nM anti-CD20 IT (left) or 0.01 nM anti-CD22 IT (right) without (black columns) or in the presence (white columns) of 10 μM pan-caspase inhibitor (Z-VAD) (**a**), 10 μM necroptosis inhibitor necrostatin-1 (Nec-1) (**b**) or 10 U/mL hydrogen peroxide scavenger catalase (CAT) (**c**). Z-VAD, Nec-1 and CAT were added 3 h before the ITs, and the viability was measured after the indicated times. Means ± S.D. of at least three independent experiments, each in triplicate, are shown as the percentage of untreated cell values. Statistical significance was determined by ANOVA/Bonferroni test. Asterisks indicate the significant difference in each experimental condition between IT alone and IT plus inhibitors/scavenger (**** $p < 0.0001$; *** $p < 0.001$; ** $p < 0.01$; * $p < 0.05$).

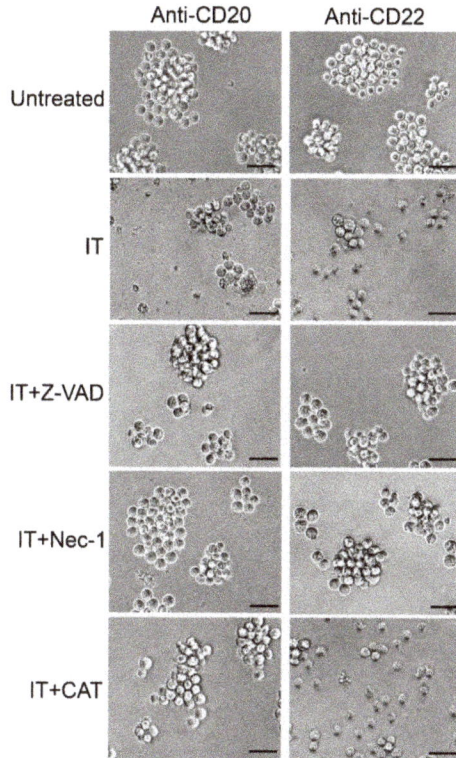

Figure 7. Morphological analysis of Raji cells assessed using phase-contrast microscopy. Cells were treated for 96 h with 1 nM anti-CD20 IT (left) or 0.01 nM anti-CD22 IT (right) alone (IT) or in the presence of 10 μM pan-caspase inhibitor (Z-VAD), 10 μM necroptosis inhibitor necrostatin-1 (Nec-1), or 10 U/mL hydrogen peroxide scavenger catalase (CAT). Z-VAD, Nec-1 and CAT were added 3 h before the ITs. Untreated cultures were grown in the absence of ITs. Magnification, 400 ×. Scale bars correspond to 50 μm.

2.6. Combined Cytotoxic Effect of ITs with the Proteasome Inhibitors MG-132 and PS-341 and with the Chemotherapeutic Drug Fludarabine

To evaluate the possibility of enhancing the cytotoxic effect of the anti-CD20 and anti-CD22 ITs on Raji cells, we tested two proteasome inhibitors, MG-132 and PS-341 (bortezomib), giving them to Raji cells as single agents or in combination with the ITs (Figure 8a,b).

The sensitivity of Raji cells to both the anti-CD20 and anti-CD22 ITs was significantly augmented by co-incubation with MG-132. The combination of MG-132 and anti-CD20 IT produced a 2.8-fold enhanced toxicity compared to the proteasome inhibitor and a 1.4-fold enhancement compared to the IT (Figure 8a, left). The combination of MG-132 with the anti-CD22 IT gave similar results, showing a 2.4-fold increase in toxicity compared to MG-132 and a 1.3-fold increase compared to IT (Figure 8a, right).

Similarly, responsiveness to the ITs was significantly augmented when cells were co-incubated with PS-341 (Figure 8b). The combination of PS-341 and the anti-CD20 IT produced a significant augmented effect in comparison to the single agents, showing a 1.7- and 1.3-fold enhanced toxicity compared to PS-341 and IT, respectively (Figure 8b, left). The combination of PS-341 with the anti-CD22

IT gave a significant 1.9- and 1.4-fold increase in toxicity compared to PS-341 and IT, respectively (Figure 8b, right).

The combination of the purine analogue fludarabine (FLU) and the anti-CD20 IT significantly reduced Raji cell viability compared to single compounds, resulting in a superadditive effect, showing a 6.6- and 2-fold enhanced toxicity compared to FLU and IT, respectively (Figure 8c, left). In combination with the anti-CD22 IT, FLU gave a 6.3- and 1.9-fold increase in cytotoxicity compared to FLU and IT, respectively (Figure 8c, right).

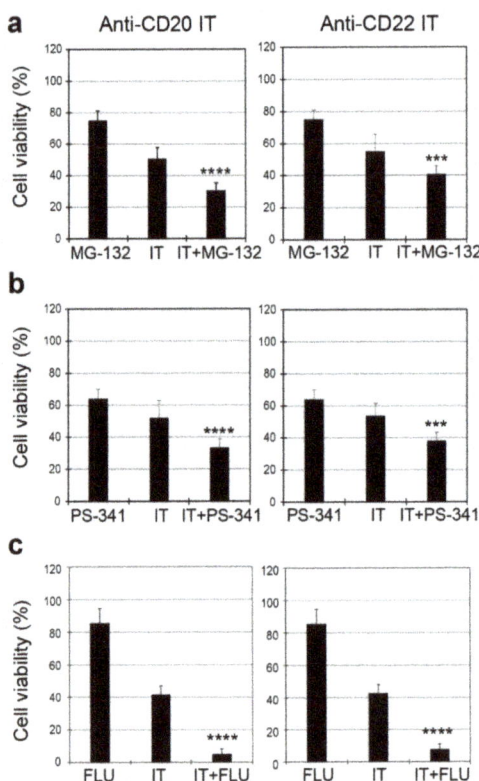

Figure 8. Viability of Raji cells treated for 96 h with 1 nM anti-CD20 IT (left) or 0.01 nM anti-CD22 IT (right) alone or in the presence of the proteasome inhibitors MG-132 (0.1 μM) (**a**) or PS-341 (1 nM) (**b**) or the purine analogue fludarabine (FLU) (0.75 μM) (**c**). Inhibitors and FLU were added 3 h before the ITs and maintained for the entire incubation time. Cell viability was determined after 96 h. Means ± S.D. of three independent experiments, each in triplicate, are showed as the percentage of the untreated cell values. Statistical significance was determined by ANOVA/Bonferroni test. Asterisks indicate the significant difference in each experimental condition between IT alone and IT plus the inhibitor or the purine analogue (**** $p < 0.0001$; *** $p < 0.001$).

3. Discussion

The specific targeting of B-cells via restricted surface antigens has already been demonstrated to be an efficacious approach for immunotherapy against antigen-positive NHLs. In the last few years, several anti-CD20 and anti-CD22 ITs composed of mAbs linked to RIPs [15,16,29], bacterial toxins [30], ribonucleases [31], drugs [32,33], or radioisotopes [1,2] have demonstrated potent anti-tumor effects in vitro and in vivo in animal models and, in some cases, also in clinical trials.

The study of the IT mechanisms of action in target cells may help the design of new immunoconjugates with higher cytotoxic potential and specificity to target cells. In our study, we tested and compared in vitro the specific cytotoxic properties and the cell death pathways triggered by two ITs defined as anti-CD20 IT, which was obtained by the chemical coupling of saporin-S6 to the anti-CD20 mAb rituximab, and anti-CD22 IT, which was produced by coupling saporin-S6 to the anti-CD22 mAb OM124.

Despite CD20 and CD22 antigens being the most utilized targets for immunotoxin based NHL therapy, very little information is available about the pathogenic mechanism of cell intoxication.

Rituximab showed a 40% higher reactivity than OM124 for the cross-linking reagent 2-iminothiolane. As a consequence, a higher saporin-S6/antibody molar ratio was obtained for the anti-CD20 IT. After the conjugation processes to anti-CD20 and anti-CD22 mAbs, saporin-S6 retained almost the same enzymatic activity as native saporin-S6 on eukaryotic ribosomes (0.06–0.09 nM).

The ability of the ITs to kill lymphoma cells was tested in the CD20-/CD22-positive Raji cell line. In both cases after conjugation the cytotoxic activity of saporin-S6 and mAbs was significantly augmented. Despite the anti-CD20 IT showing a slightly higher saporin-S6 payload than the anti-CD22 IT, the latter resulted approximately in being 30-fold more efficient in cell protein synthesis inhibition and 80-fold more toxic to Raji cells. By analyzing the cytotoxicity curves at different time points, it is possible to see that the different behavior of the two ITs is maintained over time.

It is well known from the literature that the CD20 antigen is poorly internalized after ligand binding [34]. However, in previous papers we demonstrated that rituximab increases its internalization after conjugation with the plant toxin saporin-S6, giving an IT with a good specific cytotoxicity for lymphoma target cells [16,35]. The higher specific cytotoxic effect of the anti-CD22 IT could be expected on the basis of the more rapid CD22 internalization after ligand binding and its recycling to plasma membrane [36]. Actually, our experiments confirm the rapid internalization of the anti-CD22 IT, as the complex antigen/IT was completely internalized already after 30 min at 37 °C, while with the CD20/IT complex the same result was obtained after 120 min at 37 °C.

In agreement with the results reported for other saporin-S6-containing ITs [37–39], in the present study both the anti-CD20 IT and anti-CD22 IT were found to induce apoptosis in the target cells. However, differences in timing and intensity were observed in the activation of executioner caspases that were significantly augmented after 12 or 24 h of incubation with the anti-CD20 IT and anti-CD22 IT, respectively. The anti-CD22 IT caused a slightly delayed but exponential caspase activation that, after 72–96 h, reached much higher values than that caused by anti-CD20 IT. The retard in caspase activation observed for the anti-CD22 IT, in spite of its rapid internalization, could be justified by a different and slower routing of this IT with respect to the anti-CD20 IT. Finally, the anti-CD22 IT efficiently reaches its intracellular targets resulting in a stronger caspase activation, without ROS involvement, as commented below.

Despite the differences observed in caspase activation, the pretreatment with the pan-caspase inhibitor Z-VAD gave a good, even if incomplete, cell protection (approx. 90% survival) when used with each IT. This incomplete protection prompted us to evaluate other cell death mechanisms, taking into account that: (i) saporin-S6 and some other RIPs were reported to trigger multiple death pathways [25,40] and (ii) RIP conjugation to a specific carrier can modify its intracellular routing and consequently its toxicity pattern.

Necroptosis is a recently identified programmed cell death, depending on the serine-threonine kinase receptor-interacting proteins 1 and 3. Nec-1 blocks necroptosis through the inhibition of serine-threonine kinase receptor-interacting protein 1 [41]. In our experiments, Nec-1 was able to protect Raji cells from the damage induced by the anti-CD20 and anti-CD22 ITs at a fairly similar level as Z-VAD, thus suggesting the involvement of both apoptosis and necroptosis, with a quite similar timing.

Oxidative stress was identified as being involved in cell death induced by a saporin-based conjugate directed against TfR1 [42]. Recently, it was reported that also the type 2 RIP stenodactylin is able to induce the early formation of ROS molecules in a neuroblastoma cell line and that several ROS scavengers can protect cells from RIP intoxication [40]. In our experiments, catalase significantly protected cells from damage induced by the anti-CD20 IT, but no protection was evidenced in cells treated with the anti-CD22 IT in the considered period of time. The cytotoxicity of anti-CD20 mAbs was previously demonstrated to be partially dependent on ROS generation in lymphoma cells [43,44]. Our data confirm the involvement of ROS in the cytotoxic effect of the anti-CD20 IT, but it appears to be absent for the anti-CD22 IT. This different behavior indicates that the induction of oxidative stress by ITs is not dependent on the toxic moiety but instead on the mAb/antigen interaction. The immunotoxin routing, depending mainly on the carrier moiety, can lead to RIP accumulation in specific compartments, such as the endoplasmic reticulum, where UPR can trigger ROS generation [45]. To date this is the only way to generate an oxidative stress in RIP intoxicated cells. The lack of oxidative stress in anti-CD22 IT treated cells suggests that massive endoplasmic reticulum accumulation and UPR are not involved in cell damage.

In vitro studies using rituximab-resistant cell lines showed that the development of rituximab resistance could be attributed to significant changes that occur in the CD20 antigen and in the deregulation of the ubiquitin-proteasome system [46,47]. Moreover, type 2 RIP cytotoxicity was shown to be partially reduced by proteasome activity [48], and bortezomib activity was enhanced by combination with an anti-CD22 mAb immunotherapy [49]. For these reasons, we were expecting a strongly augmented effect in the combined treatment with IT and a proteasome inhibitor. In our experiments, proteasome inhibition was able to increase cytotoxicity of both ITs, even if the combination gave a cytotoxicity lower than the sum of the single agents. The lack of superadditive effect suggests that our immunotoxins do not undergo a relevant cytoplasmic degradation.

We previously demonstrated that fludarabine strongly enhanced rituximab/saporin-S6 cytotoxicity [16]. The current experiments demonstrate a superadditive effect also when fludarabine was given in combination with the anti-CD22 IT. The ability of saporin-S6 to deadenylate poly(ADP-ribosyl)ated poly(ADP-ribose) polymerase in a cell free system was previously described [50]. If this occurs also in vivo, it would interfere with the DNA repair mechanism through the base excision repair complex inhibition, probably amplifying the effect of fludarabine. Altogether, these results indicate the advantage of a combination IT/FLU therapy with respect to proteasome inhibitor/IT therapy.

The biological properties of saporin-S6-containing ITs make them attractive molecules for the treatment of NHLs because their ability to induce cell death by more than one pathway makes the selection of tumor clones resistant to saporin-S6-induced cell death difficult.

Although some critical opinions about the clinical use of immunoconjugates that distrust their therapeutic potential have been expressed, the response rates observed in clinical trials have often been greater than those reported for conventional antiblastic drugs [8,51]. In past years, the clinical development of ITs has been hampered by several limitations, like immunogenicity and vascular leak syndrome. However, novel immunoconjugates aimed at reducing these side effects have been designed, and new and exciting opportunities for controlled drug delivery and release have been developed [52,53]. Furthermore, an anti-IL-2R IT was approved in 1999 by the U.S. Food and Drug Administration for the treatment of cutaneous T-cell lymphoma in adults [54]. Today, over thirty antibody-drug conjugates and ITs are in clinical trials, thus demonstrating that the magic bullet idea of Paul Ehrlich is still relevant [55]. The great interest in the field of institutional investigators and pharmaceutical companies is also suggested by the increasing number of patented immunoconjugates, and many researchers agree that immunoconjugates will likely become important actors in cancer therapy in the foreseeable future. The study of new ITs is today particularly interesting due to the availability of a new generation of mAbs, such as chimeric or humanized molecules, that are already at the clinical stage. The development of combination therapies may result in new effective options for the treatment of cancer, and the knowledge of the mechanism of action of the substances is mandatory in order to design proper effective protocols.

4. Materials and Methods

4.1. Immunotoxins

Saporin-S6 was purified from the seeds of *Saponaria officinalis* [19]. Anti-CD20 rituximab/saporin-S6 IT and anti-CD22 OM124/saporin-S6 IT were produced as described in [15,16]. Briefly, mAbs and saporin-S6 were dissolved in 50 mM sodium borate buffer, pH 9.0 and were derivatized by adding 2-iminothiolane (Sigma-Aldrich, St. Louis, MO, USA), as described in [56]. MAbs and the reduced RIP were allowed to react for 16 h at room temperature. The resulting conjugates were separated from RIP homopolymers and free antibody by gel filtration on a Sephacryl S-200 high-resolution column (100 cm × 2.5 cm) (GE-Healthcare, Buckinghamshire, UK), equilibrated, and eluted with phosphate-buffered saline (PBS, 0.14 M sodium chloride in 5 mM sodium phosphate buffer, pH 7.4).

The immunoconjugates were analyzed under non-reducing conditions by SDS-PAGE on a 4–15% PhastGel gradient, and then stained with Coomassie brilliant blue and analyzed, as described in [16]. Molecular weight markers were from Sigma: myosin (205 kDa), b-galactosidase (116 kDa), phosphorylase B (97 kDa), bovine serum albumin (66 kDa), ovalbumin (45 kDa), carbonic anhydrase (29 kDa).

4.2. Cells

The activity of the conjugates was assayed in the CD20-/CD22-positive Raji cell line (ATCC number CCL-86™) and in the non target Jurkat cell line (ATCC number TIB-152™). Cells were from long term culture in our department. The cells were cultured in RPMI 1640 medium supplemented with 10% heat-inactivated fetal bovine serum (FBS), 2 mM L-glutamine, 100 U/mL penicillin and 100 µg/mL streptomycin (hereafter named complete medium), cultured at 37 °C in a humidified environment with 5% CO_2 in a HeraCell Haereus incubator (Hanau, Germany) and routinely checked for the absence of Mycoplasma infection. The viability was checked before each experiment by Trypan blue (BioWhittaker, Vervies, Belgium) dye exclusion. Flasks and plates were from Falcon (Franklin Lakes, NJ, USA). All the other cell culture reagents were from Sigma-Aldrich.

4.3. Reagents and Kits

Caspase activity was evaluated using the luminescent kit Caspase-Glo™3/7 Assay (Promega Corporation, Fitchburg, Wisconsin, USA). Morphological membrane changes were detected using Annexin V-EGFP/PI detection kit (Biovision, Mt. View, CA, USA). Viability was measured using the colorimetric CellTiter 96® Aqueous One Solution Cell Proliferation Assay (Promega). The CellTiter 96® Aqueous One Solution Reagent contains the tetrazolium compound 3-(4,5-dimethylthiazol-2-yl)-5-(3-carboxymethoxyphenyl)-2-(4-sulfophenyl)-2H-tetrazolium (MTS), and an electron coupling reagent (1-methoxy phenazine methosulfate—PMS). The liquid scintillation cocktail was the Ready-Gel (Beckman Instrument, Fullerton, CA, USA). The proteasome inhibitors PS-341 and MG-132, the necroptosis inhibitor Necrostatin-1 and the pan-caspase inhibitor carbobenzoxy-valyl-alanyl-aspartyl-[O-methyl]fluoromethylketone (Z-VAD-fmk) were supplied by Vinci-Biochem (Florence, Italy). The hydrogen peroxide scavenger catalase (CAT) and fludarabine (FLU) were purchased by Sigma-Aldrich. For SDS-PAGE, precasted gels and buffer strips obtained from GE Healthcare were used. Other reagents used were from Merck (Darmstadt, Germany), Carlo Erba (Milano, Italy) and Sigma-Aldrich.

4.4. Cell Protein Synthesis Inhibition Assays

The inhibitory activity of IT and free RIP on protein synthesis was evaluated using a cell-free rabbit reticulocyte lysate system. After reduction with 20 mM 2-mercaptoethanol for 30 min at 37 °C, the IT and RIP were diluted and added to the reaction mixture as previously described [57]. Each experiment was carried out in duplicate. The concentration of IT, expressed as RIP content, causing 50% inhibition of [4,5-^{3}H]leucine incorporation (IC$_{50}$) was calculated by linear regression analysis.

The inhibitory activity of free RIP and ITs was also evaluated on Raji cells as described in [37]. Cells (4×10^4/well) were seeded in 96-well microtiter plates in 100 µL of complete medium in the presence of 100 µL of IT added at final concentrations ranging from 0.01 to 100 nM. Control samples were run with RIP alone, mAb alone, or a mixture of unconjugated anti-CD20 or anti-CD22 mAb and RIP. After 96 h, 1 µCi of L-[4,5-^{3}H]leucine was added to each well. After further 6 h, the cells were harvested with an automatic cell harvester (Skatron Instruments, Lier, Norway) onto glass-fiber diskettes. Cell-incorporated radioactivity was determined by a β-counter (Beckman Coulter, Brea, CA, USA) with Ready-Gel scintillation liquid containing 0.7% acetic acid. The IC$_{50}$ values were calculated by regression analysis.

4.5. Cell Viability Assay

Cell viability was evaluated using the colorimetric CellTiter 96® Aqueous One Solution Cell Proliferation assay. Cells (2×10^4/well) were seeded in 96-well microtiter plates in 100 µL of complete medium. After 24 h, the cells were incubated in the absence or in the presence of ITs or the mixture of unconjugated RIP and mAb at the desired concentrations in complete medium. After the indicated times, 20 µL/well of kit solution was added. After 1 h of incubation at 37 °C, the absorbance at 492 nm was measured by a microtiter plate reader Multiskan EX (Thermo Labsystems, Helsinki, Finland). Cell viability was also evaluated on non-target Jurkat cell line incubated with the immunotoxins for 96 h at the same conditions above described.

Experiments with cell death inhibitors and CAT were carried out by pretreating cells with 10 µM Z-VAD, 10 µM Nec-1, or 10 U/mL CAT (the highest concentrations found to be non-toxic to Raji cells in preliminary tests). Experiments with proteasome inhibitors and FLU were carried out by pretreating cells with 0.1 µM MG-132, 1 nM PS-341, or 0.75 µM FLU (the highest concentrations giving a toxicity for Raji cells between 15% and 35% in preliminary tests). All the above reported reagents were added to cells 3 h before the treatment with the ITs and maintained for the entire incubation times.

4.6. Cell Binding Assay

The time required for the antigen-IT complex to be internalized into Raji cells was evaluated by flow cytometry. Cells (5×10^5) were incubated in cytometer tubes in a volume of 200 µL with anti-CD20 and anti-CD22 ITs at a final concentration of 10 nM for 30 min at 0 °C to allow the binding of the IT to the antigen and avoid the internalization of the antigen/IT complex. Afterwards, the samples were kept at 37 °C for different times (0, 15, 20, 30, 60, and 120 min). Negative controls were carried out by incubating cells with complete medium alone. Other controls were not pre-incubated for 30 min at 0 °C with the IT but instead exposed to the IT only for a very short contact time of a few seconds (No inc.). After the incubation time, the cells were washed twice in PBS containing 1% FBS, and incubated again for 30 min on ice in a volume of 50 µL/vial containing 0.1 µL of anti-mouse-FITC (for the anti-CD22 IT) or 0.1 µL of anti-mouse-FITC + 0.1 µL IgG anti-human-FITC (for the anti-CD20 IT). After three washes as above, the samples were fixed with PBS containing 70% ethanol. Cells were analyzed by flow cytometry on a FACSAria BD analyzer using FACSDiva software (Franklin Lakes, NJ, USA). The amount of bound IT was expressed as the percentage of MFI values for each time point with respect to that of samples incubated for 0 min at 37 °C, which was considered as the highest value because in this case binding is not followed by internalization.

4.7. Evaluation of Apoptosis

Apoptotic cell death was assessed using a flow cytometry Annexin V-EGFP/PI detection kit and by a luminescent reagent detecting caspase activity. Before flow cytometry, cells (2×10^5/ 1 mL complete medium) were seeded in a 24-well microtiter plate, and after incubation with the ITs or unconjugated RIP + mAb, the cells were pelleted at $400 \times g$ for 5 min, washed in 2 mL of complete medium, pelleted, and resuspended in 294 μL of binding buffer. Annexin V-EGFP (3 μL) and PI (3 μL) were added. After 10 min incubation in the dark at room temperature, cells were analyzed by flow cytometry. Data were analyzed as above described. The caspase-3/7 activity was assessed by the luminescent Caspase-Glo™3/7 Assay as described in [39]. Briefly, cells (2×10^4/well) were seeded in 96-well microtiter plates in 40 μL of complete medium. Cells were treated with 40 μL of complete medium containing the ITs or unconjugated mAb and saporin-S6 to reach the desired concentration. After incubation for the indicated amounts of time, 80 μL/well of caspase kit reagent was added. The plates were shaken at 420 rpm for 1 min and then incubated for 20 min at room temperature in the dark. The luminescence was acquired (integration time 10 s) by a Fluoroskan Ascent FL (Thermo Labsystems) and values were normalized to cell viability.

The morphological analysis of the treated cells was conducted through phase contrast microscopy, directly in 96-well plate, using an inverted microscope Nikon Eclipse TS100 (Nikon, Melville, NY, USA), at the end of the above described experiments.

4.8. Statistical Analyses

Statistical analyses were conducted using XLSTAT-Pro software, version 6.1.9, 2003 (Addinsoft, Inc., Brooklyn, NY, USA). The results are presented as the means ± S.D. of three different experiments. The data were analyzed using ANOVA/Bonferroni test. The Dunnett's test was used in addiction to ANOVA, when necessary.

Acknowledgments: This work was supported by funds for selected research topics from the Alma Mater Studiorum—University of Bologna and by the Pallotti Legacies for Cancer Research.

Author Contributions: L.P. and A.B. conceived and designed the experiments; L.P., D.M., M.B., A.D. and S.M. performed the experiments; L.P., M.B., D.M., and M.G.B. analyzed the data; L.P., M.B. and D.M. wrote the paper.

Conflicts of Interest: The authors declare no conflict of interest.

References

1. Polito, L.; Mancuso, R.; Mercatelli, D.; Bortolotti, M.; Bolognesi, A. mAbs targeting CD20 and other lymphocyte CD markers in lymphoma treatment. In *Monoclonal Antibodies in Oncology*; Uckun, F.M., Ed.; Future Medicine: London, UK, 2013; pp. 6–19.
2. Polito, L.; Bortolotti, M.; Maiello, S.; Battelli, M.G.; Bolognesi, A. Rituximab and other new anti-CD20 mAbs for Non-Hodgkin's Lymphoma treatment. *Eur. Med. J.* **2014**, *2*, 63–69.
3. Govindan, S.V.; Goldenberg, D.M. New antibody conjugates in cancer therapy. *Sci. World J.* **2010**, *10*, 2070–2089. [CrossRef] [PubMed]
4. Teicher, B.A.; Chari, R.V. Antibody conjugate therapeutics: challenges and potential. *Clin. Cancer Res.* **2011**, *17*, 6389–6397. [CrossRef] [PubMed]
5. Bolognesi, A.; Polito, L. Immunotoxins and other conjugates: Pre-clinical studies. *Mini Rev. Med. Chem.* **2004**, *4*, 563–583. [CrossRef] [PubMed]
6. Gilabert-Oriol, R.; Weng, A.; von Mallinckrodt, B.; Melzig, M.F.; Fuchs, H.; Thakur, M. Immunotoxins constructed with ribosome-inactivating proteins and their enhancers: a lethal cocktail with tumor specific efficacy. *Curr. Pharm. Des.* **2014**, *20*, 6584–6643. [CrossRef] [PubMed]
7. Ferreras, J.M.; Citores, L.; Iglesias, R.; Jiménez, P.; Girbés, T. Use of ribosome-inactivating proteins from Sambucus for the construction of immunotoxins and conjugates for cancer therapy. *Toxins (Basel)* **2011**, *3*, 420–441. [CrossRef] [PubMed]
8. Frankel, A.E.; Kreitman, R.J.; Sausville, E.A. Targeted toxins. *Clin. Cancer Res.* **2000**, *6*, 326–334. [PubMed]
9. Polito, L.; Bortolotti, M.; Pedrazzi, M.; Bolognesi, A. Immunotoxins and other conjugates containing saporin-S6 for cancer therapy. *Toxins (Basel)* **2011**, *3*, 697–720. [CrossRef] [PubMed]
10. Bolognesi, A.; Bortolotti, M.; Maiello, S.; Battelli, M.G.; Polito, L. Ribosome-Inactivating Proteins from Plants: A Historical Overview. *Molecules* **2016**, *21*, E1627. [CrossRef] [PubMed]
11. Polito, L.; Djemil, A.; Bortolotti, M. Plant Toxin-Based Immunotoxins for Cancer Therapy: A Short Overview. *Biomedicines* **2016**, *4*, 12. [CrossRef] [PubMed]

12. Di Maro, A.; Citores, L.; Russo, R.; Iglesias, R.; Ferreras, J.M. Sequence comparison and phylogenetic analysis by the Maximum Likelihood method of ribosome-inactivating proteins from angiosperms. *Plant Mol. Biol.* **2014**, *85*, 575–588. [CrossRef] [PubMed]

13. Polito, L.; Bortolotti, M.; Maiello, S.; Battelli, M.G.; Bolognesi, A. Plants Producing Ribosome-Inactivating Proteins in Traditional Medicine. *Molecules* **2016**, *21*, E1560. [CrossRef] [PubMed]

14. Stirpe, F.; Battelli, M.G. Ribosome-inactivating proteins: progress and problems. *Cell. Mol. Life Sci.* **2006**, *63*, 1850–1866. [CrossRef] [PubMed]

15. Bolognesi, A.; Tazzari, P.L.; Olivieri, F.; Polito, L.; Lemoli, R.; Terenzi, A.; Pasqualucci, L.; Falini, B.; Stirpe, F. Evaluation of immunotoxins containing single-chain ribosome-inactivating proteins and an anti-CD22 monoclonal antibody (OM124): in vitro and in vivo studies. *Br. J. Haematol.* **1998**, *101*, 179–188. [CrossRef] [PubMed]

16. Polito, L.; Bolognesi, A.; Tazzari, P.L.; Farini, V.; Lubelli, C.; Zinzani, P.L.; Ricci, F.; Stirpe, F. The conjugate Rituximab/saporin-S6 completely inhibits clonogenic growth of CD20-expressing cells and produces a synergistic toxic effect with Fludarabine. *Leukemia* **2004**, *18*, 1215–1222. [CrossRef] [PubMed]

17. Bolognesi, A.; Polito, L.; Farini, V.; Bortolotti, M.; Tazzari, P.L.; Ratta, M.; Ravaioli, A.; Horenstein, A.L.; Stirpe, F.; Battelli, M.G.; et al. CD38 as a target of IB4 mAb carrying saporin-S6: design of an immunotoxin for ex vivo depletion of hematological CD38+ neoplasia. *J. Biol. Regul. Homeost. Agents* **2005**, *19*, 145–152. [PubMed]

18. Flavell, D.J.; Warnes, S.L.; Bryson, C.J.; Field, S.A.; Noss, A.L.; Packham, G.; Flavell, S.U. The anti-CD20 antibody rituximab augments the immunospecific therapeutic effectiveness of an anti-CD19 immunotoxin directed against human B-cell lymphoma. *Br. J. Haematol.* **2006**, *134*, 157–170. [CrossRef] [PubMed]

19. Polito, L.; Bortolotti, M.; Mercatelli, D.; Battelli, M.G.; Bolognesi, A. Saporin-S6: A useful tool in cancer therapy. *Toxins (Basel)* **2013**, *5*, 1698–1722. [CrossRef] [PubMed]

20. Vooijs, W.C.; Otten, H.G.; van Vliet, M.; van Dijk, A.J.; de Weger, R.A.; de Boer, M.; Bohlen, H.; Bolognesi, A.; Polito, L.; de Gast, G.C. B7-1 (CD80) as target for immunotoxin therapy for Hodgkin's disease. *Br. J. Cancer* **1997**, *76*, 1163–1169. [CrossRef] [PubMed]

21. Falini, B.A.; Bolognesi, A.; Flenghi, L.; Tazzari, P.L.; Broe, M.K.; Stein, H.; Dürkop, H.; Aversa, F.; Corneli, P.; Pizzolo, G.; et al. Response of refractory Hodgkin's disease to monoclonal anti-CD30 immunotoxin. *Lancet* **1992**, *339*, 1195–1196. [CrossRef]

22. Barbieri, L.; Valbonesi, P.; Bonora, E.; Gorini, P.; Bolognesi, A.; Stirpe, F. Polynucleotide:adenosine glycosidase activity of ribosome-inactivating proteins: effect on DNA, RNA and poly(A). *Nucleic Acids Res.* **1997**, *25*, 518–522. [CrossRef] [PubMed]

23. Barbieri, L.; Bolognesi, A.; Valbonesi, P.; Polito, L.; Olivieri, F.; Stirpe, F. Polynucleotide: adenosine glycosidase activity of immunotoxins containing ribosome-inactivating proteins. *J. Drug Target.* **2000**, *8*, 281–288. [CrossRef] [PubMed]

24. Battelli, M.G.; Barbieri, L.; Bolognesi, A.; Buonamici, L.; Valbonesi, P.; Polito, L.; van Damme, E.J.; Peumans, W.J.; Stirpe, F. Ribosome-inactivating lectins with polynucleotide:adenosine glycosidase activity. *FEBS Lett.* **1997**, *408*, 355–359. [CrossRef]

25. Polito, L.; Bortolotti, M.; Farini, V.; Battelli, M.G.; Barbieri, L.; Bolognesi, A. Saporin induces multiple death pathways in lymphoma cells with different intensity and timing as compared to ricin. *Int. J. Biochem. Cell Biol.* **2009**, *41*, 1055–1061. [CrossRef] [PubMed]

26. Vago, R.; Marsden, C.J.; Lord, J.M.; Ippoliti, R.; Flavell, D.J.; Flavell, S.U.; Ceriotti, A.; Fabbrini, M.S. Saporin and ricin A chain follow different intracellular routes to enter the cytosol of intoxicated cells. *FEBS J.* **2005**, *272*, 4983–4995. [CrossRef] [PubMed]

27. Bolognesi, A.; Polito, L.; Scicchitano, V.; Orrico, C.; Pasquinelli, G.; Musiani, S.; Santi, S.; Riccio, M.; Bortolotti, M.; Battelli, M.G. Endocytosis and intracellular localisation of the type 1 ribosome-inactivating protein saporin-S6. *J. Biol. Regul. Homeost. Agents* **2012**, *26*, 97–109. [PubMed]

28. Horrix, C.; Raviv, Z.; Flescher, E.; Voss, C.; Berger, M.R. Plant ribosome-inactivating proteins type II induce the unfolded protein response in human cancer cells. *Cell. Mol. Life Sci.* **2011**, *68*, 1269–1281. [CrossRef] [PubMed]

29. Gilabert-Oriol, R.; Thakur, M.; Haussmann, K.; Niesler, N.; Bhargava, C.; Görick, C.; Fuchs, H.; Weng, A. Saponins from *Saponaria officinalis* L. Augment the Efficacy of a Rituximab-Immunotoxin. *Planta Med.* **2016**, *82*, 1525–1531. [CrossRef] [PubMed]

30. Sullivan-Chang, L.; O'Donnell, R.T.; Tuscano, J.M. Targeting CD22 in B-cell malignancies: Current status and clinical outlook. *BioDrugs* **2013**, *27*, 293–304. [CrossRef] [PubMed]
31. Weber, T.; Mavratzas, A.; Kiesgen, S.; Haase, S.; Bötticher, B.; Exner, E.; Mier, W.; Grosse-Hovest, L.; Jäger, D.; Arndt, M.A.; et al. A Humanized Anti-CD22-Onconase Antibody-Drug Conjugate Mediates Highly Potent Destruction of Targeted Tumor Cells. *J. Immunol. Res.* **2015**, *2015*, 561814. [CrossRef] [PubMed]
32. Li, Z.H.; Zhang, Q.; Wang, H.B.; Zhang, Y.N.; Ding, D.; Pan, L.Q.; Miao, D.; Xu, S.; Zhang, C.; Luo, P.H.; et al. Preclinical studies of targeted therapies for CD20-positive B lymphoid malignancies by Ofatumumab conjugated with auristatin. *Investig. New Drugs* **2014**, *32*, 75–86. [CrossRef] [PubMed]
33. Polson, A.G.; Calemine-Fenaux, J.; Chan, P.; Chang, W.; Christensen, E.; Clark, S.; de Sauvage, F.J.; Eaton, D.; Elkins, K.; Elliott, J.M.; et al. Antibody-drug conjugates for the treatment of non-Hodgkin's lymphoma: target and linker-drug selection. *Cancer Res.* **2009**, *69*, 2358–2364. [CrossRef] [PubMed]
34. Countouriotis, A.; Moore, T.B.; Sakamoto, K.M. Cell surface antigen and molecular targeting in the treatment of hematologic malignancies. *Stem Cells* **2002**, *20*, 215–229. [CrossRef] [PubMed]
35. Bortolotti, M.; Bolognesi, A.; Battelli, M.G.; Polito, L. High in vitro anti-tumor efficacy of dimeric Rituximab/Saporin-S6 immunotoxin. *Toxins (Basel)* **2016**, *8*, E192. [CrossRef] [PubMed]
36. Sieber, T.; Schoeler, D.; Ringel, F.; Pascu, M.; Schriever, F. Selective internalization of monoclonal antibodies by B-cell chronic lymphocytic leukaemia cells. *Br. J. Haematol.* **2003**, *121*, 458–461. [CrossRef] [PubMed]
37. Bolognesi, A.; Tazzari, P.L.; Olivieri, F.; Polito, L.; Falini, B.; Stirpe, F. Induction of apoptosis by ribosome-inactivating proteins and related immuntoxins. *Int. J. Cancer* **1996**, *68*, 349–355. [CrossRef]
38. Bolognesi, A.; Polito, L.; Tazzari, P.L.; Lemoli, R.M.; Lubelli, C.; Fogli, M.; de Boer, M.; Stirpe, F. In vitro anti-tumor activity of anti-CD80 and anti-CD86 immunotoxins containing type 1 ribosome-inactivating proteins. *Br. J. Haematol.* **2000**, *110*, 351–361. [CrossRef] [PubMed]
39. Polito, L.; Bortolotti, M.; Farini, V.; Pedrazzi, M.; Tazzari, P.L.; Bolognesi, A. ATG-saporin-S6 immunotoxin: A new potent and selective drug to eliminate activated lymphocytes and lymphoma cells. *Br. J. Haematol.* **2009**, *147*, 710–718. [CrossRef] [PubMed]
40. Polito, L.; Bortolotti, M.; Pedrazzi, M.; Mercatelli, D.; Battelli, M.G.; Bolognesi, A. Apoptosis and necroptosis induced by stenodactylin in neuroblastoma cells can be completely prevented through caspase inhibition plus catalase or necrostatin-1. *Phytomedicine* **2016**, *23*, 32–41. [CrossRef] [PubMed]
41. Vandenabeele, P.; Galluzzi, L.; Vanden Berghe, T.; Kroemer, G. Molecular mechanisms of necroptosis: An ordered cellular explosion. *Nat. Rev. Mol. Cell. Biol.* **2010**, *11*, 700–714. [CrossRef] [PubMed]
42. Daniels-Wells, T.R.; Helguera, G.; Rodríguez, J.A.; Leoh, L.S.; Erb, M.A.; Diamante, G.; Casero, D.; Pellegrini, M.; Martínez-Maza, O.; Penichet, M.L. Insights into the mechanism of cell death induced by saporin delivered into cancer cells by an antibody fusion protein targeting the transferrin receptor 1. *Toxicol. In Vitro* **2013**, *27*, 220–231. [CrossRef] [PubMed]
43. Fengling, M.; Fenju, L.; Wanxin, W.; Lijia, Z.; Jiandong, T.; Zu, W.; Xin, Y.; Qingxiang, G. Rituximab sensitizes a Burkitt lymphoma cell line to cell killing byX-irradiation. *Radiat. Environ. Biophys.* **2009**, *48*, 371–378. [CrossRef] [PubMed]
44. Honeychurch, J.; Alduaij, W.; Azizyan, M.; Cheadle, E.J.; Pelicano, H.; Ivanov, A.; Huang, P.; Cragg, M.S.; Illidge, T.M. Antibody-induced nonapoptotic cell death in human lymphoma and leukemia cells is mediated through a novel reactive oxygen species-dependent pathway. *Blood* **2012**, *119*, 3523–3533. [CrossRef] [PubMed]
45. Farooqi, A.A.; Li, K.T.; Fayyaz, S.; Chang, Y.T.; Ismail, M.; Liaw, C.C.; Yuan, S.S.; Tang, J.Y.; Chang, H.W. Anticancer drugs for the modulation of endoplasmic reticulum stress and oxidative stress. *Tumour Biol.* **2015**, *36*, 5743–5752. [CrossRef] [PubMed]
46. Czuczman, M.S.; Olejniczak, S.; Gowda, A.; Kotowski, A.; Binder, A.; Kaur, H.; Knight, J.; Starostik, P.; Deans, J.; Hernandez-Ilizaliturri, F.J. Acquirement of rituximab resistance in lymphoma cell lines is associated with both global CD20 gene and protein down-regulation regulated at the pretranscriptional and posttranscriptional levels. *Clin. Cancer Res.* **2008**, *14*, 1561–1570. [CrossRef] [PubMed]
47. Tsai, P.C.; Hernandez-Ilizaliturri, F.J.; Bangia, N.; Olejniczak, S.H.; Czuczman, M.S. Regulation of CD20 in rituximab-resistant cell lines and B-cell non-Hodgkin lymphoma. *Clin. Cancer Res.* **2012**, *18*, 1039–1050. [CrossRef] [PubMed]

48. Battelli, M.G.; Scicchitano, V.; Polito, L.; Farini, V.; Barbieri, L.; Bolognesi, A. Binding and intracellular routing of the plant-toxic lectins, lanceolin and stenodactylin. *Biochim. Biophys. Acta* **2010**, *1800*, 1276–1282. [CrossRef] [PubMed]

49. Martin, S.M.; Churchill, E.; McKnight, H.; Mahaffey, C.M.; Ma, Y.; O'Donnell, R.T.; Tuscano, J.M. The HB22.7 Anti-CD22 monoclonal antibody enhances bortezomib- mediated lymphomacidal activity in a sequence dependent manner. *J. Hematol. Oncol.* **2011**, *4*, 49. [CrossRef] [PubMed]

50. Barbieri, L.; Brigotti, M.; Perocco, P.; Carnicelli, D.; Ciani, M.; Mercatali, L.; Stirpe, F. Ribosome-inactivating proteins depurinate poly(ADP-ribosyl)ated poly(ADP-ribose) polymerase and have transforming activity for 3T3 fibroblasts. *FEBS Lett.* **2003**, *538*, 178–182. [CrossRef]

51. Thrush, G.R.; Lark, L.R.; Clinchy, B.C.; Vitetta, E.S. Immunotoxins: an update. *Annu. Rev. Immunol.* **1996**, *14*, 49–71. [CrossRef] [PubMed]

52. Madhumathi, J.; Verma, R.S. Therapeutic targets and recent advances in protein immunotoxins. *Curr. Opin. Microbiol.* **2012**, *15*, 300–309. [CrossRef] [PubMed]

53. Pizzo, E.; Di Maro, A. A new age for biomedical applications of Ribosome Inactivating Proteins (RIPs): From bioconjugate to nanoconstructs. *J. Biomed. Sci.* **2016**, *23*, 54. [CrossRef] [PubMed]

54. FitzGerald, D.J.; Wayne, A.S.; Kreitman, R.J.; Pastan, I. Treatment of hematologic malignancies with immunotoxins and antibody-drug conjugates. *Cancer Res.* **2011**, *71*, 6300–6309. [CrossRef] [PubMed]

55. Dosio, F.; Stella, B.; Cerioni, S.; Gastaldi, D.; Arpicco, S. Advances in anticancer antibody-drug conjugates and immunotoxins. *Recent Pat. Anticancer Drug Discov.* **2014**, *9*, 35–65. [CrossRef] [PubMed]

56. Barbieri, L.; Bolognesi, A.; Stirpe, F. Purification and conjugation of type 1 ribosome-inactivating proteins. In *Immunotoxins: Methods and Protocols*; Hall, W.A., Ed.; Humana Press: New York City, NY, USA, 2001; Volume 166, pp. 71–85.

57. Polito, L.; Bortolotti, M.; Mercatelli, D.; Mancuso, R.; Baruzzi, G.; Faedi, W.; Bolognesi, A. Protein synthesis inhibition activity by strawberry tissue protein extracts during plant life cycle and under biotic and abiotic stresses. *Int. J. Mol. Sci.* **2013**, *14*, 15532–15545. [CrossRef] [PubMed]

Article

Improvement of the Pharmacological Properties of Maize RIP by Cysteine-Specific PEGylation

Ka-Yee Au [1], Wei-Wei Shi [1], Shuai Qian [2], Zhong Zuo [2] and Pang-Chui Shaw [1,*

[1] Centre for Protein Science and Crystallography, School of Life Sciences, The Chinese University of Hong Kong, Shatin, N.T., Hong Kong, China; aaky73@yahoo.com.hk (K.-Y.A.); Shiww@cuhk.edu.hk (W.-W.S.)
[2] School of Pharmacy, Faculty of Medicine, The Chinese University of Hong Kong, Shatin, N.T., Hong Kong, China; silence_qs@163.com (S.Q.); joanzuo@cuhk.edu.hk (Z.Z.)
* Correspondence: pcshaw@cuhk.edu.hk; Tel.: +852-3943-1363; Fax: +852-2603-7246

Academic Editors: Julien Barbier and Daniel Gillet
Received: 7 September 2016; Accepted: 11 October 2016; Published: 17 October 2016

Abstract: To improve the pharmacological properties of maize ribosome-inactivating protein (maize RIP) for targeting HIV-infected cells, the previously engineered TAT-fused active form of maize RIP (MOD) was further engineered for cysteine-directed PEGylation. In this work, two potential antigenic sites, namely Lys-78 and Lys-264, were identified. They were mutated to cysteine residue and conjugated with PEG_{5k} or PEG_{20k}. The resultant PEG derivatives of MOD variants were examined for ribosome-inactivating activity, circulating half-life and immunogenicity. Our results showed that MOD-PEG conjugates had two- to five-fold lower biological activity compared to the wild-type. Mutation of the two sites respectively did not decrease the anti-MOD IgG and IgE level in mice, but the conjugation of PEG did dramatically reduce the antigenicity. Furthermore, pharmacokinetics studies demonstrated that attachment of PEG_{20k} prolonged the plasma half-life by five-fold for MOD-K78C and 17-fold for MOD-K264C, respectively. The site-specific mutation together with PEGylation therefore generated MOD derivatives with improved pharmacological properties.

Keywords: MOD; PEGylation; antigenicity; pharmacokinetics study; circulation half-life; antibody induction

---

1. Introduction

Ribosome-inactivating proteins (RIPs) are RNA N-glycosidases that remove a specific adenine at the α-sarcin/ricin loop (SRL) of large ribosomal RNA (rRNA) and in turn cease protein synthesis [1,2].

RIPs are categorized into three types according to their structural organization. Type 1, represented by trichosanthin (TCS) [3] and pokeweed antiviral protein (PAP) [4], is a monomeric protein with full SRL depurinating activity. Type 2, such as ricin and abrin [5], is heterodimeric with an enzymatically active A chain linked to a lectin-like B chain by a disulphide bond. Maize RIP is a type 3 RIP, which is first synthesized with an extra 25-amino-acid internal segment and requires proteolytic removal upon which the precursor is activated to resume N-glycosidase function [6,7]. We have recently fused the HIV-1 TAT transduction peptide to the N-termini of maize RIP and exploited its unique activity regulatory feature of maturation by introducing HIV-1 protease recognition sequences to the internal inactivation region to sensitize maize RIP towards HIV-infected cells, in which the activation of maize RIP was triggered by HIV-1 protease [8]. Our work has demonstrated the potential inhibitory effect of maize RIP towards immunodeficiency viruses, both in HIV-infected cells [8] and in Chinese rhesus macaques [9], suggesting its potential as a novel anti-HIV agent.

The therapeutic applications of RIPs, however, have long been challenged for two reasons. First, since most RIPs are naturally produced in plants, their uses in animals are prone to trigger adverse immune responses. AIDS patients receiving PAP-derived immunotoxin showed undesirable antibody production targeting the cytotoxin moiety [10] and patients infused with trichosanthin were found to develop allergic symptoms with TCS-specific antibodies detected in vivo, and repeated administration of TCS might cause fetal anaphylaxis [11]. Second, proteins of low molecular weight are susceptible to rapid elimination from the circulation through renal filtration or degradation [12]. Typically, RIPs are of sizes ranging from 27 to 32 kDa and the 27 kDa trichosanthin was previously reported to have a plasma residence time between 8.4 and 12.7 min [13]. The short circulation retention necessitates frequent infusion to maintain an effective concentration for treatment and decreases the medical value of RIPs.

Various strategies have been adopted to improve the aforementioned shortfalls and PEGylation, and the covalent coupling with polymer polyethylene glycol (PEG) is the most common. With high hydrophilicity and low immunogenicity, PEG attachment increases the size of biopharmaceuticals for a long circulating life as well as confers a shielding effect at antigenic sites for low immune responses [14,15]. PEG-adenosine deaminase (ADA) is the first FDA-approved drug modified by PEGylation and its clinical application for treating ADA deficiency is positive and well tolerated [16]. We have coupled Icarapin, a bee venom protein, with PEG in a site-specific manner to diminish antibody generation [17], and also mono-PEGylated trichosanthin to slow down the circulation clearance and decrease the immunogenicity [18,19].

In this study, we aim to enhance the therapeutic value of TAT-fused maize RIP through conjugation with PEG. Our results demonstrated the resultant PEGylated MOD, especially PEG_{20k}-conjugated MOD shows a prolonged circulating half-life in plasma and decreased immunogenicity, while only lost half of the ribosome-inactivating activity. Our study will benefit therapeutic applications of MOD to combat HIV/AIDS and the development of effective RIP-related biomedicine as therapeutics.

2. Results

2.1. Selection of PEGylation Sites

This study aimed to improve the therapeutic potential of maize RIP by cysteine-directed PEGylation and the conjugation sites were chosen based on three criteria: (i) the site was of high antigenic index so that PEG attachment could reduce immunogenicity; (ii) it located in protrusion regions of the protein surface to enable efficient coupling; and (iii) it was away from the catalytic center and interaction site so that *N*-glycosidase activity was least impaired. A total of nine antibody epitopes were predicted using ElliPro serve [20]. Among these epitopes, two sites, namely Lys-78 and Lys-264, were selected for modification in accordance with the above criteria (Figure 1).

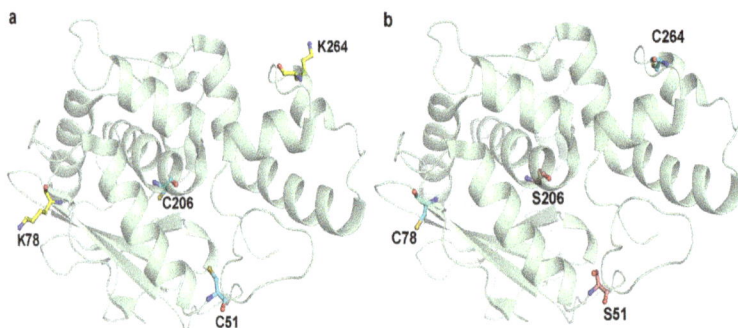

Figure 1. MOD (**a**) before and (**b**) after mutations. Cysteine, lysine and serine residues were highlighted with cyan, yellow and pink colors, respectively. C51 and C206 were mutated to serine and the selected PEGylated sites, K78 and K264, were mutated to cysteine for PEGylation.

2.2. PEGylated Maize RIP Variants only Lost Half of the Ribosome-Inactivating Activity

MOD-PEG conjugates were evaluated for N-glycosidase activity on the human T lymphocyte C8166 cell line using the quantitative PCR method. The active form His-TAT-MOD exerted over a five-fold stronger SRL-depurinating effect than the precursor His-TAT-Pro (containing the internal inactivation loop) (Figure 2). Compared to the wild-type, non-PEGylated mutants had the activity decreased for approximated 50% and PEG conjugates had the bioactivity further reduced, with MOD-PEG$_{20k}$ displayed a better depurinating effect than MOD-PEG$_{5k}$.

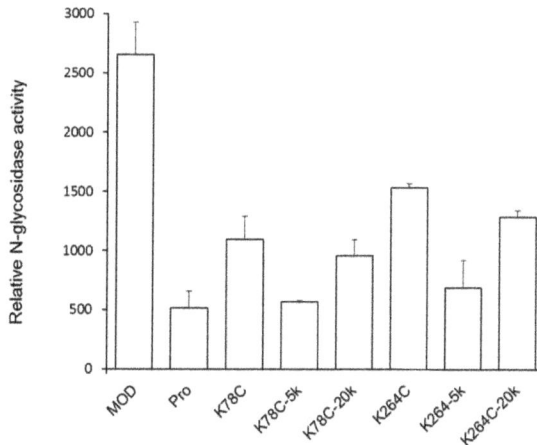

Figure 2. Relative N-glycosidase activity of maize RIP variants, mutants and PEGylated variants in human T lymphocyte (C8166). All the proteins contain the His-tag and TAT sequence at the N-termini. C8166 cells in log-phase growth were seeded at density of 1×10^6 cell/well on a six-well plate and treated with 5 µM protein samples. After incubation at 37 °C for 6 h, cells were harvested for RNA isolation, cDNA synthesis and N-glycosidase activity was determined by qPCR using primers that target the modified site as previously described [8]. The relative N-glycosidase activity was calculated as the relative amount of depurinated rRNA of protein-treated cells against that of untreated cells. Data are presented as mean ± SD ($n = 3$).

2.3. PEG$_{20k}$-Conjugated Maize RIP Variants Significantly Prolonged Circulating Half-Life in Rats

The pharmacokinetics of PEGylated maize RIP variants were examined in rats administered with a single intravenous injection of protein samples. To evaluate the plasma concentration of the variants, the corresponding antigen level was measured by ELISA (Figure 3). In both MOD-K78C and MOD-K264C, the PEG$_{20k}$ conjugates could be detected 4 h after dosing whereas all other variants had their plasma levels below the detection limit within 1 h and could not have the concentration estimated. Table 1 lists the pharmacokinetic parameters calculated using WinNonlin software (version v3, Certara, Princeton, NJ, USA). As shown, MOD-PEG$_{5k}$ conjugates had comparable plasma half-lives as the corresponding unmodified variants whereas coupling with PEG$_{20k}$ extended the plasma half-life by five-fold for MOD-K78C and 17-fold for MOD-K264C, respectively. The more prominent prolonging effects of MOD-PEG$_{20k}$ conjugates on in vivo half-life are likely attributed to the larger size of the attached PEG which makes the variants more resistant to degradation or clearance.

(a)

(b)

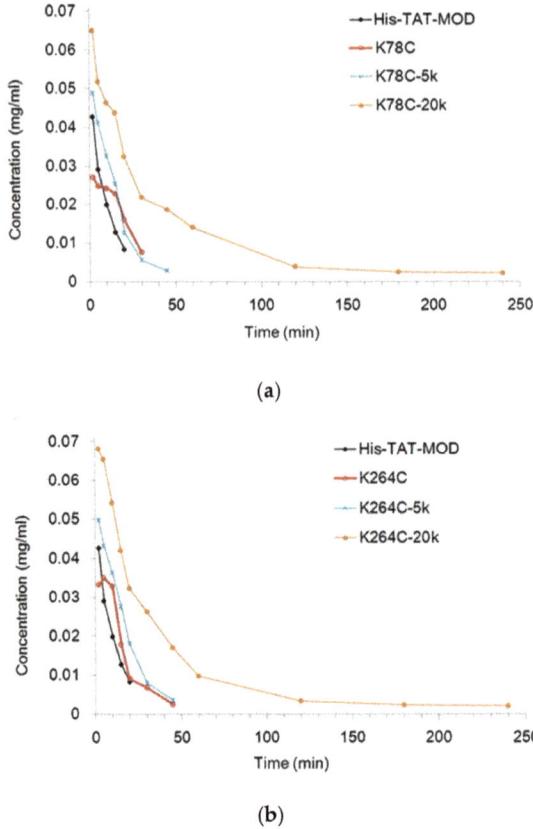

Figure 3. Plasma concentration–time profiles of maize RIP variants in rats. MOD-PEG$_{20k}$ conjugates were detected 4 h after dosing whereas the non-PEGylated variants and MOD-PEG$_{5k}$ conjugates had their concentrations below the detection limit within 1 h. (**a**) Plasma concentration–time profiles of K78C mutant and its PEGylated variants. (**b**) Plasma concentration–time profiles of K264C mutant and its PEGylated variants.

Table 1. Statistic and pharmacokinetic parameters of MOD and variants.

Treatment	AUC$_{0-t}$ (mg min/mL)	T$_{1/2}$ (min)	Fold Increased Compared to Non-PEGylated Form
His-TAT-MOD	0.46	8.0	-
MOD-K78C	0.59	9.4	-
MOD-K78C-5k	0.81	9.8	1.04
MOD-K78C-20k	2.65	47.6	5.06
MOD-K264C	0.68	10.1	-
MOD-K264C-5k	0.93	10.1	1
MOD-K264C-20k	2.57	171.1	16.94

All parameters were calculated by the WinNonlin software. AUC$_{0-t}$: the area under the curve from the time of dosing to the time of the last observation; T$_{1/2}$ is the time required for a quantity to reduce to half its initial value.

2.4. PEG$_{20k}$-Modified Maize RIP Variants Elicit Weak Immune Responses in Mice

Mice immunized with maize RIP variants were detected for IgE/IgG antibody generation to assess the effect of PEGylation on immunogenicity. By ELISA assays, no signal of maize RIP–specific

IgE was found in treated animals whereas IgG was detected and shown in Figure 4. Both variants, K78C and K264C, showed subtle differences in IgG levels compared to the wild-type MOD. However, the variants modified with PEG_{20k} conjugates were shown to trigger lower IgG levels compared to the wild-type MOD and MOD variants. The K264C-PEG_{20k} conjugate gave the most significant decrease in IgG level.

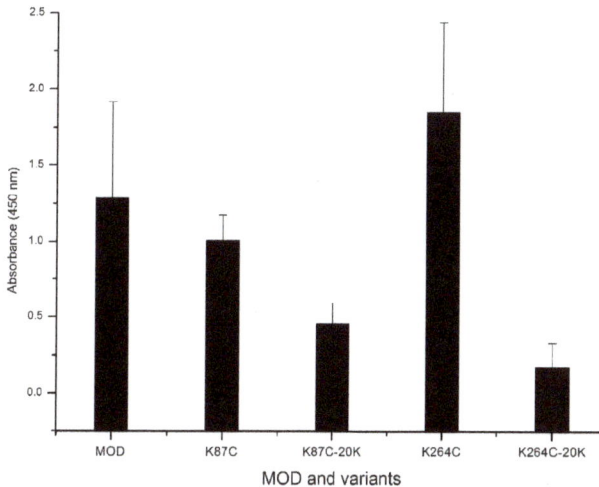

Figure 4. ELISA detection of specific IgG levels in mice serum. C57BL/6N inbred mice six to eight weeks old were randomly assigned into groups, and immunized with wild-type MOD (His-TAT-MOD), MOD mutants and PEGylated variants. ELISA assays were carried out at a 500-fold dilution of serum samples and each sample was repeated in triplicate.

3. Discussion

With high cytotoxicity and potent antiviral and anti-tumor effects, RIPs have been investigated as standalone molecules or as immunotoxins [3,11,21,22]. Even though the anti-HIV mechanism is still unclear, it has reported that several classical type I and II RIPs, such as ricin A chain [23], GAP31 [24], DAP30 [25], MAP30 [26], pokeweed antiviral protein (PAP) [10] and TCS [27], possess anti-HIV activity by inhibiting viral replication in vitro and in vivo. Our previous HIV-inhibitory activity studies showed the active maize RIP can suppress viral replication in acutely HIV-1 infected C8166 cells [8]. Recently, we assessed the anti-HIV effect of maize RIP on simian immunodeficiency virus (SIV)- and chimeric simian/human immunodeficiency viruses (SHIV)-infected macaque peripheral blood mononuclear cells (PBMC) and tested the antiviral activity of maize RIP in SHIV 89.6–infected Chinese rhesus macaque model [9]. We showed that the active recombinant maize RIP, His-TAT-MOD, can enhance PBMC cell survival and reduce the viral load in SHIV-infected macaque cells, suggesting His-TAT-MOD is a promising anti-HIV agent.

However, RIP-derived products have shortcomings such as a short plasma half-life and adverse immunogenic responses. Here, Lys-78 and Lys-264 were converted to cysteine and coupled with PEG_{5k} or PEG_{20k} to improve the above pharmacological properties. The resultant conjugates were shown to have enhanced blood circulation (Figure 3 and Table 1) and lower immunogenicity (Figure 4). We also found that MOD-PEG_{5k} conjugates showed a limited prolonging effect, whereas MOD-K78C- and MOD-264C-PEG_{20k} conjugates had the half-lives increased for five- and 17-fold, respectively. A large-sized molecule has reduced plasma clearance via renal filtration or proteolytic degradation [12]. The coupling of highly hydrophilic PEG may also facilitate the dissolution of conjugates within circulation.

RIPs are toxins that depurinate a specific adenine nucleotide at the 28S rRNA of the large ribosomal subunit, resulting in the ceasing of protein synthesis. The RIP activities of the PEGylated products were assessed by the *N*-glycosidase activity on the T lymphocyte C8166 cell line. Despite the PEGylation sites, K78 and K264, employed in this study being remote from the highly conserved catalytic center and putative interaction site with the ribosome, it was found that the non-PEGylated variants and MOD-PEG conjugates displayed decreased RIP activity compared to wild-type MOD (Figure 2). A possible reason may be that such a modification has obstructed the access of the RIP to the ribosome.

Though PEG conjugation reduces biological activity, this loss is often compensated by the prolonged plasma half-life and in turn gives enhanced in vivo efficacy [19,28]. In this case, compared to the unconjugated MOD, the pharmacological effectiveness of MOD-K264C-PEG$_{20k}$ showed an obvious advantage to prolong the half-life to five-fold while only half the *N*-glycosidase activity was lost.

Modification with inert polymers covers the protein surface to mediate the shielding effect over antibody epitopes and PEGylation has been employed to reduce the immunogenicity of several RIPs, as exemplified by the type I trichosanthin [19] and MAP30 [29]. In this study immunogenicity of maize RIP and its PEG conjugates was examined in C57BL/6N inbred mice and no detectable IgE production was found upon immunization. On the other hand, IgG levels were determined and the results showed that in both K78 and K264 cases, the lysine-to-cysteine mutation dose did not decrease the IgG response of the MOD. However, PEG$_{20k}$ coupling led to a substantial decline of the IgG level (Figure 4), suggesting K78 and K276 are located in two important antigenic epitopes of MOD, and thus PEG modification of these two residues can effectively block the immunogenicity of MOD. Figure 5 summarizes the locations of all possible antibody epitopes in MOD predicted by ElliPro and PEG attachment at K78C and K264C could help mask the protruding antigenic regions from antibody recognition, thereby causing less IgG response. In summary, with a series of biochemical and in vivo assays, we have shown that coupling with PEG helps enhance the plasma circulating life and alleviate the immune response elicited by MOD, suggesting the resultant PEGylated variants are of improved therapeutic value. Our study also benefits the rational engineering design and development of effective RIP-related biomedicine.

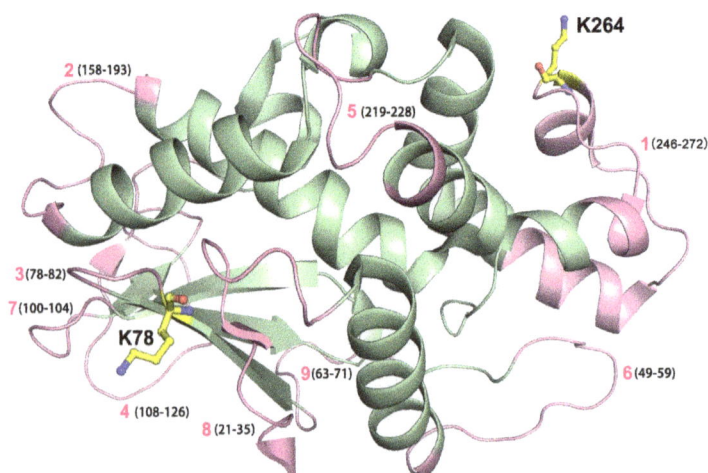

Figure 5. Predicted antibody epitopes in active form of maize RIP, MOD. Potential antibody epitopes in maize RIP predicted by the ElliPro program. These epitopes are labeled individually and the amino acid numbers are marked in brackets. The MOD molecule (PDB: 2PQI) was shown in pale green, and predicted nine epitopes are labeled in pink color. Two lysine residues, as labeled yellow sticks K78 and K264 located in epitopes 1 and 3 (pink), were mutated respectively in this study to attain cysteine-specific PEGylation.

4. Materials and Methods

4.1. Computer Modeling for PEGylation Sites

Conjugation with PEG took place specifically at cysteine residue where the polymer was covalently linked to the protein through thioether bond formation. The ElliPro method was employed to predict B-cell epitopes in maize RIP for PEGylation [30]. Computational analysis identified two lysines of significant antigenic index, namely K78 and K264, in the amino acid sequence of maize RIP which were then selected for PEG-coupling.

4.2. Cloning, Expression and Purification of Maize RIP Variants

To attain mono-PEGylation, the native cysteine residues in maize RIP, C51 and C206, were substituted with serine, followed by the lysine-to-cysteine mutations at selected modification sites. Two maize RIP variants, namely MOD-K78C and MOD-K264C, were constructed using Phusion DNA polymerase (Finnzymes) and primers containing the desired modifications (Table 2) with the recombinant plasmid His-TAT-MOD-pET3a as template. DNA products were cloned into expression vector pET3a and sequenced to confirm correct mutagenesis. All variant proteins also contain His-tag and TAT sequence at the N-termini, and they were expressed, purified, and stored in the same manner as previously described [31].

Table 2. List of primers and their corresponding sequences used for constructs. The mutated amino acid is underlined.

Primer	Sequence (5'-3')
MOD-C15S-F	GTGATCAAACAC<u>TCT</u>ACCGACC
MOD-C15S-R	GGTCGGT<u>AGA</u>GTGTTTGATCAC
MOD-C206S-F	GTGGTCATGGTG<u>TCT</u>GAGGGGCTG
MOD-C206S-R	CAGCCCCTC<u>AGA</u>CACCATGACCAC
MOD-K78C-F	ACAGAGCTC<u>TGT</u>ACTAGGACC
MOD-K78C-R	GGTCCTAGT<u>ACA</u>GAGCTCTGT
MOD-K264C-F	GACATGCAG<u>TGT</u>CTTGGCATC
MOD-K264C-R	GATGCCAAG<u>ACA</u>CTGCATGTC

4.3. Preparation of PEGylated Variants

Cysteine-specific PEGylation was carried out by incubating the maize RIP variants with methoxy PEG (5 or 20 kDa)-maleimide reagents (Nanocs) in 20 mM phosphate buffer, 10 mM EDTA, pH 6.5 overnight at 4 °C [12].

MOD-PEG$_{5k}$ reaction mixture with buffer exchanged to 20 mM phosphate buffer, 1.5 M $(NH_4)_2SO_4$ was loaded to a 5 mL HiTrap Phenyl High Performance column (GE Healthcare Biosciences, Pittsburgh, PA, USA) for hydrophobic interaction chromatography. Elution was carried out with 1.05–0.15 M $(NH_4)_2SO_4$ linear gradient and fractions with MOD-PEG$_{5k}$ conjugate were pooled and concentrated.

MOD-PEG$_{20k}$ conjugate was purified stepwise using ion-exchange and size-exclusion chromatography. MOD-PEG$_{20k}$ reaction mixture was first loaded to a 5 mL HiTrap DEAE Fast Flow column (GE Healthcare) equilibrated with 20 mM NaOAc, pH 6.5 and eluted with a 0.5–1.0 M NaCl linear gradient. Fractions of higher purity were pooled and loaded to Superdex75 (GE Healthcare Biosciences, Pittsburgh, PA, USA) equilibrated with 20 mM phosphate buffer, 0.5 M NaCl, pH 6.5 for further isolation. Fractions with pure MOD-PEG$_{20k}$ were identified by SDS-PAGE.

All variants were buffer-exchanged to 20 mM phosphate buffer, 200 mM NaCl, 5% glycerol, pH 7.4 and stored at −80 °C.

4.4. Evaluation of Sarcin-Ricin Loop Depurination Activity

Human T lymphocyte cell line (C8166) was obtained from AIDS Reagent Project, Medical Research Council, UK and maintained in RPMI-1640 medium supplemented with 10% fetal bovine serum (Invitrogen, Carlsbad, CA, USA). C8166 cells in log-phase growth were seeded at density of 1×10^6 cell/well on six-well plate and treated with 5 μM protein samples. After incubation at 37 °C for 6 h, cells were harvested for RNA isolation and cDNA synthesis as previously described [8]. Degree of SRL depurination was estimated by quantitative real-time PCR method [32] using 7500 Fast Real-Time PCR System (Applied Biosystems, Foster, CA, USA) and Power SYBR Green PCR Master Mix Kit (Applied Biosystems, Foster, CA, USA). The relative amount of depurinated rRNA was calculated as nucleic acid estimated by test primers divided by that estimated by control primers. The relative SRL depurination activity was determined as the relative amount of depurinated rRNA of treated cells against that of untreated cells.

4.5. Pharmacokinetics Study

Sprague-Dawley rats weighed 230–250 g were supplied by the Laboratory Animal Services Centre of The Chinese University of Hong Kong and experiments were performed in accordance with the CUHK Basic Principles and Guidelines (License No. (13-351)IN DH/HA&P/8/2/1 PT.31, 10 September 2013, Department of Health, Hong Kong SAR). Animals were fasted overnight with free access to water. The right jugular vein of rat was cannulated with a polyethylene tube (Braintree Scientific, Inc., MA, USA) under ketamine/xylazine (80/20 mg/kg) induced anesthesia. Prior to protein injection, rat was allowed to recover for 1 h. The cannulated animal was administered with maize RIP variant at dose of 4.5 mg/kg intravenously, followed by injecting 0.5 mL blank rat blood and then 0.5 mL heparinized normal saline (25 I.U./mL) to rinse the catheter. The blood samples, approximately 0.4 mL, were collected into heparin-rinsed tubes at 2, 5, 10, 15, 20, 30, 45, 60, 120, 180 and 240 min post dosing. Plasma was isolated by centrifugation of blood at $13{,}000 \times g$ for 5 min.

ELISA was carried out to estimate the concentration of MOD or variant in plasma for in vivo half-life determination. In brief, a 96-well ELISA plate (Thermo Fisher Scientific, Waltham, MA, USA) was pre-coated with polyclonal rabbit anti-MOD antibody in 0.05 M sodium carbonate/bicarbonate buffer, pH 9.6 overnight at 4 °C. The plate was then rinsed three times with washing buffer (PBS with 0.5% Tween 20) and blocked with 5% non-fat milk at 37 °C for 2 h. Diluted plasma samples were added and incubated at 37 °C for 2 h. After washing, biotin-labeled anti-MOD antibody was applied for detection followed by streptavidin-horseradish peroxidase conjugate (Invitrogen, Carlsbad, CA, USA). Finally, 3,3′,5,5′-tetramethylbenzidine (TMB) substrate solution (BD Bioscience, Bedford, MA, USA) was added and incubated at room temperature for 10 min. The reaction was terminated by adding 1 M H_2SO_4 and $OD_{450nm/630nm}$ was measured using an ELISA plate reader. Pharmacokinetic parameters were calculated by WinNonlin software (version 3, Certara, Princeton, NY, USA).

4.6. Immunogenicity Assay

Immunization and blood collection of mice were conducted at Guangdong Medical Laboratory Animal Centre, Foshan, China. C57BL/6N inbred mice of 6-8 week old were randomly assigned into groups of six. Wild-type or PEGylated variants were administered subcutaneously at the back with 10 μg in complete Freund's adjuvant on Day 0. Sampling for IgE detection was carried out on Day 10. Booster injection was given with incomplete Freund's adjuvant on Day 21. Sampling for IgG detection was performed 7 day after booster injection by retrobulbar puncture. Blood samples were centrifuged instantly right after collection and the isolated sera were stored at −80 °C.

IgE and IgG specific for maize RIP were detected by ELISA method. In brief, a 96-well ELISA plate (Thermo Fisher Scientific, Waltham, MA, USA) was pre-coated with antigen in 0.1 M sodium carbonate/bicarbonate buffer, pH 9.6 overnight at 4 °C. The plate was then washed and blocked with 5% non-fat milk at 37 °C for 2 h. Next, diluted serum samples were added for incubation at 37 °C for 2 h. After washing, the specific secondary detecting antibody (Goat anti-Mouse IgE Secondary Antibody-HRP conjugates, Goat anti-Mouse IgG (H + L) Secondary Antibody-HRP conjugates (Thermo Fisher Scientific, Waltham, MA, USA) was added and incubated at 37 °C for 2 h, followed by TMB substrate solution (BD Bioscience, Bedford, MA, USA). After termination, $OD_{450nm/630nm}$ was measured with an ELISA plate reader.

Acknowledgments: We thank Rebecca Boston of North Carolina State University for the clone of maize RIP. This work was supported by a One-off Funding for Research (Ref: C4045-14G) from The Chinese University of Hong Kong.

Author Contributions: All authors conceived and designed the experiments; Ka-Yee Au, Wei-Wei Shi, Shuai Qian performed the experiments; Ka-Yee Au, Wei-Wei Shi and Pang-Chui Shaw analyzed the data; Ka-Yee Au wrote the paper; Wei-Wei Shi prepared figures; Zhong Zuo and Pang-Chui Shaw supervised and revised the paper.

Conflicts of Interest: The authors declare no conflict of interest.

References

1. Endo, Y.; Tsurugi, K. The RNA *N*-glycosidase activity of ricin a-chain. The characteristics of the enzymatic activity of ricin A-chain with ribosomes and with rRNA. *J. Biol. Chem.* **1988**, *263*, 8735–8739. [PubMed]
2. Hartley, M.R.; Legname, G.; Osborn, R.; Chen, Z.; Lord, J.M. Single-chain ribosome inactivating proteins from plants depurinate *Escherichia coli* 23S ribosomal RNA. *FEBS Lett.* **1991**, *290*, 65–68. [CrossRef]
3. Shaw, P.C.; Lee, K.M.; Wong, K.B. Recent advances in trichosanthin, a ribosome-inactivating protein with multiple pharmacological properties. *Toxicon* **2005**, *45*, 683–689. [CrossRef] [PubMed]
4. Domashevskiy, A.V.; Goss, D.J. Pokeweed antiviral protein, a ribosome inactivating protein: Activity, inhibition and prospects. *Toxins (Basel)* **2015**, *7*, 274–298. [CrossRef] [PubMed]
5. Olsnes, S. The history of ricin, abrin and related toxins. *Toxicon* **2004**, *44*, 361–370. [CrossRef] [PubMed]
6. Hey, T.D.; Hartley, M.; Walsh, T.A. Maize ribosome-inactivating protein (B-32). Homologs in related species, effects on maize ribosomes, and modulation of activity by Pro-peptide deletions. *Plant Physiol.* **1995**, *107*, 1323–1332. [CrossRef] [PubMed]
7. Mak, A.N.; Wong, Y.T.; An, Y.J.; Cha, S.S.; Sze, K.H.; Au, S.W.; Wong, K.B.; Shaw, P.C. Structure-function study of maize ribosome-inactivating protein: Implications for the internal inactivation region and the sole glutamate in the active site. *Nucleic Acids Res.* **2007**, *35*, 6259–6267. [CrossRef] [PubMed]
8. Law, S.K.; Wang, R.R.; Mak, A.N.; Wong, K.B.; Zheng, Y.T.; Shaw, P.C. A switch-on mechanism to activate maize ribosome-inactivating protein for targeting HIV-infected cells. *Nucleic Acids Res.* **2010**, *38*, 6803–6812. [CrossRef] [PubMed]
9. Wang, R.R.; Au, K.Y.; Zheng, H.Y.; Gao, L.M.; Zhang, X.; Luo, R.H.; Law, S.K.; Mak, A.N.; Wong, K.B.; Zhang, M.X.; et al. The recombinant maize ribosome-inactivating protein transiently reduces viral load in SHIV89.6 infected Chinese rhesus macaques. *Toxins (Basel)* **2015**, *7*, 156–169. [CrossRef] [PubMed]
10. Uckun, F.M.; Bellomy, K.; O'Neill, K.; Messinger, Y.; Johnson, T.; Chen, C.L. Toxicity, biological activity, and pharmacokinetics of TXU (anti-CD7)-pokeweed antiviral protein in chimpanzees and adult patients infected with human immunodeficiency virus. *J. Pharmacol. Exp. Ther.* **1999**, *291*, 1301–1307. [PubMed]
11. Puri, M.; Kaur, I.; Perugini, M.A.; Gupta, R.C. Ribosome-inactivating proteins: Current status and biomedical applications. *Drug Discov. Today* **2012**, *17*, 774–783. [CrossRef] [PubMed]
12. Jevsevar, S.; Kunstelj, M.; Porekar, V.G. PEGylation of therapeutic proteins. *Biotechnol. J.* **2010**, *5*, 113–128. [CrossRef] [PubMed]
13. Byers, V.S.; Levin, A.S.; Malvino, A.; Waites, L.; Robins, R.A.; Baldwin, R.W. A phase II study of effect of addition of trichosanthin to zidovudine in patients with HIV disease and failing antiretroviral agents. *AIDS Res. Hum. Retrovir.* **1994**, *10*, 413–420. [CrossRef] [PubMed]
14. Abuchowski, A.; McCoy, J.R.; Palczuk, N.C.; van Es, T.; Davis, F.F. Effect of covalent attachment of polyethylene glycol on immunogenicity and circulating life of bovine liver catalase. *J. Biol. Chem.* **1977**, *252*, 3582–3586. [PubMed]
15. Kontermann, R.E. Strategies to extend plasma half-lives of recombinant antibodies. *BioDrugs* **2009**, *23*, 93–109. [CrossRef] [PubMed]

16. Levy, Y.; Hershfield, M.S.; Fernandez-Mejia, C.; Polmar, S.H.; Scudiery, D.; Berger, M.; Sorensen, R.U. Adenosine deaminase deficiency with late onset of recurrent infections: Response to treatment with polyethylene glycol-modified adenosine deaminase. *J. Pediatr.* **1988**, *113*, 312–317. [CrossRef]

17. Wong, K.L.; Li, H.; Wong, K.K.; Jiang, T.; Shaw, P.C. Location and reduction of Icarapin antigenicity by site specific coupling to polyethylene glycol. *Protein Pept. Lett.* **2012**, *19*, 238–243. [CrossRef] [PubMed]

18. He, X.H.; Shaw, P.C.; Xu, L.H.; Tam, S.C. Site-directed polyethylene glycol modification of trichosanthin: Effects on its biological activities, pharmacokinetics, and antigenicity. *Life Sci.* **1999**, *64*, 1163–1175. [CrossRef]

19. He, X.H.; Shaw, P.C.; Tam, S.C. Reducing the immunogenicity and improving the in vivo activity of trichosanthin by site-directed pegylation. *Life Sci.* **1999**, *65*, 355–368. [CrossRef]

20. ElliPro serve. Avaliable online: http://tools.iedb.org/ellipro/ (accessed on 10 September 2014).

21. Kaur, I.; Gupta, R.C.; Puri, M. Ribosome inactivating proteins from plants inhibiting viruses. *Virol. Sin.* **2011**, *26*, 357–365. [CrossRef] [PubMed]

22. Stirpe, F. Ribosome-inactivating proteins: From toxins to useful proteins. *Toxicon* **2013**, *67*, 12–16. [CrossRef] [PubMed]

23. Till, M.A.; Zolla-Pazner, S.; Gorny, M.K.; Patton, J.S.; Uhr, J.W.; Vitetta, E.S. Human immunodeficiency virus-infected T cells and monocytes are killed by monoclonal human anti-gp41 antibodies coupled to ricin A chain. *Proc. Natl. Acad. Sci. USA* **1989**, *86*, 1987–1991. [CrossRef] [PubMed]

24. Lee-Huang, S.; Kung, H.F.; Huang, P.L.; Bourinbaiar, A.S.; Morell, J.L.; Brown, J.H.; Huang, P.L.; Tsai, W.P.; Chen, A.Y.; Huang, H.I.; et al. Human immunodeficiency virus type 1 (HIV-1) inhibition, DNA-binding, RNA-binding, and ribosome inactivation activities in the N-terminal segments of the plant anti-HIV protein GAP31. *Proc. Natl. Acad. Sci. USA* **1994**, *91*, 12208–12212. [CrossRef] [PubMed]

25. Lee-Huang, S.; Kung, H.F.; Huang, P.L.; Huang, P.L.; Li, B.Q.; Huang, P.; Huang, H.I.; Chen, H.C. A new class of anti-HIV agents: GAP31, DAPS 30 and 32. *FEBS Lett.* **1991**, *291*, 139–144. [CrossRef]

26. Huang, P.L.; Sun, Y.; Chen, H.C.; Kung, H.F.; Lee-Huang, S. Proteolytic fragments of anti-HIV and anti-tumor proteins MAP30 and GAP31 are biologically active. *Biochem. Biophys. Res. Commun.* **1999**, *262*, 615–623. [CrossRef] [PubMed]

27. Zhao, W.L.; Feng, D.; Wu, J.; Sui, S.F. Trichosanthin inhibits integration of human immunodeficiency virus type 1 through depurinating the long-terminal repeats. *Mol. Biol. Rep.* **2010**, *37*, 2093–2098. [CrossRef] [PubMed]

28. Wang, Y.J.; Hao, S.J.; Liu, Y.D.; Hu, T.; Zhang, G.F.; Zhang, X.; Qi, Q.S.; Ma, G.H.; Su, Z.G. PEGylation markedly enhances the in vivo potency of recombinant human non-glycosylated erythropoietin: A comparison with glycosylated erythropoietin. *J. Control. Release* **2010**, *145*, 306–313. [CrossRef] [PubMed]

29. Meng, Y.; Liu, S.; Li, J.; Meng, Y.; Zhao, X. Preparation of an antitumor and antivirus agent: Chemical modification of alpha-MMC and MAP30 from *Momordica Charantia* L. with covalent conjugation of polyethyelene glycol. *Int. J. Nanomed.* **2012**, *7*, 3133–3142.

30. Ponomarenko, J.; Bui, H.H.; Li, W.; Fusseder, N.; Bourne, P.E.; Sette, A.; Peters, B. Ellipro: A new structure-based tool for the prediction of antibody epitopes. *BMC Bioinform.* **2008**, *9*, 514. [CrossRef] [PubMed]

31. Yang, Y.; Mak, A.N.; Shaw, P.C.; Sze, K.H. Solution structure of an active mutant of maize ribosome-inactivating protein (MOD) and its interaction with the ribosomal stalk protein P2. *J. Mol. Biol.* **2010**, *395*, 897–907. [CrossRef] [PubMed]

32. Melchior, W.B., Jr.; Tolleson, W.H. A functional quantitative polymerase chain reaction assay for ricin, Shiga toxin, and related ribosome-inactivating proteins. *Anal. Biochem.* **2010**, *396*, 204–211. [CrossRef] [PubMed]

MDPI

St. Alban-Anlage 66

4052 Basel

Switzerland

Tel. +41 61 683 77 34

Fax +41 61 302 89 18

www.mdpi.com

Toxins Editorial Office

E-mail: toxins@mdpi.com

www.mdpi.com/journal/toxins